THE TETTIGONIIDAE:
Biology, Systematics and Evolution

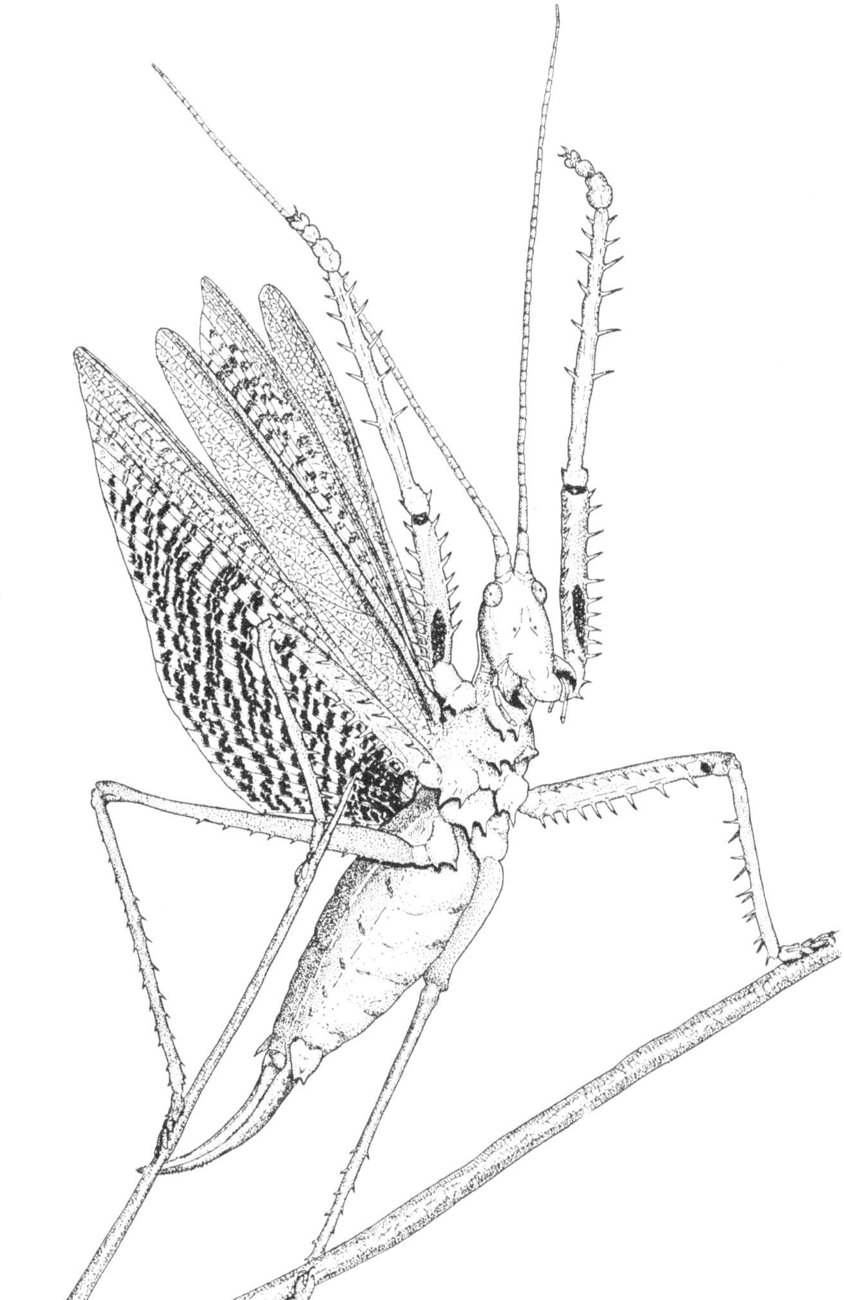

THE TETTIGONIIDAE:
Biology, Systematics and Evolution

W.J. Bailey
and
D.C.F. Rentz
[Eds.]

Springer-Verlag Berlin Heidelberg GmbH

Published by
Springer-Verlag

Originally published by Springer-Verlag Berlin Heidelberg New York in 1990
Softcover reprint of the hardcover 1st edition 1990

Distribution rights for Australia, New Zealand and Papua New Guinea:
Crawford House Press. For all other countries:
Springer-Verlag Berlin Heidelberg GmbH.

ISBN 978-3-662-02594-9 ISBN 978-3-662-02592-5 (eBook)
DOI 10.1007/978-3-662-02592-5

Designed and produced by
Crawford House Press
Panorama Avenue
Bathurst NSW 2795 Australia

10 9 8 7 6 5 4 3 2 1

CONTENTS

CONTRIBUTORS

Dr W.J. Bailey
Department of Zoology
University of Western Australia
Western Australia 6009

Dr W. Barendse
CSIRO Tropical Cattle Research Centre
Box 5545
Rockhampton Mail Centre
Queensland 4702
Australia

Dr J.J. Belwood
Department of Entomology
University of Florida
3103 McCarty Hall
Gainesville
Florida 32611
USA

Dr D.H. Colless
Divison of Entomology
CSIRO
PO Box 7600
Canberra ACT 2601
Australia

Dr I. R. Dadour
Entomology Division
State Department of Agriculture,
South Perth,
Western Australia 6151

Dr Michael Greenfield
Department of Biology
University of California
Los Angeles
California 90024
USA

Dr Darryl Gwynne
Department of Biology
University of Toronto
Erindale Campus
Mississauga
Ontario L5L 1C6
Canada

Dr J. C. Hartley
Department of Zoology
University of Nottingham
Nottingham NG7 2RD
United Kingdom

Dr Klaus-Gerhard Heller
Institute fir Zoologie II
Universität Erlangen-Nurnberg
Staudstrasse 5
8520 Erlangen
West Germany

Professor Alfred P. Kaltenbach
Naturhistorisches Museum Wien
Zweite Zoologische Abteilung (Insekten)
Burgring 7 – Postfach 417
A-1014 Wien
Austria

Dr W. Kaiser
Fachbereich Biologie-Zoologie
Philipps Universität
D-3550 Marburg/Lahn
West Germany

Dr A. Keuper
Fachbereeich Biologie-Zoologie
Philipps Universität
D-3550 Marburg/Lahn
West Germany

Professor Dr Klaus Kalmring
Fachbereich Biologie-Zoologie
Philipps Universität
D-3550 Marburg/Lahn
West Germany

Dr R. Lakes
Zoologisches Institut
Universität Göttingen
Berliner Strs. 28
3400 Göttingen
West Germany

Dr D.C.F. Rentz
Division of Entomology
CSIRO
GPO Box 1700
Canberra ACT 2601
Australia

Professor Dr J. Rheinlaender
Zoologisches Institut
J.W. Goethe Universität Frankfurt
D-6 Frankfurt
West Germany

Dr David Robinson
Department of Biology
The Open University
Walton Hall
Milton Keynes MK7 6AA
United Kingdom

Dr H. Römer
Lehrstuhl für Allgemeine Zoologie
Ruhr Universität Bochum
Postfach 102148
4630 Bochum 1
West Germany

Dr A. Schatral
Department of Zoology
University of Western Australia
Western Australia 6009

Dr T. Schikorski
AK Neurochemie
J.W. Goethe Universität Frankfurt
Postfach 111 932
6000 Frankfurt 11
West Germany

Dr N. Ueshima
Department of Biology
Matsusaka University
1846 Kubo-cho
Matsuka
Mie 515
Japan

DR DAVID RAGGE
An Appreciation

David Ragge was born in Surrey, England in 1930 and was educated at Trinity School and Imperial College, London University. He developed an early interest in natural history and by his teens was a keen bird-watcher, bug-hunter and botanist. He applied to Imperial College to read Botany at a time when this interest was uppermost, but later changed to Zoology and specialised in Entomology. At that time the Zoology Department at Imperial College organised summer camps in the New Forest for its students and it was during these that Ragge developed a serious interest in Orthoptera, becoming especially fascinated by their behaviour and characteristic calls.

After graduating at the age of 20 he spent two years working for a PhD on the wing-venation of Orthoptera, and it was during this time that he was exposed to the rich Orthopteran fauna of southern Europe. The British Museum was only a short walk from Imperial College, and there he met the renowned B.P. (later Sir Boris) Uvarov, who, although never on the Museum staff, had played a major part in building up the Orthoptera collection. Uvarov persuaded Ragge to join the permanent staff in 1952 and Ragge lost no time in convincing the Museum authorities that field studies, particularly on the calling songs, would be needed for detecting and establishing the limits of real species of sound-producing Orthoptera, as opposed to the rather artificial species resulting from work on dried specimens alone.

The BMNH collection was particularly rich in African material, as a result of the strong British links with that continent, and, since the Acridoidea were already well catered for by the taxonomic unit of the then Anti-Locust Research Centre, Ragge chose the African Tettigoniidae for most of these early studies. He also extended his wing-venation studies to the Phasmida, and realised an early amibition by writing a book, as a private project, on the British Orthoptera.

Ragge was surprised to find on his appointment to the BMNH that the Museum's collection of European Orthoptera was decidedly poor, as a result of the Museum's traditional emphasis on tropical faunas. Being particularly interested in the European fauna, and believing that this deficiency was in any case undesirable, Ragge set about remedying it by collecting in many parts of Europe himself. In tracking down this European material he was able to associate with almost all the European orthopterists, and considered himself fortunate to have met such early workers as Richard Ebner and Malcolm Burr shortly before they died.

He became involved in corporate entomology in the role of Secretary (and later Vice-President) of the Royal Entomological Society of London. During this period he was able to travel to the USA in order to visit the American Museum of Natural History, New York; Academy of Natural Sciences of Philadelphia; US National Museum, Washington; University of Michigan, Ann Arbor; and California Academy of Sciences, San Francisco. In 1968 he was appointed to a partly administrative job as Deputy Keeper of Entomology (his time for research being from then on much restricted) and in 1973 was made a Research Fellow of the Academy of Natural Sciences of Philadelphia.

During the sixties an opportunity at last arose for the construction of an acoustic laboratory at the Museum. Having worked closely with the BBC Natural History Unit, both as a broadcaster and in collaborating with the Unit in tape recording the songs of the British Orthoptera, he was able to make use of the BBC's expertise on building and equipping an acoustic laboratory which finally came into use in 1973. With W.J. Reynolds, Ragge began work on a comprehensive review of the songs of the European Orthoptera in relation to their taxonomy. This project has naturally required extensive field-work in southern Europe and the making of over 1000 field and studio recordings of Orthoptera songs. Most of the field-work has now been done (giving rise to about 10 preliminary research papers on groups urgently needing such studies) and the completion of the review is scheduled for the early nineties. Although due to retire in 1990, Ragge has made it clear that, far from dropping the project, he intends to spend more of his time on it from 1990 onwards.

Outside natural history Ragge's main interests are music and art. As a young man, when more leisure time allowed such luxuries, he wrote keyboard music, sketched and painted. What better way to combine his musical and entomological interests in later life than by studying the songs of insects!

A complete list of publications by David Ragge appears at the end of this volume.

David Ragge

1. THE TETTIGONIIDAE – THEIR DIVERSITY AND IMPORTANCE IN BIOLOGICAL RESEARCH

W.J. Bailey and D.C.F. Rentz

Why study tettigoniids ?

A perusal of the contents of this book will reveal a diversity of scientific study that relies on tettigoniids. Why is this so? The reasons are probably many. For taxonomists, they are obvious: the fauna is relatively large and diverse and in most parts of the world is poorly known. For the behaviourist the array of lifestyles and broad range of ecological niches is reflected in the abundant behaviour patterns awaiting study. Ecologists are just beginning to investigate the many and unusual ways tettigoniids interact with each other and members of other communities. Functional anatomists and physiologists have used the unique features of the group's obvious auditory communication to investigate processes of sound production and reception.

A tool for the evolutionary biologist

Evolution in action is an attractive concept for the experimental biologist but is always elusive. Darwin held dear the comparative approach where characters would evolve at different rates in related taxa providing in extant species (or indeed the paleontological record) a spectrum of morphology and behaviour. A number of authors have attempted to provide a comparative view of their specific area of study using as wide a range of taxa as possible, and although not all have evolution as their *modus operandi*, the wide sweep of a review provides the data for further comparisons. These comparative chapters either directly or indirectly focus on reproductive behaviour, through the process of mate attraction, mate location, the mating system or the biology of the egg.

Mate attraction is almost entirely by song and the complexity of the song of tettigoniids, unlike the more repeated notes of cicadas and crickets is often bizarre. K-G Heller shows some exquisite examples of this extreme in one Mediterranean group of phaneropterine species. Although social signals have this primary role of attracting mates from a distance, their complexity may have evolved in response to a number of selective influences that in addition to the recognition of like, may also include the avoidance of predation and the enhancement of male characters through sexual selection. Like the complex structure of the genitalia (Eberhard 1985), song structure seems almost "too complex" for species isolation alone. And in recognizing the role of reproductive isolation we are less ready to accept the role of sexual selection in shaping such complexity. Why can we accept that the elaborate colours of butterflies, the horns of beetles or the plumage of tropical birds could have evolved through sexual selection, but remain conservatively fixed on the species paradigm for insect song? Like the bird's plumage, the song acts to attract the female, and as in many birds the female is quite selective of which male she takes as a mate: selection on call structure may be through female choice. In addition where there are more males than females ready to mate at any location, these males will compete and in this process the song may have an essential function of outsinging competitors or distancing competing males from a strategic calling site.

Although frequency and temporal structure of the song may have a demonstrable role in maintaining species isolation, there is one example where sympatric congenerics are isolated by other means than the song. Chapters by Dadour and Bailey and more explicitly by Barendse examine the biology and evolution of an indigenous West Australian species of Copiphorinae (*Mygalopsis marki*), where between two recently formed species, perhaps dating from a late Pleistocene separation of the parent population, the songs do not provide a barrier to mating: the songs are indistinguishable and the F1 generation of hybrids are sterile. However, other, perhaps older, species of this genus do show good song differences and there is evidence that the song may provide a barrier to mating at one parapatric boundary.

Where the male provides the female with a nuptial gift that both enhances the chances of successful copulation as well as the increased survival of the offspring (Gwynne 1988), selection will act on the mating system to increase female choice on the one hand, and a reluctance on the part of the male to give up his offering on the other. Tettigoniids are renowned for the production of massive male gifts in the form of the spematophylax, and this protein mass has a profound influence on mating behaviour. One hypothesis is that its sole function is to protect the sperm ampula from being consumed by the mated female. However, although accepting this as an initial function, Gwynne traces the evolution of this nuptial gift through the ensiferan Orthoptera, suggesting that its function extends beyond this and that it has evolved through sexual selection.

The evolution of song

Stridulation may have evolved as a by-product of flight in that the wings use the same muscles for stridulation as used for flight (Elsner & Popov 1978). Although such an explanation may be parsimonious other routes are equally possible. Thus in crickets the motion of the wings not only produces sound, but in many species the vortices of disturbed air around the moving wings are detected by the highly sensitive anal cerci of a nearby insect (Heinzel & Dambach 1987). If the wings were initially used in a nocturnal blattid-gryllid ancestor for the projection of pheromones rather like the raised wings of a moth, then air movement may have added social information via the anal cerci. It is a small step for these wing movements to produce airborne sounds that could again augment the patterned pheromone/air-pulsed signal. Processes involved in detecting and interpreting these signals would have co-evolved with sound production, one system driving the other.

The ears of both ensiferan and caeliferan Orthoptera have a far wider role than social communication – they serve as an early warning system detecting an approaching predator. In crickets the distinction between conspecific detection and the cries of searching bats is quite straightforward. The ear is able to detect both the sonic social signals as well as the ultrasound of bats, and interneurons are able to separate these signals giving an appropriate phonotactic behaviour: attraction to the conspecific call and repulsion from the bat cry. Although less is known of the role of the tettigoniid ear in detecting the sounds of predators, unlike the cricket, the male's call usually covers a wide frequency extending well into the ultrasonic range and overlapping the frequencies used by bats.

The role of predation may also have shaped both the signal itself and the manner in which it is produced. Any intraspecific signal not only acts as an attractant for the conspecific, but also may be used by predators to actively locate the prey. Examples in other taxa are predation by fireflies on calling heterospecifics (Lloyd 1984) or the location of calling crickets by tachiniid flies (Cade 1975). Sexual selection theory would predict that if there were any risks involved in mating then the sex offering the highest investment should take fewer of them. Given these factors and an inherent risk of detection by predators of calling individuals, males are more likely to signal than females (Thornhill 1979). Further, if the quantity (length and/ or loudness) of the acoustic signal were correlated with the risks of exposure, we may predict that insects calling from exposed perches, or in areas of high predation, would use calls with a low duty cycle (the calls are short and widely spaced in time). Belwood provides an extensive account of defensive strategies used by tettigoniids in the tropical rainforests of Central and South America. Her discussion includes both cryptic strategies in avoiding day-time predators and also signalling strategies avoiding those operating at night.

As the signal shortens so selection is transferred from the caller to the searching female, and male location may become more risky than calling itself. At this juncture it may be advantageous for the female to call in reply to the male. This role reversal in signalling often results in duetting where the male calls and the female replies within a very narrow time window. The male now searches for the female, taking on both risks of calling and moving. The detection of like species may now revolve around a time specific window between caller and respondent not only announcing the presence of a conspecific but also its proximity. The nature of duetting has attracted the attention of numerous workers not only in the tettigoniids, evidenced in chapters by Robinson and Heller, but also in taxa ranging from the Neuroptera (Henry 1980) through the Hemiptera (Claridge 1985) to acridid grasshoppers (Helversen & Helversen 1983).

The interaction of other noisy conspecifics and/or heterospecifics inevitably leads to a change in the behaviour of the calling insect. Dadour and Bailey describe the tactics used by a male to avoid interacting with a close neighbour by progressively shifting its position on the bush. When sound intensities reach too high a level the male moves off the bush to another singing perch. Many species, however, call in a chorus, and chorusing appears to be stimulated by the presence of calling neighbours. The often incessant choruses of North American coneheads are a perfect example of this type of interaction. Greenfield analyses chorusing behaviour in this group, suggesting that calling strategies may have evolved to allow dominant males the opportunity to monitor calling neighbours through acoustic windows in their near continuous song.

Given the propensity of males to react to others, it is perhaps of little surprise to find that sympatric congenerics, and even heterospecifics, evoke a response in the caller. Schatral has developed the argument of reactive and non-reactive species that compete for calling time within the same habitat. Although the chapter is centered on a number of European studies she adds new data from a set of interacting species in Western Australia.

THE EAR

Signal reduction can be extreme where predation is high, and this is suggested by the modification of signalling strategies in many tettigoniids of the tropical rainforests of central and south America (Morris 1980 and Belwood chapter 2). In some species the acoustic signal has been replaced by substrate vibration. The male and female determine their identity and location by vibrating the substrate in a tremulation behaviour. Here species specific signals are detected by the subgenual organs of the leg. These organs, slung on curtains of tissue behind the knees are highly sensitive to the slightest disturbance of the substrate. In rather specialised cases, where tremulation as used alone, the temporal pattern of vibration will be interpreted by the nervous system as the species' template. The receptor neurones of the foreleg open into the same auditory neuropile as those for airborne signals. That this system is widespread throughout the tettigoniids is evidenced by the comparative treatment provided by Kalmring and his co-workers. His research group has already demonstrated that when substrate vibration, carrying the same temporal structure of the song, is perceived by both the subgenual organ and the *crista acustica* the information is amplified within the central nervous system – one input reinforces the other (Kalmring & Kuhne 1980).

Inevitably, advantage through reproductive success leads to evolved behaviours, and these may be associated with changes in structure. The ear is one such structure that must respond to demands of increased sensitivity to absolute sound levels, selected frequencies within the song, or importantly the direction of the sound's source. By reviewing the anatomy of the ear in a comparative manner Bailey suggests how structure may reflect behavioural pressures from both the detection of the conspecific and the sensitivity to noises generated by predators. This review of the gross anatomy is supported by a much wider study of the neuroanatomy of the mechanoreceptor system of which the ear is part. By using *in vitro* staining, Lakes and Schikorsky not only display the intricate neural anatomy of the tettigoniid mechanoreceptor system, but for the first time give a picture of the ear of a tettigoniid derived from an intricate knowledge of its function.

How does the ear work? Rheinlaender and Römer discuss distance perception and auditory acuity: the way in which an insect is able to locate a sound source. By using novel techniques of recording nerve activity in field preparations, they have exposed some of the real problems insects face in sorting out relevant signals from those of environmental noise, and also changes in the insect's ability to locate the sound as it moves within the vegetation. By a system of cross inhibition, the tettigoniid is able to sharpen its acuity by constant comparison of sound fields either side of its mid-line. They include some comparative data on a quite differently evolved system in the acridids.

The egg, predatory life styles and systematics

Hartley provides a comprehensive review of the tettigoniid egg, linking structure, egg diapause and embryology with the biology of the adult. His classification of egg development supports aspects of the systematic organisation of the tettigoniids set by Rentz in 1979.

The small but impressive subfamily of predatory tettigoniids, the Saginae, are dealt with by A. Kaltenbach in a worldwide review of all forms excepting those described from Australia which he regards as different. These large insects are primarily distributed in southern Africa. Kaltenbach reviews the taxonomic history of the group, the relationships of genera and species and comments on the array of genera which have been included in the Saginae from time to time. He also reviews the known biology and ecology of some species, and presents extensive notes on their behaviour.

Ueshima and Rentz present the first comprehensive cytological analysis of the Australian shield-backed tettigoniids. The group was revised by Rentz (1985), and the cytological component for the specimens he reviewed is presented here. The chromosome complement for nearly every species is illustrated and described and a scheme of chromosome evolution for the group is discussed. This is the first comprehensive cytological treatment for any major group of Australian tettigoniids and illustrates the importance of this kind of study in understanding the evolution of a group. Tettigoniids are very amenable to chromosomal analysis and yet this form of investigation is not at all commonplace. Rentz and Colless' classification of the shield-backs of the world, based on an exhaustive set of characters, extend our understanding of this group. The study is opportune since virtually all major regions of the world inhabited by these insects have been recently revised. They compared cladistic and phenetic analyses and demonstrate that the latter can reflect relationships (based on zoogeographical parameters) better than the former, a result that is bound to elicit considerable discussion. Phenograms are used to compare the results.

The Higher Classification of the Tettigoniidae

The higher classification of organisms is always a topic that provokes argument among taxonomists, first, because of its general lack of objectivity, and second, by the outrage of non-taxonomists because the latter can never seem to settle on a scheme. The beginner is always befuddled and frequently emerges with a negative view of taxonomists and their science.

The situation regarding ensiferan (the more-or-less traditional name of the long-horned Orthoptera) is no different. Nearly every author at some point presents a scheme of higher classification. Some follow such schemes without ever having to present the rationale in arriving at them. Within the Tettigoniidae there is generally no accepted classification. Some specialists regard long-established subfamilies as families, others raise tribes to familial status and others are die-hard traditionalists, resisting change at all costs. As a general observation, it is probably safe to say that most recent classifications have 'evolved' from an escalation in rank of previously well-known taxa. Such changes are based on opinion rather than fact. It is one editor's (DCFR) opinion that there is little to be gained in the general escalation of rank of taxa for several reasons. In splitting taxa there are generally so many intermediate forms that the boundaries between them are blurred resulting in some taxa relegated to a 'limbo' state. Larger groupings would accommodate these. Additionally, our state of knowledge is such that attempts at higher classification are generally outdated because of the many new taxa that are constantly being discovered. Conservative classifications are more appropriate at this time.

Probably the most conservative in present-day classifications of the Tettigoniidae is that espoused by Rentz (1979) who generally follows that of Key (1970). Readers should consult Kevan (1976, 1977, 1982) for examples of an expanded or escalated classification of the Tettigoniidae. The latest classification is that of Gorochov (1988) (see Fig. 1) who incorporated fossil and traditional morphological characters to arrive at a scheme that is very conservative but still controversial. Certain subfamilies (such as the Phaneropterinae, Bradyporinae and Conocephalinae) which Kevan and others have raised to family rank are considered as subfamilies of the Tettigoniidae. Discussion is bound to result from Gorochov's aligning the Conocephalinae with the Hetrodinae and the Meconematinae with the Phasmodinae. His uncertainty of the relationships of the Apteropedetinae are discussed in the chapter by Rentz and Colless. Rentz (1979) felt that the Microtettigoniinae were related to the Conocephalinae.

An Orthopteran down-turn

It is curious and somewhat sad to note that with interest in tettigoniid-science more intense now than ever before, and important in understanding the role of organisms in the diminishing natural ecosystems of the world, that certain institutions have decided to eliminate or reduce their programmes in Orthoptera. Thus the University of Michigan, housing one of the most important Orthoptera collections in the world and Quebec's MacDonald College, home of Orthoptera for more than 30 years are loosing without replacement their leading systematicists. Europe has suffered a similar fate with the retirement of Dr A Kaltenbach's from the Zoologisches Museum in Vienna, and with the retirement of the man we honour, Dr D.R. Ragge, the Orthoptera Section of the British Museum (Natural History) will virtually go out of existence.

What's in a name

One of the early problems we encountered in putting together this edited work was to find an acceptable common name for tettigoniids. People with European backgrounds generally call them bushcrickets and to those of North American origins they are known as katydids. Within a very short time, we discovered considerable resistance to consensus for either term.

North Americans, brought up on an onomatopoeic word reflecting the loud call of the most common species of tettigoniid in the eastern portion of the continent, *Pterophylla camellifolia* (Fabricius), prefer katydid. Males call with the cry "katydid katydid" or "katy-she-did". However, despite the British-born editor, after listening to *P. camellifolia* in central Florida for some hours, being unconvinced that "katy-did" anything, not so unconvinced was Harris (1842) in his work on the Insects of New England, cited by Darwin (1871) who noted that "groves resound with the call of the "katydid-she-did the live long night". Even though the term katydid is used throughout the United States and Canada, the distribution of *P. camellifolia* would mean that many who use the term have never had a chance to hear the originating species. However, there is a strong aversion on the part of naturalists to use any term that may confuse tettigoniids with the true crickets – the Gryllidae.

Across the Atlantic, Europeans are familiar with the ubiquitous and equally noisy *Tettigonia viridissima* (Linnaeus) with its range across central and northern Europe. This insect provides the common namesake of the "great green bushcricket" which describes its size, colour and to some extent its habitat, and to most the term "bushcricket" has become generic to all tettigoniids. In Germany, the direct translation again applies in "Laubhauschrecken" – the foliage grasshopper. Australia with its highly multicultural society uses both names interchangeably, and as if to make some statement of independence some workers, particularly in applied fields, have used yet another term "the longhorned grasshopper".

A used name is "common" and therefore by definition remains the common name! We have decided to allow both terms in this volume – a reflection of the international complexion of the contributors. Undoubtedly the safest term is tettigoniid – a compromise taken in this chapter.

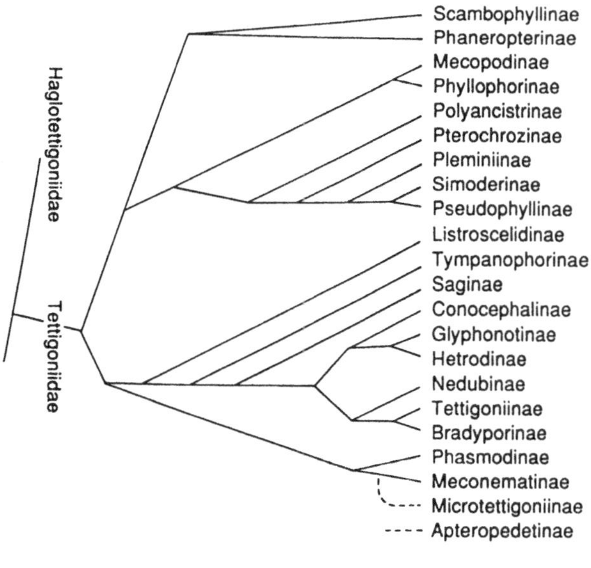

Fig. 1. A recent classification of the Tettigoniidae. This scheme by Gorochov (1988) is generally conservative but will provoke some discussion and controversy. His indecision regarding the Apteropedetinae has been dealt with by Rentz and Colless in chapter 17. The Microtettigoniinae may be more related to the conocephaline branch than the Phasmodine-Meconematine branch.

References

Cade W.H. 1975. Acoustically orienting parasitoids: fly phonotaxis to cricket song. Science 190: 1312-1313.

Claridge M.F. 1985. Acoustic signals in the Homoptera: Behavior, Taxonomy, and Evolution. Ann.Rev.Ent. 30: 297-317.

Darwin C. 1871. The descent of man, and sexual selection in relation to sex. Murray. London.

Eberhard W.G. 1985 Sexual Selection and Animal Genitalia. Harvard University Press pp 244.

Elsner N. & A.V. Popov 1978. Neuroethology of acoustic communication. Adv.Insect Physiol. 13:229-355.

Gorochov A.V. 1988. (In Russian). 4. The classification and phylogeny of grasshoppers (Gryllida-Orthoptera, Tettigonioidea). In: The Cretaceous Biocoenotic Crisis and the Evolution of Insects. Ed. A. Pomerenko. "Hayka", Moscow, pp. 145-190.

Gwynne D.T. 1988. Courtship feeding and the fitness of female katydids (Orthoptera: Tettigoniidae). Evolution 42: 545-555.

Heinzel H-G. & M.Dambach 1987. Travelling air vortex rings as potential communication signals in a cricket. J.Comp.Physiol. 160: 79-88.

Helversen D. v. & O.v.Helversen 1983. Species recognition and acoustic localization in acridid grasshoppers. A behavioural approach. In: Huber F & H.Markl (eds.), Neuroethology and behavioural physiology. Springer, Berlin, Heidelberg, New York, Tokyo, pp 95-102.

Henry C.S. 1980. The importance of low frequency, substrate-borne sounds in lacewing communication (Neuroptera: Chrysopidae) Ann. Ent. Soc. Amer. 73: 617-621.

Kalmring K. & R. Kuhne 1980. The coding of airborne-sound and vibrational signals in bimodal ventral nerve cord neurons of the grasshopper *Tettigonia cantans*. J. Comp. Physiol. 139: 267-275.

Kevan D.K.McE. 1976. The higher classification of the Orthopteroid insects. Lyman Ento. Mus. & Res. Lab., 4, Special publ. 12: 1-31.

Kevan D.K.McE. 1977. Suprafamilial classification of "Orthopteroid" insects, applying the principles of symbolic logic–a draft scheme for discussion and consideration. Notes from the Lyman Ento. Mus. & Res. Lab., 2 (second edition):1-27.

Kevan D.K.McE. 1982. Orthoptera. In: Synopsis and Classification of Living Organisms. Vol. 2. Ed. S.P. Parker. McGraw Hill, New York, pp. 252-379.

Key K.H.L. 1970. Orthoptera (Grasshoppers, locusts, crickets). In: The Insects of Australia. A Textbook for Students and Research Workers. Ed. CSIRO. Melbourne University Press, Melbourne, pp. 323-347.

Lloyd J.E. 1984. Occurrence of aggressive mimicry in fireflies. Florida Entomologist. 67: 368-376.

Morris G.K. 1980. Calling display and mating behaviour of *Copiphora rhinoceros* Pictet (Orthoptera: Tettigoniidae). Anim.Behav. 28: 42-518.

Rentz D.C.F. 1979. Comments on the classification of the orthopteran family Tettigoniidae, with a key to subfamilies and description of two new subfamilies. Aust. Jour. Zool. 27: 991-1013.

Rentz D.C.F. 1985. A Monograph of the Tettigoniidae of Australia. Volume 1. The Tettigoniinae. With an Appendix by D.H. Colless. CSIRO, Canberra; E.J. Brill, Leiden.

Thornhill R. 1979. Male and female sexual selection and the evolution of mating strategies in insects. In: Sexual selection and reproductive competition in Insects. Eds. M.S. Blum & N.A. Blum. Academic Press, New York. pp 81-121

2. ANTI-PREDATOR DEFENCES AND ECOLOGY OF NEOTROPICAL FOREST KATYDIDS, ESPECIALLY THE PSEUDOPHYLLINAE

Jacqueline J. Belwood

Introduction

Protection from predators is an important part of every animal's biology. Katydids, especially in the tropics, are well known for their striking resemblance to leaves and for their abilities to blend in with lichen-covered tree surfaces (e.g., Cott 1957; Nickle 1988; Plates 1-3). As such, they offer many opportunities for the study of anti-predator adaptations. It is surprising then, that with few exceptions (Robinson 1968; 1969a,b; Rentz 1975; Morris 1980; Morris and Beier 1982; Belwood 1988) katydids have been ignored in the context of predator-prey studies. Reasons for this are likely to include the relative inaccessibility of many tropical species for long-term studies, their nocturnal activity cycles, and a paucity of information on exactly which predators feed on these insects.

In addition to their cryptic appearance, katydids are good candidates for studies of anti-predator defences for several reasons. First, they number more than 3000 species worldwide (Nickle, pers. comm.) and are a dominant component of both the Old and New World tropics. In the neotropics alone there are over 2200 species (ibid.), including about 1000 described pseudophyllines (Beier 1960, 1962). Second, many katydids are large (25 to > 100 mm in body length), slow-moving insects that are easy to see and study at night on vegetation in the forest understory (Morris 1980; Rentz 1987; Belwood 1988). Third, katydids can be collected with little difficulty (by hand or with a long-handled insect net), and many species (particularly the pseudophyllines) do well in captivity, are long-lived, and can be maintained for study in large cages for months at a time (Belwood 1988). For unknown reasons, phaneropterine katydids are more difficult to maintain and usually die after about a week, but with care and patience they can also be studied in the laboratory.

This paper discusses anti-predator adaptations in Latin American forest katydids. It is based on the results of two ongoing (in 1989) long-term (> 2 years) field and laboratory studies. The first is on Barro Colorado Island (BCI), Panama (9° 09'N, 79°52'W) (Belwood & Morris 1987; Belwood 1988; Belwood, unpublished), the second at three sites near Iquitos, Peru, (approximately 40 and 80 km NNE from Iquitos on the Amazon River, and 160 km NNE on the Napo River, respectively). Results of the Peruvian study (Nickle, Castner and Belwood, unpublished), will only be discussed briefly as detailed results will be presented elsewhere.

The katydids in these studies, and those discussed in the greatest detail here, are understory forest species that can be observed directly. They are mainly pseudophyllines. The phaneropterines receive less attention because they are uncommon at ground level. Although they are regular visitors to ultraviolet lights, phaneropterines appear to spend most of their time in the forest canopy. The species discussed are potential prey for the foliage-gleaning bats (FGB) and birds and other predators that occur in the study areas.

Katydid Predators

Recent studies show that katydids are food for many of the animals, particularly vertebrates, that inhabit tropical New World forests. In Panama, some understory FGB such as *Micronycteris hirsuta*, *Tonatia silvicola* (Plate 7), and *Trachops cirrhosus*) (Phyllostomidae: Phyllostominae) include large numbers of pseudophylline katydids in their diets (Wilson 1971; Belwood 1988). These bats use the airborne calling songs of males, rather than echolocation, as prey-finding cues (Tuttle *et al.* 1985; Belwood & Morris 1987; Belwood 1989), but also feed heavily on females (Belwood 1988). How females, which are silent, or sporadically calling males (see below) are located at night in the dark is a mystery.

Presently, the extent to which FGB over large geographic areas feed on katydids is not known. Long-term studies on the diets of these bats, which are diverse and abundant in the tropics, have been scarce. The results of one study, however, show that bat-katydid interactions can be very specific (Belwood 1988). Over a two-year period, 42% by number, and 61% by weight of the diet of *M. hirsuta*, a 15 g bat on BCI, consisted of members of only six species: *Cocconotus wheeleri*, *Docidocercus gigliotosi*, *Xestoptera cornea*, *Idiarthron major*, and two unknowns. All are pseudophyllines. Near Iquitos, another FGB was found that appears to specialize on phaneropterines, but this bat could not be captured for identification (Belwood, unpublished).

Other studies on BCI (Gradwohl & Greenberg 1982, 1984) and in Peru, in the Tambopata Wildlife Reserve (12°50'S, 69°17'W) (Rosenberg, in press) and Manu National Park (Remsen and Parker 1984), show that some tropical antbirds (Formicariidae), ovenbirds (Furnariidae), and other birds (Plate 8) also consume large numbers of katydids. At Tambopata, Rosenberg (pers. comm.) has seen members of at least 22 bird species from 8 families catch or eat katydids in the field. These are taken from a variety of substrates, including live foliage, palms, bark surfaces, and at army ant swarms. Orthoptera also dominate in the stomach contents of many of these birds and appear to represent the single most important food source for many species (ibid.). Some of these birds forage exclusively by pulling arthropods from 'aerial leaf litter' (*sensu* Gradwohl & Greenberg 1982), the curled dead leaves that hang from vegetation in the forest canopy. It is noteworthy that many of the katydids consumed by antwrens in this habitat on BCI (Gradwohl & Greenberg 1984; Gradwohl, pers. comm.), are the same species that are fed on at night by FGB (e.g., *Cocconotus wheeleri*, *D. gigliotosi*, *Copiphora brevirostris*) (Belwood 1988).

Other katydid predators include small primates (e.g., titi monkeys, *Callicebus* spp.; Plate 9) (Belwood, unpublished), spiders (Gradwohl & Greenberg 1982; Belwood 1988:162; Plate 10), and Tropical Screech Owls, *Otus choliba*, in Peru (Belwood unpublished). Carnivorous katydids (e.g., *Copiphora brevirostris*) and mantids are probably also important katydid predators, because they have ears and can probably hear calling individuals (Belwood, unpublished; Yager & Hoy 1986, respectively).

Katydid Defences

There are two main types of animal defences (Robinson 1969a,b; Edmunds 1974): primary or indirect defences (that operate regardless of the presence of a predator and decrease the chances of an encounter taking place), and secondary or direct defences (that operate only after a predator encounters a potential prey animal). Primary defences are often based on morphology or overall appearance, and on the behavioural traits that enhance these features. These function either to prevent predators from detecting prey (crypsis), or if prey are detected, to prevent the recognition of the latter as edible objects (aposematism and batesian mimicry). Secondary defences increase the prey's chances of survival during a predator's attack (Edmunds 1974).

Anti-predator defences in neotropical katydids are closely tied to overall appearance. Four basic groups can be recognized, although these are not always mutually exclusive. For example, bark-mimicking katydids, which are in group 2 (below), are fusiform like the katydids in group 4.

1) Leaf-Mimicking Katydids (LMK)

These include members of the Phaneropterinae and the Pseudophyllinae (tribe Pterochrozini).

Primary defences

The phaneropterines are usually green, but sometimes brown or mottled (Rentz 1983; Castner 1989), and their wings resemble simple oval leaves (Plates 1, 11; Nickle, in press). The Pterochrozini occur in a greater array of colours (greens, yellows, and browns, and combinations of these) and have more elaborate leaf shapes (Plate 2). Their wings also often have drip-tips and/or irregular margins and clear hyaline patches that mimic necrotic leaf tissue. Others have white patches on them reminiscent of bird droppings (Castner 1989).

The survival value of crypsis in LMK is contingent on the correct choice of background substrate (Edmunds 1974:16). These insects should actively choose diurnal resting environments whose colours match their own wings and bodies. That this occurs was observed on BCI where several hundred green and brown LMK were maintained individually in wood and screen cages (at least 50 x 50 x 65 cm) that each contained a "mini tropical rainforest", i.e., live plants, pieces of tree bark, twigs and branches, rotting hunks of wood, and dead-leaf litter. These katydids were marked with white paint and a waterproof number that enabled them to be found and followed in their cages for a variety of observations.

Katydids were checked each morning at sunrise as part of their regular maintenance. At this time, green LMK were usually found on green leaves and brown LMK on brown leaves, including leaf litter. Large wooden surfaces and tree bark were not used as diurnal resting places. The insects were always in place *before* sunrise. How they located appropriately coloured backgrounds in the dark is not known.

To maximize cryptic behaviour, a LMK should minimize diurnal movement or other behaviours that make its presence known to potential predators (Edmunds 1974:19). Once in place on leaves, the LMK remained fully exposed on the upper or lower surfaces of leaves, but were motionless for hours at a time. The phaneropterines usually kept their legs close to their bodies, but in general did not adopt special resting attitudes. An exception, *Aegimia elongata*, rests on foliage with its head, which is flattened and resembles the petiole of a leaf, apposed to the substrate (Plate 11). This enhances the insect's leaf-like appearance. Resting LM pseudophyllines were also motionless, but most had special resting postures in which their legs were extended away from their bodies. This posture efficiently camouflages the legs, which are usually brown, and emphasizes the leaf-like wings (Robinson 1969a; Castner 1989). If disturbed during the day, some katydids such as *Aegimia elongata* rock and teeter as they walk (Robinson 1966), which resembles a leaf moving in a breeze.

In addition to being studied in the laboratory, LMK were also observed in the field at the Peruvian study site (Castner, in preparation). Over a three-month interval, 20 individuals were located on vegetation along trails one or two hours before dawn. (The species involved have not yet been identified.) Their locations at sunrise were flagged and the katydids were marked with a small dot of fast-drying green or yellow typewriter correction fluid. During the day, there was little diurnal movement by these insects, which is consistent with laboratory observations and with the cryptic appearance of these insects as a primary anti-predator defence.

LMK in the field moved more at night, but still cover relatively small areas. At dusk, observers at the Peruvian site were positioned near marked insects whose movements were followed throughout the night. Most of these katydids had not moved since they were marked during the day. Fourteen such katydids were watched and none moved more than one m. Similarly, on BCI a solitary male *Mimetica mortuifolia* sang from the same small tree for seven consecutive evenings.

There is evidence that lack of diurnal movement in LMK is adaptive against bird predation. Rosenberg (pers. comm.) reports that a captive Black-Capped Antshrike (*Thamnophilus schistaceus*) did not locate a brown LMK placed in its cage as long as the insect did not move. This was true even when the insect was only 3-5 cm from the bird. In a later experiment, the bird pursued and ate the katydid, but only after it was flushed from its resting place.

Secondary defences

The LMK studied have a variety of secondary defences. Most phaneropterines jump and/or fly away from an object that disturbs them. Most pseudophyllines and some phaneropterines "freeze", but grip the substrate tightly when attempts are made to lift them from a leaf surface (Rentz 1975). This behaviour is carried to an extreme by *Scopiorinus fragilis*, which flatten their bodies and hug the substrate when disturbed (Robinson 1969b).

Several LM pseudophyllines (e.g. *Tanusia* sp., *Pterochroza ocellata*, *Typophyllum mortuifolia*) have brightly coloured hindwings with eye spots that are exposed in response to tactile stimulation (Cott 1957: Plate 39; Linsenmaier 1972: Figs. 17/16 and 78/1; Morris *et al.* 1989, respectively). The hindwings remain exposed for about a minute after the insect has been disturbed. Presumably, this display intimidates or startles a predator due to the katydid's increased size or the sudden appearance of the eye spots (Edmunds 1974). This behaviour is most adaptive against diurnal predators, during the day when sunlight highlights the colours of the hindwings (Lisenmaier 1972). It should not be effective against nocturnal predators.

Members of most species regurgitate a darkish fluid from the crop when handled. Some species also bite, or attempt to bite. This response is best developed in the larger-bodied carnivorous species that have well developed mandibles (e.g. *Copiphora brevirostris*). Members of other large-bodied species (e.g., *Steirodon* spp.) kick when handled. The robust hindlegs are particularly effective for this because they are covered with large spines. The spines can draw blood from a human handler and could probably injure the eyes of a small bat or bird that has the insect in its mouth (Plate 7). Members of other katydid species, such as *Cnemidophyllum eximium*, flail their legs rapidly in an up and down motion when handled. This produces a loud raspy sound, which is startling the first few times it is heard. Male *Steirodon careovirgulatum*, *Cnemidophyllum eximium*, and other species also produced a loud startling disturbance sound with the sound producing structures on the tegmina. This is a noisy, broad-band signal (Belwood, unpublished), as are most insect disturbance calls (Masters 1980). In *Xestoptera cornea*, *Scopiorinus fragilis*, and *Parascopioricus exarmatus* females also produce disturbance sounds (Belwood 1988).

2) Bark-Mimicking Katydids (BMK)

BMK (Plate 3) are large (20-85 mm), mottled (grey, brown, green and white) pseudophylline katydids and include *Acanthodis curvidens* and *Pristonotus tuberosus* on BCI and several unidentified species in Peru. Some are lichenose. They are uncommon and constitute only a small percentage of the katydids found at night. During the day they are difficult to find, because they resemble the tree bark on which they rest.

Primary defences

In Peru, several marked individuals were observed in a large walk-in cage (2.5 x 1.5 x 1.5 m) that was equipped with vegetation as described on p. 10. During the day, these were usually found on tree bark (in 60.5% of cases; n = 35) or other woody surfaces (in 15.7% of the cases). They often rested in close contact with the substrate in shallow depressions, with the antennae and forelegs apposed and facing forward as described by Robinson (1969b) for *Acanthodis curvidens*. At night, most activities including calling in males and oviposition in females also occurred on wood, suggesting that a large number of these insects' activities take place on this substrate. Although several BMK at the Peruvian site were collected at night on large trees, only one individual was watched in the field. Except for a short period when it fed on a leaf, it remained on wood throughout its observation period.

Secondary defences

Like their leaf-mimicking counterparts, BMK are motionless during the day, they grip the substrate tightly when handled, and some (*Acanthodis curvidens*, *Pristonotus tuberosus*) produce disturbance sounds. *Acanthodis curvidens* also bite and kick, and have a conspicuous startle display involving exposure of the wings and a red-coloured abdomen. This is described in detail by Robinson (1969b).

3) Aposematic Katydids (including batesian mimics)

Aposematic katydids

Many katydids have large cuticular spines that are sharp and should cause serious damage to predators [see Nickle (in press) for figures of most of the species mentioned in this section]. In *Panacanthus* sp. (Plate 4), these structures cover large portions of the body, but this morphotype is rare. Instead, spines are usually restricted to the legs, and are best developed on the hindfemora where they point backwards and are effective as defensive weapons when the insects kick (Belwood, unpublished). Some species also have conspicuous or exaggerated spines on the prothorax (*Markia hystrix*, *Steirodon careovirgulatum*) or head (*Markia hystrix*, *Copiphora rhinoceros*, *Acanthoremus* sp.).

Most katydids are nocturnal and are cryptically rather than conspicuously coloured. Therefore, whether or not spines are aposematic signals to diurnal predators is not known. In some cases (*Markia hystrix*) the spines also help camouflage the insect (Nickle, pers. comm.) At night, however, thoracic spines appear to offer some protection from predation by FGB such as *Tonatia silvicola*, which usually grab and hold their insect prey by the thorax (Plate 7). In the laboratory these bats refused to eat the *Steirodon careovirgulatum* they were offered as food, although the insects were taken if their spines were removed. This was observed under low-intensity red light, indicating that the bats perceived the spines using echolocation, rather than vision.

In the field, spine detection in this manner would require close contact between bat and insect. In such a scenario, it is reasonable to suggest that the calling songs of spiny insects should be distinctive and function as aposematic signals. The evolution of such songs would decrease attacks on spiny katydids that call and would protect predators from potentially dangerous encounters. Morris (pers. comm.) reports that the calls of spiny *Panacanthus* in

Ecuador appear to fit this description. Frog-eating FGB (*Trachops cirrhosus*) have been shown to distinguish between palatable and poisonous prey based on call alone (Tuttle and Ryan 1981), but this has not yet been investigated with insect songs.

Some katydids such as *Harroweria gloriosa* on BCI and *Euceraia rufovariega* in Peru appear to be truly aposematic (Wickler 1968). When placed on green or brown backgrounds their bodies are bright and conspicuous. Such species are rare, however, and nothing is known about their biologies, or about their usual diurnal resting places.

Wasp-mimicking katydids (WMK)

The most striking katydids are the wasp-mimicking forms. These are a glaring exception to the notion that most tettigoniids are large, slow-moving nocturnal insects that resemble leaves or tree bark. WMK are harmless insects that appear to gain protection from predators as batesian mimics (Wickler 1968) of wasps. In the New World tropics, two phaneropterine genera (*Aganacris* and *Scaphura*) have been described, and *Aganacris* consists of five known species (Grant 1958; Nickle, pers. comm.). On BCI, the most common species, *Aganacris insectivora*, resembles the sympatric diurnal pompilid *Hemipepsis mexicana*. *Aganacris insectivora* males are about 35 mm long, have shiny black heads and bodies, and translucent yellow-orange wings (Plate 5); females are similar in appearance, but are slightly larger. Their antennae are short, thick, and orange with black tips and bases. *Aganacris insectivora* also resemble wasps in behaviour; when active they move about incessantly with rapid jerking motions.

As is usually the case with mimics (Wickler 1968), *Aganacris insectivora* on BCI are relatively rare; few individuals are found at one time. They are present from May to August and have two daily activity periods (Belwood, submitted). In late afternoon (ca. 1600 h), they can be seen in the field moving in vegetation 1-2 m off the ground in sunlit treefall gaps (where lone pompilids also occur). At night, between about 0130 and 0430, males can be collected at ultraviolet lights.

The bimodal circadian activity cycle of these insects is unusual because most katydids are active either at night or during the day. It is also unusual because the protective advantage afforded by an aposematic appearance is lost at night. The activity cycles of these insects were confirmed in the lab. Under natural photoperiod, and in the absence of ultraviolet light, caged *Aganacris insectivora* were active during the day between 1200 and 1700 h and again at night between 0130 and 0530.

Why these insects have two distinct active periods is not known, but may be related to the finding that singing for mates in these insects occurs only in the early pre-dawn hours. Some predators use acoustic cues to locate singing insects (Burk 1981). There is evidence that such predation pressure may be responsible for the aggregation during song production of some insects (cicadas), the combined intensities of which may repel birds (Simmons *et al.* 1971). Because wasp-mimicking katydids are rare, they are not likely to aggregate in large groups to chorus. In spite of aposematic colouration, the best anti-predator defence for these and other solitary diurnal katydids that call to attract mates may be to adopt a nocturnal singing schedule. At night, these insects sing within a very narrow time band (0300-0500 h), a period that coincides with the nightly lull in the feeding behaviour of katydid-eating FGB on BCI. This appears to be an effective anti-bat adaptation as these katydids are not fed on by FGB (Belwood, unpublished).

Clearly, more information is needed on the biology of *Aganacris insectivora* and other WMK in order to better understand various aspects of their behaviour. At least one other species, *Aganacris pseudosphex*, has been collected at night near Iquitos (Nickle, Castner &

Belwood, unpublished). This indicates that partial nocturnality may be the rule rather than the exception in wasp-mimicking species.

4) Fusiform (Spindle-Shaped) Katydids (excluding bark-mimicking species)

The great majority of pseudophylline and copiphorine katydids are long, slender, fusiform insects (Plate 6). They do not resemble leaves and are usually a mottled, nondescript brown or olive colour; a few (*Copiphora* spp.) are bright green. Spots and variously coloured patches can also adorn the body. The most distinctive features of these katydids are their antennae (Plates 13, 14), which are two to three times body length, and their faces (Plates 14-17), which are usually not body coloured, but bright blue or green, black, striped, or spotted. The two *Copiphora* species on BCI have black and tan faces, respectively.

Primary Defences

These katydids are unusual in that most do not appear to dayroost in the open on vegetation. *Ischnomela pulchripennis*, which roost on large spiny bromeliads (*Aechmea magdalanae*), may be an exception to this. Rather, most of these insects spend the day concealed from predators in a variety of tubular nooks and crannies in the forest. Stout-bodied species such as *Balboa tibialis* and *Idiarthron major* occur in cavities or notches in wood, while smaller-bodied species (*Cocconotus wheeleri*, *D. gigliotosi*, *Melanonotus bradleyi*) roost in the curled dead leaves (Plate 12) that accumulate in dense foliage in the forest understory. Green species such as *Copiphora brevirostris* inhabit live curled leaves near the ground, such as furled *Heliconia*.

Several interesting features are associated with the leaf- and crevice-roosting mode. First, some species (e.g., *Cocconotus wheeleri*, *Docidocercus gigliotosi*, and *Melanonotus bradleyi* on BCI) appear to be unique to this microhabitat (Gradwohl & Greenberg 1982). Second, these katydids are always positioned with the long axis of their bodies parallel to the long axis of their leaf or crevice. Third, individuals usually roost in leaves that closely match their own body colours (Rosenberg, pers. comm.). Last, individuals appear to use the same leaf-roosts day after day. This was first suspected by Gradwohl (pers. comm.), who regularly examined the arthropods in leaves as part of a feeding study on antwrens. Belwood (1988) also noticed this in field and in captive animals. This phenomenon was further investigated in the Peruvian study, where 14 marked individuals were observed over 53 study nights in the large cage described on p. 10.

Under red light, the insects were checked several times per night to ascertain that leaf-roosts are indeed vacated after dark. Leaves were checked one hour after dusk (at which time most roosts were empty), at midnight (when all were empty), and one and a half hours before dawn, when individuals were several meters from their roosts and began to seek daytime shelter. The last observation period lasted until dawn. Results confirm that some individuals return to a former roost; four individuals used the same roosting sites for more than one day, three for more than two days, and two for more than three days. The record for consecutive use of the same roost by a captive animal was twenty days by a female *Schedocentrus* sp.

Katydids in the field at the Peruvian site showed similar roost site fidelity. Members of seven leaf-roosting species (as yet unidentified) were located along trails, on leaves, half an hour to an hour before dawn. Under red light, each was followed to its diurnal retreat, which usually consisted of a curled dead leaf. Each was visited once during the day of its original sighting and was marked. This was done with care, so as not to cause an insect to abandon its leaf.

About half an hour before sunset, a pair of observers was positioned near a known leaf-

1

2

3

Plates 1-6. Basic morphotypes on which most katydid defences are based: leaf-shaped bodies characteristic of (1) phaneropterine (*Orophus conspersus*) and (2) pseudophylline (*Mimetica mortuifolia*) (previous page); (3) bark-mimicking katydid (*Acanthodis curvidens*), (4) spine-covered aposematic form (*Panacanthus* sp.), (5) wasp-mimicking katydid (*Aganacris insectivora*), and (6) fusiform katydid (*Ischnomela pulchripennis*).

4

5

6

7

8

Plates 7-10. Neotropical katydid predators: (7) foliage gleaning bat *Tonatia silvicola* (Phyllostomidae) from Barro Colorado Island BCI), Panama, (8) Plain Brown Woodcreeper, *Dendrocincla fuliginosa* (Dendrocolaptidae), from Costa Rica, (9) titi monkey from lowland Peru (*Callicebus* sp.), and (10) ctenid spider from BCI (species unknown).

9

10

Plates 11-13. Katydid defence tactics: (11) diurnal resting attitude of *Aegimia elongata* in which the flattened head touches substrate enhancing leaf-like appearance, (12) leaf-roosting in pseudophylline, and (13) position of *Melanonotus bradleyi* about to tremulate on leaf. Note katydid's antennae protruding from leaf in (12).

11

12

13

14

15

Plates 14-17. Facial patterns characteristic of leaf-roosting katydids. (14) blue face of *Docidocercus gigliotosi*, (15) striped face of *Cocconotus wheeleri*, (16) black and striped face of *Eucoconotus camerani*, and (17) black face of *Copiphora* nr. *brevirostris*. (14) to (16) are pseudophyllines, (17) is a copiphorine.

16

17

Plate 18. *Mygalopsis thielei* Bailey. This is one of the larger species in the genus occurring along the south-central coast of Western Australia. Speciation in a related species in the genus is discussed in Chapter 6.

Plate 19. Male tettigoniine *Chlorodectes baldersoni* Rentz from eastern Australia is similar in many respects to species of *Tettigonia* but appearances are deceiving. The phenogram presented in Chapter 17 (Fig. 2) indicates that the two genera are not related.

Plate 20. *Microtettigonia kangaroo* Rentz, female from Kangaroo Island, South Australia, is one of the smallest tettigoniids in the world. This specimen barely attains 1 cm in length.

roost. After the insect emerged, it was followed with red light as it moved, usually upward and outward into the canopy. It could usually be followed for four to six hours before visual contact was lost in dense vegetation. In spite of this, observers remained in position to document the insect's possible return, which occurred about an hour before dawn. At this time the individual reappeared abruptly, usually only a short distance from its leaf. It moved rapidly back to its roost and was usually in place 15 to 20 minutes before dawn. The longest record of this circadian behaviour in the field was six days for a female *Schedocentrus* sp. At Tambopata, Rosenberg (in press) also marked dead leaves and found that 15 of 20 katydids observed, which were not marked but which were presumably the same individuals, were in their roosts one day, 3 were present for 2 days, 1 for 3 days, and 1 for 9 consecutive days.

How insects find their way 'home', in the dark after covering long distances in dense forest, in some cases > 10 m, is not known. However, it seems likely that a trail of chemical cues, good spatial memory (i.e., the formation of a cognitive map, or route-specific memory), or a combination of these may be used. On several occasions, captive *Ischnomela pulchripennis* in the laboratory on BCI have been observed to lay trails of clear saliva droplets as they walk in a large glass aquarium. Also, at the Peruvian site, large numbers of wild and captive katydids have been seen to 'mouth', but not feed on, the plants on which they travel. It is tempting to speculate that these behaviours allow particular paths to be walked and then retraced. Whether substrate trails can be perceived when katydids move back to their leaves is not known. In the absence of this, some form of spatial memory would appear to play a role in re-locating roosts.

Dayroosts are probably very important for katydid survival. In addition to roost fidelity, this is indicated by the observation that leaves are actively defended against potential intruders, which can include other katydids, crickets, cockroaches, spiders, ants, and small wasps (Gradwohl and Greenberg 1982; Rosenberg, in press; Belwood, unpublished). This was documented in captivity at the Peruvian site on at least three occasions when one katydid attempted to enter a leaf that was already occupied. Without leaving it, the resident shook vigorously and repeatedly in its roost, causing the intruder to depart. In spite of this observation, some 'roost-sharing' can occur. Rosenberg (pers. comm.) has observed individual katydids in the field roosting in the same leaf as another katydid, a cockroach, a spider, and one or more ants.

Leaf- or cavity-roosting should protect katydids from diurnal predators such as foliage-gleaning birds. These katydids, however, are still potential prey for birds that specialize on the arthropods that occupy curled leaves (Gradwohl & Greenberg 1982, 1984; Remsen & Parker 1984; Rosenberg, in press). These birds systematically peer into individual leaves as they look for prey. Leaves of all shapes and sizes are checked, but curled ones receive the most attention (Remsen & Parker 1984). With this in mind, it is tempting to speculate that the facial patterns and colours that characterize leaf-roosting insects help conceal them within their roosts. To the human observer, spots and stripes help break up the shape of the face while blue, tan, and black colours help the insect to blend in with leaves.

Secondary Defences

Secondary defences in the leaf-roosting katydids are similar to those described for the BMK. Particularly well developed in these insects is their ability to "hold-on tightly" to their substrates when disturbed inside curled leaves. Rosenberg (pers. comm.) has observed small antwrens in feeding experiments play "tug-of-war" with katydids inside leaves, for up to 12 minutes, as they try to extract them. Under such circumstances many of the birds give up their pursuit of the insect in question. Rosenberg has also observed katydids jump out of their leaves when disturbed. Some birds, however, actively pursue such escaping insects, and even appear to specialize in this "flush-pursuit" hunting technique.

Calling and Anti-Predator Behaviour

Acoustic behaviour (sound production and audition) has been studied in detail in relatively few katydid species. In this respect, temperate species that are plentiful, easily accessible,

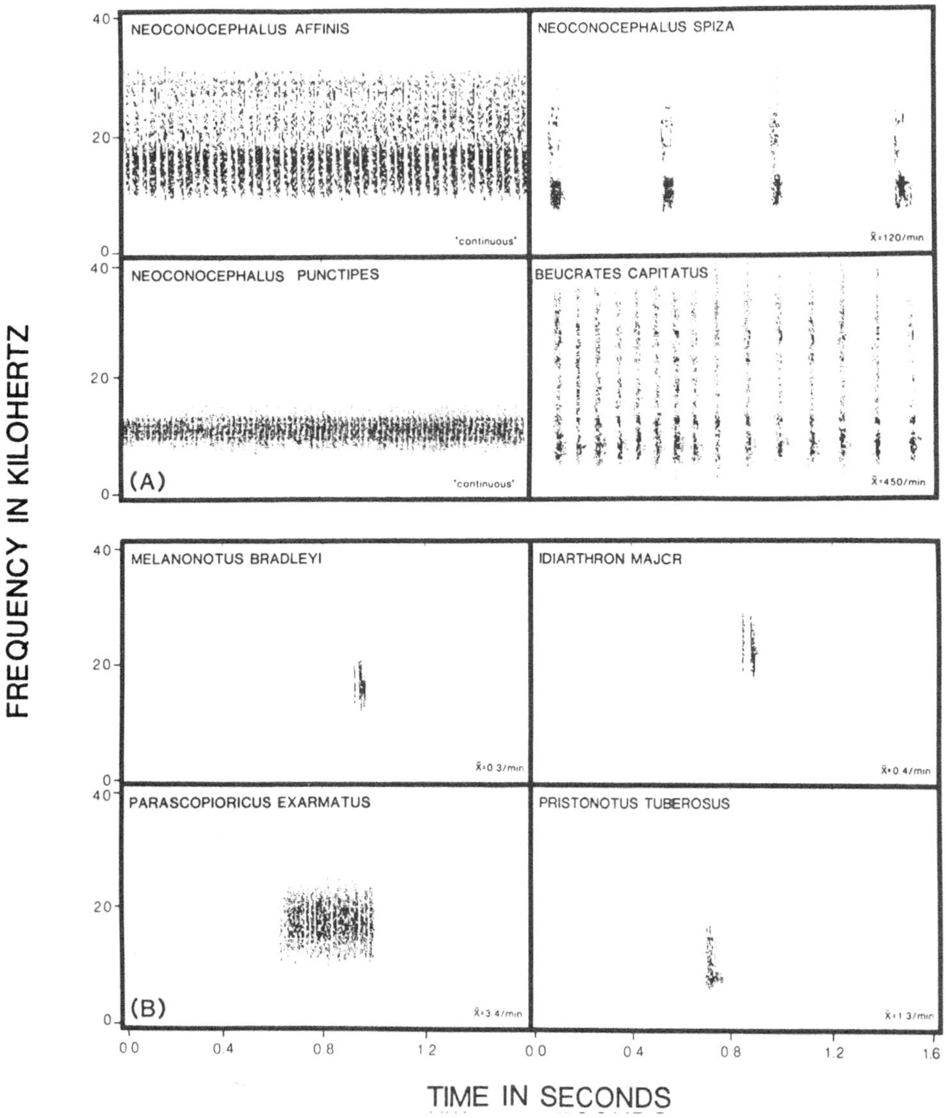

Fig. 1. Airborne calling songs (a) of members of four dominant clearing-katydid species on Barro Colorado Island (BCI), Panama. These insects are not preyed on by bats. Their calls emphasize lower frequencies, are broadband, and are produced frequently. X = average call repetition rate per minute for 3 to 5 individuals. Audiospectrograms (b) of airborne calling songs of members of four forest-inhabiting katydid species on BCI. These calls are short, infrequent, and relatively high and pure in tone, making them difficult for mammalian ears to localize. They typify katydid songs in areas inhabited by tropical New World foliage gleaning bats (Phyllostomidae: Phyllostominae). Figure from Belwood and Morris (1987). Copyright 1987 by the AAAS.

active before midnight, easy to maintain in the laboratory, and that have conspicuous (audible and sustained) airborne calls have received the most attention [e.g., the numerous studies cited in Alexander (1960), Otte (1977), Gwynne & Morris (1983), Kalmring and Elsner (1985), and Bellman (1988)]. The calls of some tropical species that share these characteristics have also been examined (e.g., Walker & Greenfield 1983; Greenfield 1983, 1988).

Results of these studies have made katydids universally well known for their song producing abilities. Consequently, they are used extensively to investigate acoustical pair formation and acoustical physiology in insects (Otte 1977; Gwynne & Morris 1983; Lewis 1983; Kalmring & Elsner 1985). For this reason, the finding that calling songs of neotropical forest katydids differ significantly from those of temperate species is surprising (Rentz 1975).

The calls of most neotropical katydids studied to date are characterized by several unusual features. They are short (usually < 1 s) and infrequently produced (every few seconds to every few minutes) (Fig. 1b). They emphasize high and/or ultrasonic frequencies [often > 20 kHz and as high as 108 kHz (Morris *et al.*, in prep.)], and their song duty cycles (percent of song period spent singing) are extremely low. In 17 species studied on BCI, the proportion of time male katydids spent singing ranged from only 0.06% to 4.55%. Short sporadic calls, low duty cycles, and high-pitched carrier frequencies also typify the calls of most of the 10 Costa Rican species described by Morris and Beier (1982), most of the nine Ecuadorian species described by Morris *et al.* (1989), and all but one of about 20 Peruvian pseudophyllines, which have been recorded, but are not yet analyzed. In contrast, the duty cycles of most temperate species are much higher [see figures in Pierce (1948), Alexander (1960), Haskell (1961), and Keuper *et al.* (1985)].

Short, sporadic, airborne calls in neotropical pseudophyllines may be widespread anti-predator adaptations to the acoustically orienting FGB that are plentiful and sympatric with these katydids and that use the airborne calling songs of these insects to locate prey. Laboratory studies (Belwood & Morris 1987) show that FGB have difficulty locating sporadically-calling katydids.

On BCI, the large pseudophylline *Ischnomela pulchripennis* appears to be the one 'exception to the rule' that forest katydids call sporadically (Belwood 1988). Its duty cycle is about 50%. This insect sings from large (> 2 m), terrestrial bromeliads, which are covered with long (> 1 cm), sharp spines that could injure or kill a FGB (Gillette and Kimbrough 1970). They also sing from other protected sites. In contrast, most of BCI's other forest katydids usually occur on, and sing from, more exposed locations. For example, the medium-sized *Copiphora brevirostris*, *Copiphora* nr. *brevirostris*, *D. gigliotosi*, and *Melanonotus bradleyi* are found at night on the upper surfaces of leaves. In captivity these are the sites that are used for calling by these insects. Larger-bodied species such as *Acanthodis curvidens*, *Balboa tibialis*, *Eucocconotus camerani*, *Idiarthron major*, and *Pristonotus tuberosus* are usually found fully exposed on sturdier substrates such as large branches and tree trunks; these are the only sites from which these insects will sing in captivity. The short high frequency calls of the katydids listed above appear to allow them to sing from these exposed areas with decreased threat of predation.

On BCI, the short sporadic calls of forest katydids (Belwood & Morris 1987) contrast sharply with the "loud penetrating" calls of *Neoconocephalus* spp. (Walker and Greenfield 1983; Greenfield 1983, 1988; Fig. 1a) that are found in grassy clearings where FGB do not occur. Mistnets placed over patches of grass containing *Neoconocephalus* spp. and *Bucrates capitatus*, another regularly calling copiphorine, captured only bats that fly through, or feed in, cleared areas (Belwood, unpublished). These are frugivores and 'aerial insectivore' bats that feed on small flying insects, respectively. This suggests that katydids in areas without FGB can sing longer, more conspicuous songs than their forest counterparts.

In Ecuador, *Typophyllum mortuifolium* recorded by Morris *et al.* (1989) also have songs with high duty cycles and that are conspicuous to the human ear. These insects are common in disturbed forest habitats in both Ecuador and Peru (Morris *et al.* 1989; Nickle, Castner & Belwood, unpublished, respectively), where FGB are uncommon, but also occur in climax forest. How they avoid predation by FGB in the latter is not known.

Neotropical lowland katydids that chorus or duet (Greenfield & Shaw 1983) in the forest canopy (e.g., *Xestoptera cornea* on BCI) also have calls that are conspicuous to the human ear. The calls of these insects are loud, broad-band signals that are produced frequently and in response to the calls of other individuals (Belwood 1988). Chorusing has been suggested to reduce predation on group singers either by protecting centrally located individuals, saturating the appetites of predators, or by decreasing the ease with which single katydids can be located using acoustic cues (Greenfield & Shaw 1983). Because these katydids occur in the forest canopy, their predator-prey interactions are difficult to study. Therefore, which, if any, of these mechanisms is in operation is not known. Recent evidence (Belwood 1988), however, shows that *X. cornea* are not protected from predation by FGB. Over two years, these katydids made up 17% (by number) and 26% (by weight) of the diets of *Micronycteris hirsuta*.

Structure of Airborne Calls

For the most part, the airborne songs of the Panamanian katydid species studied consisted of only one or a few short pulses that were produced erratically and sporadically. These occurred either singly (in *Acanthodis curvidens*, *Balboa tibialis*, *Copiphora brevirostris*, *Eucocconotus camerani*, *Idiarthron major*, *Melanonotus bradleyi*, *Parascopioricus exarmatus*, *Pristonotus tuberosus*) or in single pairs (in *D. gigliotosi*, and individual *X. cornea*). In a few species, the single or paired notes were repeated up to ten times in a short period (in *Copiphora* nr. *brevirostris*, *Mimetica mortuifolia*, and one unknown species). In only the minority of cases were the short calls produced more or less continuously and at regular intervals (*Cocconotus wheeleri*, *Scopiorinus fragilis*). However, song duty cycles of these insects were still very low (< 5%).

Principal carrier frequencies used by BCI's forest katydids were bimodal in distribution. About half used frequencies between 13 and 18 kHz [(*Acanthodis curvidens* (13-17 kHz), *Balboa tibialis* (approx. 13 kHz when the calls of several individuals are examined), *Ischnomela pulchripennis* (approx. 14 kHz), *Melanonotus bradleyi* (approx. 17.5 kHz), *Mimetica mortuifolia* (approx. 16 kHz), plus two unknown species (17.2 and 17.4 kHz, respectively)], while the others *Cocconotus wheeleri* (26.0 kHz), *D. gigliotosi* (22.4 kHz), *Idiarthron major* (23-27 kHz), *Scopiorinus fragilis* (approx. 25 kHz), *X. cornea* (27.5 kHz) used ones between 23 and 27.5 kHz. Calls from Costa Rican (Morris & Beier 1982) and Ecuadorian species (Morris *et al.* 1989) seem to fit a similar pattern, although there are exceptions (Morris, pers. comm.). It is noteworthy that the calls of the two forest copiphorines studied on BCI, *Copiphora brevirostris*, and *Copiphora* nr. *brevirostris*, have bimodal peak frequencies (approx. 15 and 31, and 14.5 and 28.7 kHz, respectively). Each falls within the boundaries of one of the two frequency categories cited above.

Four of the six katydid species used as food by the bat *M. hirsuta* on BCI sing in the 23-27 kHz range. This suggests that bat audition may be a factor that determines which species are taken as prey. However, hearing curves in these bats have not yet been investigated. Therefore, the significance, if any, of the species taken cannot be appreciated.

Temporal Aspects of Calling

Contrary to popular belief, not all katydid species become active and start to call at sunset (Walker 1983). In the Panamanian fauna studied, different species were active, and called, at different times of the night. For example, *Cocconotus wheeleri*, *D. gigliotosi*, *Idiarthron major*, and *X. cornea* were active early in the evening (between 1830 and 2400 h), while *Acanthodis curvidens* and *Balboa tibialis* were active several hours later (2000 to 0200 h). *Copiphora brevirostris*, *Copiphora* nr. *brevirostris*, and *Aganacris insectivora* were active between 2300 and 0500 h. These patterns appear to be important in determining which species are taken as prey by bats.

Katydid species that are active shortly after sunset, when bat foraging intensity is the greatest, make up the bulk of FGB prey on BCI. Species active later in the evening are conspicuously absent from the diets. This is particularly apparent in the case of *Copiphora brevirostris*. It was taken by FGB, but only in small numbers even though it is the most common katydid on BCI and was extremely palatable to bats in feeding experiments. Thus, it appears that a late active period may protect this species from bat predation.

Tremulation

The short, sporadic calls of the neotropical pseudophyllines described above appear to lessen some of the predation risks associated with airborne calling, but should also increase localization difficulties for female katydids that are attracted to males. In apparent response to this, several aspects of song structure and ear morphology in these katydids appear designed to improve localization of the short, sporadic airborne calls (Morris *et al.* 1989). In addition, males of most of the pseudophylline and copiphorine species studied to date (in the studies cited above) supplement their shortened calls with complex, species-specific body movements that generate vibratory signals that are inaudible to bats, but travel to conspecifics through shared plant substrates (Fig. 2).

These movements are called tremulations and appear as a rapid succession(s) of vertical oscillations of the body that are performed when the insect's six legs are planted firmly on a substrate (Plate 13). These movements are forceful and visible to the naked eye as a bouncing up and down of the insect's body; peak-to-peak oscillation amplitude of the abdomen can exceed one cm during tremulation. According to Morris (1980), who reviewed the occurrence of substrate signalling in Ensiferan Orthoptera (katydids and crickets), tremulation on vegetation imparts forces to a plant through the insect's feet that are propagated as waves along various plant parts (stem, leaf, petiole).

All katydids are believed to produce faint vibrations that can be transmitted to other organisms via the substrate when they stridulate (Keuper *et al.* 1985). Females use these in conjunction with airborne calls to locate singing males (Kalmring & Kühne 1980; Latimer and Schatral 1983). The vibrations produced by the tremulations discussed here differ from those produced during stridulation: they are more forceful and are produced independent of airborne calls. In many of the neotropical katydids observed to date (Belwood, unpublished), the production of airborne and substrate-borne signals are separate events that occur in seemingly random fashion within the same block of time. They are produced in the absence of females and appear to be true long distance communication signals that circumvent the airborne channel and acoustically orienting predators. Tremulations are analogous to the mating signals used by some spiders (Rovner and Barth 1981) and should function well as long distance communication cues since katydids have sensitive vibration receptors (subgenual organs) (Hutchings & Lewis 1983).

Tremulations, as described here, have been observed only in katydids with the fusiform bodies typical of non-pterochrozine pseudophyllines and copiphorines; this body shape appears to enhance the ability to oscillate the body forcefully. These katydids are also the same species that day roost in curled leaves (and use body vibrations to defend them) and other cylindrical hiding places. They also include many bark-mimicking katydids, which are fusiform.

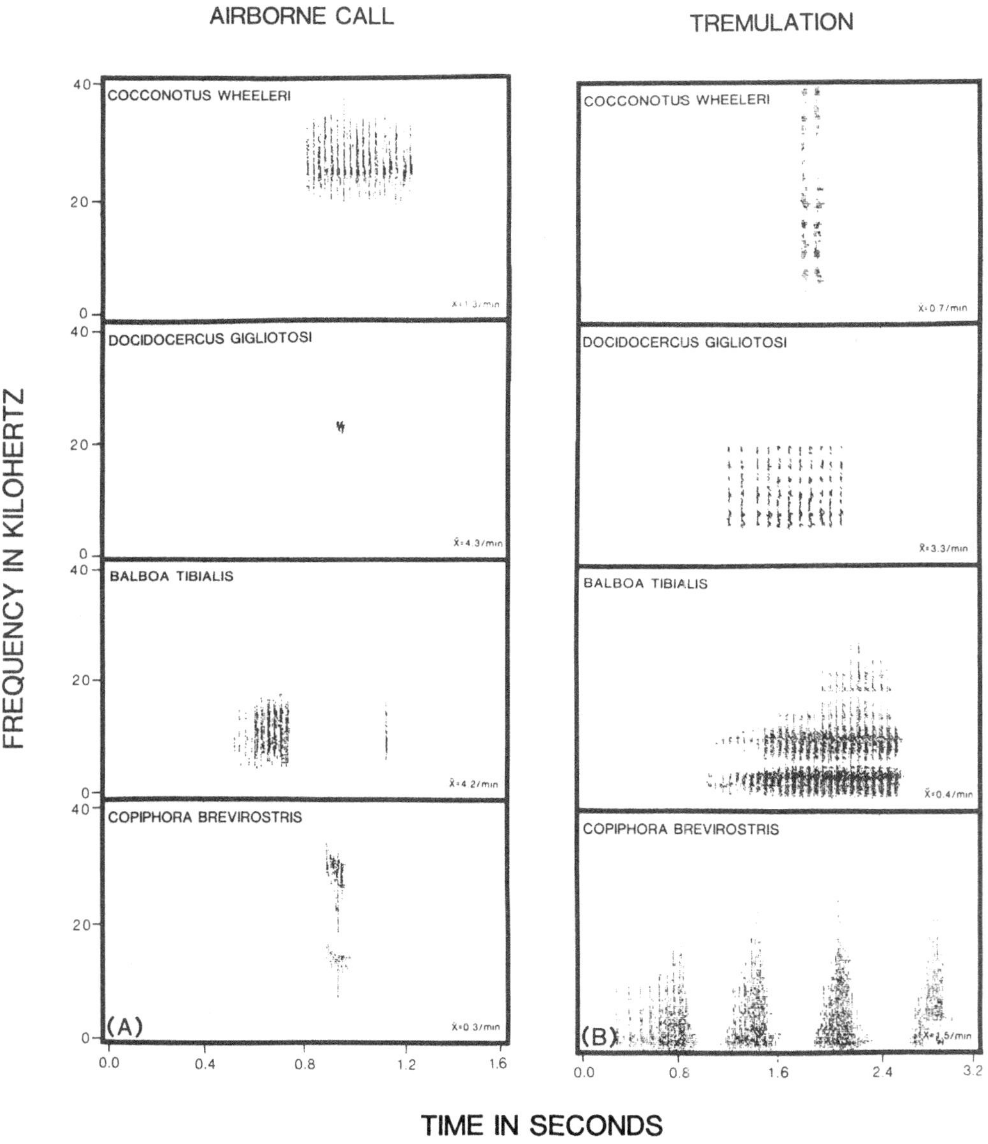

Fig. 2. Patterned, species-specific, male-produced airborne (a) and substrateborne (b) calling songs of members of four forest-katydid species on Barro Colorado Island, Panama. Individual tremulation pulses in the last three pulse sequences of *Copiphora brevirostris* 's call occur too rapidly to be differentiated by the recording method used (see Belwood and Morris 1987). However, they can be seen in the first pulse sequence. N.B., tremulations are not audible; the spectral component of the patterns is an artifact of the method used to record the signals. Also, the scales differ for the two types of signals. Figure from Belwood and Morris (1987). Copyright 1987 by the AAAS.

The vibrations recorded to date appear to be highly stereotyped, patterned, and species-specific (Belwood & Morris 1987; Belwood 1988; Fig. 2). Presumably, this helps distinguish one species from another to promote reproductive isolation. Patterned signals – ones that are unique with respect to frequency, duration, regularity, and amplitude and frequency modulation – should also help katydids distinguish their vibrations from ones produced by abiotic factors (e.g., wind, and leaves and rain that fall on vegetation). Such parameters allow some spiders to discriminate between the vibrations of potential prey and inanimate objects (Bleckmann 1985).

Temporal aspects of airborne and substrate-borne calls were compared in five katydid species on BCI. In four of the species, the duty cycle of the tremulation signal exceeded that of the airborne signal by a factor of at least 2. Substrateborne-to-airborne call ratios for *Balboa tibialis*, *Acanthodis curvidens*, *D. gigliotosi*, and *Copiphora brevirostris* were 2.2 to 1, 2.8 to 1, 12.0 to 1, and 95.3 to 1, respectively. This suggests that tremulation is an important component of the advertisement call. However, in some species (e.g., *Balboa tibialis*, *Copiphora brevirostris*) the vigorous, sustained, relatively long-lasting body vibrations of tremulation appear to be energetically more expensive to produce than the relatively short song pulses of stridulation. This means that forest katydids may pay a high price in terms of energy expenditure to avoid bat predation. This should be examined in future studies.

Although the function of tremulation has not yet been studied in detail, preliminary observations indicate that it is also important as an integral part of courtship in at least five species *Copiphora rhinoceros* (Morris 1980); *Copiphora brevirostris*, *D. gigliotosi*, and *Cocconotus wheeleri* on BCI, and *Schedocentrus* sp. in Peru (Belwood, unpublished). In the laboratory, females of these species responded to conspecific male tremulations with tremulations of their own, and this behaviour continued between males and females for at least an hour prior to copulation. This is reminiscent of precopulatory behaviour in ephiggerine katydids (Busnel *et al.* 1955) and is also analogous to the way phaneropterine females respond acoustically to the airborne calls of conspecific males (Spooner 1964, 1968).

Katydids observed in the laboratory were extremely sensitive to vibration. Without fail, they stopped call production airborne or substrateborne) if an observer or other organism (e.g., cockroach) accidently touched their cages. This may be adaptive, especially in the case of tremulation, as individuals probably increase their susceptibility to predation by animals with vibration receptors when they tremulate. For example, some ctenid spiders have repeatedly been observed feeding on katydids in the field, particularly on *Copiphora brevirostris* (Belwood 1988:162), which has the longest and most complex tremulation signal of all BCI's katydids. When pseudophyllines are active, their long antennae are in constant circular motion around their bodies (Plates 13, 14). This behaviour may help sense approaching spiders.

Audition in Katydids

Insectivorous bats and their prey have long co-evolutionary histories, and bats have influenced the development of anti-predator behaviour in moths (Roeder 1967; Spangler 1988), green lacewings (Miller & Olesen 1979), crickets (Moiseff *et al.* 1978), and mantids (Yager & Hoy 1986). In all these cases, insect defences are based on the ability of the potential prey insects to hear echolocating bats.

Tropical New World forest katydids are exposed to predation pressure from FGB whose echolocation calls contain high frequency ultrasonics (peak frequencies on BCI range from 75 to 110 kHz) (Belwood 1989). Because these insects produce very high-pitched sounds themselves and have ears, they should be preadapted to detect the calls of the bats with which they are sympatric. Whether this actually occurs is not known, as audition has not been studied in neotropical species.

Recent studies indicate that katydids in other geographic locations may indeed respond to bat echolocation calls. Spangler (1984) reports that phaneropterines in the southwestern U.S.A. (*Insara covilleae*, *Insara elegans*) stop singing in the presence of flying bats. Sales and Pye (1974:121) report similar results for stridulating *Conocephalus conocephalus* and *Conocephalus maculatus* in Nigeria. Similarly, McKay (1969, 1970) has found auditory neurons in *Homorocoryphus* (= *Ruspolia*) katydids, which are believed to be preyed on by bats, that respond to sounds greater than 30 kHz. Ultrasonic sensitivity in katydids, to frequencies as high as 100 kHz, has also been reported by several workers in Europe (Kalmring and Kühne 1980, 1983:281; Rheinlaender & Römer 1980; Hutchings & Lewis 1983:201).

Whether katydids on BCI can perceive the sounds of echolocating bats is not known. It is noteworthy, however, that in the laboratory and field, at least some of BCI's katydids (*Acanthodis curvidens*, *Balboa tibialis*, *Ischnomela pulchripennis*, *Parascopioricus exarmatus*, and *Scopiorinus fragilis*) continue to sing in the presence of the bats that are attracted to their calls (Belwood, unpublished). Also, the preponderance of katydids in the diets of FGB indicate that the bats are not perceived by at least some katydid species.

Conclusions

An attempt has been made here to review and organize the little that is known about anti-predator behaviour and ecology in tropical New World katydids. Needless to say, much of this information is anecdotal and more studies are needed in this area as well as on all aspects of katydid biology. With respect to katydid-predator interactions, several points made here should be useful in future studies:

1) A large number of predators, both nocturnal and diurnal, can feed on members of the same katydid species. *Copiphora brevirostris* is a good example of this, as it is fed on by nocturnal FGB (Belwood 1988), spiders (Plate 10), and diurnal birds (Hebard 1927:139). In response, katydids can have a suite of characteristics that help protect them from a wide range of predators, each of which has a different foraging style. The use of tremulation by *Copiphora brevirostris* as a long distance communication signal that circumvents bat hearing is facilitated by a fusiform body. This body shape also allows the insects to roost in curled dead leaves and to lessen predation by foliage-gleaning birds. However, anti-predator defences against one type of predator can increase threats by other predators. Tremulation and leaf-roosting may render katydids susceptible to predation by spiders and dead-leaf-searching birds, respectively.

2) Katydid morphology and anti-predator behaviour are intimately related. For example, the effectiveness of leaf-mimicking, bark-mimicking, wasp-mimicking, and fusiform bodies as anti-predator adaptations is contingent on factors such as the correct choice of background substrates or roosts, activity and movement patterns, and other aspects of behaviour.

3) Anti-predator defence mechanisms should be studied in the context of real ecological settings. In other words, it is important to know precisely which predators feed on which prey, because the predator-prey interactions are likely to be very specific.

4) There is no such thing as "the" katydid, or the "typical" katydid. Just as all sympatric species are unique with respect to their calling songs, activity cycles and other aspects of biology and behaviour are also unique. This includes predator-prey interactions.

5) Although some katydids may be difficult to find and observe in the field, many can be studied in the laboratory, and much can be learned in this way about their ecology and

behaviour.

6) Katydids can be added to the long list of insects with anti-predator adaptations against bats.

7) There are several potentially fruitful areas for future predator-prey studies on katydids. These include studies on katydid orientation, roost localization, the energetics of song production as it compares to tremulation production, site fidelity in the leaf-roosting species, selection of appropriate backgrounds by leaf mimics, spider-katydid interactions, and anti-predator (bat) audition. Interactions between katydids and the many species of diurnal birds that feed on them should also be studied.

Acknowledgments

Reading Dave Rentz's (1975) paper on Costa Rican *Melanonotus* was pivotal in my decision to pursue the study of neotropical bat-katydid interactions. Therefore, it is a great pleasure to be able to thank him and W.J.Bailey for the opportunity to be included in this volume. Both are also thanked for their patience regarding the late submission of this chapter. Field work in Panama was supported by the Smithsonian Institution, the Smithsonian Tropical Research Institute, the (U.S.) National Academy of Sciences, the (U.S.) National Science Foundation, Sigma Xi, the American Society of Mammalogists, the Florida Entomological Society, and the Exxon Corporation. The Center for Field Research, Watertown, Massachusetts, U.S.A., supported the work in Peru, and Peter Jensen and his staff at Explorama, Iquitos, provided continual logistical support. In Panama much field help was provided by Robin Abbey, Aixa Cabrera, Roberto Cambra, and Elizabeth Stockwell. The work in Peru would not have been possible without the help of 52 wonderful Earthwatch volunteers. Judy Gradwohl first introduced me to dead-leaf-roosting insects on BCI, and Ken Rosenberg generously made available his unpublished manuscript on this subject. He also went over the manuscript with a fine-toothed comb and provided a large number of unpublished observations. J.L. Castner provided Plate 3 and Doug Levey Plate 8. This manuscript was also read by G.K.Morris (who also contributed unpublished observations), W.J.Bailey, D.C.Rentz, and D.A. Nickle and J.L. Castner (both of whom are equally responsible for the results presented here from Peru). This paper was written during a Smithsonian postdoctoral fellowship.

References

Alexander R.D. 1960. Sound communication in Orthoptera and Cicadidae. In: Animal Sounds and Communication. Eds. W.E. Lanyon & W.A. Tavolga. AIBS, Washington D.C. pp 38-92.

Beier M. 1960. Orthoptera, Tettigoniidae, Pseudophyllinae. II. Das Tierreich 74: 1-396.

Beier M. 1962. Orthoptera, Tettigoniidae, Pseudophyllinae. I. Das Tierreich 73: 1-468.

Bellman H. 1988. A field guide to the grasshoppers and crickets of Britain and northern Europe. William Collins Sons & Co., London. 213 pp.

Belwood J.J. 1988. The influence of bat predation on calling behavior in neotropical forest katydids (Insecta: Orthoptera: Tettigoniidae). PhD Thesis, University of Florida, Gainesville, Florida, U.S.A. 286 pp.

Belwood J.J. 1989. Foraging behavior, prey selection, and echolocation in phyllostomine bats (Phyllostomidae). In: Animal Sonar: Processes and Performance. Eds. P.E. Nachtigall & P.W.B. Moore. Plenum Press, New York. pp 601-605.

Belwood J.J. Behavior and acoustics of the wasp-mimicking katydid *Aganacris insectivora* (Tettigoniidae: Phaneropterinae). MS in review.

Belwood J.J. & G.K. Morris. 1987. Bat predation and its influence on calling behavior in neotropical katydids. Science 238: 64-67.

Bleckmann H. 1985. Discrimination between prey and non-prey wave signals in the fishing spider *Dolomedes triton* (Pisauridae). In: Acoustic and Vibrational Communication in Insects. Eds. K. Kalmring & N. Elsner. Parey Verlag, Hamburg. pp 215-222.

Burk T. 1981. Evolutionary significance of predation on sexually signalling males. Fla. Entomol. 65: 90-104.

Busnel R.-G., F. Pasquinelly & B. Dumortier. 1955. La trémulation du corps et la transmission aux supports des vibrations en résultant comme moyen d'information a courte portées des Ephippigères mâle et female. Bull. Soc. Zool. Fr. 80: 18-22.

Castner J.L. 1989. Life as a leaf. Int. Wildl. 19(2): 18-20.

Cott H.B. 1957. Adaptive Coloration in Animals, 2nd ed. Metheun & Co., London. 508 pp.

Edmunds M. 1974. Defence in Animals. Longman Group Ltd., Harlow, Essex, England. 357 pp.

Gillette D.D. & J.D. Kimbrough. 1970. Chiropteran mortality. In: About Bats: A Chiropteran Symposium. Eds. B.H. Slaughter & D.W. Walton. Southern Methodist University Press, Dallas, Texas. pp 262-283.

Gradwohl J. & R. Greenberg. 1982. The effect of a single species of avian predator on the arthropods of aerial leaf litter. Ecology 63: 581-583.

Gradwohl J. & R. Greenberg. 1984. Search behavior of the checker-throated antwren foraging in aerial leaf litter. Behav. Ecol. Sociobiol. 15: 281-285.

Grant H.J. 1958. A revision of the genus *Aganacris* (Orthoptera; Tettigoniidae; Phaneropterinae). Not. Nat. 306: 1-11.

Greenfield M.D. 1983. Unsynchronized chorusing in the coneheaded katydid *Neoconocephalus affinis* (Beauvois). Anim. Behav. 31: 102-112.

Greenfield M.D. 1988. Interspecific acoustic interactions among katydids *Neoconocephalus*: inhibition-induced shifts in diel periodicity. Anim. Behav. 36: 684-695.

Greenfield M.D. & K.C. Shaw. 1983. Adaptive significance of chorusing with special reference to the Orthoptera. In: Orthopteran Mating Systems. Eds. D.T. Gwynne & G.K. Morris. Westview Press, Boulder, Colorado. pp 1-27.

Gwynne D.T. & G.K. Morris. Eds. 1983. Orthopteran mating systems: Sexual Competition in a Diverse Group of Insects. Westview Press, Boulder, Colorado. 376 pp.

Haskell P.T. 1961. Insect Sounds. Quadrangle Books, Chicago. 189 pp.

Hebard M. 1927. Studies in the Tettigoniidae of Panama (Orthoptera). Trans. Am. Entomol. Soc. 53: 79-156.

Hutchings M. & B. Lewis. 1983. Insect sound and vibration receptors. In: Bioacoustics: A Comparative Approach. Ed. B. Lewis. Academic Press, London. pp 181-205.

Kalmring K. & N. Elsner. Eds. 1985. Acoustic and Vibrational Communication in Insects. Parey Verlag, Hamburg. 230 pp.

Kalmring K. & R. Kühne. 1980. The coding of airborne-sound and vibration signals in bimodal ventral-nerve cord neurons of the grasshopper *Tettigonia cantans*. J. Comp. Physiol. 139: 267-275.

Kalmring K. & R. Kühne. 1983. The processing of acoustic and vibrational information in insects. In: Bioacoustics: A Comparative Approach. Ed. B. Lewis. Academic Press, London. pp 261-282.

Keuper A., C. Otto, W. Latimer & A. Schatral. 1985. Airborne sound and vibration signals of bushcrickets and locusts; their importance for the behaviour in the biotope. In: Acoustic and Vibrational Communication in Insects. Eds. K. Kalmring & N. Elsner. Parey Verlag, Hamburg. pp 135-142.

Latimer W. & A. Schatral. 1983. The acoustic behaviour of the bushcricket *Tettigonia cantans*. I. Behavioural responses to sound and vibration. Behav. Proc. 8: 113-124.
Lewis B. Ed. 1983. Bioacoustics: A comparative approach. Academic Press, London.
Lisenmaier W. 1972. Insects of the World. McGraw-Hill, New York. 392 pp.
Masters W.M. 1980. Insect disturbance stridulation: characterization of airborne and vibrational components of the sound. J. Comp. Physiol. 135: 259-268.
McKay J.M. 1969. The auditory system of *Homorocoryphus* (Tettigonioidea, Orthoptera). J. Exp. Biol. 51: 787-802.
McKay J.M. 1970. Central control of an insect auditory interneurone. J. Exp. Biol. 53: 137-146.
Miller L.A. & J. Olesen. 1979. Avoidance behavior in green lacewings. I. Behavior of free flying green lacewings to hunting bats and ultrasound. J. Comp. Physiol. 131: 113-120.
Moiseff A., G. Pollack & R.R. Hoy. 1978. Steering responses of flying crickets to sound and ultrasound: mate attraction and predator avoidance. Proc. Natl. Acad. Sci. U.S.A. 75: 4052-4056.
Morris G.K. 1980. Calling display and mating behaviour of *Copiphora rhinoceros* Pictet (Orthoptera: Tettigoniidae). Anim. Behav. 28: 42-51.
Morris G.K. & M. Beier. 1982. Song structure and description of some Costa Rican katydids (Orthoptera: Tettigoniidae). Trans. Am. Entomol. Soc. 108: 287-314.
Morris G.K., A.C. Mason, J.J. Belwood & P. Wall. High ultrasonics in Tettigoniidae. MS in prep.
Morris G.K., D. Klimas & D.A. Nickle. 1989. Acoustic signals and systematics of false leaf katydids from Ecuador (Orthoptera, Tettigoniidae, Pseudophyllinae). Trans. Am. Entomol. Soc. 114: 215-264.
Nickle D.A. 1988. Singing in the rainforest: the katydids of the Peruvian Amazon. Orion 7(3): 42-47.
Nickle D.A. The katydids of Panama (Orthoptera: Tettigoniidae). In: The Insects of Meso-America. Oxford University Press. In Press.
Otte D. 1977. Communication in Orthoptera. In: How Animals Communicate. Ed. T.A. Sebeok. Indiana University Press, Bloomington, Indiana. pp 334-361.
Pierce G.W. 1948. The Songs of Insects. Harvard University Press, Cambridge, Massachusetts. 329 pp.
Remsen J.V.Jr. & T.A. Parker. 1984. Arboreal dead-leaf searching birds of the neotropics. Condor 86: 36-41.
Rentz D.C. 1975. Two new katydids of the genus *Melanonotus* from Costa Rica with comments on their life history strategies (Tettigoniidae: Pseudophyllinae). Entomol. News 86: 129-140.
Rentz D.C. 1983. *Orophus conspersus* (esperanza, bushcricket). In: Costa Rican Natural History. Ed. D.H. Janzen. University of Chicago Press, Chicago, Illinois. pp 749-750.
Rentz D.C. 1987. Techniques and approaches in studying an unknown fauna: the Tettigoniidae of Australia. In: Evolutionary Biology of Orthopteroid Insects. Ed. B. Bacetti. Ellis Horwood Ltd., Halstead Press, John Wiley & Sons, New York. pp 427-432.
Rheinlaender J. & H. Römer. 1980. Bilateral coding of sound direction in the CNS of the bushcricket *Tettigonia viridissima* L. (Orthoptera, Tettigoniidae). J. Comp. Physiol. 140: 101-111.
Robinson M.H. 1966. Anti-predator adaptations in stick and leaf-mimicking insects. PhD Thesis, Oxford University, Oxford.
Robinson M.H. 1968. The startle display of *Balboa tibialis* (Brunner) (Orthoptera:

Tettigoniidae). Entomol. Mon. Mag. 104: 88-90.

Robinson M.H. 1969a. Defenses against visually hunting predators. In: Evolutionary Biology, Vol. 3. Eds. T. Dobzhansky, M.K. Hecht & W.C. Steere. Appleton, Century, Crofts, New York. pp 225-259.

Robinson M.H. 1969b. The defensive behaviour of some orthopteroid insects from Panama. Trans. R. Entomol. Soc. Lond. 121: 281-303.

Roeder K.D. 1967. Nerve Cells and Insect Behavior. Harvard University Press, Cambridge, Massachusetts. 188 pp.

Rosenberg K.V. Dead-leaf foraging specialization in tropical forest birds: measuring resource availability and use. In: Studies in Avian Biology. In press.

Rovner J.S. & F.G. Barth. 1981. Vibratory communication through living plants by a tropical wandering spider. Science 214:464-466.

Sales G. & D. Pye. 1974. Ultrasonic Communication by Animals. Chapman and Hall, London. 281 pp.

Simmons J.E., E.G. Wever & J.M. Pylka. 1971. Periodical cicada: sound production and hearing. Science 171: 212-213.

Spangler H.G. 1984. Silence as a defense against predatory bats in two species of calling insects. Southwest Nat. 29: 481-488.

Spangler H.G. 1988. Moth hearing, defense, and communication. Ann. Rev. Entomol. 33: 59-81.

Spooner J.D. 1964. The Texas bush katydid – its sounds and their significance. Anim. Behav. 12: 235-244.

Spooner J.D. 1968. Pair-forming acoustic systems of phaneropterine katydids (Orthoptera, Tettigoniidae). Anim. Behav. 16: 197-212.

Tuttle M.D. & M.J. Ryan. 1981. Bat predation and the evolution of frog vocalizations in the neotropics. Science 214: 677-678.

Tuttle M.D., M.J. Ryan & J.J. Belwood. 1985. Acoustical resource partitioning by two species of phyllostomid bats (*Trachops cirrhosus* and *Tonatia silvicola*). Anim. Behav. 33: 1369-1371.

Walker T.J. 1983. Diel patterns of calling in nocturnal Orthoptera. In: Orthopteran Mating Systems: Sexual Competition in a Diverse Group of Insects. Eds. D.T. Gwynne & G.K. Morris. Westview Press, Boulder Colorado. pp 45-72.

Walker T.J. & M.D. Greenfield. 1983. Songs and systematics of Caribbean *Neoconocephalus* (Orthoptera: Tettigoniidae). Trans. Am. Entomol. Soc. 109: 357-389.

Wickler W. 1968. Mimicry in Plants and Animals. World University Library, McGraw-Hill, New York. 255 pp.

Wilson D.E. 1971. Food habits of *Micronycteris hirsuta* (Chiroptera: Phyllostomatidae). Mammalia 35: 107-110.

Yager D.D. & R.R. Hoy. 1986. The cyclopean ear: a new sense for the praying mantis. Science 231: 727-729.

3. THE KATYDID SPERMATOPHORE: Evolution of a Parental Investment

Darryl T. Gwynne

Introduction

In most insects insemination is achieved through the transfer of an ejaculate packaged in a spermatophore. Since the insect spermatophore is thought to be an adaptation to a terrestrial mode of life, sperm packaging is considered to be ancestral to the transfer of a free ejaculate through an intromittent organ (Davey, 1960).

Virtually all orthopterans transfer sperm in a spermatophore (but see Ono *et al.* 1989) and a variety of different types of spermatophore have been observed in this group. During copulation male grasshoppers (suborder Caelifera) insert the long neck of a specialized spermatophore, the body of which is retained in the male terminalia throughout mating (Davey, 1960). All katydids (Tettigoniidae) and crickets (Gryllidae) (suborder Ensifera) that have been examined have a spermatophore which is transferred entirely to the female at mating (Alexander and Otte, 1967). The tettigoniid spermatophore is more complex than that of most Ensifera. It consists of several parts that are moulded together during copulation. This contrasts with the more simple gryllid spermatophore which is fully-formed prior to mating (Boldyrev, 1915, 1928; Ander, 1939). Of particular interest to the present paper is the fact that in virtually all katydids the spermatophore exhibits a gelatinous mass, the spermatophylax. This mass does not function in transporting sperm and, since it is eaten by the female after mating (Boldyrev, 1915; Gwynne 1983), appears to be analogous to "courtship feeding" of prey items as observed in other animals (Lack, 1940; Thornhill, 1976a). A spermatophylax is seen in only a few other Ensifera (Gryllidae, Prophalangopsidae and Raphidophoridae) (Boldyrev, 1927; Ander, 1939; Alexander & Otte, 1967) and a neuropteran (*Osmylus chrysops*, Withycombe, 1922).

The main purpose of this chapter is to summarize evidence that suggests a unique function for the provision of a spermatophylax meal by male katydids. The results suggest that selective pressures leading to the origin of this 'nuptial gift' were quite different from the pressures that currently act to maintain the form of this structure. The complex katydid spermatophore is assumed to have evolved from a more simple sperm package. I begin the paper by reviewing knowledge of the spermatophore structure of extant tettigoniids and related Ensiferans in order to trace the likely evolutionary steps in the elaboration of the katydid spermatophore from a more simple ancestral type.

A possible evolutionary history of the katydid spermatophore

The katydid spermatophore is very different both in size and complexity from that of related insect groups. In orthopteroid insects other than the suborder Ensifera (suborder Caelifera and orders Blattodea, Mantodea and Phasmatodea) the structure consists of a small single sperm-containing reservoir (ampulla) which in some groups is retained by the male after copulation. In others it is positioned in or outside of the female genital tract (see Mann, (1984), for a recent review). Most ensiferans possess a sperm ampulla that after mating is positioned outside the female gonopore. The Gryllidae are a typical example. In this speciose family, the spermatophore has a long neck which is inserted into the female with the ampulla remaining outside. Since this is the most common form of spermatophore, it seems reasonable to assume that this "simple" sperm ampulla is the ancestral state for the Ensifera.

Figure 1 presents a suggested evolution of the spermatophores of 11 of 19 recognized tettigoniid subfamilies and most of the common subfamilies (see Rentz, 1979). Information

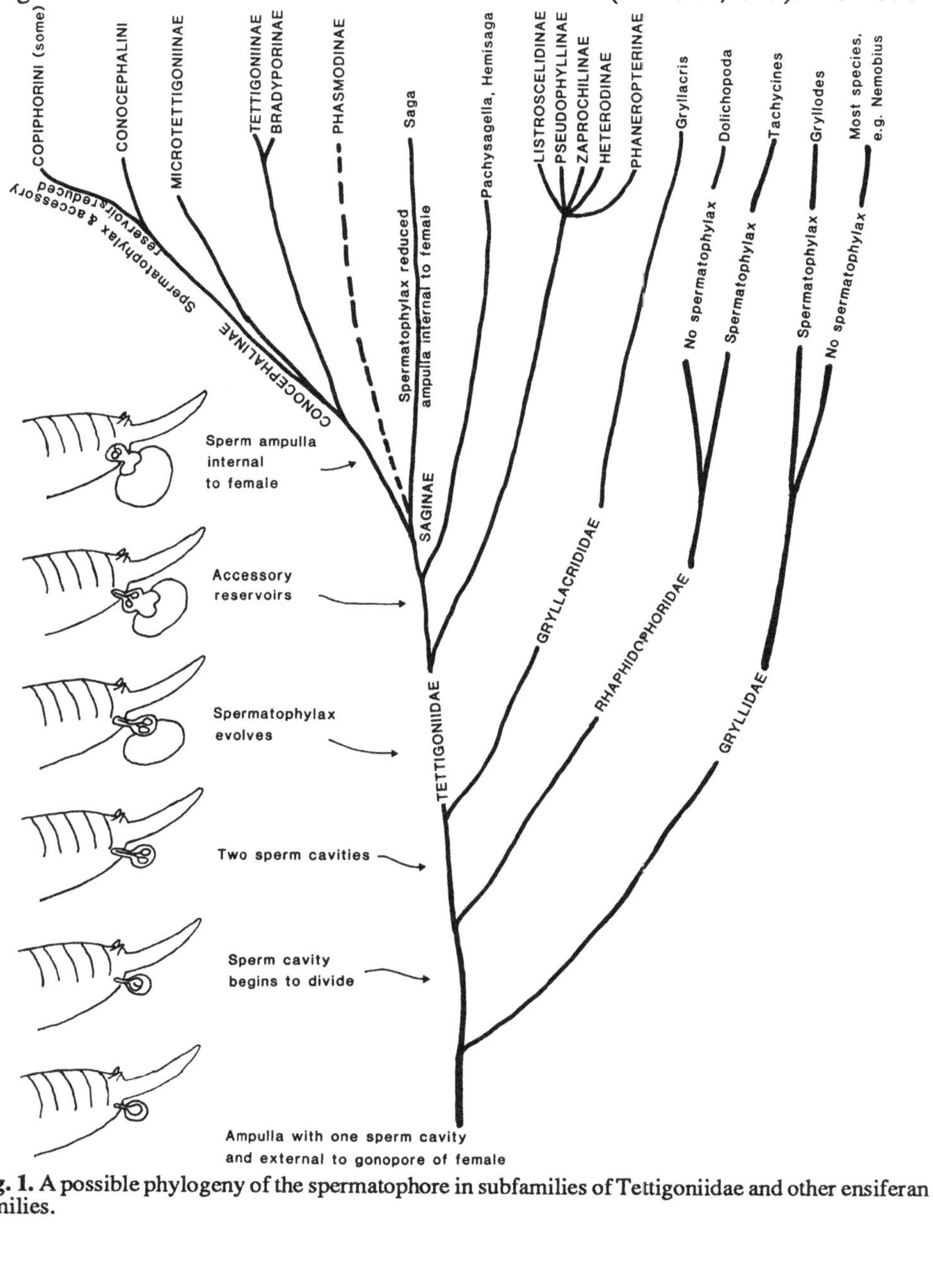

Fig. 1. A possible phylogeny of the spermatophore in subfamilies of Tettigoniidae and other ensiferan families.

on several other extant ensiferans are included in the figure to represent putative early steps in the evolution of the tettigoniid spermatophore. Details of species names and spermatophore structure are presented in Table 1.

The putative ancestral ensiferan spermatophore is represented by extant species of *Nemobius* and *Gryllotalpa* (Gryllidae) (Fig. 2). The spermatophore structure of certain species of Rhaphidophoridae represents the next suggested stage (Fig. 4). Only tettigoniids, gryllacridids (*Gryllacris*) and possibly the prophalangopsids (*Cyphoderris*), show a sperm ampulla with two sperm cavities (Ander, 1939). This appears therefore to be a derived trait shared by these groups. Species of Rhaphidophoridae may represent an early stage in the evolution of a double sperm vesicle in that "the ridge in the hind wall (of the ampulla) suggests the first beginning of a separating wall" (Ander, 1939). Any adaptive significance of two sperm cavities is unknown. *Gryllacris* represents the next step; in this species the ampulla is fully divided into two sperm cavities. *Gryllacris* does not have a spermatophylax.

The next step is the evolution of a spermatophylax. Although the ancestral katydid probably featured a spermatophylax, since virtually all extant species examined thus far do, it appears that this structure evolved independently in other ensiferans including several Gryllidae (Boldyrev, 1927 [Fig. 3]; Alexander & Otte, 1967; Sakaluk, 1984) and a rhaphidophorid *Tachycines* (*Diestrammena*) (Gerhardt, 1913) (Fig. 1).

The tettigoniid subfamilies Pseudophyllinae, Phaneropterinae, Zaprochilinae, Heterodinae and Listroscelidinae have a spermatophore that is probably the ancestral type for the family. The spermatophores of these groups appear to be primitive, relative to other katydids, since other groups show further elaborations (see below). Gerhardt's (1913) figure of the spermatophore of the rhaphidophorid *Tachycines* (Fig. 4), shows it to be very similar to the suggested primitive katydid spermatophore (Figs. 6, 8,9,13) (the difference being that *Tachycines* has a single-chambered ampulla).

Most of the remaining tettigoniids have elaborations of the sperm ampulla that, although referred to by Boldyrev (1915) as "accessory reservoirs," have an unknown function (Boldyrev, 1928). The Australian sagines *Hemisaga* and *Pachysagella* (*Pachysaga*) have accessory reservoirs and ampullae that are positioned external to the female. They appear to be more primitive in spermatophore structure than certain Conocephalinae (tribe: Conocephalini) (Fig. 10), Tettigoniinae (Figs. 7 & 12), Bradyporinae and Microtettigoniinae in that in the latter groups the sperm ampulla is positioned internal to the female in an enlarged genital chamber. The European sagine *Saga ephippigera* (Fig. 5), however, differs from the antipodean Saginae (see below).

The main difference among the tettigoniids that have both accessory reservoirs and an ampulla internal to the female, appears to be in the shape of the sperm cavities, particularly in the length of their ejaculatory ducts. All studied Tettigoniinae (tribes Decticini and Tettigoniini) have a shorter ejaculatory duct than the Conocephalini and the single bradyporine species. The spermatophore of Microtettigoniinae (*Microtettigonia*) has a very short ejaculatory duct. The spermatophore anatomy of this recently-described subfamily groups it with the Conocephalinae and Tettigoniinae. Rentz (1979) suggested that the subfamily appeared to be closely related to the Conocephalinae.

The three remaining groups of Tettigoniidae for which there is some knowledge of spermatophore anatomy are *Saga* (Saginae), the Phasmodinae (*Phasmodes ranatriformis*) and copiphorines that inhabit grassy areas (Conocephalinae, *Ruspolia* and *Belocephalus*) (Fig. 11). The spermatophores of these katydids have very small spermatophylaxes (covered by the female subgenital plate [see Fig. 5]), ampullae positioned inside the female, and very short ejaculatory ducts. Given the widespread occurrence of the large spermatophylax in

Table 1. Characters of ensiferan spermatophores used in constructing figure 1.

Taxon	Spermatophylax size[1]	Number of sperm cavities	Location of ampullae[2]	Presence of accessory reservoirs	Length of ejaculatory duct
Tettigoniidae					
Conocephalinae:					
Conocephalini;					
Orchelimum delicatum Brunner (1)	large	2	internal	+	long
Conocephalus fuscus (F) (2)	large	2	internal	+	long
Copiphorini;					
Ruspolia nitidula (Scop)(2)	small	2	internal	-	short
Belocephalus subapterus (Scud.)(1)	small	2	internal	-	short
Microtettigoniinae:					
Microtettigonia sp.(1)	large	2	internal	+	short
Saginae:					
Saga ephippigera Fisch.(2)	small	2	internal	+	short
Hemisaga denticulata White (1)	large	2	external	+	long
(Pachysaga) Pachysagella(1)	large	2	external	+	long
Tettigoniinae:					
Tettigoniini;					
Tettigonia viridissima (L)	large	2	internal	+	medium
T. cantans (Fuessly)(2)	large	2	internal	+	medium
Anabrus simplex Haldeman (1)	large	2	internal	+	medium
Decticus albifrons Fabr. (2)	large	2	internal	+	medium
D. verricivorous L (2)	large	2	internal	+	medium
Platycleis vittata Charp. (2)	large	2	internal	+	medium
Metrioptera roeseli (Hagenb.)(2)	large	2	internal	+	medium
Bradyporinae;					
Bradyporus multituberculatus FW(2)	large	2	internal	+	long
Phasmodinae:					
Phasmodes ranatriformis West.(1)	small	2	internal	-	short
Zaprochilinae:					
Gen. Nov. 22[3](1)	large	2	external	-	short
Listroscelidinae;					
Requena verticalis Walker (1)	large	2	external	-	medium
Phaneropterinae					
Phaneroptera quadripunctata Brunner (2)	large	2	external	-	long
Tylopsis thymifolia (2)	large	2	external	-	long

Table 1. Characters of ensiferan spermatophores used in constructing figure 1 (continued)

	Spermatophylax size[1]	Number of sperm cavities	Location of ampullae[2]	Presence of accessory reservoirs	Length of ejaculatory duct
Isophya (2)	large	2	external	-	long
Poecilimon bosphoricus Ret.(2)	large	2	external	-	long
Leptophyes punctatissima Bosc (2)	large	2	external	-	long
Polichne sp.(1)	large	2	external	-	long
Acanthoplus bechuanus Per.(3)	large	2	external	-	long
Pseudophyllinae:					
Cocconotus insularis (Brun.)(4)	large	2	external	-	?
Zabalius apicalis Bolivar (8)	large	2	external	-	?
Adenes sp. (8)	large	2	external	-	medium
Haglidae: (=Prophalangopsidae)					
Cyphoderris strepitans Morris & Gwynne (1,4)	large	2	external	?	-
Rhaphidophoridae:					
Tachycines sp.(2)	large	1+[4]	external	-	long
Dolichopoda euxina Sem.(2)	large	1	external	-	short
Gryllacrididae					
Gryllacris sp.(5)	none	2	external	-	?
Gryllidae:					
Gryllotalpidae					
Gryllotalpa sp.(2)	none	1	external	-	long
Gryllinae:					
Gryllus frontalis Fieb.(2)	none	1	external	-	long
Gryllodes supplicans Walker (6)	medium	1	external	-	long
Gryllomorpha dalmatina (Oesk)(7)	medium	1	external	-	long
Nemobiinae:					
Nemobius sp.(2)	none	1	external	-	long

[1]A large spermatophylax is much larger than the ampulla, a medium one is about the same size as the ampulla and a small one is barely visible.
[2]This refers to the position of the ampulla with respect to the female's gonopore.
[3]This species is in an undescribed genus (D.C. Rentz, pers. comm.).
[4]This species has a partial dividing wall (see text).
The numbers in brackets after each taxon refer to the sources for the information: (1) Gwynne, unpublished, (2) Boldyrev (1915) and Gerhardt (1913, 1914), (3) Power (1958), (4) Leroy (1969), (5) Ander (1939), (6) Alexander and Otte (1967), (7) Boldyrev (1927), and (8) Eluwa (1979).

2

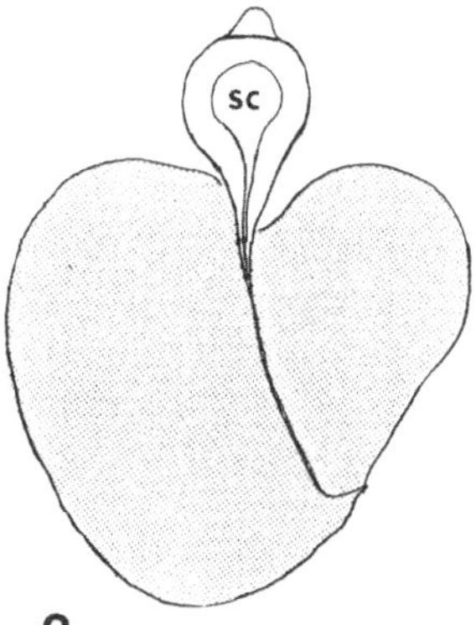

3

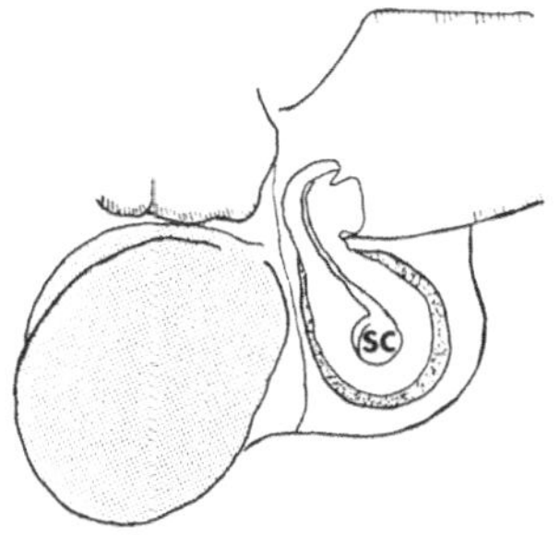

4

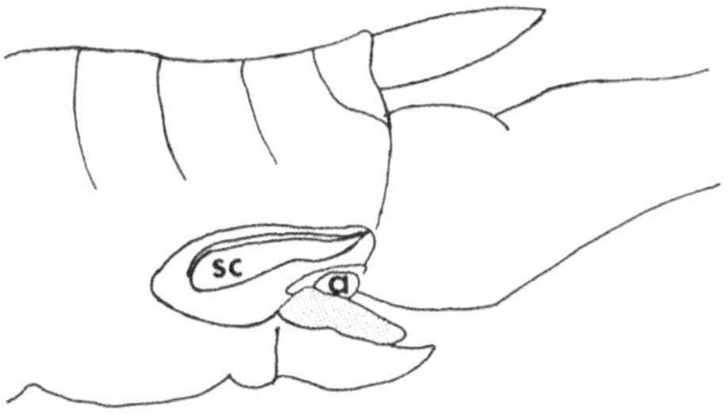

5

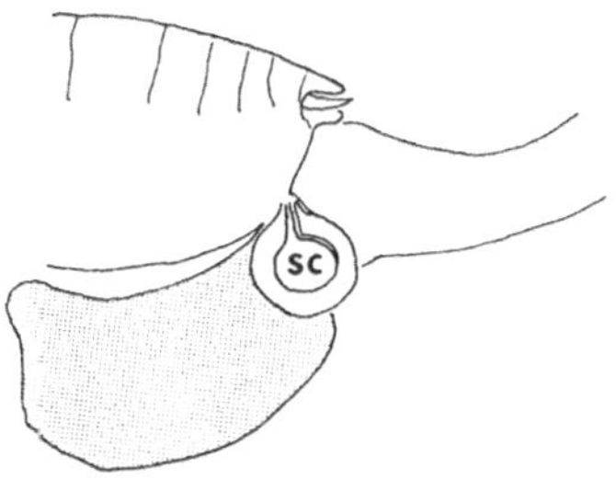

6

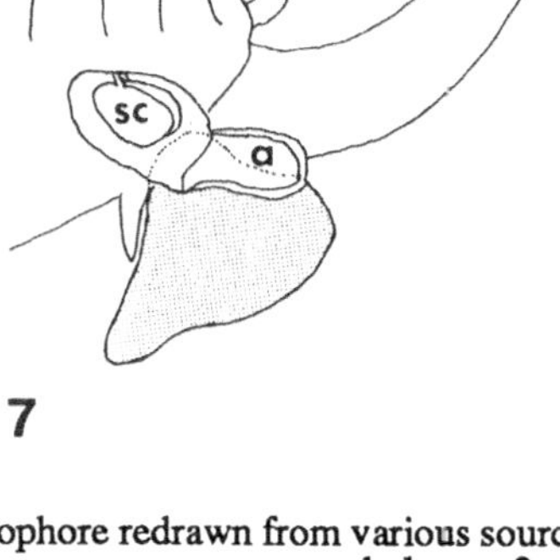

7

Figs. 2-7. A sampler of ensiferan spermatophore redrawn from various sources: sc = sperm cavity, a = accessory reservoir. The stippled areas represent spermatophylaxes. **2.** *Gryllotalpa gryllotalpa* (Gryllidae) (Boldyrev, 1915). **3.** *Gryllomorpha dalmatina* (Gryllidae) (Boldyrev, 1927). **4.** *Tachycines* sp. (Rhaphidophoridae) (Gerhardt, 1913). **5.** *Saga ephippigera* (Tettigoniidae) (Boldyrev, 1915). **6.** *Isophya acuminata* (Tettigoniidae) (Boldyrev, 1915). **7.** *Metrioptera roeselii* (Tettigoniidae) (Gerhardt, 1913; Boldyrev, 1915). The dotted line in figure 7 indicates where the spermatophylax was removed to reveal the accessory reservoir.

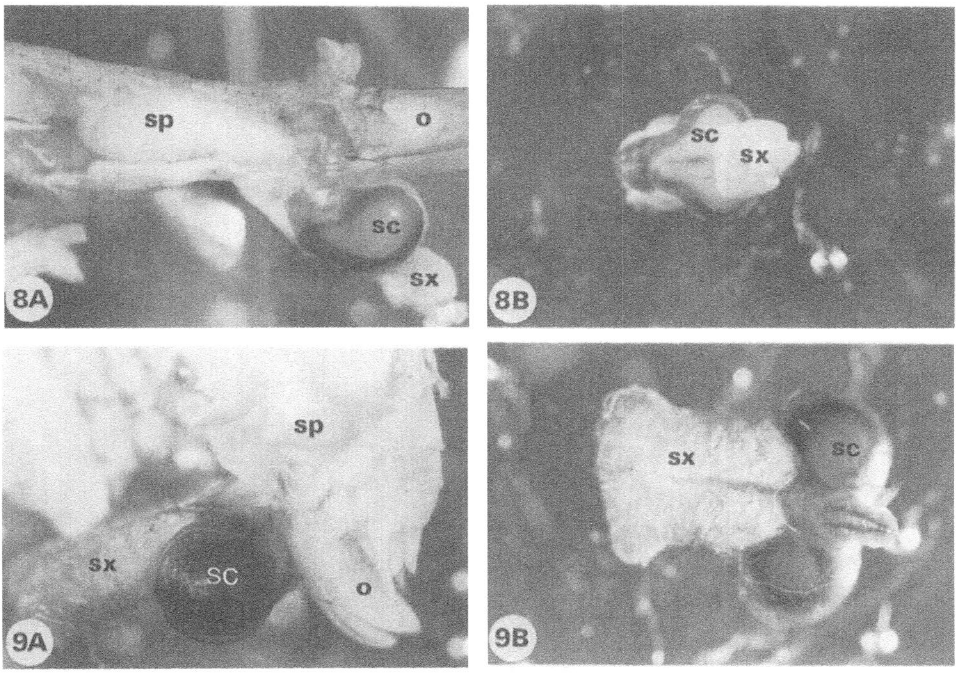

Figs. 8-9. A sampler of tettigoniid spermatophores: o = ovipositor, sc = sperm cavity, a = accessory reservoir, sx = spermatophylax, sp = spermatheca. **8A.** An unnamed zaprochiline: lateral dissection of mated female showing spermatophore. **8B.** Ventral view of spermatophore. **9A.** *Polichne* sp. (Phaneropterinae): lateral dissection of female. **9B** Dorsal view of spermatophore.

katydids it appears that the small spermatophylax of these groups is a derived trait. *Saga* is similar to the other sagines in having accessory reservoirs but *Ruspolia* and *Belocephalus* do not have these structures (although they are present in other Conocephalinae). This suggests a secondary loss of accessory reservoirs accompanied the reduction in spermatophylax size in these copiphorine genera (*Copiphora rhinoceros*, a copiphorine that inhabits rainforest exhibits a large spermatophylax (Morris, 1980)).

Differences in the structure of the spermatophores supports the separation of Zaprochilinae and Phasmodinae into separate subfamilies as advocated by Ander (1939) and supported by Rentz (1979). Unlike the Zaprochilinae, Phasmodinae fall into the 'derived' tettigoniid spermatophore group: the ampulla of the phasmodine spermatophore is positioned inside the female. The relationship of this spermatophore within this group is somewhat uncertain. The phasmodine spermatheca has 'spermatodoses,' separate sperm vesicles within the spermatheca. Since this character is shared with the Tettigoniinae and Saginae (Boldyrev, 1915), I have placed the Phasmodinae with these two subfamilies (Fig. 1). This arrangement assumes that the similarity of the phasmodine and *Belocephalus – Ruspolia* copiphorine spermatophores (i.e. internal to the female, with no accessory reservoir and a reduced spermatophylax) is due to convergence. An alternative hypothesis is that the phasmodine spermatophore is closely-related to that of copiphorine katydids.

THE FUNCTION OF THE SPERMATOPHYLAX

What are the selection pressures that may have led to the evolution of the elaborate katydid spermatophore from a simple cricket-like structure? The tettigoniid spermatophylax is large

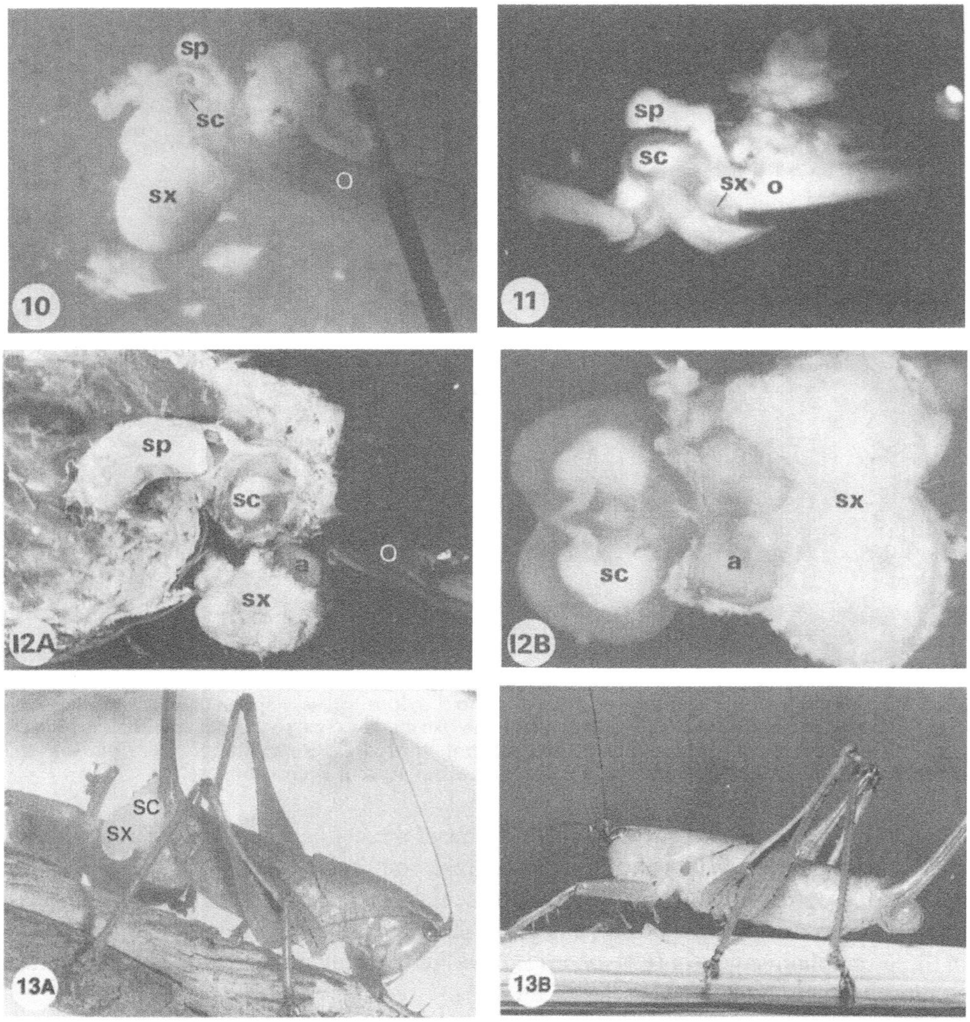

10-13. *Orchelimum delicatum* (Conocephalinae, Conocephalini): lateral dissection of female. **11.** *Belocephalus subapterus* (Conocephalinae, Copiphorini): lateral dissection of female. **12A.** *Anabrus simplex* (Tettigoniinae): lateral dissection of female. **12B.** Dorsal view of spermatophore. **13A.** *Requena verticalis* (Listroscelidinae) female with spermatophore after mating. **13B.** Female consuming the spermatophylax.

and costly (Davies & Dadour, 1989) so is undoubtedly not a result of random evolution (e.g genetic drift) or "correlated effects." Presentation of a food gift at mating is an elaborate trait. It is such traits that arouse curiosity about their evolved function (see Thornhill & Alcock, 1983).

Theoretical background

Below I review evidence for the function of the spermatophylax. By "function" I refer to "evolved function," i.e. the present role of an adaptive feature shaped by natural selection for this role (Williams, 1966).

There have been a number of recent discussions in the literature about the function of courtship gifts. I will review these arguments before presenting empirical data. Several authors have argued that these male donations are maintained by sexual selection on males to acquire fertilizations by providing nutrition in exchange for gamete transfer (and ultimately fertilizations of eggs). An alternative hypothesis is that the current selective advantage of spermatophylax meal is a result of natural selection for parental investment, a source of nutrition which increases the fitness of the male's offspring (Trivers, 1972).

On theoretical grounds Alexander and Borgia (1979), Gwynne (1984a) and Wickler (1985) argued that courtship gifts were probably a result of sexual selection since males are unlikely to have evolved to provide parental investment before fertilization. The suggestion is that, in contrast to females, males never have complete assurance of parentage or control over the fate of their gametes. Thus, relative to females, nutrients donated to the next generation by males seem less likely to result in benefits to their own progeny. In contrast to these arguments, other authors present the view that courtship feeding in insects can represent investment by the male in the fitness of their progeny. Since Trivers (1972) defined parental investment in terms of its effect on offspring fitness (i.e. an implicit inclusion of both function and incidental effect), Zeh and Smith (1985) and Sakaluk (1986) stated that it is appropriate to refer to male gifts donated before fertilization as parental investment, even if this is not the selective context for the evolution of these gifts.[1] Sakaluk (1986) pointed to the fact that most studies of insects have shown mixed paternity of offspring with these data predicting that courtship-feeding males would father at least some of the offspring containing nutrients from their donations.

A point that should be emphasized is that, even if defined in terms of beneficial effect, courtship feeding by males must enhance the fitness of offspring if these contributions represent parental investment in the sense of Trivers (1972). There have been several studies in which nutritional donations by male insects before fertilization takes place increase the fitness of the mated female by increasing the number of eggs she lays (in a scorpionfly, Thornhill, 1981; a grasshopper, Butlin *et al.*, 1987; and a butterfly, Rutowski *et al.*, 1987). However, since none of these studies reported any change in fitness parameters of individual offspring (e.g. egg size), it is unclear to what extent that these offspring benefit (see Wickler, 1985). One argument might be that since an increase in the number of eggs produced results in newly-created offspring, these offspring obviously receive a fitness increase. It is not clear, however, whether the fitness of all offspring is increased. In fact, it remains a possibility that an increase in the size of a clutch may adversely affect the fitness of individual newly-hatched offspring such as by increased competition for food resources (e.g. Simmons, 1987).

Some suggestions about the function of the katydid spermatophylax preceded the above arguments. Boldyrev (1915, 1928) and Alexander and Otte (1977) provided the equivalent of a sexual selection hypothesis by arguing that this nuptial meal in crickets and katydids functioned to distract the female and thus to prevent her from interfering with insemination (i.e. by removing the sperm ampulla prematurely). Presumably, females that interfered with sperm transfer obtained nutritional or other benefits from their manipulation of the ejaculate (Gwynne *et al.* 1984).

[1]Quinn and Sakaluk (1986), suggested that this original concept of parental investment is the most suitable one for understanding the relationship between parental investment and the intensity of sexual selection on the sexes. It is important to note, however, that this question of the consequences of male nutritional donations is separate from that of the adaptive significance of these traits (Gwynne, 1986a; Quinn and Sakaluk, 1986). Both questions are relevant to the evolutionary biology of nuptial feeding bahaviour.

Empirical support

Support for the "ejaculate-protection" hypothesis was provided by a study of a spermatophylax-bearing cricket *Gryllodes supplicans*. In those insects exhibiting courtship feeding, transfer of the ejaculate ends when the female finishes feeding. Sakaluk (1984) showed that the average time taken by female *G. supplicans* to consume the spermatophylax approximated the time taken for the full complement of sperm to be transferred into the female reproductive tract. Thus, males providing small offerings are less likely to transfer full-sized ejaculates. A parallel result was obtained for a courtship-feeding scorpionfly *Hylobittacus apicalis*. By allowing a fixed length of feeding time by females males provided no more of the nuptial gift (an arthropod prey item) than necessary to transfer a full ejaculate. Furthermore, due to female rejection, ejaculate transfer was incomplete if males provided undersized prey (Thornhill, 1976b).

A series of experiments were undertaken to investigate the function of the spermatophylax in a katydid, *Requena verticalis* (Listroscelidinae) (Gwynne *et al.* 1984, Gwynne, 1984b, 1986b, 1988a,b) (Fig. 13). Initial experiments with this species suggested that the spermatophylax had a similar function to that of *Gryllodes* in enabling the male to transfer a full ejaculate; female *R. verticalis* remove the spermatophylax within a couple of minutes after mating and, when experimentally-deprived of this male contribution, females removed and ate the sperm ampulla and prevented ejaculate transfer (Gwynne *et al.*, 1984). Additional experiments indicate, however, that in this species ingestion of spermatophylax nutrients by females increases the fitness of offspring fathered by the investing male (Gwynne, 1988a).

Given this apparent "dual role" of the spermatophylax it appeared to be difficult to determine whether ensuring insemination or investment in the male's offspring is the selective context currently maintaining the form of this structure. In fact, Gwynne (1984a) considered that the two hypotheses may be impossible to separate, and Quinn and Sakaluk (1986) suggested that classification of this sort of male effort according to 'primary function' should be avoided since these difficulties may result in erroneous conclusions.

However, the katydid spermatophylax, like any other trait, may well have a 'primary function,' i.e. a selection pressure mainly responsible for maintaining the trait. Given the above theory, three conditions appear to be necessary to provide a reasonable demonstration that this nuptial gift functions as male parental investment. First is a refutation of the most plausible alternative – that variation in spermatophylax size is related to variation in the ability of males successfully to transfer ejaculates. Secondly, there should be evidence that male contributions benefit offspring and lastly, that the courtship-feeding male fathers these offspring.

With regard to the first condition, in contrast to initial observations mentioned above, later results revealed that the spermatophylax of *R. verticalis* does not appear to function in ensuring transfer of a complete ejaculate; variation in the size of this food gift does not affect variation in the ability of males to inseminate females. The female takes an average of 5.25 hours to consume all the spermatophylax material and remove the ampulla, yet both a full complement of sperm and apparent substances that induce a full sexual refractory period in the female have been transferred by 2.5 hours after mating. All males in the experiment produced a spermatophylax of a size that was more than sufficient to ensure ejaculate transfer (Gwynne, 1986b).

Gwynne (1984b, 1988a) provided evidence for the second condition, that spermatophylax nutrition affects the fitness of offspring produced by the courtship-feeding female. Females that consumed larger nuptial meals produced larger eggs but there was no significant effect on the number of eggs produced (Gwynne, 1988a). An increase in the number of spermatophylaxes eaten by females, such as would be obtained by multiple mating, resulted in an increase in both the size and number of eggs produced (Gwynne, 1984b, 1988a). Since egg size is known to affect offspring fitness (e.g. Capinera, 1979), Gwynne (1988a) examined the progeny of experimental females. Females that produced larger eggs had significantly more offspring surviving the subsequent winter (eggs are laid in late summer and autumn). Furthermore, there was some evidence that spermatophylax nutrients influence the fitness of progeny after maturity: sons of females that had received additional courtship meals had a significantly longer development time before their moult to adulthood. Development time in turn was positively correlated with adult body size. Assuming that the cost of additional development time is not too large, larger males would probably enjoy increased reproductive success: larger male *R. verticalis* are more successful than small ones in sexual competition (Schatral, unpublished).

Although courtship feeding in *R. verticalis* appears to influence offspring fitness, does the mating male actually invest parentally by fathering these offspring? Gwynne (1988b) examined this question by individually radio- labelling the spermatophore proteins of two male *R. verticalis* of different genotypes (determined electrophoretically) that were mated to the same female. Matings were four days apart, a time period approximating the length of the sexual refractory period of the female (Gwynne, 1986b). Radioactive proteins (14C and 3H) were used as labels and a polymorphic enzyme (esterase) was used to determine genotypes. The results of the experiment showed that the first male to mate did indeed invest parentally in offspring by fathering almost all the progeny, many of which contained more of his nutrients than those of the other male.

THE ORIGINS OF THE SPERMATOPHYLAX: A NUPTIAL GIFT THAT HAS CHANGED IN FUNCTION?

In contrast to the selection for the maintenance of the form of the spermatophylax being mainly natural selection for male parental investment, it is likely that sexual selection – i.e. protection of the ejaculate to succeed in sexual (sperm) competition with other males for egg fertilizations – was responsible for the origin of the katydid spermatophylax. Females of the ancestral species may have obtained nutritional benefits by mating with a number of males and consuming the sperm ampullae without using the enclosed sperm to fertilize eggs. Boldyrev (1915) describes the simple sperm ampullae of several gryllid species and provides an account of one species which may employ such a foraging tactic. Females of this species do mate a number of times and remove the ampulla quickly after mating, often before the ejaculate has emptied. Boldyrev (1915) suggested that ensiferans may have evolved several different ways to ensure complete insemination: in some species males 'guard' the female, preventing her from removing the ampulla; and in others males 'distract' females with edible secretions, either from external glands or from a sticky mass, the spermatophylax, added to the sperm ampulla.

The nutritional benefits to ancestral females of consuming the proteinaceous sperm ampullae (Khalifa, 1949) may in fact be similar to those obtained by the courtship-feeding female katydid; in crickets (*Gryllus bimaculatus*) nutrition from sperm ampullae effects an increase in the size of eggs produced by the mated female (Simmons, 1988). Simmons states that male nutritional investment in the *Gryllus* ampulla is unlikely to function as parental investment since many more ampullae than a single male can supply are necessary to affect a detectable increase in egg size.

Thus, female consumption of the ampulla before insemination was complete, may have been the original selection pressure that produced the ancestral spermatophylax. Sakaluk

(1984) stated that the 'protective' spermatophylax of the gryllid *G. supplicans* may parallel such a stage in the evolution of the katydid spermatophylax (see also Quinn and Sakaluk, 1986). Material which comprised the spermatophylax was probably derived from glandular tissue similar to that producing the sperm ampullae. The nutritional effects of such material (Simmons, 1988) would have likely preadapted ancestral katydid males for the elaboration of the spermatophylax into a source of nutrition which benefitted their offspring.

Summary

In most species of Tettigoniidae the spermatophore is a large, elaborate structure. It consists of two main parts, each with a different function. The sperm ampulla has two sperm cavities which contain the ejaculate. The spermatophylax is a sperm-free mass which is eaten by the female after mating. In most species the spermatophylax accounts for the majority of the spermatophore's mass. In the first part of this chapter I use details of spermatophore anatomy in extant species of Ensifera to outline possible steps in the evolution of this structure in katydids, and to suggest relationships among the different types of spermatophores found within the Tettigoniidae. Since the spermatophores of most orthopteroid insects consist of a sperm ampulla with a single sperm cavity, I assume that the complex katydid spermatophore has evolved from a more simple ancestral type. Of particular interest is the evolution of the spermatophylax. The second part of this chapter reviews the literature on the evolved function of this food item provided by males to their mates. The spermatophylax appears to have originated in a sexual selection context to facilitate the transfer of the ejaculate by distracting the female with a courtship 'meal.' The katydid spermatophylax appears, however, to have undergone a change in function. I summarize information suggesting that, rather than in facilitating ejaculate transfer, its large size is currently maintained by natural selection for provision of nutritional benefits by males to their offspring.

Acknowledgements

Thanks to W. J. Bailey, D. C. Rentz and S. Sakaluk for comments.

References

Alexander R.D. & D. Otte. 1967. The evolution of genitalia and mating behavior in crickets (Gryllidae) and other Orthoptera. Misc. Publ. Mus. Zool. Univ. Mich. 133: 1-62.

Alexander R.D. & G. Borgia. 1979. On the origin and basis of the male-female phenomenon. In: Sexual Selection and Reproductive Competition in the Insects. Eds. M. Blum & N. Blum. Academic Press, New York. pp 417-440.

Ander K. 1939. Comparative anatomical and phylogenetic studies on the Ensifera (Saltatoria). Opusc. Entomol. Suppl. II.

Boldyrev B.T. 1915. Contributions à l'étude de la structure des spermatophores et des particularités de la copulation chez Locustodea et Gryllodea. Horae Entomol. Soc. Rossicae 41: 1-245.

Boldyrev B.T. 1927. Copulation and spermatophores of *Gryllomorpha dalmatina* (Oesk.) (Orth., Gryllidae). Eos 3: 279-288.

Boldyrev B.T. 1928. Biological studies on *Bradyporus multituberculatus* F.W. (Orth., Tettig.). Eos 4: 15-57.

Butlin R.K., C.W. Woodhatch & G.M. Hewitt. 1987. Male spermatophore investment increases female fecundity in a grasshopper. Evolution 41: 221-225.

Capinera J. 1979. Qualitative variation in plants and insects: effect of propagule size on ecological plasticity. Am. Nat. 114: 350-361.

Davey K. 1960. The evolution of spermatophores in insects. Proc. R. Entomol. Soc. Lond. (A)35: 107-113.

Davies P.M. & I.R. Dadour. 1989. A cost of mating by male *Requena verticalis* (Orthoptera: Tettigoniidae). Ecol.Ent. 14:467-469.

Eluwa M.C. 1979. Observations on the spermatophore and post-copulatory behaviour of three West African bush-crickets (Orth., Tettigoniidae). Entomol. Mon. Mag. 114: 1-7.

Gerhardt U. 1913. Copulation und Spermatophoren von Grylliden und Locustiden I. Zool. Jahrb. (Syst.) 35: 415-532.

Gerhardt U. 1914. Copulation und Spermatophoren von Grylliden und Locustiden. Zool. Jahrb. (Syst.) 37: 1-64.

Gwynne D.T. 1983. Male nutritional investment and the evolution of sexual differences in the Tettigoniidae and other Orthoptera. In: Orthopteran Mating Systems: Sexual Competition in a Diverse Group of Insects. Eds. D.T. Gwynne & G.K. Morris. Westview Press, Boulder, Colorado.

Gwynne D.T. 1984a. Male mating effort, confidence of paternity and insect sperm competition. In: Sperm Competition and the Evolution of Animal Mating Systems. Ed. R.L. Smith. Academic Press, New York. pp. 117-149.

Gwynne D.T. 1984b. Courtship feeding increases female reproductive success in bushcrickets. Nature 307: 361-363.

Gwynne D.T. 1986a. Reply to: stepfathers in insects and their pseudo-parental investment. Ethology 71: 74-77.

Gwynne D.T. 1986b. Courtship feeding in katydids (Orthoptera: Tettigoniidae): investment in offspring or in obtaining fertilizations? Am. Nat. 128: 342-352.

Gwynne D.T. 1988a. Courtship feeding and the fitness of female katydids (Orthoptera: Tettigoniidae). Evolution 42: 545-555.

Gwynne D.T. 1988b. Courtship feeding in katydids benefits the mating male's offspring. Behav. Ecol. Sociobiol. 23: 373-377.

Gwynne D.T., B.J. Bowen & C.G. Codd. 1984. The function of the katydid spermatophore and its role in fecundity and insemination. Aust. J. Zool. 32: 15-22.

Khalifa A. 1949. The mechanism of insemination and mode of action of the spermatophore in *Gryllus domesticus*. Q. J. Microsc. Sci. 90: 281-294.

Lack D. 1940. Courtship feeding in birds. Auk 58: 56.

Leroy Y. 1969. Quelques aspects de la reproduction des Tettigonioidea de Trinidad (Orth.). Ann. Soc. Fr. (N.S.) 5: 775-798.

Morris G.K. 1980. Calling display and mating behaviour of *Copiphora rhinoceros* Pictet (Orthoptera: Tettigoniidae). Anim. Behav. 28: 42-51.

Mann T. 1984. Spermatophores. Springer Verlag.

Ono T., M. Siva-Jothy & A. Kato. 1989. Removal and subsequent ingestion of rivals' semen during copulation in a tree cricket. Physiol.Ent. 14:195-202.

Power J.H. 1958. On the biology of *Acanthoplus bechuanus* Per. (Orthoptera: Tettigoniidae). J. Entomol. Soc. South. Afr. 21: 376-381.

Quinn J.S. & S.K. Sakaluk. 1986. Prezygotic male reproductive effort in insects: why do males provide more than sperm? Fla. Entomol. 69: 84-94.

Rentz D.C. 1979. Comments on the classification of the orthopteran family Tettigoniidae, with a key to subfamilies and description of two new subfamilies. Aust. J. Zool. 27: 991-1013.

Rutowski R.L., G.W. Gilchrist & B. Terkanian. 1987. Female butterflies mated with recently-mated males show reduced reproductive output. Behav. Ecol. Sociobiol. 20:

319-322.
Sakaluk S.K. 1984. Male crickets feed females to ensure complete sperm transfer. Science 223: 609-610.
Sakaluk S.K. 1986. Is courtship feeding by male insects parental investment? Ethology 73: 161-166.
Simmons L.W. 1987. Competition between larvae of the field cricket, *Gryllus bimaculatus* (Orthoptera: Gryllidae) and its effects on some life-history components of fitness. J. Anim. Ecol. 56: 1015-1027.
Simmons L.W. 1988. The contribution of multiple mating and spermatophore consumption to the lifetime reproductive success of female field crickets (*Gryllus bimaculatus* De Geer). Ecol. Entomol. 13: 57-69.
Thornhill R. 1976a. Sexual selection and paternal investment in insects. Am. Nat. 110: 153-163.
Thornhill R. 1976b. Sexual selection and nuptial feeding behavior in *Bittacus apicalis* (Insecta: Mecoptera). Am. Nat. 110: 529-548.
Thornhill R. 1981. *Panorpa* (Mecoptera: Panorpidae) scorpionflies: systems for understanding resource-defense polygyny and alternative male reproductive effort. Annu. Rev. Ecol. Syst. 12: 355-386.
Thornhill R. & J. Alcock. 1983. The evolution of insect mating systems. Harvard, Cambridge, MA.
Trivers R.L. 1972. Parental investment and sexual selection. In: Sexual Selection and the Descent of Man, 1871-1971. Ed. B. Campbell. Aldine, Chicago. pp 136-179.
Wickler W. 1985. Stepfathers in insects and their pseudo-parental investment. Z. Tierpsychol. 69: 72-78.
Williams G.C. 1966. Adaptation and natural selection. Princeton Univ., Princeton, N.J.
Withycombe C.L. 1922. Notes on the biology of some British Neuroptera (Plannipenna). Trans. R. Entomol. Soc. Lond. 1922: 501-594.
Zeh D.W. & R.L. Smith. 1985. Paternal investment by terrestrial arthropods. Am. Zool. 25: 785-805.

4. EGG BIOLOGY OF THE TETTIGONIIDAE

J.C. Hartley

INTRODUCTION

In the life of the Tettigoniidae the biology of the egg stage assumes an over-riding importance since, at least in temperate climates, most species spend far more time in the egg than in the other stages. In a large number of temperate species the egg stage persists for rather more than a year, thus the egg has to be adapted to withstand all the adversities of the year as well as to complete its own development and hatch at the right time. Therefore it is possible to use the egg alone as a device for testing the ability of a species to survive in a particular location; if the egg cannot synchronize and develop under the conditions of that place then the species would be unable to survive. On the other hand if the egg is capable of surviving there then it is likely that the mobile active stages, which can adjust to local changes in the prevailing conditions, would be able to do so too.

This extraordinary feature of the very long egg life span distinguishes the Tettigoniidae from most other insect families. It is this degree of uniqueness that makes their egg biology so interesting.

Nearly 100 years ago the American entomologist W.M. Wheeler (1893) in his "Contribution to Insect Embryology", introduced the word "diapause" to insect development. It so happened that he was looking at the embryology of conocephaline bushcrickets or meadow grasshoppers. For a long time afterwards there was little further study of the developmental biology of these insects. One reason for this paucity seems to have been because the few attempts to hatch the eggs were rather unsuccessful and the family acquired a reputation for being difficult. The major break through came when we discovered that many of the species had multiple diapause and very precise hatching temperature requirements (Hartley & Warne, 1972). Since then much has been published on the development of the European species, and we now know that the eggs of the Tettigoniidae exhibit a far greater diversity of structural features, embryological patterns, and adaptations to environmental conditions than those shown within most other insect families. Although the concept of diapause has changed somewhat from Wheeler's usage, diapause, as we now know it, plays a major role in the life history patterns and the habitat occupancy displayed by the Tettigoniidae.

The term, biology of the egg stage, is used to cover all aspects of development from the time when the egg is laid to when it hatches. While information on many species mentioned here has been published previously, I have accumulated a fair amount of unpublished data on other species, and additional data for some of those for which information already exists. This is generally referred to as Hartley (unpublished). Most of these new data conform to existing published patterns and therefore scarcely warrants separate detailed treatment, however, where I have unpublished data that contribute new information or information on very different species, this is included in an appendix and is referenced in the text as (Appendix 1). I am also indebted to Dr Sigfrid Ingrisch for further unpublished information. The list of species for which egg biological data is known to me is given in Table 1.

The classification of the Tettigoniidae used in this section follows that of Rentz (1979) in that I am including the Copiphorinae as a tribe within the subfamily Conocephalinae as a tribe Copiphorini. I would also include the Decticinae within the Tettigoniinae were it not for the fact that much of the information in this account stems from European decticines and that to include them within the Tettigoniinae at this stage might be unnecessarily confusing to many readers familiar with the old system. This topic will be discussed further towards the end of the chapter.

Table 1. List of Tettigoniidae of which I have some knowledge of the embryonic development

	Development rate	Initial diapause	Embryonic diapause type	Source
Phaneropterinae				
Phaneroptera falcata (Poda)	fast	-	-	1
" "		-	3d	4,3
Phaneroptera nana Fieb.	fast	-	?2d	6
" "		-	3d/-	4,29
Acrometopa macropoda (Burm.)	slow	++	?	29
Acrometopa syriaca Br.	slow	+	?	29
Trigonocorypha unicolor (Stoll)	fast	-	-	25
Tylopsis lilifolia (Fab.)	slow	?	3d	1
" "	fast	+	?	29
Amblycorypha oblongifolia (De Geer)	?	?	?	8
Caedicia simplex (Walker)	fast	-	-	2
Isophya pyrenea (Serv.)	slow	?	3d	29
Isophya taurica Br.	?	?	?	7
Barbitistes serricaudus (Fab.)	slow	+	3d	2,29
Barbitistes ocskayi (Charp.)	slow	-	3d	2
Barbitistes fischeri Yers.	slow	?	3d	1
Andreiniimon nuptialis (Karny)	slow	+	?	29
Leptophyes albovittata (Koll.)	slow	+	3d	2,4
" "	med.	+	3d	29
Leptophyes punctatissima (Bosc)	slow	+	3d	1,20,4
Leptophyes boscii Br.	slow	?	3d	2
Leptophyes laticauda (Friv.)	slow	+	3d	2
Poecilimon cretensis Wern.	slow	+	3d	2
Poecilimon gracilis (Fieb.)	slow	?	?	29
Poecilimon hamatus Br.	slow	+	?	29
Poecilimon schmidti (Fieb.)	slow	+	3d	2
Poecilimon ornatus Schmidt	slow	+	3d	2
Polysarcus denticaudus Charp.	slow	+	3d	23,2
Polysarcus scutatus Br.	slow	+	3d	1
Meconematinae				
Meconema thalassinum (De Geer)	med.	?	2a	1,3,4
Pseudophyllinae				
Jamaicana flava Caudell	fast	-	-	2
Jamaicana subguttata Walker	fast	-	-	2
Zabalius apicalis Bolivar	fast	-	-	14
Mecopodinae				
Corycoides kraussi (Kirby)	fast	-	-	13
Vetralla quadrata Walker	slow	-	-	2

Table 1. List of Tettigoniidae of which I have some knowledge of the embryonic development

	Development rate	Initial diapause	Embryonic diapause type	Source
Euthypoda acutipennis Karsch	slow	-	-	12
Segestes decoratus Redtenbacher	slow	-	-	27
Segestidea uniformis (Willemse)	slow	-	-	28
Conocephalinae, Conocephalini				
Conocephalus discolor Thunb.	fast	-	2b	1,3,4,5
Conocephalus dorsalis (Latr.)	fast	-	2b	1,3,4,5
Conocephalus conocephalus (L.)	fast	-	-	26
Conocephalus semivittatus (Walker)	fast	-	-	2
Conocephalus bilineatus (Erichson)	?	-	2a	2
Conocephalus ensiferum Scudder	?	-	2a	11
(As *Xiphidium ensiferum*)				
Phlesirtes merumontanus Sjostedt	fast	-	-	2
Conocephalinae, Copiphorini				
Banza parvula Walker	slow	-	?	30
Ruspolia nitidula (Scop.)	fast	-	2a	1,21
Ruspolia differens (Serville)	fast	-	-	22,21
Euconocephalus thunbergi Redtenb.	fast	-	-	19
(=*E.nasutus* (Thunberg))				
Conocephalinae, Agraeciini				
Anelytra nov.sp.	fast	-	-	29
Tympanophorinae				
Lipotactes nov.sp.	slow	-	-	29
Listroscelidinae				
Requena verticalis Walker	fast	++	-	2
Tettigoniinae				
Tettigonia viridissima L.	slow	++	3d	1,3,4
Tettigonia cantans (Fuessly)	slow	++	3d	1,3,4
Tettigonia caudata (Charp.)	slow	++	3d	4
Tettigonia hispanica (Bol.)	slow	++	3d	29
Decticinae				
Anabrus simplex Haldeman	slow	+	3d	9
Decticus verrucivorus (L.)	slow	++	3d	1,24
Decticus albifrons (Fab.)	slow	++	3d	1
Platycleis grisea (F.)	med.	-	2d	6,29
Platycleis albopunctata (Goeze)	med.	-	3d	1,3,4
Platycleis sabulosa Azam	med.	-	3c	1
Platycleis intermedia (Serv.)	med.	-	3d	1
Platycleis falx (F.)	med.	-	3d	1
Platycleis affinis Fieb.	med.	-	3d	1
Platycleis romana Rme.	med.	-	?	5,29
Platycleis montana Koll.	slow	?	?	29
Platycleis modesta (Fieb.)	slow	?	+	29
Platycleis holoptera (Ramme)	med.	?	?	1,2
(as *Platycleis tessellata* (Charp.))				
Platycleis veyseli Kocak	?	-	?	5
(as *Platycleis vittata* (Charp.))				
Platycleis sepium (Yers.)	slow	-	3d	1,5
(as *Metrioptera sepium*)				
Metrioptera brachyptera (L.)	slow	+	3d	1,29
Metrioptera saussuriana (Frey-Gess.)	slow	++	3d	1,4
Metrioptera buyssoni (Saulcy)	slow	?	3d	2
Metrioptera tadzhica Bei-Bienko	slow	?	?	7

Table 1. List of Tettigoniidae of which I have some knowledge of the embryonic development

	Development rate	Initial diapause	Embryonic diapause type	Source
Metrioptera hime Furukawa	slow	-	3d	17
Metrioptera bicolor (Phil.)	slow	-	3d	1,5
" "	med.	+	3d	4,29
Metrioptera roeselii (Hgb.)	slow	-	3d	1,5
" "	med.	+	2d	3,4,29
Metrioptera abbreviata (Serv.)	slow	-	3d	1
Metrioptera kuntzeni Rme.	slow	?	?	29
Pholidoptera aptera (F.)	slow	+	3d	2
Pholidoptera littoralis (Fieb.)	slow	+	3d	2,5
Pholidoptera femorata (Fieb.)	slow	+	3d	1
Pholidoptera fallax (Fisch.)	slow	?	3d	2
Pholidoptera griseoaptera (De Geer)	slow	+	3d	1,4
Pholidoptera pustulipes (F.W.)	slow	?	?	7
Eupholidoptera chabrieri (Charp.)	slow	+	3d	1,5
Eupholidoptera smyrnensis (Br.)	slow	+	2c,3	4
Psorodonotus illyricus Ebner	slow	+	3d	2
Anadrymadusa retowskii (Adel.)	slow	?	?	7
Anonconotus alpinus (Yers.)	slow	?	3c	1
Yersinella raymondi (Yers.)	slow	+	3c	1
Pachytrachis striolatus (Fieb.)	?	?	?	5
Rhacocleis germanica (H.S)	slow	?	3c	2,5
Thyreonotus corsicus (Ramb.)	slow	++	3c	1
Antaxius pedestris (F.)	slow	+	3c	1
Antaxius chopardi Mor.Ag.	slow	+	3c	1
Antaxius capellei Caz.	slow	-	3c	1,2
Antaxius sorrezensis (Marqu.)	slow	+	3c	2,29
Antaxius hispanicus Bol.	slow	?	3c	1
Gampsocleis glabra (Herbst)	slow	+	3c	2
Gampsocleis buergeri (De Haan)	slow	?	?	18
Saginae				
Saga pedo (Pallas)	slow	++	3d	1,2
Saga rhodiensis Salfi	slow	++	3d	4
Saga natoliae Serville	slow	++	3d	4
Saga hellenica Kaltenbach	slow	++	3d	4
Saga campbelli Uvarov	slow	++	3d	29
Saga rammei Kaltenbach	slow	++	3d	29
Ephippigerinae				
Ephippiger ephippiger (Fieb.)	slow	++	3d	1,16
Ephippiger perforatas (Rossi)	slow	+	3d	2
Ephippiger provincialis (Yers.)	slow	++	3d	1
Ephippiger terrestris (Yers.)	slow	++	3d	1
Ephippiger discoidalis (Fieb.)	slow	++	3d	2,5
Ephippigerida areolaria (Bol.)	slow	+	3d	1
Ephippigerida zapateri (Bol.)	slow	++	3d	1
Ephippigerida saussureiana (Bol.)	slow	++	3d	2
Steropleurus stali (Bol.)	slow	+	3d	1
Steropleurus brunneri (Bol.)	slow	++	3d	1
Steropleurus nobrei (Bol.)	slow	+	3d	2
Steropleurus catalaunicus Bol.	slow	+	3d	1
Uromenus rugosicollis (Serv.)	slow	?	3d	1
Callicrania monticola (Serv.)	slow	?	3d	1,2
Callicrania seoanei (Bol.)	slow	+	3d	2

Table 1. List of Tettigoniidae of which I have some knowledge of the embryonic development

	Development rate	Initial diapause	Embryonic diapause type	Source
Platystolus obvius (Nav.)	slow	+	3d	1
Platystolus faberi Harz	slow	+	3d	2
Bradyporinae				
Bradyporus dasypus (Illig.)	slow	++	3d	29
Callimenus multituberculatus (F.W.)	slow	++	3d	15
Deracanthinae				
Deracantha sp	slow	++	3d	2
Zichyia sp	slow	++	3d	2
Hetrodinae				
Acanthoplus bechuanus Peringuey (=*A.discoidalis* (Walker))	slow	?	?	10

Key: - Not known to diapause at this stage; + Some may diapause at this stage; ++ Most diapause at this stage; ? Probably diapause at this stage.

Key to authors:

1 Hartley & Warne 1972	2 Hartley unpublished	3 Ingrisch 1985
4 Ingrisch 1986	5 Helfert & Sanger 1976	6 Helfert 1980
7 Chetyrkina 1966,1971	8 Chopard 1938	9 Cowan 1929
10 Power 1958	11 Wheeler 1893	12 Eluwa 1970
13 Eluwa 1971	14 Eluwa 1975	15 Feodorov 1962
16 Dumortier 1967	17 Arai 1977	18 Arai 1979
19 Ando unpublished	20 Deura & Hartley 1982	21 Hartley & Ando 1988
22 Hartley 1971	23 Hartley 1986	24 Ingrisch 1984
25 Wagan 1985	26 Warne unpublished	27 Young 1985
28 Froggatt & O'Connor 1940	29 Ingrisch 1987(p.com.)	30 Strazanac 1986

Embryology

The morphological stages through which tettigoniid embryos pass were described by Warne (1972). His scheme was developed from a comparative approach based on a study of over 40 species and revealed three contrasting patterns. Ingrisch (1984b) re-defined, slightly extended and gave in more detail the stages for *Decticus verrucivorus*. Although there are some minor differences between these authors there is general agreement on the phases of the embryology which are displayed schematically in Fig.1 and outlined, with the stages, in Table 2. The system employed here basically follows that of Warne, but I am inclined to agree with Ingrisch that stage 10 is better regarded as the end of anatrepsis and that stage 18 should be in phase IV, as the end of catatrepsis, rather than in phase V as the start of dorsal closure. I have also put stages 25 and 26 together in Phase VII, hatching, since it appears that, once the spines start to darken, the embryonic moult has commenced and the hatching sequence proceeds as a continuous process.

The three contrasting patterns found by Warne (1972) occur in the precatreptic development of the embryos and are each characteristic of a subfamily. The basic pattern, is perhaps that of the decticines – *Pholidoptera griseoaptera*, described by Warne and *Decticus verrucivorus*, described by Ingrisch (1984b). The embryology of the Ephippigerinae, Saginae, Tettigoniinae and Hetrodinae is similar, as is also that of *Requena*, (Listroscelidinae) (Hartley, unpublished). In this pattern the embryonic primordium develops dorsally near the posterior pole of the egg (Fig.1,a,II) and during anatrepsis grows backwards on the dorsal surface of the yolk, where it remains during mesentrepses (Fig.1). During catatrepsis (Fig.1,b) the embryo grows forward around the posterior pole to adopt its proper orientation with the head pointing towards the anterior end, it then grows forward (Fig.1,c) to eventually fill the whole of egg space. I am not regarding the pattern exhibited by the Phaneropterinae as materially different. Although the early embryo develops to one side of the mid line, this is a feature induced by the laterally compressed nature of the egg. Otherwise the embryology is identical with the decticine pattern and from catatrepsis onward the embryo becomes central.

The embryonic primordium in *Ruspolia* (described under the previous name of *Homorocoryphus*) forms with the head facing anteriorly on the mid ventral side of the yolk (Fig.1), during anatrepsis the embryo moves backwards through the yolk to lie near the dorsal

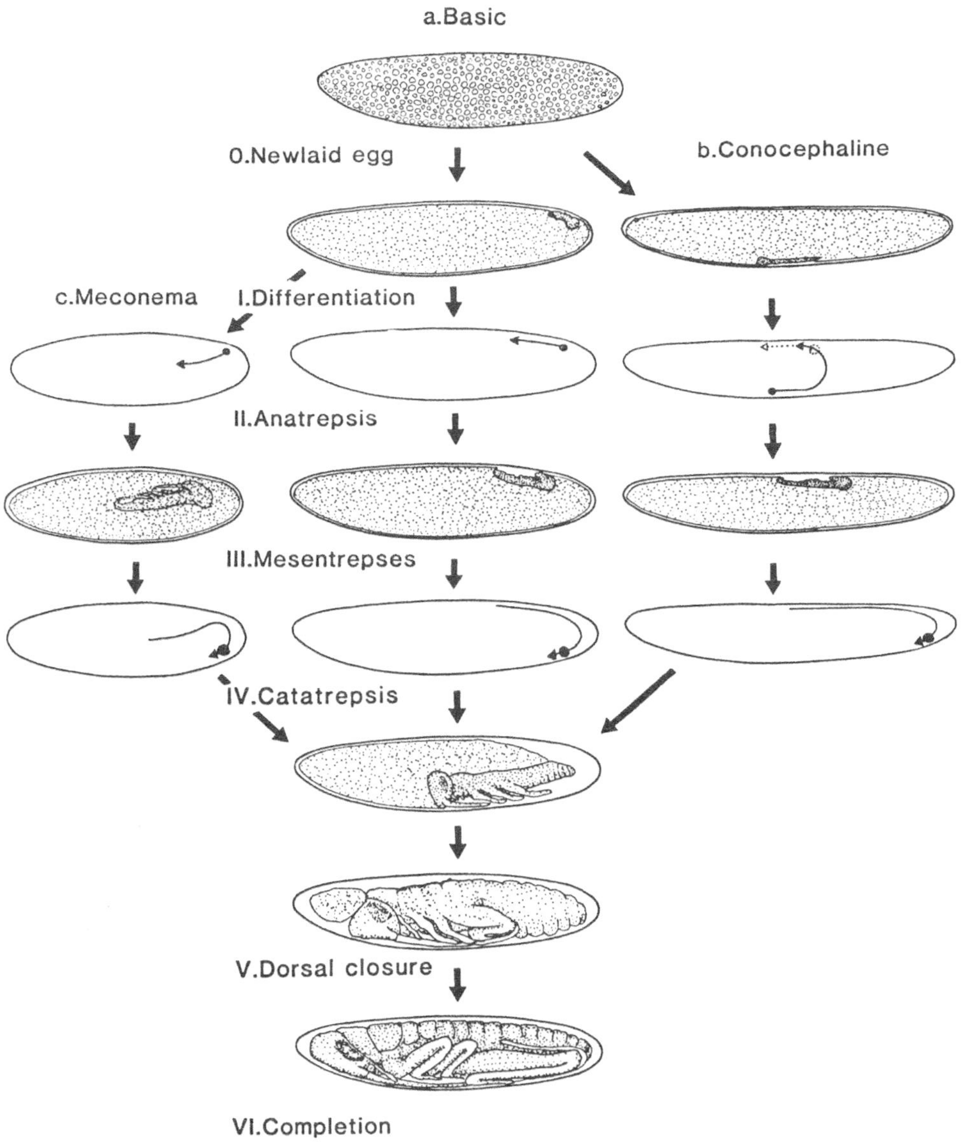

Fig. 1. Embryonic phases of the Tettigoniidae. Further details are given in Table 2.

surface, still in the middle of the egg but with the head now facing posteriorly (Fig.1). Catatrepsis is normal (Fig.1,d,III) with the embryo moving around the posterior pole. Similar movements were described by Wheeler (1893) for *Xiphidium* (=*Conocephalus*) *ensiferum* and another conocephaline, *Orchelimum vulgare*. This was confirmed for the European *Conocephalus* spp by Warne (1972). The two tribes of conocephalines, Conocephalini and Copiphorini, have eggs of very similar appearance and shell structure (Hartley 1964, 1971) and the pseudophylline *Jamaicana flava* although with a very different egg shell structure has a similar embryology (Warne, 1972). *Meconema thalassinum* (Meconematinae) is quite different again, here the embryonic primordium forms near the posterior pole as in the Decticinae, but during anatrepsis it sinks backwards into the yolk only emerging again as a much larger embryo during catatrepsis (Fig.1,e).

In comparison with other Orthoptera, the Decticinae are not unlike the Acrididae with similar embryonic positions and movements. The New Zealand weta *Deinacrida* (Stenopelmatidae) (Richards 1973), also has a pattern like that of the Decticinae. On the other hand *Meconema* is more like the Gryllidae, where much of the precatatreptic development also occurs within the yolk. The Conocephalinae – Pseudophyllinae pattern seems to be unique amongst the Orthoptera.

Table 2. Embryonic phases of the Tettigoniidae with abridged stages

	Phase		Stages
0	Blastoderm formation	1-3.	Blastoderm and yolk cells formed
I	Embryonic primordium formed	4-6.	Differentiation of embryonic primordium and differentiated into protocephalic and protocormic regions
II	Anatrepsis	7-10.	Growth of embryo backwards, segmentation developed, limb buds formed
III	Mesentrepses	11-13.	Embryo broadens but remains stationary, limb segmentation begins
IV	Catatrepsis	14-18.	Forward rotation of embryo around posterior pole, embryo less than half length of egg
V	Dorsal closure	19-21.	Embryo grows around and encloses yolk, limbs small
VI	Completion	22.	Embryo entire, limbs small
		23.	Limbs half extended
		24.	Limbs fully extended, pigmentation develops
VII	Hatching	25.	Spines darken, fully pigmented
		26.	Hatches

Egg development rates

Like many other insects the egg development of the Tettigoniidae is temperature dependent. Eggs of most species will develop over a range of temperatures without added stimuli. This was clearly demonstrated by Hartley and Ando (1988) for *Ruspolia differens* where there was a very high correlation and very small standard error between development rate and temperature over a range of temperatures, which can be termed the "Developmental Temperature Range" (DTR). In other words, over the DTR there is an almost linear relationship between the developmental rate and temperature. For *R. differens* at constant temperatures the DTR is from 18 to 30°C. Below 18°C, although some development may start, embryos usually die at an early stage. At higher temperatures (31-33°C) there is a high mortality and no further reduction of development time. A similarly linear regression was obtained for the locust (Hunter-Jones, 1970). The temperature development curve and its regression line y=bx+c for *R. differens* are shown in Fig.2. The regression line y=bx+c over the DTR (18-30°C) has values of b=0.00397 and c= -0.0589. This expression can be converted into the thermal summation equation K=D(T-t) where D = duration of development, T = incubation temperature, t = c/b = lower thermal threshold and K = 1/b = thermal constant in day°C. Thus this species has a thermal constant K = 252 day°C above a lower threshold t = 14.8°C. These constants can be used to define species characteristics. It was possible to make these accurate determinations with *R. differens* because development is direct and there is no intervening diapause. Data are available on three other non-diapausing species, and these are given in Table 3. Also given in this table are the development times of the species at two temperatures.

Table 3. Comparison of developmental characteristics of four species.

Species	Origin	K day°C	t °C	Days to hatch at 20°C	25°C	Source
Ruspolia differens	Uganda	252	14.8	48	25	
Euconocephalus thunbergi	Japan	320	13.5	47	27	(1)
Jamaicana subguttata	Jamaica	344	13.5	53	30	(2)
Jamaicana flava	Jamaica	423	13.5	65	37	(2)

(1) Ando (Pers.comm.), (2) Hartley, unpublished).

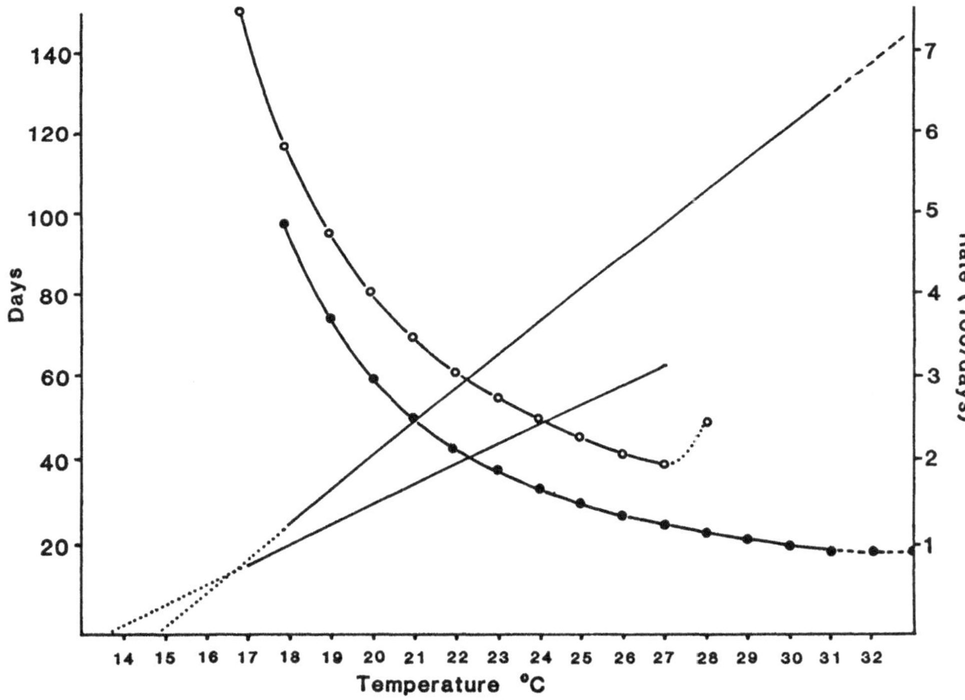

Fig. 2. Development times and rates for two species, *R. differens* and *J. flava*.

Hence, it is possible to use the thermal constant as a comparative developmental parameter, the larger it is the slower is the development. The developmental curves for the slowest of these species, *J. flava*, is also shown in Fig.2 for comparison. Where diapause intervenes it becomes more difficult to produce a value for K for the whole development period, but Hartley and Ando (1988) were able to produce a partial value of K for *Ruspolia nitidula*, a species with an early embryonic diapause (see below), and to extrapolate this to give estimated total values K = 320 day°C, t=13.5°C.

Unfortunately these values have not been determined for other species, most of the available information on them relates to just one or two temperatures. *Conocephalus* species also seem to have rapid development rates, although many species diapause, some during mesentrepses and others at late catatrepsis. *C. dorsalis*, which normally diapauses at late catatrepsis can be prevented from so doing by prior cooling, and will then develop and hatch in 45 days at 20°C. Normal post diapause development times are given as 16-28 days at temperatures of 20-24°C (Hartley & Warne, 1972; Helfert & Sanger, 1976; Ingrisch, 1985). *C. discolor* appears to require slightly longer. Non-diapausing eggs of *C. semivittatus* from New Zealand hatched in 32 days at 25°C (Appendix 1). In the phaneropterine *Leptophyes punctatissima* the developmental pattern is very dependent on the eggs history (Deura & Hartley, 1982). New laid eggs subject to high temperature (30°C) develop immediately and produce complete diapausing embryos in a mean time of 49 days at a rate of 5.6%/day. Eggs will also develop at 20°C but only after a long delay (Table 4). If eggs are kept for 3 or more weeks at 20°C, and then placed at 30°C the start of development will be retarded, development then proceeds more slowly and usually ceases before the embryo is fully formed. Such embryos eventually die unless transferred to a lower temperature (<26°C), when they are able to complete embryogenesis normally. At 25°C new eggs start development sporadically, but once started the standard 25°C development rate is maintained. These development features are summarized in Table 4.

Table 4. Constant temperature prediapause development of *Leptophyes punctatissima*

Temp. °C	Min.days to mesentrepses	Mean days to complete embryo	±1sd	Rate %/day
30	10	49	6.1	5.6
25	25	93	22.6	1.5
20	103	196	28.3	1.3

At lower temperatures no development is initiated, but keeping new eggs at low temperatures (e.g.8°C) for 40 days or more greatly facilitates subsequent development. Thus, previously cooled eggs placed at constant 25°C had a mean development time of 60 days at a rate of 3.3%/day, and at 20°C development was also slightly enhanced. However, similarly treated eggs mostly failed to complete development if kept at 30°C. Eggs of *L. punctatissima* that were already developing could continue development at temperatures below the normal DTR. Thus 70% of eggs with embryos starting catatrepsis were able to complete when transferred to 12°C as did all post catatreptic embryos. There is, therefore, in this species evidence that the DTR shifts downwards with egg age. The changing DTR and problems caused by diapause, which are discussed below, make it virtually impossible to obtain a thermal summation constant for this species and others like it.

The species discussed so far tend to be fairly synchronous with little variation between the development rates for individual eggs in any one batch. This is not so for many slow developing species. Thus in *Ephippiger ephippiger* (Dean & Hartley, 1977a) there is a considerable asynchrony in the start of development and also the developmental rates between eggs of any one batch vary considerably. Again, there is an indication of a downward shift in the DTR. This is also so, although to a lesser extent, in *Pholidoptera griseoaptera* where Hartley & Warne (1973) estimated a thermal constant of about 800 day°C above 10°C. *E. ephippiger* certainly requires a heat input in excess of this, but as with *L. punctatissima* it depends on the previous history of the eggs. Downward shifting DTRs are not so evident in those species which have their last diapause at an intermediate stage, e.g. *Antaxius*, *Conocephalus* (Hartley & Warne,1972).

With *E. ephippiger* Bugren (1980) found that daily fluctuating temperatures 10-30°C and 20-40°C resulted in development rates nearer the upper temperature. In the 20-40°C regime development often ceased during catatrepsis in embryos of southern races but many embryos of northern races were able to complete development. The normal upper temperature limit for *E. ephippiger* in the later part of embryogenesis is about 28°C.

One of the features of temperate species is that the developmental cycle will have evolved to fit the climatic cycle of their region. A slow developing European species for example requires much of the heat input of the summer months, but if the season is poor or truncated, the downward shifting DTR will allow development to continue. This also has an effect on the diapause patterns discussed below. The slow developing species have another feature in common. There is often a long period of inactivity, 10-30 days after the eggs are laid, before the yolk blocks and any embryo starts to form.

Slow developing eggs are not just associated with temperate species. All the known examples of the Mecopodinae are slow developers, thus the Nigerian *Euthypoda acutipennis* had an incubation period of 78-190 days at a temperature fluctuating between 22-30°C (Eluwa, 1970).

The width of the DTR is also variable between species. For *R. differens* it is 12°C between 18 and 30°C and it is at least 12°C in *P. griseoaptera* (14-26°C). *E. ephippiger* is very like *P. griseoaptera* but with a slightly higher range(18-30°C). *J. flava*, with a range of only 9°C between 17-26°C, should be considered a stenothermal species. *R. differens* is more tolerant at high temperatures and a few eggs can succeed at 33°C but *J. flava* fails completely above 27°C. *L. punctatissima* can develop through to late embryonic diapause at 30°C and can also do so at 15°C, but not within a batch of eggs of the same age group, it can also continue development at 12°C. This makes *L. punctatissima* clearly a eurythermal example of the Tettigoniidae.

The picture that emerges from the information available is that most development rates fall into either of two categories: (a) fast with a thermal requirement of less than 420 day°C, or needing less than 40 days at 25°C, and (b) slow requiring more than 700 day°C, or 60-90days or more at 25°C. A few species are able to complete embryogenesis in an intermediate time. For example *Platycleis albopunctata* can complete embryogenesis in less than 50 days at 24°C when it enters diapause, and although it still requires a further 20-30 days after diapause to hatch, it is clearly much faster than *P. griseoaptera*. On the other hand it is distinctly slower than *Conocephalus*. Such rates are typical of the genus *Platycleis* and some other species. Although they would normally rate as slow in my classification it might be better to designate them as belonging to an intermediate category. The categories, fast, slow and intermediate (med.) are indicated in column 2 of Table 1.

In general the division between fast and slow follows phylogenetic lines, with Conocephalinae and Pseudophyllinae both fast and Saginae, Tettigoniinae, Ephippigerinae etc all slow. Since the first group of subfamilies are all essentially tropical with a few species that have invaded temperate regions and conversely the others are basically temperate, it might appear that fast development is associated with a tropical environment. However, the Mecopodinae are slow and tropical. Phaneropterinae have both fast and slow species in temperate regions. The fast developing species of Phaneropterinae, which are all long winged, might like *R. nitidula* be considered recent invaders of temperate regions. The brachypterous species all seem to be slow developers. It would be particularly useful to have information on Decticinae and Phaneropterinae from other parts of the world. Information is too scant on other subfamilies to predict their characteristics.

In the intermediate rated decticines, the species of *Platycleis* are long winged. Like the phaneropterines, they also may be evolving toward a faster pattern as part of their role as mobile colonisers.

In comparative terms most known examples of the Acrididae and Gryllidae are rapid developers. I know of none that has a slow development rate similar to that of the Tettigoniidae.

WATER REQUIREMENTS

All tettigoniid eggs require water during the course of development. Hartley (1971) recorded a weight increase of 150% in *Ruspolia differens*, which was accompanied by an anisometric expansion of the egg shell, the length increasing by 14% and the girth by 50%. Warne (1970) showed that there were similar weight increases, ranging from 90% to 160%, for a number of species and that most of the increase occurred after diapause, regardless of whether the diapause was early or late. The weight increase of four species exhibiting different patterns of water uptake is shown in Fig.4. The course of weight increase was examined in detail for *Platycleis grisea* and *Phaneroptera nana* by Helfert (1980) for a variety of different developmental regimes. She found a 150% increase in *P. grisea* but only about a 100%

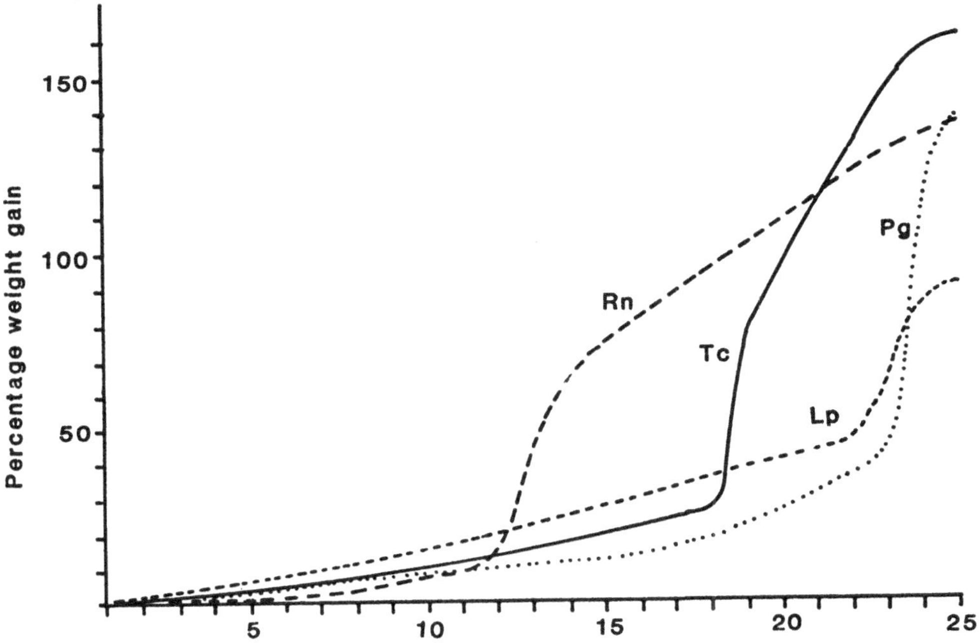

Fig. 3. Weight changes with developmental stages of four tettigoniid eggs. Rn = *Ruspolia nitidula*; Lp = *Leptophyes punctatissima*; Tc = *Tettigonia cantans*; Pg = *Platycleis grisea*. (After Warne 1970)

increase in *P. nana*. In these species, where there is no obligate diapause but where quiescence may be induced at lower temperatures, water uptake occurs steadily, except during periods of quiescence or when eggs are kept at 5°C. Likewise Warne (1970) found phaneropterine eggs only increased in weight by no more than 100% whereas others tended to increase by 160%. Warne considered the flattened shape of the phaneropterine egg limited its potential to expand by water uptake, since, although they dilate considerably, they never become completely cylindrical as do the others.

Ingrisch (1986c) studied the consequences of drought at different times in the development of a number of European species. He found that development proceeds more slowly if eggs are deprived of contact water. Without contact water embryogenesis cannot proceed beyond a certain stage, although high humidities will often allow development most of the way up to the whole embryo stage or to the onset of embryonic diapause. Eggs deprived of water when in an active developmental phase will usually die if the period of low humidity is too long. Thus 20 weeks at a humidity of 75% or less produced a high mortality in *Pholidoptera griseopatera, Decticus verrucivorus, Leptophyes punctatissima, Tettigonia viridissima*, and *Tettigonia cantans*.

Desiccation immediately prior to the hatching period resulted in very high mortality. On the other hand, drought at an early developmental stage usually only delayed development, but with some species, e.g. *Tettigonia caudata*, desiccation is necessary to overcome the initial diapause (qv). Warne (1970) found development of *R. differens* was inhibited during mesentrepses at less than 100% humidity. Such eggs when wetted, rapidly absorb water and development proceeds normally.

Eggs of species with little or no initial diapause may behave as though there is an intial diapause if kept dry after laying. Thus Ingrisch (1986c) found *Metrioptera roeselii* and *M. bicolor*, which normally hatch after a cooling period following 4 weeks incubation at 24°C, only hatched after a second incubation and cooling cycle if kept dry during the first warm period. *Conocephalus dorsalis* kept 5 months at 24°C and at a humidity >75% after the first period of cooling stayed in diapause and then hatched successfully in moist conditions after a second cooling period. Desiccation induced quiescence is known in Acrididae as for example in *Locustana* (Matthee, 1951).

Diapause patterns

Diapause in the egg stage is defined here as an arrest of active development or embryogenesis caused by intrinsic factors at a time when conditions are still favourable. It serves to hold eggs at a particular predetermined stage, thereby preventing precocious development or hatching at a time when adversity could strike the next stages. At the same time it can add an element of synchrony to the population. The diapause stage also may be physiologically better able to survive the oncoming period of adversity. Diapause is considered to be a physiological state, requiring a particular set of conditions. It can be completed by a period of time in the adverse conditions, hot, cold or dry. After diapause completion the embryo enters a quiescent state and is able to resume development when temperature or other conditions permit. In some instances diapause at a later stage may be prevented by subjecting the eggs to adverse conditions at an earlier stage. This is often referred to as diapause aversion.

Eggs of some species may cease development if too cold or dry, survive quite well a period of cold or drought and resume development satisfactorily when temperatures rise or water becomes available; such enforced quiescence is a response to the immediate conditions and not a programmed arrest, it does not require a reactivation process and is therefore not regarded as diapause.

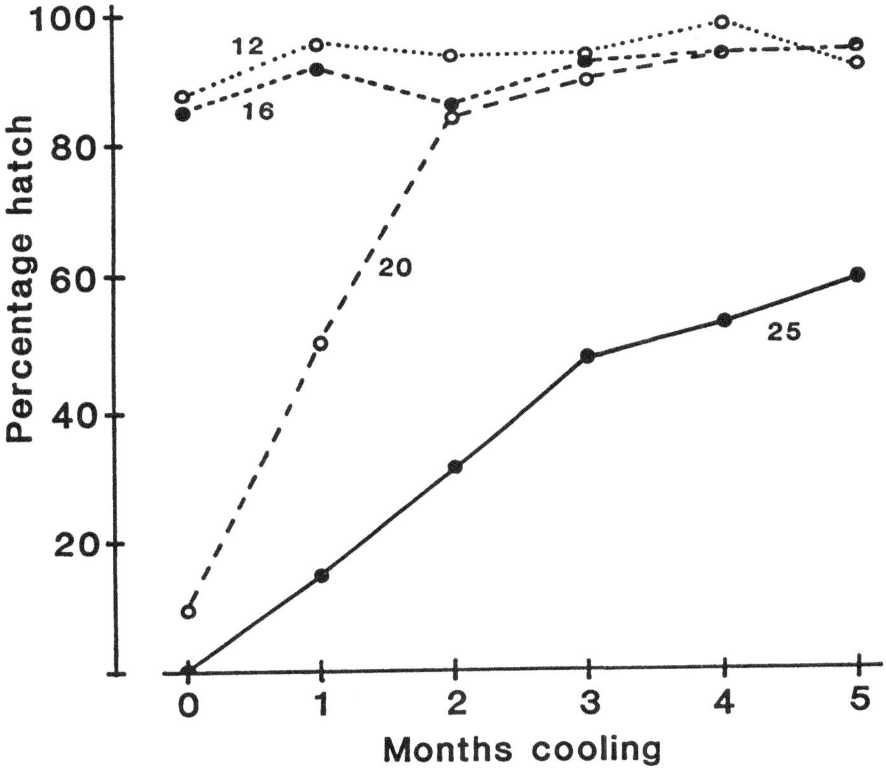

Fig. 4. Hatching success in *Ephippiger* after various periods of cooling at different temperatures.

The diapause patterns in the Tettigoniidae are quite variable and often complex. Many species exhibit two diapause periods in the egg stage (Hartley & Warne, 1972) although often where this occurs it may be more correct to consider the first as an egg diapause, that is a diapause affecting the egg at an early stage before there is any significant embryonic component present, and the second as an embryonic diapause, when there is well established embryo. Examples of the different diapause patterns will now be considered. The final embryonic diapause will be considered first since it is the least ambiguous of the diapause types found in the Tettigoniidae and perhaps the best known.

Final embryonic diapause (Phase VI diapause)

The most obvious type of diapause is the final embryonic diapause of many temperate species. It was investigated in detail by Dean and Hartley (1977b) for *Ephippiger cruciger* (now referred to as *Ephippigerida ephippiger* (Hartley & Warne, 1982)). Visible embryonic activity stops at stage 23 when embryogenesis is complete and the embryo occupies the whole egg space, but before the limbs are fully extended. The most obvious feature is that the embryo's heart stops, or only beats spasmodically. If embryos at this stage continue to be kept at the previous development temperatures they will gradually shrink and eventually die, although they may tolerate a month or more under these conditions. With *Ephippiger* a period of cooling enables the embryo to become active again when placed at a suitable temperature. The post-cooling response is not immediate, and it is only after a quiescent period of many days that full activity is resumed. The first observable change is an increase in heart activity, followed by limb extension and gradual cuticular pigmentation. Finally tanning of spines occurs a day or so before hatching.

Diapause development, that is the phsyiological processes that end diapause and enables the embryo to respond when conditions are suitable, is a function of time and temperature. Dean and Hartley (1977b) showed that with *Ephippiger*, cooling at temperatures between 4 and 12°C for a period of 3 months enabled almost all eggs to hatch at temperatures up to 20°C, further cooling only slightly reduced the mean hatch time. Hatching at higher temperatures (25°C) was never very satisfactory. Shorter cooling periods resulted in fewer eggs hatching at 20 and 25°C (Fig.4). However successful hatches of better than 80% were obtained at 12 and 16°C regardless of cooling period. This is because diapause development was possible at temperatures up to 20°C, although with less success at 20°C. In general for *Ephippiger* cooling periods of up to 3 months tend to reduce mean hatching time and to reduce the hatching spread. With *Pholidoptera griseoaptera* (Hartley & Warne, 1972), *Phaneroptera falcata*, *Platycleis albopunctata* and *Metrioptera roeselii* (Ingrisch, 1985) there was a significant reduction in hatching time obtained by increasing the cooling time to 4 months. Ingrisch also found a similar progressive reduction for up to 6 months cooling in *Tettigonia viridissima*. The assessment of completion of the final embryonic diapause is based on hatching results, which in turn are also dependent on hatching temperature. This will be further discussed below under hatching.

In nearly all known examples where diapause occurs at this stage (Table 1, Col.4,3d), its elimination is brought about by about 3-4 months at low temperature, i.e. one winter. However, it is possible that it could be a wet/dry response as is, perhaps, implied for *Acanthoplus* (Power, 1958).

Platycleis grisea presents something of a problem. Helfert (1980) found a transient delay, in the stage 23 embryo, of 60-120 days in embryogenesis in eggs kept at constant 20°C. However eggs laid under short day conditions (L:D::8:16), and incubated under short-day lighting, developed without delay. Helfert also found that eggs of this species when kept cool for 90 days after laying, were able to develop directly and hatch without further delay when placed at 20-28°C. Hartley and Warne (1972) found a final diapause in the very closely related *P. albopunctata* that was only present in eggs kept at 20°C or above. At 15°C hatching occurred sporadically after a total incubation period of 280 days. This might suggest a weak diapause to be present. Some hatching even occurred at constant 20°C. Ingrisch (1986a) also confirmed diapause in *P. albopunctata* at this stage. Hartley and Warne did not test the effect of precooling on this species.

In the final diapause, of the sort that responds to cold, extending the period of cooling beyond about 6 months results in loss of viability. Ingrisch (1987) has shown that metabolic activity may not be greatly suppressed during this diapause except by low temperature. After diapause development is completed the embryo may continue to use its reserves, so that, if it is at too low a temperature to hatch, it may gradually become exhausted and die. This may show, firstly, as a failure to hatch properly, and secondly by a reduced percentage attempting to hatch at all. In *Ephippiger* failure to hatch properly is a feature particularly evident in eggs that start to hatch at low temperatures. Here the embryo escapes from the egg but then is unable to shed the embryonic cuticle (Dean & Hartley, 1977b). A reduced percentage hatch also occurs after 6 months cooling with *P. falcata* and *M. roeselii* (Ingrisch, 1985). The intensity of the final embryonic diapause is independent of the earlier history of the egg or embryo. Dean & Hartley (1977b) found no significant difference in the hatching performance of *Ephippiger* following 2, 3 or 4 months incubation at 25°C before cooling, nor was there any appreciable difference in the hatching of *Metrioptera roeselii* following one or two incubation and cooling cycles (Ingrisch, 1985).

Initial diapause (Phase I diapause)

Diapause occurring at a very early stage of embryogenesis is referred to as initial diapause or egg diapause, since it is often controlled by factors carried over from the female and operates before embryonic control is possible. In *Ephippiger* it was found that a variable percentage of newly laid eggs would not develop beyond the embryonic primordium stage at temperatures favourable to morphogenesis however long they were kept at such temperatures (Hartley & Warne, 1972; Dean & Hartley, 1977a). Elimination of this initial diapause in *Ephippiger* follows a different course from that found in many other embryonic diapauses. Dean and Hartley (1977a) found that about one month of cooling was just as effective as three months in the percentage of eggs that would start to develop on return to incubation temperatures. Although cooling for longer periods did not increase the number of eggs which developed it did affect the reactivation time. Surprisingly, increasing the cooling period also increased the period of time before development started when the eggs were placed at a suitable temperature. This is shown in Table 5.

Table 5. Time for appearance of phase III embryos (anatrepsis) in 50% of eggs of *Ephippiger* that were in initial diapause after cooling for the time indicated.

Days cooling	Mean days to anatrepsis	Total percentage developing
30	15	76
60	57	85
90	81	79
120	87	84

They also found that excessively long cooling periods reduced the ability of embryos to complete embryogenesis at 25°C, although they could do so at lower temperatures. Hartley and Warne (1972) found with some ephippigerines that a similar percentage of initially diapausing eggs would start development on each reincubation cycle for at least four cooling periods.

Ingrisch (1986b) showed that initial diapause in some species varied with locality. Thus more northerly populations tended to produce eggs with less initial diapause, so that eggs of *Decticus verrucivorus* from Babin Potok, Yugoslavia, had 100% initial diapause at 24°C whereas those from Aachen, Germany, laid under long day conditions, only 40% had initial diapause. I can substantiate this with new data given in table 5. In addition the occurrence of initial diapause is also variable within a local population, and may be under some form of maternal control. This is discussed below.

Manifestation of diapause at this stage is often dependent on incubation temperature. High temperature (30°C), for example, facilitated direct development in *Leptophyes punctatissima*, but intermediate temperatures 25-16°C induced an initial diapause (Deura & Hartley, 1982). This diapause was not evident in eggs that were kept from the time of laying at low temperatures (8-10°C) for 40 days. This could be regarded as diapause aversion. However, after induction of initial diapause by intermediate temperatures, 100 or more days at low temperature were then required to eliminate it. All eggs of this species developed in the second incubation cycle. The effect of initial incubation temperature is even more striking

Table 6. Percentage initial development in eggs of *Decticus verrucivorus* from Fussen, Bavaria, Germany and Lozere, France 1987, at different incubation temperatures.

Origin	Temp.°C		
	35/15*	30	25
Germany	32 (776)	18 (1140)	0.5 (834)
France	49 (466)	4 (449)	0 (450)

Total numbers in brackets.
*This was a 12hr diurnal switch between 35°C and 15°C

in *Decticus verrucivorus* as shown in Table 6. The French population had a much higher incidence of initial diapause than the German at 30°C, but with the alternating temperature this was reversed. It would seem that the 35°C component was able to overcome the initial diapause. Constant 35°C was lethal.

No development was induced in the eggs of the Australian *Requena verticalis* kept for three months at 25 and 30°C. After three months cooling a proportion of eggs then developed at both 25 and 30°C, others appeared to remain in diapause (Appendix 1).

Initial diapause may persist in some eggs of some species for several years. Diapauses persisting for at least 5 winters have been recorded in eggs of *Tettigonia viridissima* kept under natural field conditions (Ingrisch, 1986a; Hartley, unpublished), and up to seven in a number of species (Ingrisch, 1986a). Ingrisch has coined the term plurennial for such species. Ingrisch (1986c) has shown that the initial diapause responds not to cold but to dryness in *Tettigonia caudata* from Greece, and that in general recruitment to development in the next favourable season was proportional to the water lost previously. Eggs of *Poecilimon cretensis* which were kept continually at 30°C from laying, showed no development within the first 2 months. They were kept dry for a further 2 months at 30°C and then when moistened they started to develop (Appendix 1).

Initial diapause is clearly a very variable feature and in some cases is indistinguishable from a high initial thermal threshold. Obviously where in a batch of eggs some start development after one cooling and warming cycle and others start after a further identical sequence, there is a good case for calling the initial resistance to development a genuine diapause. The only problem here is that a gradually decreasing threshold, which falls at different rates in different eggs, would produce an identical result. Therefore, although it is possible to categorize the possible types of initial diapause, as is done below, it is considered that we do not have enough precise information to allot the possible categories to more than two or three species. Consequently, only the fact that initial diapause can occur is indicated in table 1 (+ in column 3).

Early intermediate diapause (Phase III, mesentrepses diapause)

This term is employed for the diapause occurring after anatrepsis, exhibited by the copiphorine *Ruspolia nitidula*. Here Hartley and Ando (1988) found that this diapause may be eliminated by time as well as by low temperature. Embryogenesis ceased in eggs kept at normal developmental temperatures when the embryos reached mesentrepses. At 20-30°C this stage was reached within 30 days after laying. Eggs kept continuously at these temperatures erratically resumed morphogenesis and, since there is no further diapause in this species, they proceeded to hatch, as shown in Table 7.

Table 7. Days to hatch at constant temperature in *Ruspolia nitidula*. The hatching rate is the average rate between the 25 and 75% hatch.

Temp.°C	Mean days to whole embryo	±1sd	Rate %/day
30	93	23.1	1.5
25	97	17.2	2.0
20	121	7.1	4.8

When diapausing eggs of this species were cooled for increasing periods, up to 5 weeks, they progressively completed development sooner and hatched more synchronously on return to higher temperatures. Thus after 8 weeks cooling hatching occurred in 21 + 1.5 days at 25°C, with a hatching rate of 26%/day. So although this diapause disappears with time it can be eliminated more effectively at low temperature. The more synchronous hatch at 20°C, as indicated by the lower sd and higher hatching rate, suggests diapause development followed by normal post diapause development. This diapause can probably be eliminated by prior cooling, as with *Concephalus* described below. Diapause at this stage is also found in a number of conocephaline species including *Xiphidium* (*Conocephalus*) *ensiferum* described by Wheeler (1893) and in some eggs of *C. semivittatus* (Appendix 1).

Mid intermediate diapause (Phase IV, late catatrepsis diapause)

In some *Conocephalus* species, such as *C. dorsalis*, diapause normally sets in just after the end of catatrepsis. It usually persists until the egg is subject to a period at low temperature, but a few embryos may resume development after a while without temperature change (Hartley & Warne, 1972; Hartley, unpublished). In another species, *C. discolor*, there is a greater tendency for this to happen. Ingrisch (1985) found that the length of the cooling period, 3-6 months, had no effect on the subsequent time to hatching in either species and Hartley & Warne found that it could be as short as 18 days. Eggs of *Conocephalus* species can also be cooled before the onset of diapause and they will then develop without further delay at suitable temperatures. Diapause can also be averted by dryness. Newly laid *C. dorsalis* eggs that were kept dry for several weeks at developmental temperatures mostly produced embryos that had ceased growth during mesentrepses. When moistened they developed through to hatching without further delay (Appendix 1), thus showing dryness as a diapause eliminating factor as with the intial diapause found in *Tettigonia caudata* referred to under that section.

Late intermediate diapause (Phase V, dorsal closure diapause)

A diapause occurs just before dorsal closure is complete in a number of decticine species. *Antaxius* sp. diapaused at stage 20 when about 3/4 grown. This diapause was obligate and is very similar to the final diapause in response. After this diapause there was no further arrest. Similar situations were found in *Platycleis sabulosa*, *Thyreonotus corsicus*, and *Gampsocleis glabra*. Ingrisch (1986a) found an arrest in *Eupholidoptera smyrnensis* at the same stage in eggs at 24°C. In this species development in some individuals resumed without change of conditions after 20 weeks, although it was facilitated by lowering the temperature below 20°C. However, a normal final diapause occurs in this species.

Summer diapause

A number of tettigoniid species show an arrest in embryogenesis if kept at high temperatures.

With the exception of the intermediate diapause found in *Eupholidoptera smyrnensis* referred to above these delays are probably more in the nature of high temperature inhibition perhaps caused by a decreasing DTR as in *L. punctatissima* rather than a true diapause.

Classification of diapause types

The diapause types of the Tettigoniidae can now be classified, with examples, as follows:-

1. Initial or egg diapause, effective by stage 4, at the beginning of phase I;

i. facultative, influenced by maternal environment, type 1a; *Metrioptera roeselii*

ii. facultative, influenced by the environment of the new laid egg, type 1b; *Leptophyes punctatissima*

iii. obligate, not affected by changing maternal environment or the present environment on the new laid egg, type 1c; *Requena verticalis.*

2. Mutable embryonic diapauses, usually obligate but perhaps sometimes maternally influenced, often transient, can be eliminated before they become effective;

i. Early, phase III, mesentrepses, type 2a; *Ruspolia differens*

ii. Mid, phase IV, late catatrepsis, type 2b; *Conocephalus dorsalis*

iii. Late intermediate, phase V, dorsal closure, type 2c; *Eupholidoptera smyrnensis*

iv. Final, phase VI, whole embryo, type 2d; *Platycleis grisea.*

3. Immutable embryonic diapauses, obligate only responding to changed conditions while in diapause;

i. Late intermediate, phase V, dorsal closure, type 3c; *Antaxius pedestris*

ii. Final, phase VI, whole embryo, type 3d; *Pholidoptera griseoaptera.*

These categories where known are also included in Table 1 in column 4.

As previously mentioned there is a problem with the facultative initial or egg diapauses in that often only a proportion of the eggs in any one batch may diapause. Factors such as incubation temperature, maternal history and geographical source all have some effect as detailed elsewhere. Therefore most species cannot as yet be allocated to a particular category. I expect that as further data are obtained more and more species will fit both categories 1a and 1b. Similarly some early embryonic diapauses may be affected by maternal history and present conditions.

The other Orthoptera appear to be much more conservative in their diapause patterns. Hence the normal egg diapausing stage of the Acrididae is during mesentrepses (Uvarov, 1966, 1977), although *Melanoplus* species are reported as diapausing at the whole embryo stage. In the Gryllidae, where egg diapause occurs, it can be an initial diapause as in the Tettigoniidae or at mesentrepses (Tanaka 1984). Diapause in the Acrididae requires a period of cooling for its completion, whereas that of the Gryllidae may be terminated by a short high temperature impulse, or often ends just with the passage of time (Masaki *et al.*, 1979), much in the way that it does in *Ruspolia* (Hartley & Ando, 1988). There is also some evidence that some of the diapauses occurring in the Acrididae can be prevented by high temperature incubation. Extended diapauses also occur in some Acrididae (Ushatinskaya, 1984).

Maternal effects

Helfert (1980) found that the diapause manifestation within a population depends on the photoperiod experienced maternally in *Platycleis grisea* and *Phaneroptera nana*. Eggs laid under long day conditions by *P. grisea* had a longer quiescent or diapause period than those laid during short days, but there was a tendency for the opposite situation with *P. nana*. This was further complicated by the actual day length experienced by the developing eggs. She found eggs exposed to short days tended to have an accelerated development. There was a distinct photoperiod effect in *Metrioptera roeselii*. Here Ingrisch (1986b) found a switch from less than 10% initial diapause in eggs laid under a long day 16 hrs light – 8 hrs dark regime, through 50% at around 13.5-14 hrs light to about 90% at a L:D::12:12 photoperiod as shown in Fig.5. In *Eupholidoptera smyrnensis* the change was even more abrupt at about 14 hrs light (Fig.5). *Pholidoptera griseoaptera* on the other hand responded differently with some initial diapause in eggs laid at 16 hrs daylight but none at 14 hrs.

The incidence of initial diapause in several other species is also under maternal control. Ingrisch (1984,1986b) found in several *Metrioptera* species that with oviposition at L:D::16:8 most eggs developed directly at 24°C but at L:D::14:10 there was a much greater resistance to development in the first 4 weeks. This was even more pronounced in *Decticus verrucivorus* as in *Eupholidoptera smyrnensis*. *Platycleis albopunctata* and *M. saussuriana* were not susceptible to changes in day length, *M. saussuriana* having a well defined and obligate initial diapause and *P. albopunctata* being without one at all. There is also a possibility that temperature thresholds are set maternally as a response to the temperature experienced by the female at time of egg formation, so that eggs maturated under warmer conditions have a

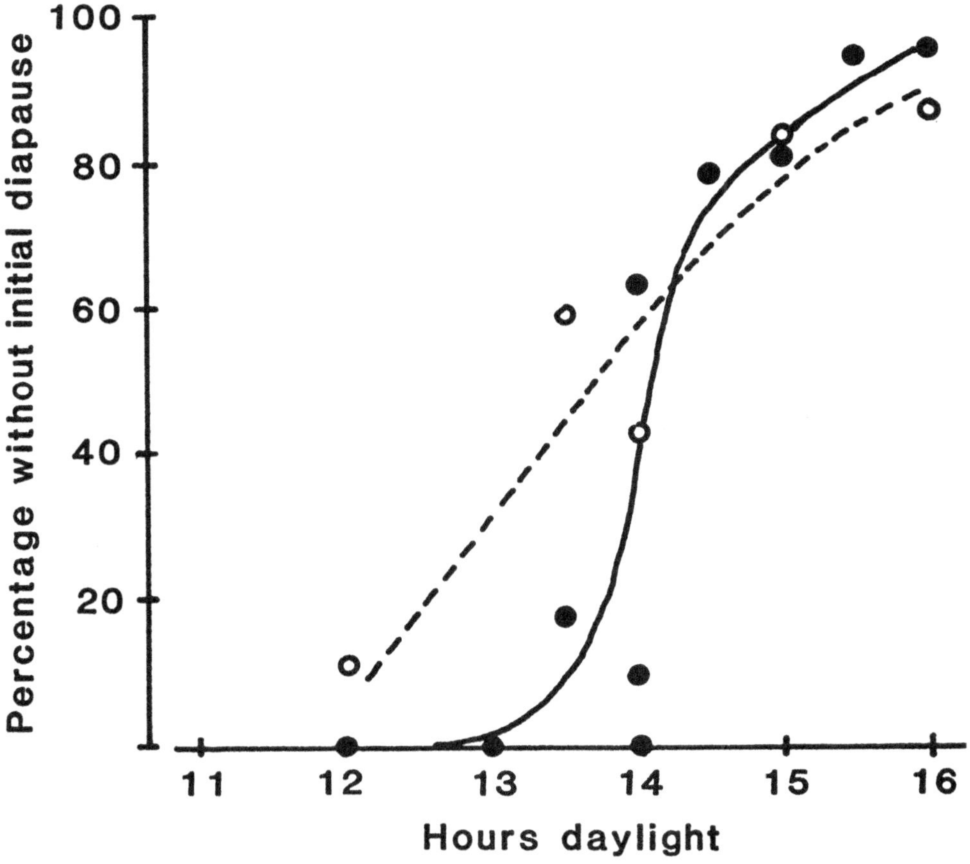

Fig. 5. Change in levels of initial diapause as a response to different photoperiods (Modified after Ingrisch, 1986b).

higher threshold than eggs produced when cooler (Dean & Hartley, 1977a).

Response to day length is fairly rapid, Ingrisch (1986b) found *E. smyrnensis* changed from laying eggs without initial diapause to eggs with initial diapause 2-3 weeks after changing day length from 16 to 10 hrs light. Ingrisch also suspects that maternal age may affect initial diapause, in *E. smyrnensis* eggs from young females had only a low percentage of initial diapause, but eggs laid 2 weeks after the start of the oviposition period now exhibited intial diapause and this situation persisted for the rest of the females lives. Since the environmental conditions were not changed this suggests that for this species aging causes an increased diapause factor.

As already indicated under initial diapause (see above) not all populations of one species respond in the same way. Thus while Ingrisch (1986) found *D. verrucivorus* from Aachen, Germany, responded to long days with a lower incidence of initial diapause, he was unable to induce any shift from the 100% initial diapause in eggs of this species from Babin Potok, Yugoslavia.

Hatching

Since hatching is able to occur at temperatures between 8 and 25°C there can be a considerable overlap of both diapause and post diapause development. This creates difficulty in separating diapause and the following reactivation processes. In *Ephippiger ephippiger*, for example, embryos cooled for 2 months require some 540 day°C above 9°C to hatch, whereas after 3 months they require only 425 day°C (calculated from Dean & Hartley, 1977b). Further cooling of up to 5 months at 4-8°C made little difference to the hatching time, particularly at 20°C. However, there was a shortening of the mean time to hatch, at 20°C, after 4 months at intermediate temperatures (12-16°C), where the diapause and post diapause processes overlap. Hatching also eventually occurred below 9°C. Eggs with diapausing embryos, kept continuously at 8°C, started hatching in 300-340 days and had a mean hatching time of nearly 400 days. However hatching at 8°C following a period at 4°C began about 300 days from start of the cooling period. Here the mean hatch time was about 370 days regardless of the period of 1-5 months at 4°C. Thus the slight rise in temperature reduced the overall time to hatch. It also shortened the hatching period and increased the success from 35% to 60%.

Although hatching under such conditions cannot be considered a natural phenomenon, these results suggest that after diapause is completed at low temperature, reactivation may occur and the hatching programme start, albeit very slowly. The ability to hatch at 8°C is further evidence of temperature thresholds changing with time. Ingrisch (1985) also found the mean hatch time became earlier in Tettigonia species as the cooling period was extended to 6 months, but in *Platycleis albopunctata* and *Phaneroptera falcata* there was no further significant decrease in time to hatch if cooling was extended beyond 3 months.

For many species the hatching temperature range (HTR) is rather lower than the DTR (Hartley & Warne, 1972). This varies very considerably from species to species, but is generally true of species diapausing on completion (phase VI diapause). In the Pholidopterini, for example, *Pholidoptera griseoaptera* was able to hatch from 9 to 19°C with best results between 11-16°C. No hatching occurred at 20°C. In *Eupholidoptera chabrieri* the HTR was 8-20°C, but with *P. femorata* and *P. aptera* hatching only occurred at or below 12xC and at 15°C the embryos became exhausted and eventually died (Hartley & Warne, 1972; Hartley, unpublished). This is probably the cause of the failure recorded by Helfert & Sanger (1976) for *P. littoralis*. Many of the early problems Hartley & Warne (1972) had were due to attempts to induce hatching at too high a temperature.

In those species without an embryonic diapause, or where diapause occurs before embryogenesis is complete, i.e. those with early or mid diapause, the HTR is obviously the same as the DTR. Thus *Ruspolia nitidula* and *Conocephalus discolor* both hatch as continuous processes from embryogenesis and so also does *Requena* (Appendix 1). Some Decticinae with a late mid diapause (phase V) hatched well at 20°C and above suggesting that these species do not have the lower HTR found in those with a final embryonic (phase VI) diapause. Thus Helfert & Sanger (1976) found *Pachytrachis striolatus, Tessellana vittata* and *Platycleis romana* hatched well at 30°C, but they were unsuccessful with *Rhacocleis germanica* above 20°C. Hartley and Warne (1972) found that *Antaxius* species, diapausing during dorsal closure (phase V), were able to resume development at a lower temperatures after diapause than before, but it is not known whether the upper limits had fallen below 20°C. Following diapause treatment *Antaxius* is able to resume embryogenesis without delay and is ready to start hatching about a month after being moved to warm conditions. On the other hand in *E. ephippiger*, where embryogenesis was completed before diapause set in and the only post diapause changes are limb extension and cuticular tanning, it also requires about a month to start hatching. This long reactivation time, which is found in other species with late diapause, e.g. *Pholidoptera* species, is probably important in preventing hatching too soon in the spring before conditions are stable.

In all species, of which I have experience, hatching appears to occur around dawn. This was recorded by Hartley (1971) for *Ruspolia differens* and has been observed more casually for other species. Arai (1977, 1979a,b,c) proved dawn hatching under experimental conditions for *Metrioptera hime* and *Gampsocleis buergeri* and found that synchrony was lost under constant light or dark conditions. There appears to be a different response pattern in *M. hime* in comparison with *G. buergeri*, the former showed an hour-glass response to light-off enhanced by light-on of the next dawn, but the second, light-on, stimulus was not found in *G. buergeri*. Here hatching usually shortly preceded dawn, and diurnal temperature changes produced a similar effect. Under constant darkness the thermoperiod response for *M. hime* was 12hrs after temperature-down, but in constant light it was only 4-6 hrs. *G.buergeri* hatched 8-10hrs after temperature-down in the dark but lost synchrony in light.

Egg batches transferred from cyclical to constant conditions lost synchrony after a day. Arai also achieved hatching at high temperatures (20,25,30°C) in both species when provided with a 12:12 light-dark photoperiod, although with lower percentage hatch, and with 12:12 thermoperiod between 7.5 and 20-30°C. Some of these differences relate to the egg laying site, for instance *G. buergeri* lays eggs underground where photoperiod is unlikely to be perceived whereas *M. hime* inserts eggs into surface material.

One characteristic of many species of Tettigoniidae is the extended period of hatching. For *E. ephippiger* under laboratory conditions Dean and Hartley (1977b) found a minimum spread of 25 days, but frequently the hatching period spread over 50 days. Very much extended hatches were usually due to less than optimal treatment beforehand. Ingrisch (1985) recorded field hatching in *Tettigonia viridissima* and *T. cantans* in different years. He found the period ranged from 10 to 27 days. The longer hatching periods were associated with earlier first hatching dates, however the minimum temperatures were also lower.

It is likely that field hatching conditions will normally truncate the hatching spread, since as temperatures rise the less advanced embryos will be accelerated. Further temperature rises will, of course, tend to be inhibitory and suppress extended hatching. This could mean that a proportion of the eggs may never hatch, but even if they were to hatch much later their chances of survival might well be poor. However such eggs could provide a reservoir if there were an abnormally early spring followed by a cold spell. Although the long hatching periods found in the laboratory may be something of an artifact, the ability to have an extended hatching period is probably another important survival strategy.

Ingrisch (1985), considering hatching only, suggested that there were two patterns. In late hatching species (e.g. *Conocephalus*, *Phaneroptera*) hatching date was not affected by the length of the cooling period, but in early hatching species (e.g. *Metrioptera*, *Tettigonia*) extension of the cooling period advanced the hatching date. He considered the latter strategy as an adaptation, for such species, to the necessity of hatching as early in the year as possible. So that, for early hatching species, hatching would be relatively earlier in a late spring, thus enabling them to recover some of the lost time. Late hatching species do not have the same seasonal constraints.

Discussion

The extraordinary diversity of developmental and diapause patterns in the eggs of the Tettigoniidae has profound effects on their habitat occupancy and biogeography. For instance the biennial developmental cycle can result in populations being extremely abundant, even to pest proportions, in alternate years; or conversely local populations may be absent every other year. Whether diapause has evolved more than once in the Tettigoniidae is a difficult question to answer. The diversity of diapauses patterns would suggest that diapause could be of multiple origin, but it may be that there was a fundamental necessity for slow developing eggs to have an overwintering arrest and that this evolved into present diversity of diapause patterns. This possible evolutionary story would fit a very ancient family of insects that had probably formed distinct subfamilies before the breakup of the world supercontinent during the Jurrasic. On the other hand the intermediate diapauses of the Conocephalinae are of perhaps rather different nature and origin.

Most of the adaptations described in this account have evolved along with these insects, and enable them to live in particular environments. Some of the various diapause patterns achieve a common objective but approach it from different directions. Thus the embryonic diapauses serve to hold and synchronize the eggs over winter, or other periods of adversity, and enable the resulting post-embryonic population to make best use of the coming favourable season. The same applies to the short term initial diapauses (1.a & b).

Prolonged initial diapause (1.c), which Ushatinskaya (1984) has termed superdiapause, performs a rather different function. It must be considered as a device for maintaining the species over a number of years in unpredictable habitats by providing an egg bank, just as many plants have a seed bank. This is particularly important for insects with low mobility. It is, therefore, not surprising to find the initial diapause so well developed in, for example, some ephippigerines living in the hot dry Mediterranean areas of France and Spain. Here it is quite possible for local summer fires to sweep over an area killing all above ground stages, or for a severe drought to desiccate the environment with the same effect. Prolonged initial diapause should only be found in species that lay eggs in the ground, and as far as is known this is so. It might also be expected to occur more in flightless species, but this does not seem to be obviously so. Such diapause ought to be widespread in Tettigoniidae from other areas where fire is a frequent hazard, such as parts of Africa and Australia.

Similar ideas were expressed by Ingrisch (1986c). He was also critical of the view expressed by Tauber *et al.* (1984) that diapause was not appropriate for unpredictable adversities. Tauber *et al.* (1986) now recognize prolonged diapause as a bet-hedging technique in seasonally unpredictable environments, although they do not show it in either their Table 2.1 or Fig. 2.1.

That initial diapauses may be subject to maternal control is scarcely novel. However, in this family, with their slow and variable developmental patterns, closely related to local environmental, correct synchronisation is obviously very important. If eggs are laid early in the season when sufficiently favourable conditions are still present, it may be desirable that part of the population should develop directly. On the other hand, if there is any possibility that the favourable period could be truncated, then the bet-hedging strategy will require that a substantial part of the population is held dormant. Day length as experienced by the female provides a very reliable cue but temperature experienced by the egg can be used. In general long days, or decreasing temperature after oviposition induce initial diapause, although there are exceptions.

The weaker diapauses may well be a response to species moving into, or away from, seasonal climates. Thus in *Ruspolia nitidula* the development of its diapause could be a device that has enabled it to move into the Palaearctic. The same argument can be applied to *Phaneroptera* species. The genus *Platycleis* presents more of a problem. As a decticine it should have a fairly intense diapause system but several species obviously do not. It could well be that the genus, in its history, has in part been in a situation where diapause was not necessary but is now moving back into diapause requiring habitats. All these species are fully winged and quite mobile.

As previously indicated there seems to be some subfamily characteristics that influence these patterns. For instance it appears that there is no extended initial diapause in the Phaneropterinae, although a number of phaneropterines exhibit a one winter initial diapause. The Conocephalinae have only an intermediate diapause which, is often of a temporary nature. Indeed most of the development patterns of the Pseudophyllinae and Conocephalinae are in contrast with the other Tettigoniidae. The most definite and persistent diapauses are found in the palaearctic Tettigoniinae, Decticinae, Saginae, Ephippigerinae and closely related subfamilies. There are obviously severe gaps in our knowledge, but it is likely that other predominantly temperate subfamilies will show similar characteristics. The tropical Pseudophyllinae and Conocephalinae – Cophiphorini appear to be without any diapause.

Hatching patterns show differences that can be related to habitat. In the Pholidopterini *Eupholidoptera chabrieri* has perhaps the widest HTR as might be expected from a mountain species and *P. femorata* one of the narrowest. The low HTR of *P. femorata* enables hatching early in a habitat that can suffer from desiccation later in the season (Le Midi of France), the adults can better survive on dry foods than the earlier instars. Thus in habitats where this species and *P. griseoaptera* overlap *P. femorata* becomes adult while *P. griseoaptera* is still in the penultimate larval instar. Similar early hatching occurs in *P. aptera* and *P. littoralis*, both species with a southern distribution, although *P. aptera* has extended into the Alps.

Ingrisch (1986a-c) attempted a categorization of egg development patterns for Western Palearctic Tettigoniidae. He proposed three categories which are: 1. annual (without initial diapause); 2. annual or biennial depending of whether laid early or late; and 3. biennial or many year life cycle (up to 8 years due to a prolonged initial diapause). The latter category he has termed plurennial. The annual species hatch in the next favourable season (usually spring); biennial species normally develop in the summer of the year after oviposition and hatch in the following spring; several year species are as the biennial species but require an appropriate sequence of stimuli (often summer followed by winter) to initiate development. Ingrisch's scheme however, does not cover tropical species in which there is no diapause and is inevitably imprecise on biennial species. I consider there is a real difference between annual species and species that can be annual or biennial. A modified scheme is therefore presented here to try to incorporate such patterns and consider the strategy from the embryo's activity as well as an environmental response. These categories are:

A) Non seasonal, without diapause (*Ruspolia differens*).

B) Seasonal, annual species, present population directly related to previous years population:

i. Without diapause, but seasonality induced by adverse period eliminating all but egg stage, facultative annual (*Phaneroptera nana*, *Caedica simplex*).

ii. With one diapause, obligate annual (*Ruspolia nitidula*, *Conocephalus dorsalis*, *Platycleis albopunctata*).

C) Normally biennial species, all with embryonic diapause (2c,d or 3c,d), present population related to population two years ago:

i. No initial diapause, could be annual if laid early enough and thus fit B) ii. (*Metrioptera roeselii*).

ii. With facultative initial diapause (1a,b) which tends to be stronger in early laid eggs forcing bienniality (*Metrioptera bicolor*, *Pholidoptera griseoaptera*, *Polysarcus denticaudus*).

iii. With obligate initial diapause strictly biennial (*Metrioptera saussuriana*, *Pholidoptera femorata*).

D) Extended seasonal, plurennial species, with prolonged initial diapause that may last several seasons, present population related more to climatic conditions of previous years than to the previous populations:

i. With no second embryonic diapause (*Requena verticalis*).

ii. With a second embryonic diapause (3c or 3d) (*Gampsocleis glabra*, *Tettigonia viridissima*, *Ephippiger ephippiger*).

Figure 6 is a diagrammatic representation of these patterns.

Some species do not strictly fit within this scheme. For example, *Poecilimon cretensis*, with an initial and late (3d) diapause, should belong to category C) iii. but is probably in fact an annual species. However, if one considers Crete to have two favourable seasons, spring and autumn, separated by two adverse seasons, hot dry summer and cool winter, then it hatches after the second adverse season. Therefore, it might be better in some instances to replace biennial with alternate favourable seasons. Such life history patterns may well be found in other areas with similar climatic features.

The distinction between categories is not always clear. It is, for example, quite possible

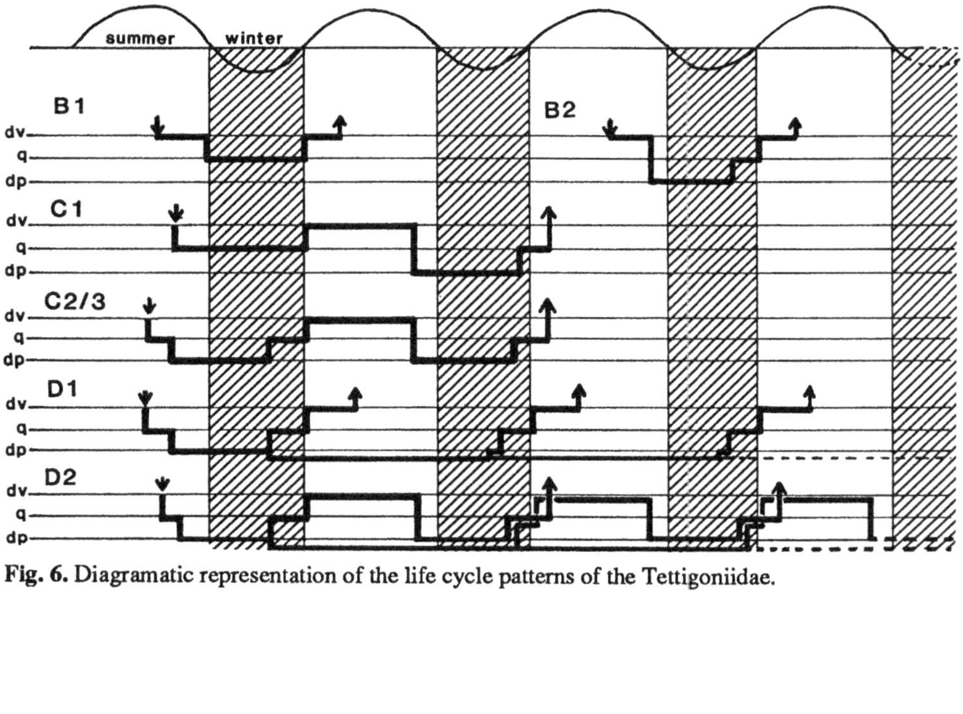

Fig. 6. Diagramatic representation of the life cycle patterns of the Tettigoniidae.

for a species like *Metrioptera roeselii* to be annual if the eggs are laid early to have developed sufficiently enough, which in this species beyond catatrepsis, before the onset of winter (Ingrisch 1986b). Indeed *M. roeselii* appears to be programmed for just such an eventuality in that eggs laid during long days will have low incidence of initial diapause, as shown in Fig. 7. Although over most of its northern and central range it is unlikely to be egg laying until late summer when there is no longer enough heat left in autumn for this to be possible, but southern populations, if they have the same initial diapause pattern, should be annual. *Eupholidoptera smyrnensis* from Rhodes, with a similar initial diapause pattern (Ingrisch, 1986b) also could well be annual as with *Poecilimon cretensis*. However, most species where this is a possibility have developed a facultative initial diapause. Hence *Pholidoptera griseoaptera* exposed to long daylight, during egg maturation, tends to produce a greater proportion of initially diapausing eggs. I have also found a strong regional difference in this species, populations from the southern part of their range also have a much greater level of initial diapause. Here where there is much more heat available in the later part of the year, it becomes more necessary to enforce bienniality with an initial diapause rather than to allow partial development in one autumn and have fully developed embryos, which are more susceptible to desiccation, produced early in the following summer.

The distinction between Cii, Ciii and Dii can also be vague. A C-type biennial population has no component with an initial diapause that will extend beyond the first winter so that they can never be more than biennial. In a Dii species, part could be biennial, but at least some will be carried over to another year. An exception to this rule was found in a Ci population of *Pholidoptera griseoaptera* near its northern limit in England. Eggs placed in the natural habitat were found in a cool summer to have had insufficient heat to complete embryogenesis that year, they survived the winter, completed embryogenesis the next summer, diapaused and hatched the year after, in the third spring. However, this should be regarded as an aberrant situation of a freak year. So if there is a capability of the initial diapause being extended beyond one adverse period, as in *Tettigonia*, then it is category D, if the initial diapause only holds for one adverse season, as in *Polysarcus*, then it must be category C.

Unfortunately there is a shortage of field data to corroborate these predictions. It is recorded that *Polysarcus denticaudus* occurred as a biennial pest in part of the Alpes Maritimes, France in 1964, 1966 and 1968 (Commeau & Malabre 1969) and there is reference to a two year cycle in mountain populations of *Anabrus simplex* (Cowan &

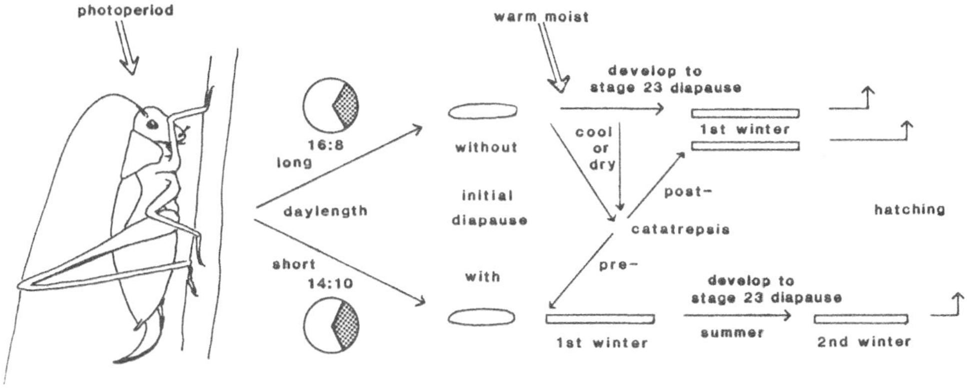

Fig. 7. Life cycle patterns for *Metrioptera roeselii*. Modified after Ingrisch (1984a).

Shipman 1940). Dumortier (1967) and Ingrisch (1985) have recorded hatching of *Ephippiger ephippiger* and *Tettigonia* sp under outside conditions and shown at least a two year cycle. It is likely to be only with positively biennial species (i.e. category C), that bienniality will be observed in the wild.

Features like the staggered hatching of a batch of eggs, variable development rates, and the extended initial diapause seem to characterize the temperate Tettigoniidae. Even some tropical species seem to have a variably extended developmental pattern. These strategies might be all considered as part of a bet-hedging policy. They all point towards evolution of life history patterns designed to ensure the continuing presence of a species in an area over the long term even if habitat occupation, in any one season, is minimal. This would be particularly useful for species of low mobility in undependable or hazardous habitats.

The diapause ability of the conocephalines, together with their propensity for laying eggs in plant stems, may well have contributed to their success as island colonisers. Warne and Hartley (1975) considered the potential of *Conocephalus dorsalis* to be carried as eggs in flotsam as the only way in which the discontinuous distribution in Britain could be explained. Ingrisch (1986c) found that these eggs would stay in diapause if kept at low humidity such as would be caused by immersion in sea water. This would also offer an alternative explanation to flight theory of Pitkin (1980) for the wide distribution for insects of this genus, which are not strong fliers, on the islands throughout the Pacific. A similar situation appears possible for the appearance of *Banza* in the Hawaiian Islands since this Copiphorine has a 2 month plus incubation period which can be extended by dryness (Strazanac, 1986 & pers. comm.).

The biological studies of the egg stage show that there is considerable similarity within any subfamily. This can provide further support for the merging of subfamilies suggested by Rentz (1979). The old Copiphorinae and Conocephalinae have many features in common, diapause stages, embryology and egg shell structure (Hartley, 1964, 1971) and are obviously closely related. As regards placing the Decticinae within the Tettigoniinae there are no features in the egg biology against it. I have a feeling that further studies of egg shell structures might provide additional useful information here. Certainly studies on the embryology of the other subfamilies could be very revealing.

Appendix

New data

Unless otherwise stated the insects were collected by the author. In all species eggs were kept moist during incubation and cooling unless otherwise stated.

Caedicia simplex. North Island, New Zealand, April 1987. Eggs laid on scuffed surface of bark, etc. Shape unlike typical phaneropterine in that the eggs are triangular in cross section with a flattened dorsal surface on which they are gummed to the substrate. The eggs are sometimes embedded into the scuffed bark and partially covered with debris. About 100 eggs, which had been laid during the previous month, were incubated at 25°C and hatching started 23 days later.

Poecilimon cretensis. Crete, April 1986 collected as half grown nymphs. Eggs laid in groups in soil at the beginning of May. Shape typical phaneropterine. About 30 eggs kept moist at 30°C for 6 weeks, with no development, then kept dry at 30°C for a further 8 weeks. The eggs were remoistened and after 5 weeks 12 had half embryos. These were placed at 25°C and had whole embryos in diapause 5 weeks later. They were cooled for 3 months at 8°C and hatching started 20 days after transfer to 15°C. Some previously non-developing eggs then started to develop.

Other Phaneropterinae. Collected at various times, Europe. Eggs laid in groups in soil *Poecilimon* sp., *Barbitistes* sp. or singly in fibrous material (*Leptophyes*). Weak initial diapause in some easily suppressed by early incubation at high temperatures as in *L. punctatissima* (Deura & Hartley, 1982). Late embryonic diapause in whole embryo, all can be reactivated after 3 months' cooling.

Jamaicana flava. One female from bananas, 1967. Many hundreds of eggs over several generations. Shape: elongate ovoid with anterior projection bearing respiratory surface, length 8 mm. Development direct, 32 days at 26°C, 75 at 19°C. K=414 day°C, t=15.24°C. (Kirby, 1979).

Jamaicana subguttata. As *J. flava*, smaller species, several generations handled, K=344 day°C, t=15.5°C.

Vetralla quadrata. From culture at London Zoo. Eggs laid in groups in moss, development 2-3 months, hatching erratic, no diapause.

Conocephalus dorsalis. Gloucestershire, 1984. About 150 eggs kept dry for 4 weeks after laying contained embryos in mesentrepses. When incubated at 25°C, 75% developed directly and started hatching in 16 days with a mean hatch at 21 days. 30% of eggs of the same initial batch incubated at 20°C developed directly and started hatching in 42 days. Eggs of this stock kept moist diapaused as usual at the end of catatrepsis.

Conocephalus semivittatus. North Island, New Zealand, April 1987. Typical conocephaline appearance. About 20 eggs laid in grass leaf sheaths, kept at 25°C, developed directly and started hatching in 32 days. Second generation eggs at 20°C were in diapause at late anatrepsis after 2 months.

Conocephalus bilineatus. New Zealand as above. Eggs incubated at 25°C entered diapause during mesentrepses.

Requena verticalis. From D.T. Gwynne, Western Australia, 1985. Small *Metrioptera* shape. Laid in sand at 25°C, up to one month, airmailed, incubated at 30°C, no development in 3 months, cooled 5 months at 8°C, then placed at 25°C, resulting in rapid development and hatching starting 37 days later. Some (1/3 approx) eggs remained in initial diapause.

Metrioptera buyssoni. Pyrenees, 1976. In all respects very similar to *M. brachyptera*.

Pholidoptera aptera. Bavaria and Austria, 1974. Several hundred eggs laid in soil. Development very similar to *P. femorata*, some initial diapause at 25°C, optimum hatching at 8°C, inhibited above 10°C. *P. littoralis* from Northern Italy very similar. *P. fallax* from Northern Yugoslavia similar to *P. griseoaptera*.

Psorodonotus illyricus. Near Rejika, Yugoslavia, 1984. Eggs like *Decticus*, about 300 laid in soil, initial diapause in less than half at 25°C, whole embryos entered final diapause. No success in hatching following cooling for 4 months at 8°C. Ingrisch (personal communication) believes lower temperature required.

Rhacocleis germanica. Northern Italy, Yugoslavia, 1981. Eggs laid in soil. At 25°C developed without initial diapause to embryonic diapause at dorsal closure stage.

Antaxius sorrezensis. Herault, France, 1976. Eggs laid in soil. Initial diapause below 20°C. Late embryonic diapause at dorsal closure.

Gampsoscleis glabra. Cevennes, France 1978. About 200 eggs laid in soil. *Decticus* shaped. Initial diapause in more than half at 25°C. Developing eggs entered diapause at dorsal closure, poor hatch at 20°C after 3-4 months' cooling at 8°C.

Deracanthines. Eggs from assorted adults collected around Ulan Bator, Mongolia, 1974, by R. Alexander, were sent to Dr. Ragge, British Museum (Natural History), by Mrs W.M. Taylor. Dark shell, *Decticus* shape. When received these eggs did not respond to high temperature (25-30°C), they were then treated as though in initial diapause. Two developed at 25°C after one cold treatment and another two after a second treatment. The fully grown embryos entered late dispause and three hatched at 15°C after 4 months chilling (8°C). Other eggs, which still appeared to be in initial diapause. failed to respond to further warm and cold cycles. Some gradually died and others were lost through mishandling. The previous treatment of these eggs was probably inappropriate. The insects arising from the eggs that hatched proved to be *Zichyia* sp. Some of the eggs were known to have been laid by *Deracantha* sp. It is probable that these species have a very intense initial diapause that may well require lower and longer low temperature treatment.

Other Ephippigerinae. *Ephippiger perforatas*, *E. discoidalis*, *Ephippigerida saussureiana*, *Steropleurus nobrei*, *Callicrania seoanei*, *Platystolus faberi*, collected at various times, Europe. Eggs laid in soil, except *E. perforatus* which also lays in fibrous matter. All show variable initial diapause, having a late embryonic diapause requiring 3-4 months' cooling, and hatch afterwards in about one month at 15-20°C.

References

Arai T. 1977. Effects of the daily cycle of light and temperature on hatchability and hatching time in *Metrioptera hime* Furukawa (Orthoptera, Tettigonidae). Kontyu 45: 107-120.

Arai T. 1979a. Effects of light-on and light-off on the hatching time in *Metrioptera hime* Furukawa (Orthoptera: Tettigonidae). Kontyu 47: 66-77.

Arai T. 1979b. Effects of temperature shifts on the hatching time in *Metrioptera hime* Furukawa (Orthoptera: Tettigonidae). Kontyu 47: 517-529.

Arai T. 1979c. Effects of daily cycles of light and temperature on hatching in *Gampsocleis buergeri* De Hann (Orthoptera: Tettigonidae). Jpn. J. Ecol. 29: 49- 55.

Bugren M.M. 1980. Comparative physiology, ecology and behaviour of some races of the *Ephippiger ephippiger* group. PhD Thesis, University of Nottingham.

Chetyrkina I.A. 1966,1971 (quoted from Ushatinskaya 1984)

Chopard L. 1938. La biologie des Orthoptères. Encycl. Entomol. 20: Paris.

Commeau J. & A.M. Malabre. 1969. Les invasions *d'Orphania scutata* ans les Alpes Maritimes. Phytoma 205: 41-43.

Cowan F.T. 1929. Life history, habits and control of the Mormon cricket. Tech. Bull. U.S. Dept. Agric. 161: 28pp.

Cowan F.T. & H.J. Shipman. 1940. Preliminary observations on two-year cycle Mormon crickets on the Big Horn mountains in Wyoming. 3rd Q. Rep. U.S. Bur. Entomol. Plant Quar.

Dean R.L. & J.C. Hartley. 1977a. Egg diapause in *Ephippiger cruciger* (Orthoptera: Tettigoniidae). I. The incidence, variable duration and elimination of the initial diapause. J. Exp. Biol. 66: 173-183.

Dean R.L. & J.C. Hartley. 1977b. Egg diapause in *Ephippiger cruciger* (Orthoptera: Tettigoniidae). II. The intensity and elimination of the final egg diapause. J. Exp. Biol. 66: 185-195.

Deura K. & J.C. Hartley. 1982. Initial diapause and embryonic development in the speckled bush-cricket, *Leptophyes punctatissima*. Physiol. Entomol. 7: 253- 262.

Dumortier B. 1967. Essais in vitro pour la rupture de la diapause embryonnaire chez quelques Tettigonioides Insectes. Orthoptères. Ann. Epiphyt. 18: 387-400.

Eluwa M.C. 1970. The biology of the West African bush-cricket, *Euthypoda acutipennis* Karsch (Orthoptera: Tettigoniidae). Biol. J. Linn. Soc. 2: 1-39.

Eluwa M.C. 1971. Notes on the biology of *Corycoides kraussi* (Orthoptera: Tettigoniidae). Bull. Entomol. Soc. Nigeria 3: 32-36.

Eluwa M.C. 1975. Studies on the life history of the African bush-cricket *Zabalius apicalis* Bolivar (Orthoptera: Tettigoniidae). J. Nat. Hist. 9: 33-39.

Feodorov S.M. 1962. (quoted from Ushatinskaya 1984)

Froggatt J.L. & B.A. O'Connor. 1940. Insects associated with the coco-nut palm. N. Guinea Agric. Gaz. 6: 16-32.

Hartley J.C. 1964. The structure of the eggs of the British Tettigoniidae (Orthoptera). Proc. R. Entomol. Soc. Lond. 39: 111-117.

Hartley J.C. 1966. Laboratory culture of a tettigoniid, *Homorocoryphus nitidulus vicinus* (Wlk) (Orthoptera). Bull. Entomol. Res. 57: 203-205.

Hartley J.C. 1971. The respiratory system of the egg-shell of *Homorocoryphus nitidulus vicinus* (Orthoptera, Tettigoniidae). J. Exp. Biol. 55: 165-176.

Hartley J.C. & Y. Ando. 1988. Egg development patterns in diapausing and non- diapausing species of *Ruspolia* (Orthoptera, Tettigoniidae). Entomol. Exp. Appl. 49: 203-212.

Hartley J.C. & A.C. Warne. 1972. The developmental biology of the egg stage of Western European Tettigoniidae (Orthoptera). J. Zool. Lond. 168: 267-298.

Hartley J.C. & A.C. Warne. 1973. The distribution of *Pholidoptera griseoaptera* (DeGeer) (Orthoptera: Tettigoniidae) in England and Wales related to accumulated temperatures. J. Anim. Ecol. 42: 531-537.

Hartley J.C. & A.C. Warne. 1984. Taxonomy of the *Ephippiger ephippiger* complex (*ephippiger*, *cruciger* and *cunii*), with special reference to the mechanics of copulation. Eos 60: 43-54.

Helfert B. 1980. Die regulative Wirkung von Photoperiode und temperatur auf den Lebenszyklus okologisch unterschiedlicher Tettigoniiden-Arten (Orthoptera, Saltatoria). Zool. Jb. Syst 107: 449-500.

Helfert B. & K. Sanger. 1976. Vergleichende Untersuchungen uber die Temperatursummierung von Laubheuschrecken (Orthoptera: Tettigoniidae) wahrend der Embryogenese. Zool. Anz. 196: 43-60.

Hunter-Jones P. 1970. The effect of constant temperature on egg development in the desert locust, *Schistocerca gregaria*. Bull. Entomol. Res. 59: 707-718.

Ingrisch S. 1984a. The influence of environmental factors on dormancy and duration of egg development in *Metrioptera roeselii* (Orthoptera: Tettigoniidae). Oecologia 61: 254-258.

Ingrisch S. 1984b. Embryonic development of *Decticus verrucivorus* (Orthoptera: Tettigoniidae). Entomol. Gen. 10: 1-9.

Ingrisch S. 1985. Effect of hibernation length on termination of diapause in European Tettigoniidae (Insecta: Orthoptera). Oecologia 65: 376-381.

Ingrisch S. 1986a. The plurennial life cycles of the European Tettigoniidae (Insecta: Orthoptera). 1. The effect of temperature on development and hatching. Oecologia 70: 606-616.

Ingrisch S. 1986b. The plurennial life cycles of the European Tettigoniidae (Insecta: Orthoptera). 2. The effect of photoperiod on the induction of an initial diapause. Oecologia 70: 617-623.

Ingrisch S. 1986c. The plurennial life cycles of the European Tettigoniidae (Insecta: Orthoptera). 3. The effect of drought and the variable duration of the initial diapause. Oecologia 70: 624-630.

Ingrisch S. 1987. Oxygen consumption by developing and diapausing eggs of *Eupholidoptera smyrnensis* (Orthoptera: Tettigoniidae). J. Insect Physiol. 33: 861-865.

Masaki S., Y. Ando & A. Watanabe. 1979. High temperature and diapause termination in the eggs of *Teleogryllus commodus* (Orthoptera: Gryllidae). Kontyu 47: 493-504.

Masaki S. 1965. Geographical variation in the intrinsic incubation period: a physiological cline in the emma field cricket (Orthoptera: Gryllidae: *Teleogryllus*). Bull. Fac. Agric. Hirosaki Univ. 11: 59-90.

Matthee J.J. 1951. The structure and physiology of the egg of *Locustana pardalina* (Walk.). Sci. Bull. Dep. Agric. S. Africa. 316: 83 pp

Pitkin L.M. 1980. A revision of the Pacific species of *Conocephalus* Thunberg (Orthoptera: Tettigoniidae). Bull. Br. Mus. (Nat. Hist.) 41: 315-355.

Power J.H. 1958. On the biology of *Acanthoplus bechuannus*. J. Entomol. Soc. South Afr. 21: 376-381.

Rentz D.C.F. 1979. Comments on the classification of the orthopteran family Tettigoniidae, with a key to subfamilies and description of two new subfamilies. Aust. J. Zool. 27: 991-1013.

Richards A.M. 1973. A comparative study of the giant wetas *Deinacrida heteracantha* and *D. fallai* (Orthoptera Henicidae) from New Zealand. J. Zool. 169: 195-236.

Strazanac J.S. 1986. Some preliminary results on the status of the Oahu species of *Banza* (Orthoptera: Tettigoniidae: Copiphorinae). Proc. Hawaii. Entomol. Soc. 27: 147-152.

Tanaka S. 1984. Seasonal variation in embryonic diapause of the striped ground cricket, *Allonemobius fasciatus*. Physiol. Entomol. 9: 97-105.

Tauber M.J., C.A. Tauber & S. Masaki. 1984. Adaptations to hazardous seasonal conditions: dormancy, migration and polyphenism. In: Ecological Entomology. Eds. Haffaker & Rabb. Wiley-Interscience, New York. pp 149-184.

Tauber M.J, C.A. Tauber & S. Masaki. 1986. Seasonal adaptations of insects. Oxford University Press. 411 pp.

Ushatinskaya R.S. 1984. A critical review of the superdiapause in insects. Ann. Zool. 21: 3-30.

Uvarov B. 1966. Grasshoppers and locusts. I. Cambridge University Press.

Uvarov B. 1977. Grasshoppers and locusts. II. Centre for Overseas Pest Research, London.

Wagan M.S. 1985. Some observations on the developmental biology of *Trigonocorypha unicolor* (Phaneropterinae: Tettigoniidae). Pak. J. Zool. 17: 110-111.

Warne A.C. 1970. The biology of the eggs of the Tettigoniidae. PhD Thesis, University of Nottingham.

Warne A.C. 1972. Embryonic development and the systematics of the Tettigoniidae (Orthoptera: Saltatoria). Int. J. Insect Morphol. Embryol. 1: 267-287.

Warne A.C. & J.C. Hartley. 1975. The distribution and dispersal of *Conocephalus dorsalis* (Latreille) (Tettigoniidae) in the British Isles. Entomol. Gaz. 26: 127-132.

Wheeler W.M. 1893. A contribution to insect embryology. J. Morph. 8: 1-160.

Young G.R. 1985. Observations on the biology of *Segestes decoratus* Redtenbacher (Orthoptera: Tettigoniidae), a pest of coconut in Papua New Guinea. Gen. Appl. Entomol. 17: 57-64.

5. EVOLUTION OF ACOUSTIC COMMUNICATION IN THE GENUS *NEOCONOCEPHALUS*: Discontinuous Songs, Synchrony, and Interspecific Interactions

Michael D. Greenfield

Introduction

Among the musicians of the insect world, cone-headed katydids of the genus *Neoconocephalus* (Conocephalinae: Copiphorini) stand out for the loud and incessant nature of their calling songs. Males of several North American species produce songs that are clearly audible to the human ear at distances greater than 300 meters (Alexander 1956), and, overall, *Neoconocephalus* calls represent a major component of the nocturnal "acoustic" environment in mesic grassland areas of the Nearctic and Neotropical regions. This conspicuous sexual signalling, usually occurring near the ground, has led various workers to study chorusing (Greenfield 1983, 1988; Meixner & Shaw 1986), spacing and territoriality (Meixner & Shaw 1978, Shaw *et al.* 1982; Brush *et al.* 1985), mechanics and energetics of stridulation (Heath & Josephson 1970; Walker 1975; Josephson 1984, 1985), and environmental influences on song parameters (Whitesell & Walker 1978) in these insects. In this chapter, I present the basic biology of *Neoconocephalus* katydids as a background to their acoustic communication. I then formulate a novel hypothesis, based on intra-sexual selection, for the evolution of the different songs produced by the various species and for the intra- and inter-specific acoustic interactions between these songs. In particular, I focus on explaining the origins of discontinuous songs as occurring in the genus, chorusing (various forms of synchrony) in species producing these discontinuous songs, and the inhibition of singing in certain of these species as caused by heterospecific calls. Moreover, these features are not peculiar to *Neoconocephalus*. By introducing data from other acoustic orthopterans, I demonstrate the prevalence of such phenomena and also marshal evidence in support of the broad applicability of the hypothesis.

Systematics and Biogeography of *Neoconocephalus*

The genus *Neoconocephalus* comprises 25 New World species of cone-headed katydids (Tettigoniidae: Conocephalinae: Copophorini) that have been defined according to current taxonomic considerations. Males of all 25 species may be separated either on the basis of the fastigium, stridulatory files or fields, or calling songs. Based on the species I have examined, females can be distinguished from one another by the fastigium. Twelve of the 25 species are restricted to the Nearctic, three are found in both the Neotropics and the Nearctic (two of these, however, are essentially Neotropical and only reach North America in southern Florida), and the remaining ten are entirely Neotropical.

The Nearctic Fauna

Treatment of the Nearctic species has been relatively stable for over 70 years: this may partly reflect the ease with which males of most species can be identified in the field by their calling songs. By 1915, 11 of the species (*N. caudellianus*, *N. ensiger*, *N. exiliscanorus*, *N. lyristes*, *N. melanorhinus*, *N. nebrascensis*, *N. palustris*, *N. retusus*, *N. robustus*, *N. triops*, and *N. velox*) had been formally described – on gross morphological bases – and an additional subspecies (*N. robustus crepitans*) was recognized (Rehn & Hebard 1915). The use of magnetic tape recordings after 1950 to analyze more precisely the acoustic parameters of calling songs confirmed the validity of the above 11 species (see Alexander 1956 and Whitesell 1969) and indicated furthermore that *N. robustus crepitans* was a distinct species; this former subspecies has since been duly elevated to *N. bivocatus* (Walker *et al.* 1973). More recently, analysis of tape recordings revealed that individuals formerly considered to be members of a disjunct (Florida) population of *N. lyristes* actually comprise a distinct, local species (*N. pahayokee*; see Walker & Whitesell 1978a), and exploration of southern Florida showed that two widespread, Neotropical species (*N. affinis* and *N. maxillosus*) occur there as well (see Walker & Whitesell 1978b and Walker & Greenfield 1983).

All but two of the Nearctic species are restricted to areas of North America east of the 100th meridian (Whitesell 1969). The two exceptions, *N. robustus* and *N. triops*, occur widely in eastern North America, but have disjunct populations in California and Arizona. It is uncertain whether these isolated, western populations originated via natural dispersal or resulted from human introduction: the excellent flight of these two species makes the former suggestion a legitimate possibility. In eastern North America, *Neoconocephalus* are found from Florida to Texas and New Mexico, and from Maine and Ontario to Minnesota and North Dakota. Many species are associated with wetland vegetation, either coastal or freshwater (e.g., marshes, bogs, poorly drained agricultural fields, roadside ditches), and species distributions are centered along the Atlantic and Gulf coasts and in the Mississippi Valley: ten species occur in New Jersey and nine occur in Illinois. Two species, *N. lyristes* and *N. palustris*, have curious, disjunct distributions, with populations both along the Atlantic coast and in central North America. This situation may have resulted from a dispersal of coastal individuals at the end of the last Pleistocene glaciation (Thomas 1933).

The Neotropical Fauna

Unfortunately, our knowledge of the Neotropical *Neoconocephalus* is as incomplete as the treatment of the Nearctic ones is stable. Due to some uncritical work during the nineteenth century and failure to correct these errors by early twentieth century cataloguers, we are presently left with a multitude of species names from this region, many of which should probably be relegated to *nomina dubia* or synonymised. This is suggested by a recent taxonomic treatment of the *Neoconocephalus* of the Caribbean Islands and Panama (Walker & Greenfield 1983), the only Neotropical areas studied intensively during the past 50 years. Currently, 13 species are recognized from these areas, whereas 36 names had been proposed prior to 1900 (see Redtenbacher 1891 and Saussure & Pictet 1898). Notably, though, 5 of the 13 species had previously been undescribed.

Of the 13 Caribbean and Panamanian species, three (*N. affinis*, *N. maxillosus*, and *N. triops*) are widespread and occur throughout the Caribbean Islands and the adjoining mainland of North, Central, and South America (*N. triops* is found as far north as Ohio in North America), and one (*N. punctipes*) is known from certain mainland areas and continental islands (Panama, Surinam, Trinidad) and various Caribbean islands nearest the mainland (St. Vincent, Jamaica), suggesting limited dispersal into the Caribbean from two points (Honduras and Venezuela). The remaining nine are either known from restricted mainland areas of continental and nearby islands (*N. saturatus* – Trinidad, Grenada, St. Vincent; *N. spiza* – Panama; *N. susurrator* – Trinidad) or are endemic to one or two Caribbean islands (*N.*

aduncus – Cuba; *N. carbonarius* – Cuba, Grand Cayman; *N. occidentalis* – Hispaniola; *N. pinicola* – Hispaniola; *N. pipulus* – Jamaica; *N. retusiformis* – Puerto Rico, Mona). It is likely that exploration of Central America (north of Panama) and South America, supplemented by a thorough examination of type specimens and species descriptions, will extend the known ranges of several species (*N. punctipes*, *N. saturatus*, *N. spiza*, *N. susurrator*). Extrapolating from the above study, it is also likely that such exploration will yield several new species and reveal many *nomina dubia* and synonyms. Because the recent systematic study of Caribbean and Panamanian *Neoconocephalus* shows that most previous work is of little value, in this review I choose to ignore the existing taxonomy for other Neotropical areas, all of which had been created by the same authors of the earlier treatments of Caribbean *Neoconocephalus*.

Phylogenetic Relationships with other Copiphorini

The closest relatives of *Neoconocephalus* are undoubtedly the 21 species of *Ruspolia* found in Africa. In fact, Bailey (1975) considers that *Neoconocephalus* and *Ruspolia* are congeneric, as the two genera are not separable morphologically. If this assessment is eventually proposed formally and then accepted, all species of *Neoconcephalus* would be assigned to the older genus, *Ruspolia*. The cosmopolitan distribution of these cone-headed katydids may result from their occurrence on both the African and South American plates prior to tectonic separation during the Cretaceous period. Alternatively, it should be noted that adults in some species of *Ruspolia* form 'swarming flights," and members of swarms had landed on ships in the Atlantic Ocean 1000 km off the African coast (Ragge 1972; Bailey & McCrae 1978): distribution on both sides of the Atlantic could be the consequence of dispersal and founder events. Of uncertain interpretation in this regard is the observation that the most favoured food plants of some Panamanian *Neoconcephalus* are species of grasses introduced from Africa after 1900 (M. Greenfield, pers. obs.).

The distributions of two of the African species of *Ruspolia* extend to Central Europe and the Arabian Peninsula (Bailey 1972, 1975), respectively, and additional members of the genus occur in the Oriental and Australian regions. Bailey (1979) suggests that Australian members of the genera *Euconocephalus* (one species) and *Pseudorhynchus* (six species) might also be assigned properly to *Ruspolia*. Finally, the Hawaiian *Conocephaloides remotus* appears to be *Neoconcophelus* (or *Ruspolia*), based on its description (as given in Perkins 1899 and Zimmerman 1948; also see Bailey 1979 regarding the decision). Other close relatives of *Neoconcophelus* (or *Ruspolia*) are three genera of flightless, brachypterous copiphorines found in southwestern Australia (*Mygalopsis*; see Bailey 1979), Hawaii (*Banza*; see Zimmerman 1948), and Florida (*Belocephalus*; see Blatchley 1920).

NATURAL HISTORY OF *NEOCONOCEPHALUS*

Host-Plant Associations

Neoconocephalus katydids are primarily grassland species. Host-plants have been reported for seven species, and in all cases, save one (Gangwere (1961) observed *N. ensiger* feeding on plantain and dock), the insects were consuming blades, seeds, or flowers of various common grasses; e.g., *Agrostis*, *Andropogon*, *Bromus*, *Hyparhenia*, *Panicum*, *Paspalum*, *Phragmites*, *Saccharum* (see Whitesell 1969, Meixner & Shaw 1979, Greenfield 1983, 1988). Oviposition, seen in four species, is also associated with the host-plants, eggs being inserted into the grass culms (Whitesell 1969; M. Greenfield, pers. obs.). As expected, these katydids are never found far from strands of grasses. Specific habitats differ somewhat though, and some species are characterized as being more xerophilic (upland dwelling) than others (Whitesell 1969). Whether such preferences originate from host plant specificity is unknown.

The insects generally remain sedentary and close to the ground during the day but migrate upward in the vegetation after nightfall (Whitesell 1969; Greenfield 1983), when they are occasionally observed feeding. While in their daytime resting positions, they are exceedingly cryptic (to a human observer). This applies equally well to green and brown colour morphs, both of which probably occur in all species. The camouflage may be enhanced by the insects' habit of lying closely appressed to, and parallel to, the axes of grass blades and stems.

Life History and Seasonality

Most of the Nearctic species of *Neoconocephalus* are univoltine (Whitesell 1969). Overwintering is accomplished by diapausing eggs, and adults are seldom present prior to July but may be found until October. *N. pahayokee*, which is endemic to peninsular Florida, and populations of *N. triops* in the southeastern United States, however, are bivoltine. Adults of *N. pahayokee* occur during both May and September (Walker & Whitesell 1978a), and a diapausing, winter generation of *N. triops* adults exists in addition to a non-diapausing summer one (Whitesell 1974, Whitesell & Walker 1978). The winter adults are the progeny of either the previous winter or summer generation. Progeny of the winter generation had failed to mature completely by August, and they then entered a photoperiodically-induced reproductive (adult) diapause which ended by February. Offspring of the summer generation also spent the early winter in diapause. Summer adults, on the other hand, are all offspring of the previous winter generation. Remarkably, the song of the winter males (produced after the termination of diapause), regardless of their ancestry, differs from that of the summer individuals. It can be generated experimentally by subjecting developing (summer) insects to short-day photoperiods which induce diapause. In more northern locations, only diapausing, winter adults are found.

Voltinism of the Neotropical *Neoconocephalus* is unknown, but in the species that have been observed regularly (*N. affinis*, *N. punctipes*, *N. spiza*, *N. triops*), adults have been seen year round with no seasonal peaks in abundance (M. Greenfield, pers. obs.). Probably, several overlapping generations occur. *N. triops* males in the Caribbean Islands and Panama all produce the "summer song" (Walker & Greenfield 1983), indicating an absence of diapause.

Adult longevity in *Neoconocephalus* may be relatively prolonged. Non-diapausing *N. triops* males have been kept for over one month in cages (Whitesell 1974), and the monitoring of marked *N. affinis* males in the field has revealed lifespans greater than 60 days (Greenfield 1983).

Reproductive Behaviour – Calling

Pairing is presumably facilitated by the loud, species-specific calls of the males, but actual demonstration, via playback experiments, of the long-range attractiveness of these songs to conspecific females has only been accomplished in *N. triops* (Whitesell 1974). In most *Neoconocephalus* species, production of these calls is nocturnal (Alexander 1956; Whitesell 1969). Moreover, in those species that have been monitored intensively, the activity is largely restricted to the first half of the night (Greenfield 1988). Because these observations were made on Neotropical species, this temporal pattern probably reflects actual diel periodicity rather than more inactivity during the cooler temperatures of the latter half of the night (see Walker & Whitesell 1982). Nonetheless, several exceptions to the above generalization do exist. Two Nearctic species, *N. exiliscanorus* and *N. melanorhinus*, are crepuscular (Whitesell 1969), and the Neotropical *N. spiza* sings during virtually all hours (Greenfield 1988). Additionally, some species that do adhere to the general pattern of early nocturnal activity produce sporadic songs during the day (Alexander 1956; Whitesell 1969; Greenfield 1983).

Males in many species remain stationary and on relatively exposed perches during calling bouts. Perches are located within the preferred habitat but are not necessarily on a host plant. For example, *N. nebrascensis*, *N. robustus* and *N. triops* often sing in trees. (Thomas 1933; & Whitesell 1969). In general, males are affected by substrate vibration, and they respond to it by ceasing to call and either flying or dropping into denser foliage nearer the ground. Whitesell (1969) reports that downward escaping is more common in species that typically perch in concealed locations amid tangled vegetation. Some of these species (*N. melanorhinus*, *N. retusus*, *N. spiza*) exhibit the interesting habit of tremulation (shaking the substrate vegetation) during or immediately after stridulation. Whitesell (1969) proposed that this behaviour is adaptive in such microhabitats in that the male advertises his precise location to females, potentially situated nearby, who would otherwise be unable to determine the specific stem on which he is perched (see also Morris 1980). Tremulation may also mediate agonistic interactions between males (K.C. Shaw, pers. comm.).

The extreme sensitivity of calling males, at least in some species, to substrate vibration suggests that singing is attractive to natural enemies. However, predation has seldom been witnessed on these insects. Furthermore, in Panama I have only observed visually-hunting predators (salticid spiders) feeding on *Neoconocephalus*, and no records exist anywhere of specifically phonotactic predators such as bats (see also Belwood & Morris 1987). It is likely, though, that male calling does attract phonotactic parasitoids. Several *Neoconocephalus* species are regularly infested by the larvae of tachiniid flies (Nutting 1953), and when records have been kept, nearly all of the parasitized insects were found to be male (Whitesell 1969, Burk 1982).

Courtship and Copulation

Close-range sexual interactions have been observed in *N. ensiger*, *N. nebrascensis*, and *N. retusus*. Gwynne (1977) witnessed *N. ensiger* females flying to within 10 cm of singing males, with the males then closing the final gap to the females. In *N. retusus* the mutual orientation is aided by both sexes tremulating (Whitesell 1969); this activity has also been seen in *N. nebrascensis* (Meixner & Shaw 1986). After initial contact with the female, which is usually made via antennation, *N. nebrascensis* males emit a low-intensity "precopulatory song" (Meixner & Shaw 1986). Copulation in all observed cases lasted between 9 and 40 minutes. (Whitesell 1969; Meixner 1976; Gwynne 1977). The spermatophore transferred is internal and small, suggesting that paternal (nutritional) investment is negligible. Males resume calling shortly after the completion of copulation.

Spacing, Movement and Aggression

Field studies on several species (*N. affinis*, *N. ensiger*, *N. nebrascensis*) that typically occur in relatively high population densities showed that calling males were regularly spaced within "local" areas (Meixner & Shaw 1979; Shaw *et al.* 1982; Greenfield 1983). Moreover, in *N. affinis* most males exhibit a considerable degree of site fidelity, some individuals remaining within 2x2-m areas for several weeks (Greenfield 1983). Agonistic encounters have occasionally been reported in several species (*N. affinis*, *N. melanorhinus*, *N. nebrascensis*, *N. velox*) when calling males approached one another (Whitesell 1969; Greenfield 1983), and in *N. nebrascensis* a specific "contact song" was associated with such encounters (Meixner & Shaw 1986). An experimental study on *N. affinis*, employing playback of recorded songs in the field, demonstrated clearly that calling males did not "tolerate" other singing conspecifics within close range: males "attacked" a loudspeaker broadcasting song above a threshold sound intensity (Brush *et al.* 1985). These responses suggest that uniform spacing in *N. affinis* is affected by the calling song (see Bailey & Thiele (1983) for similar findings in *Mygalopsis*) and that many of the males are territorial (*sensu* Fretwell 1970). At the same time, *N. affinis* populations also include silent, non-territorial (= satellite) males that are found adjacent to calling individuals (Greenfield 1983). Satellite males also exist in *N. nebrascensis* (Meixner & Shaw 1986) and *N. ensiger* (K.C. Shaw, pers. comm.).

In *N. affinis*, *N. ensiger*, and *N. nebrascensis*, the regular spacing pattern of calling males appeared to be nested within an overall contiguous distribution (Meixner & Shaw 1979; Shaw *et al.* 1982; Greenfield 1983). Most members of local populations were found within a limited fraction of an apparently suitable habitat. It is uncertain whether such restriction represents mutual phonotaxis and lek (or, more precisely, "resource-based lek" (*sensu* Alexander 1975)) formation, and/or whether the insects are responding to spatial variation in resource quality unrecognized by the investigators. Possibly, territorial males are defending effective calling stations and/or dense stands of the host plants that are most preferred by females (see Greenfield *et al.* 1987 for an acridid example).

Males in other species of *Neoconocephalus* are decidedly more vagile. *N. bivocatus*, *N. exiliscanorus*, and *N. melanorhinus* all walk during calling, and *N. melanorhinus* males usually fly 5-10 m between calling bouts (Whitesell 1969).

Song Characteristics

Amplitude Modulation

I describe *Neoconocephalus* songs onomatopoetically as either continuous "buzzes" or as

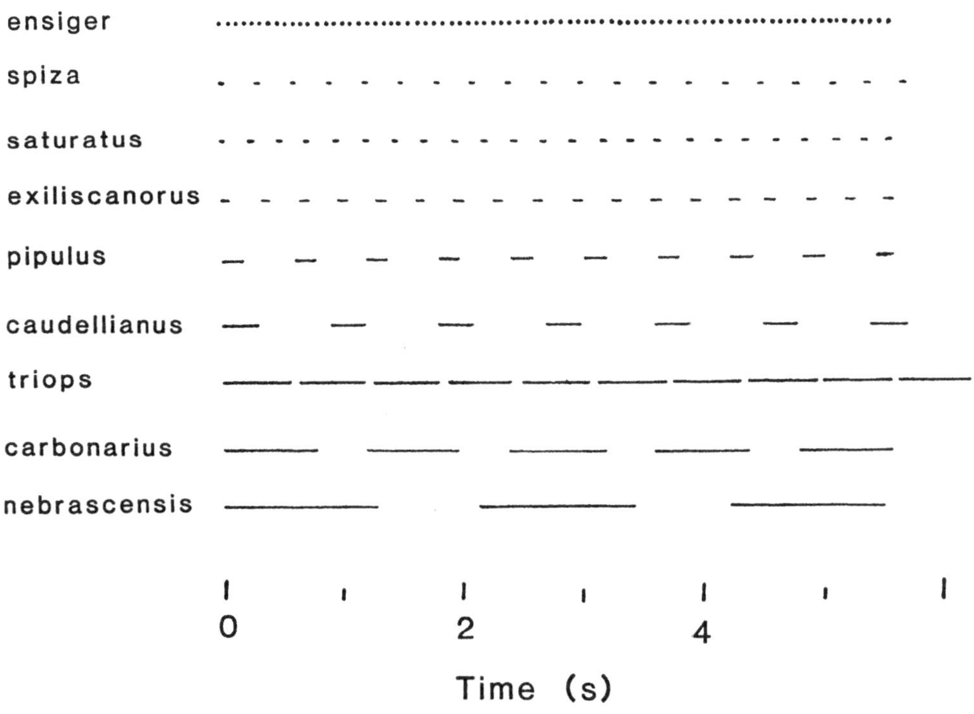

Fig. 1. Temporal pattern of chirps or interrupted buzzes of the nine *Neoconocephalus* species producing discontinuous songs.

discontinuous "chirps" or "interrupted buzzes". Oscillograms of the songs reveal the duration of chirps and inter-chirp intervals as well as the "phonatomes" (*sensu* Walker & Dew 1972) or "pulses" produced by each complete cycle of wing movement (I use phonatome in this chapter because of its unambiguous meaning). Closer inspection of these oscillograms, supplemented by high-speed, sound-synchronised cinemaphotography of wingstrokes during calling, show finer details within the phonatomes of several species and the wing movement responsible for such detail.

Calling songs have been recorded from 23 of the 25 *Neoconocephalus* species (Table I, Fig. 1). Tapes of a 24th species, *N. pinicola*, had been lost, but the general nature of the songs was recollected. The song of *N. aduncus*, endemic to Cuba, remains unknown.

Table 1. Song characteristics of *Neoconocephalus*

Species	song[a] type	phonatome[b] rate (s^{-1})	chirp or interrupted buzz[b] length (ms)	rate (s^{-1})	source[c]
aduncus	U				
affinis	C	11.7±0.7[1]			7
bivocatus	C	78[1]	(57-88;24°C)		9
carbonarius	D	198[2]	800[2]	O.8[2]	7
caudellianus	D	192[2]	300[3]	1.1[3]10	
ensiger	D	14[3]	10-30	14.0[3]	2,5,10
exiliscanorus	D	3.1[3]	50-112	3.1[3]5	
lyristes	C	130[3](23°C)			1
maxillosus	C	133.6±6.3[1]			7
melanorhinus	C	140[3]			10
nebrascensis	D	165[3] (21°C)	1296±132	0.40-0.55	3,4
occidentalis	C	49±3			7
pahayokee	C	155[3] (23°C)-23°C			8
palustris	C	225[3,4]			10
pinicola	C	[5]			6
pipulus	D	267±8[2]	175[2]	1.65[2]	7
punctipes	C	245±10.7[1]			7
retusiformis	C	49±6			7
retusus	C	170[3,4]			5
robustus	C	198[1] (168-230;25.5°C)			9
saturatus	D	161±19	43-62[4]	4.8[6]	7
spiza	D	153±17	40-60[4]	3.5[6]	7
susurrator	C	118[2]			7
triops	C	89±7			5,7
	D	114.4±4.9[1,7]	585;27°C[3]	1.6;27°C[3]	
velox	C	195[3]			10

([a]) U: song unknown; C: continuous song – regular silent intervals >0.25 x phonatome length or >5 ms not present; D: discontinuous song – silent intervals >2 x phonatome length and >40 ms present with regularity.

([b]) Rates and lengths are given for (or interpolated to) 25°C unless otherwise noted and are expressed as X± SD or range wherever possible. 1: data averaged from several localities; 2: data from 1 insect only; 3: variation not available; 4: temperature unknown; 5: tape lost; 6: chirping rate can be irregular, and the maximum rate is shown; 7: approximately 5 phonatomes delivered at 100 s^{-1} are produced at the beginning of each interrupted buzz.

([c]) Sources: 1: Alexander 1956; 2: Counter & Henke 1977; 3: Meixner 1976; 4: Meixner & Shaw 1986; 5: Walker 1975; 6: T.J. Walker, pers. obs.; 7: Walker & Greenfield 1983; 8: Walker & Whitesell 1978a; 9: Walker *et al.* 1973; 10: Whitesell 1969.

Sixteen species of *Neoconocephalus* produce continuous buzzes. Males in these species sing for indeterminate lengths of time that vary, both between and within individual insects, from several seconds to many minutes (60 min. in *N. affinis* (M. Greenfield, unpubl. data) and *N. robustus* (Heath & Josephson 1970)). They then remain silent for another unspecified interval before resuming their song. This cycle of alternating periods of singing and silence is typically repeated many times during any one night.

Nine *Neoconocephalus* species sing in a discontinuous fashion with brief uniform durations of sound alternating with intervals of silence. Discontinuous songs in *Neoconocephalus* range from the 30-ms chirps, produced at a rate of 14 s^{-1}, of *N. ensiger* to the 1300-ms interrupted buzzes, separated by 800-ms silent intervals, of *N. nebrascensis*. In several species there exists a moderate amount of variation in the length of the chirp (or interrupted buzz) and/or the inter-chirp interval (Table I). As with continuous singers, discontinuously singing species intersperse bouts of regular chirping (or emission of interrupted buzzes) with silent periods. (The discrepancy in the number of species, namely songs of 24 species are known but sixteen and nine species are listed in the two respective categories, derives from *N. triops*, in which diapausing, winter generation males produce continuous songs whereas non-diapausing, summer generation males sing discontinuously (Whitesell & Walker 1978)).

Phonatome rates in *Neoconocephalus* extend from 3.1 s^{-1} (*N. exiliscanorus*) to 240 s^{-1} (*N. pipulus*, *N. punctipes*) and show little intraspecific variation at a given temperature (Table I). *N. bivocatus* and *N. robustus*, however, deviate somewhat from this generalization (Walker *et al.* 1973), and the song of non-diapausing *N. triops* is another exception in that two different types of phonatomes occur during each interrupted buzz (Walker 1975). In discontinuous singers a chirp may consist of a single phonatome, as in *N. ensiger* and *N. exiliscanorus*, or several, as in *N. saturatus* and *N. spiza*. Sounds termed interrupted buzzes (in *N. carbonarius*, *N. caudellianus*, *N. nebrascensis*, *N. pipulus* and *N. triops*) differ markedly from chirps by consisting of trains of 50 or more phonatomes. Unless otherwise stated, all temporal parameters given were obtained from (or interpolated to) insects singing at 25°C.

Kinesiology of Stridulation

Filming of stridulatory movements in eight species of Nearctic *Neoconocephalus* showed that a phonatome generally corresponded to a brief opening of the wings followed by a more prolonged closing that tended to produce a louder sound (Walker 1975). Several species (*N. bivocatus*, *N. triops*) display the phenomenon of "paired pulses" in which an entire phonatome is produced by four movements: a full opening, a partial closing, a partial opening, and a full closing. Although not established by filming, paired pulses are likely to occur in *N. affinis* and *N. maxillosus* as well (see oscillograms in Walker & Greenfield 1983). The song of *N. exiliscanorus*, 50-112-ms single phonatome chirps repeated every 325 ms, represents a more extreme version of this stridulatory feature: the wings are partially closed and opened from three to eight times during each complete cycle of movement, creating a phonatome that consists of four to nine 12.5 ms "elements".

Wing movements producing stridulation in *Neoconocephalus*, even in those species with phonatome rates of 200 s^{-1}, are affected by synchronous muscles (Heath & Josephson 1970). The very high contraction frequencies of these muscles are achieved via unusually short twitch durations (Josephson 1984) and are sustained by endothermy: body temperatures may be elevated as much as 10°C above ambient prior to stridulation (Heath & Josephson 1970). The same muscles are responsible for flight, wherein they operate at much lower frequencies. In *N. robustus*, for example, the reduction is 90% (Josephson 1984).

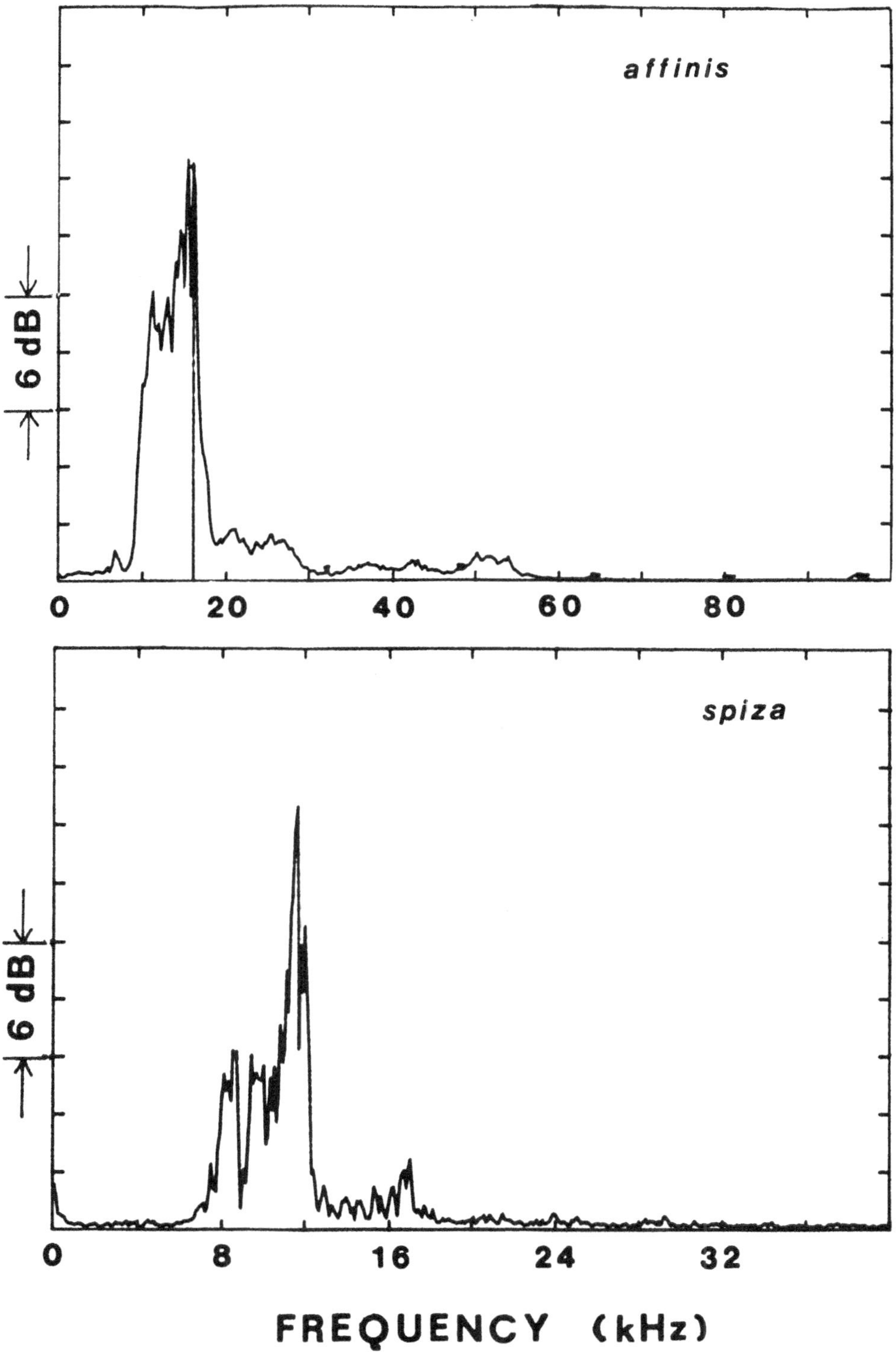

Fig. 2. Frequency spectrograms of the songs of *N. affinis* and *N. spiza*. Recordings were made with a Lafayette tape recorder operated at 76cps and a Bruel & Kjaer microphone; frequency analysis was accomplished with a Nicolet 660a Fourier spectrum analyzer.

Carrier Frequency and Sound Intensity

The principal carrier frequencies of the songs of *Neoconocephalus* tend to be broad-banded and ranging from 7 to 16 kHz (Alexander 1956; Walker & Greenfield 1988). Because nearly all calls were recorded with equipment restricted to sounds beneath 20 kHz, the existence of high-frequency harmonics in some songs remains possible. Nonetheless, the few recordings (of *N. affinis* and *N. spiza* only) that were made with equipment capable of gathering sound up to 60 kHz revealed very little sound energy above 20 kHz (Fig. 2). Counter (1977) and Counter & Henke (1977) made detailed studies of the songs of *N. robustus* and *N. ensiger*, and claimed that the principal carrier frequencies were derived from the rate of file-scraper contact. They also reported, based on neurophysiological work, that hearing in both males and females was maximally tuned to these frequencies.

Intensities of several of the louder species (*N. affinis*, *N. robustus*, *N. triops*) have been measured, and values averaging 100 dB (0 dB re 20 pa), at 10 cm dorsal aspect, were obtained (Counter 1977; Greenfield 1983 unpubl. data). In continuous singers, intensities are relatively low at the beginning of an uninterrupted song, gradually rising over a period of seconds to a value that remains fairly constant until the end of the song (M. Greenfield, unpubl. data).

Song interactions

Chorusing

Statistically significant temporal interactions between the songs of neighbouring conspecifics may occur in all *Neoconocephalus* species when population densities are sufficiently high. I refer to these interactions as choruses (*sensu* Greenfield & Shaw 1983). Choruses are most likely the effects of acoustic competition between adjacent males (Alexander 1975). By this argument, the large-scale choruses observed in some species would merely represent the summation of many pairwise competitions. The actual acoustic interactions assume different forms depending on the nature of solo singing in a species. At the simplest level, males may exhibit a nearly simultaneous onset of song at the beginning of the evening (Greenfield & Shaw 1983). On several occasions in Panama, I have heard four or five (non-diapausing) *N. triops* males, grouped within a 100 m radius, all initiate singing during a 1 min period approximately 1 h after sunset. Alexander (1956) reported comparable events in *N. nebrascensis*. Through either mutual phonoresponse, or a very precise and invariant photoperiodic mechanism, these males formed a "spree" (*sensu* Walker 1983), the temporal equivalent of a lek. A more advanced form of chorusing, either termed "unsynchronized

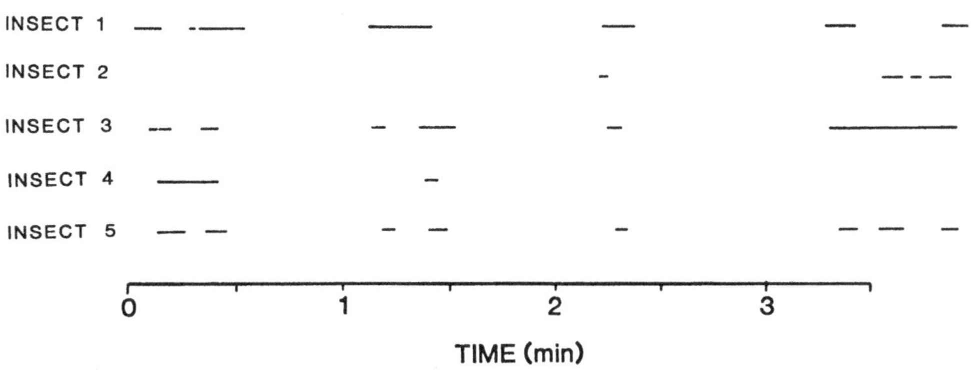

Fig. 3. Temporal pattern of song segments of five *N. affinis* males engaged in unison bout singing. See Greenfield (1983) for recording method.

chorusing" (Otte 1977) or "unison bout singing" (Greenfield & Shaw 1983), occurs regularly in several *Neoconocephalus* species (*N. affinis*, *N. melanorhinus*, *N. palustris*) that emit continuous songs (Whitesell 1969; Greenfield 1983). This interaction entails a crude synchrony of the singing bouts of neighbours. In *N. affinis*, for example, males within a 30 m radius may all sing collectively for 30 s to 5 min, remain silent for a comparable period, and then repeat the cycle many times during the evening (Fig. 3). Occasionally, more extensive groups of *N. affinis* exhibit these bouts of collective singing, perhaps facilitated by a "chain-reaction effect." Within these singing bouts, "leaders", individuals who habitually sing earliest during a collective bout and who, overall, sing during a relatively high proportion of the nocturnal calling period, can be distinguished.

The finest level of chorusing in *Neoconocephalus* is the synchrony of individual chirps or interrupted buzzes of neighbours. In his review of the acoustic insects of eastern North America, Alexander (1956) noted that those *Neoconocephalus* species that chirp, synchronise. Of the nine species that produce discontinuous songs, synchrony has been recorded in four, notably *N. caudellianus*, *N. exiliscanorus*, *N. nebrascensis*, and *N. spiza*. Closer observation of several Neotropical species may reveal additional cases.

In both *N. caudellianus* and *N. exiliscanorous* various observers have noted that the intensity of choruses of synchronising males periodically rises and falls (Rehn & Hebard 1915; Allard 1918; Fulton 1934). This effect is caused by insects that intersperse bouts of regular chirping with silent periods; i.e., a case of unison bout singing may overlie the synchrony of individual chirps. Synchrony in *N. nebrascensis* and *N. spiza* has been studied in more detail (Meixner & Shaw 1986; Greenfield 1988). Analyses reveal an imperfect synchrony in which "leaders" (in synchrony, leaders are individuals that begin their interrupted buzzes or chirps first) and "followers" exist. In *N. nebrascensis*, individuals who are leaders during synchrony are those who produce faster rates of interrupted buzzes/min while singing alone. During synchrony leaders produce longer interrupted buzzes than followers so that buzzes of these males are often completely overlapped. Leader and follower roles are seldom reversed during a singing bout in *N. nebrascensis*, but reversals are common in *N. spiza*.

Interspecific Inhibition

Several *Neoconocephalus* species that produce discontinuous songs are inhibited from singing during the more continuous songs of congeneric species. These responses probably represent an effect or by-product of the chorusing that normally occurs within the discontinuously singing species. An alternative explanation is that elements in the songs of affected species critical to effective communication with conspecific females, and males, are obscured or distorted by the sounds of inhibitors. By abstaining from singing when such interference is present, males avoid the production of essentially useless signals.

In North America *N. ensiger* and *N. exiliscanorus* refrain from singing during the long, interrupted buzzes of any nearby *N. nebrascensis* (Meixner 1976; T. Walker, pers. comm.). Both of the inhibited species sing in unison with *N. nebrascensis* by inserting several chirps during each interval between the interrupted buzzes of the latter species (Fig. 4a).

In Panama a more severe case of inhibition occurs in which *N. spiza* completely ceases to sing during the (continuous) songs of all sympatric congeners (*N. affinis*, *N. punctipes*, *N. triops*; Fig. 4b; Greenfield 1988). Playback experiments using tape recordings of *N. affinis* song or various pure-tone frequencies showed that inhibition was caused by any continuous sound above 40 dB SPL and between 8 and 16 kHz in carrier frequency. Pulsed sounds with these intensity and frequency characteristics also caused inhibition provided that the inter-pulse intervals were sufficiently short (220 ms). That *N. spiza* is inhibited only by congeners reflects the fact that few animals other than *Neoconocephalus* produce continuous, loud sounds at such high carrier frequencies. *N. spiza* exhibit two levels of adaptation to the acoustic interference originating from continuously singing species. The inhibitory responses described above represent one of these, a short-term temporal adjustment. The second adaptation is a long-term temporal adjustment which involves an actual shift in the diel periodicity of singing in populations co-occupying habitats with inhibitory species (Greenfield 1988; Meixner (1976) suggests that this adjustment may also occur in *N. ensiger* populations in response to *N. nebrascensis*). Populations of *N. spiza* that are isolated from congeners, and therefore acoustic interference, sing primarily during the evening, but where confronted by congeneric species the males call diurnally. This can be seen either by a comparison of different populations or, more dramatically, by the removal of the congeneric species: this results in an immediate shift to nocturnal singing.

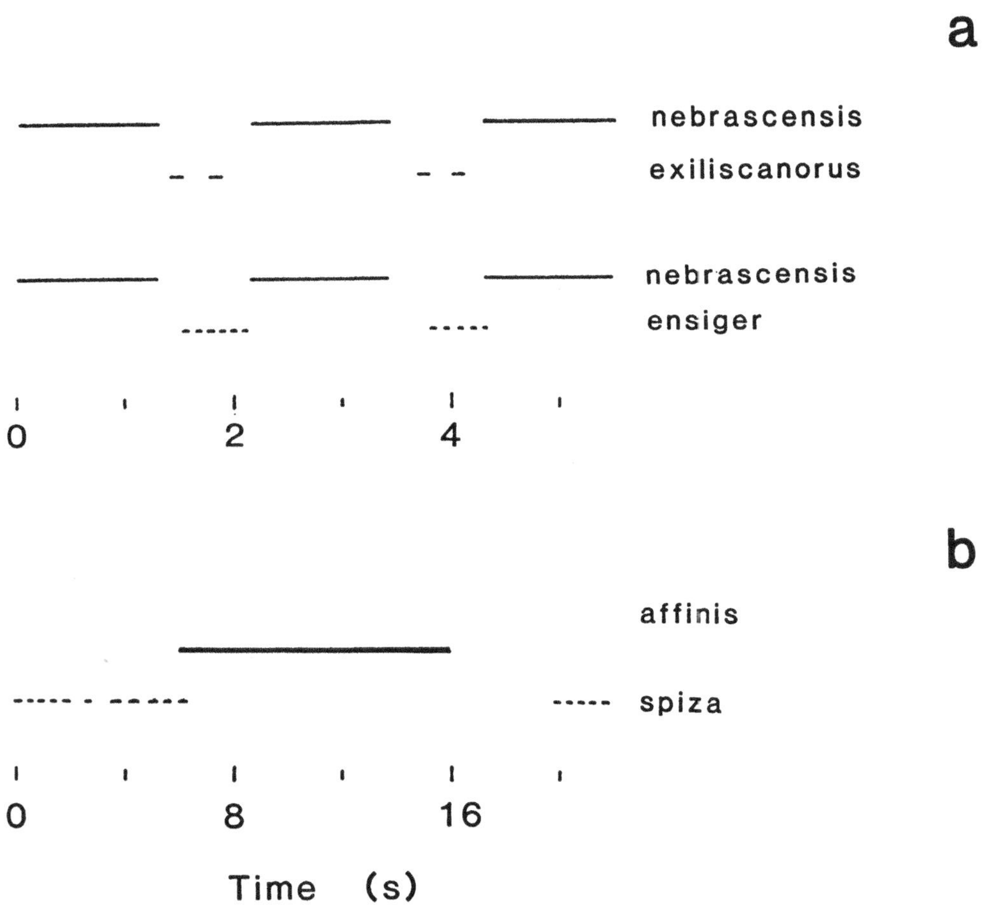

Fig. 4. Generalized temporal patterns of interspecific inhibition of discontinuously singing *Neoconocephalus* species by more continuously singing congeners. a. *N. nebrascensis* inhibits (Meixner 1976) and *N. exiliscanorous* (T. Walker, pers. comm.) from singing during its long, interrupted buzzes. b. *N. affinis* inhibits *N. spiza* from singing during its continuous songs and for a brief refractory period afterward (Greenfield 1988).

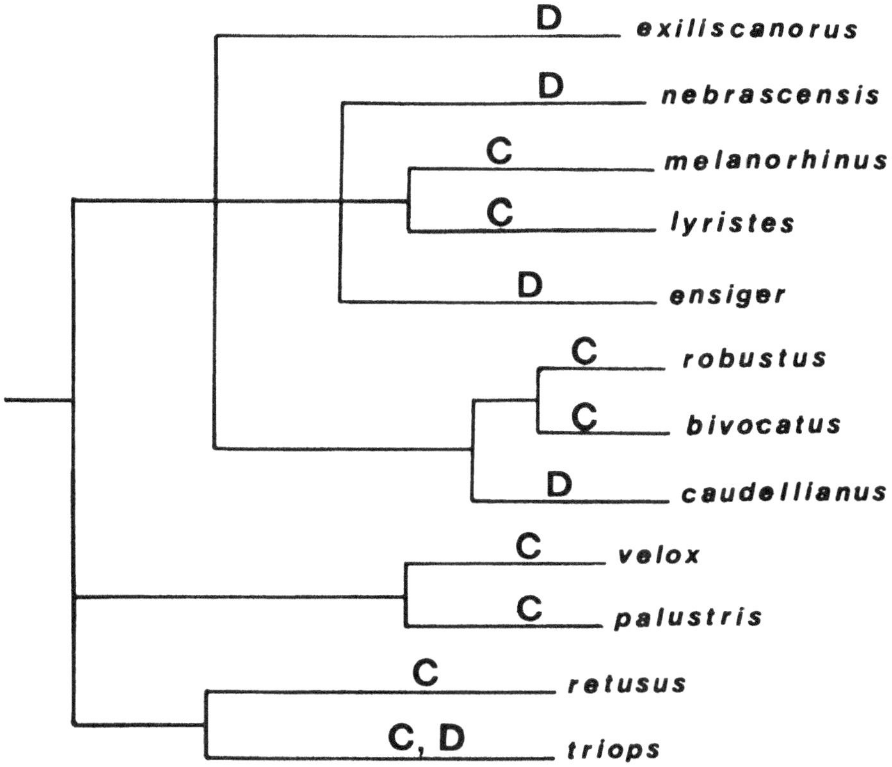

Fig. 5. Phylogeny of Nearctic *Neoconocephalus* species as constructed by Rehn & Hebard (1915) with song type (C: continuous; D: discontinuous) of each species noted.

The Evolution of Discontinuous Songs: A requirement for "acoustic windows"

As in many acoustic Orthoptera (see Alexander 1956, 1962, and Love & Walker 1979), the calling songs of *Neoconocephalus* species can be separated into two categories, those in which phonatomes are produced continuously and those in which phonatomes are grouped as discontinuous songs. I now propose that discontinuous songs have evolved largely because of the importance to a singing male of hearing clearly the calls of its neighbours so that it can perform various actions, either acoustic (chorusing), spacing, or agonistic. This hypothesis is based on the expectation that occasions during which a male would be selected to execute such actions would most often arise in dense populations, and a survey of *Neoconocephalus* (and other) species demonstrates that discontinuous singers more frequently do experience high density than continuous ones. Below, I first entertain five alternative hypotheses, none of which are as likely.

1. Song Characteristics Reflect Phylogeny

It is possible that discontinuous singers represent a separate lineage and that chirping songs occur for historical reasons. However, a morphologically-based phylogeny constructed for the Nearctic species of *Neoconocephalus* by Rehn and Hebard (1915) suggests otherwise. Continuous and discontinuous singers are apportioned among the species groups (Fig. 5), and, even allowing for tentative alterations in this initial phylogeny due to the introduction of modern systematic techniques, such as allozyme analysis or the cladistic arrangement of taxa, it is hard to foresee a change that would cluster all discontinuous singers together. Further evidence against this hypothesis is the production of both types of songs by one species, *N. triops* (see Whitesell & Walker 1978).

Continuous and discontinuous songs also co-occur within the Old World genus *Ruspolia* (Bailey 1972, 1976) and the endemic Hawaiian genus *Banza* (T. Shelly and M. Greenfield, unpubl. data). The songs of seven out of a total of 21 *Ruspolia* species have been recorded, and discontinuous, chirping songs occur in two of these (*R. differens*, *R. nitidula*). In *Banza*, songs of five of the ten species are known, and two of them are discontinuous. Comparisons with the Nearctic Gryllidae shows that variation, as described above, occurs within no fewer than seven genera (data compiled from: Fulton 1956; Alexander 1957a, 1957b; Alexander & Thomas 1959; Walker 1962, 1963, 1969b, 1969c, 1973; Love & Walker 1979; Weissman *et al.* 1981). These surveys indicate that chirping and continuous songs have arisen together within many taxa, implying the polyphyletic origins of the two song types.

2. Discontinuous Songs Result from Physiological Constraints

Species that sing at particular phonatome rates or that produce certain types of phonatomes may be constrained to insert regular gaps in their songs, possibly for energetic reasons (see Ewing 1984:234). However, a comparison of phonatome rates with song types shows that no relationship exists between these two factors; i.e., both singing categories include species with fast and slow phonatome rates. Sound-synchronised cinematography (Walker 1975) shows that several of the species emitting interrupted buzzes (*N. nebrascensis*, non-diapausing *N. triops*) produce phonatomes via wingstrokes similar in form to the wing movements of continuously singing species with comparable phonatome rates.

3. Environmental Influences

Habitat differences, however subtle, between *Neoconocephalus* species may cause certain types of songs to be transmitted effectively in some species but not in others. This could be due to the influences of specific vegetation structure or atmospheric conditions on the attenuation of sound and/or the local presence of phonotactic parasitoids who might detect or orient most efficiently to certain sounds. However, the extreme similarity in calling perches among species producing different types of songs argues against this hypothesis. For example, in Panama both *N. punctipes* (a continuous singer) and *N. spiza* (discontinuous) often call from adjacent sites on grass stems. In North America, *N. robustus* (continuous) and *N. ensiger* (discontinuous) have been heard calling simultaneously in the same fields (Counter & Henke 1977), as have *N. retusus*, *N. velox* (both continuous), and *N. triops* (discontinuous; Whitesell 1969). Additionally, both continuously and discontinuously singing *N. triops* are attacked by the phonotactic tachiniid parasite *Ormia lineifrons* equally well (Burk 1982).

4. Reproductive Isolation

The incorporation of regularly spaced, silent intervals within songs introduces a second dimension in the male's song that may act in species identification by conspecific females: interspecific variation along this dimension may have arisen within the context of reproductive isolation (see Whitesell 1969). *N. pipulus* and *N. punctipes*, two species sympatric in Jamaica whose 250 phonatome/s songs differ only in that those of the former include periodic 470 ms intervals (Table I, Fig. 1), lend support to this notion. However, various species producing discontinuous songs (*N. carbonarius*, *N. ensiger*, *N. exiliscanorus*, *N. spiza*) do so at phonatome rates unused by any sympatric species (Table I, Fig. 1), and it is unlikely that the chirping nature of their songs evolved solely, if at all, because of selection pressure to attain reproductive isolation. Additionally, the variation in chirp and inter-chirp interval duration that is evident in some species (Table I) suggests that chirp rhythm, as opposed to phonatome rate, is not always a reliable indicator of a male's identity.

5. Sexual Selection – Female Choice

One current theory suggest that interspecific variation in sexual signalling is indirectly the outcome of female choice: signals diverge under the influence of differences in such choice that arise, perhaps by chance, in isolated populations which ultimately become distinct species (West-Eberhard 1983). Thus, sexual selection can augment various factors that may lead to song divergence. For this process to explain the occurrence of continuous versus discontinuous songs, one would expect females in discontinuously singing species to prefer songs with the particular chirping pattern of conspecific males.

Unfortunately, the study of female preferences has been neglected in *Neoconocephalus*, but in the one species investigated, *N. triops*, non-diapausing females showed no preference for the discontinuous song of non-diapausing males over the continuous song presented at identical phonatome rates (Whitesell 1974). Likewise, winter generation females did not prefer the continuous songs of winter males over discontinuous songs. Comparable results have been obtained in the European gryllid *Gryllus campestris*, a chirping species (Thorson *et al.* 1982). Females responded equally well to natural songs or to continuous ones synthesized with the correct phonatome rate. On the other hand, Doherty & Hoy (1985) note various Orthoptera in which the temporal pattern of chirping is in fact critical. Their examples, nonetheless, only suggest that female choice might be responsible for specific chirp rhythms in some cases but not in others, and the origin of the discontinuous character itself is not addressed.

6. Inter-Male Competition

Various studies, beginning with Fabre's (1912:263) finding that a cannon blast did not silence singing cicadas, indicate that certain acoustic insect species do not hear during their own stridulation (see also Pringle 1954). The best evidence for this phenomenon comes from recent work by Wolf & von Helversen (1986) on the acridid *Chorthippus biguttulus*. They discovered a specific mechanism in the CNS that switches hearing on and off and that may function to prevent sensory overload (self-induced habituation) and/or the triggering of inappropriate attack/escape responses. In *Neoconocephalus*, Counter (1977) studied the question of hearing during stridulation in *N. robustus* and obtained equivocal neurophysiological results, but Brush *et al.* (1985) found behavioural evidence for hearing being switched off during singing in *N. affinis*. *N. affinis* males respond aggressively to playback of the conspecific song when silent, but display no response while singing.

If hearing and stridulation are generally incompatible, singing males in species producing continuous songs would be unable to detect the approach of calling conspecific males. Males thus incapacitated could not make various selected responses: approaching or deviating from the intruder (see Bailey & Thiele 1983; Dadour & Bailey 1985), or phonoresponding with a call reserved specifically for contact or aggression Consequently, I propose that chirps, interrupted buzzes, and the indefinite gaps in buzzes and trills of continuous singers have evolved to provide males with brief "acoustic windows" through which they can listen to neighbours. Recognizing that this hypothesis and the female choice hypothesis (#5) are not mutually exclusive, I submit that inter-male competition has been most influential.

The strongest evidence in support of the intrasexual selection hypothesis comes from two

Ruspolia species, *R. differens* and *R. nitidula*. *R. differens* often forms dense swarms, and at such times the male song changes from a continuous, 140 phonatome/s buzz to interrupted buzzes, each approximately 50 phonatomes in length, which are synchronised with neighbours (Bailey 1976; Bailey & McCrae 1978). A comparable shift in song type occurs in *R. nitidula* when neighbouring males are present (Bailey 1972).

In *Neoconocephalus*, non-systematic observations and other anecdotal reports suggest that discontinuously singing species are likely to occur in high density. For example, four species, *N. affinis*, *N. punctipes*, *N. spiza*, and *N. triops*, are prevalent in central Panama, and *N. spiza*, the most discontinuous singer, is the most abundant one throughout the lowland savannas. As many as ten males may be found singing within a 200 m^2 site, more than double the maximum density observed in any of the other species (M. Greenfield, unpubl. data).

In North America, studies of spacing in *N. ensiger* and *N. nebrascensis*, which are both discontinuous singers, revealed that males often sang in small clusters and were separated by 4 m (Meixner & Shaw 1979; Shaw *et al.* 1982). Two other discontinuous singers, *N. caudellianus* and *N. exiliscanorus*, can occur at even higher densities (Whitesell 1969). Thirty *N. exiliscanorus* males have been observed singing within an 80 m^2 clump of *Phragmites* (reeds). These insects may have been aggregated because of the patchy distribution of their preferred habitat: thus, a potential link between habitat distribution, population density, and song type can be inferred. Comparably high densities have not been reported for any of the continuously singing species of Nearctic *Neoconocephalus* (Alexander 1956; Whitesell 1969). Because the occasions during which a singing male would benefit from hearing conspecifics would occur more frequently in species that commonly exist at high density, I propose that the apparent relationship between density and song type in *Neoconocephalus* and *Ruspolia* points strongly toward the inter-male competition hypothesis. To a lesser extent, males in other species would also benefit from hearing conspecifics, and the intermittent, short gaps in their (continuous) songs may also represent windows for hearing neighbours. In *N. affinis*, 250 ms gaps occur sporadically during prolonged singing bouts in many individuals (Brush *et al.* 1985). During these pauses the forewings are not lowered to their normal resting position, and the male is able to sing at its maximum intensity immediately upon recommencing its call (M. Greenfield, unpubl. data). Similar gaps are inserted in the songs of many species of gryllids, and their songs are then usually termed "broken trills". In the genus *Oecanthus* the interruptions in broken trills were found to be unimportant for female response (Walker 1957).

Based on my earlier arguments, I propose the following evolutionary scenario. Initially, pauses in continuous songs may have arisen under the pressure of intrasexual selection. In certain species for which acoustic windows were more essential, the pauses eventually became more common, and discontinuous singing appeared. Female choice operating in species that had evolved discontinuous singing may have then led to certain specific chirping patterns. Once specific chirping patterns arose, the evolution of fine levels of chorusing proceeded in some cases (see following section). Alternatively, discontinuous songs may have been ancestral. In those cases where population densities remained or became low, continuous songs may have developed in the presence of strong intersexual selection for persistent singing (see Hedrick (1986) for an example of such female choice in gryllids).

Synchrony: a "game" of followers and leaders

Awareness of the precise synchrony between neighbouring males in certain discontinuously singing orthopteran species predates the advent of magnetic tape recording and can be traced back to several 19th century observers (Allard 1918). Early in the present century Fulton

(1928) demonstrated that such synchrony was facilitated by the ability to hear. He used *N. nebrascensis* in some of his experiments and found that deafened males sang asynchronously. However, an understanding of the evolution of the interaction has been harder to acquire. Explanations for this curious phenomenon are varied and have included: (1) it is an "apparently useless trait" (Fulton 1934), (2) individual males are forced to sing synchronously because to avoid doing so would destroy the species-specific chirp rate and prevent females from reaching any male (Walker 1969a), and (3) phonotactic parasitoids and predators are relatively unable to orient toward any one calling male when all calls are synchronized (Otte 1977). I now propose a fourth explanation for synchrony, derived from Alexander's (1975) general conception that all formats of chorusing are by-products of intra- and inter-sexual selection, and maintain that this specific format may be most typical of species in which the "envelope" of the chirp contains higher intensities toward the end.

My proposal of yet another explanation for the origin of synchrony was stimulated, in part, by judgements that the three pre-existing ones were inadequate. First, it is doubtful whether a behaviour as complex as synchronous chorusing would be "useless", and, while not advocating a staunchly adaptationist perspective, considering of alternatives to uselessness should be *de rigueur*. Second, whereas preservation of chirp rhythm may be a "goal" in species with uniform chirping rates (e.g., *Oecanthus fultoni*), variable rates (see Table I) in certain synchronising *Neoconocephalus* species (*N. exiliscanorus*, *N. nebrascensis*, *N. spiza*) suggest that rhythm preservation is not always a necessity for female attraction. Finally, the suggestion that synchrony may "confuse" phonotactic natural enemies had admittedly been offered on the basis of human subjectivity, and it is not grounded in the sensory perception of dipteran parasitoids or other potential enemies. (The one reported instance where predators are thwarted by synchronised calls involves phonotactic frog-eating bats who are deterred from finding their prey when adjacent frogs overlap their brief (500 ms), sporadic (1.7/min) calls (Tuttle & Ryan 1982)).

In arriving at the hypothesis that synchrony is a special manifestation of sexual selection, I was influenced heavily by two features of orthopteran and, in particular, *Neoconocephalus* synchrony. First, the cases of *Neoconocephalus* synchrony that have been analyzed with any precision do not reveal a perfect temporal overlap of chirps. Rather, there is a partial overlap in which followers and leaders occur. Second, while continuous and discontinuous songs co-occur within many genera of acoustic orthopterans, formats of chorusing (in discontinuous singers) are typically consistent within genera or even higher taxa. For example, all chorusing known in *Neoconocephalus* is of the synchronous variety, and this is also true for the gryllid genera *Cyrtoxipha* (Walker 1969b) and *Oecanthus* (Walker 1962). Analogously, only alternation of chirps is known in the chorusing of pseudophylline (*Pterophylla*, *Lea*; see Shaw 1968; Greenfield & Shaw 1983; Shaw & Galliart 1987) and ephippigerine tettigoniids (*Ephippiger*; see Busnel 1967).

The Importance of "Exposed" Chirp Endings

The existence of followers and leaders hints that synchrony of chirps or interrupted buzzes arises in cases where the latter part of the chirp (or buzz) is most critical for communication to females and other males. That is, if endings are crucial, individuals would be expected to sing in a manner conductive to the "exposure" of their own chirp endings while at the same time "concealing" the chirp endings of neighbours (see Otte & Smiley 1977 and Otte & Loftus-Hills 1979 for additional consideration of such factors); these criteria could only be met through initiating one's chirp immediately after the onset of a neighbour's chirp. When two neighbouring insects both behave according to the above "rules", an imperfect synchrony of chirps would arise in which the preferred, or dominant, role of follower rotates frequently between the individuals during successive chirps. This prediction is an accurate description of the synchronous chorusing in *N. spiza*.

Potential mechanisms that allow an individual to assume the role of a follower include reducing its own chirp rate immediately after being in the role of a leader (see Walker 1969a) and/or phonoresponding to the onset of a neighbour's chirp within a brief (30 ms) interval. The rapid phonoresponses known in male-female "dialogues" in certain tettigoniids (Heller & von Helversen 1986; Robinson *et al.* 1986; see Robinson and Heller chapters 7 and 8 this volume) and many occasions when a male *N. spiza* follows its neighbour's chirp by 10-30 ms after both had been silent for several seconds (Greenfield 1988) attest to the existence of the latter mechanism in some species.

The above emphasis on exposed chirp endings should not imply that this is the only song feature critical to species such as *N. spiza*. If it were, successively reduced chirp rates would be expected as chorusing individuals vie repeatedly for the (dominant) follower role. Such decreases are not evident, probably because excessively slow rates of chirping are ineffective in communication. Perhaps, delays in chirping occur up to a point beyond which the value of a high chirp rate takes precedence over the inevitability of several concealed chirp endings.

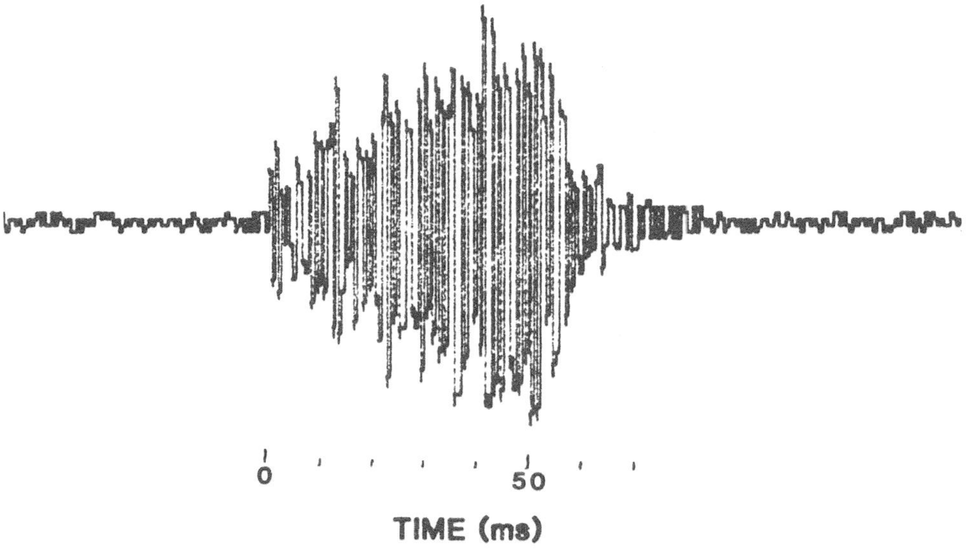

Fig. 6. Sound envelope of a six-phonatome *N. spiza* chirp showing the typical increase in intensity toward the ending. Recording was made with a Sony TC-D5M stereo cassette recorder and a Sennheiser ME-88 microphone; sound analysis was accomplished with a Bruel & Kjaer Model 2033 high resolution signal analyzer.

Constancy of a chorusing format within any one genus suggests a dependency of the format (i.e., synchrony versus alternation) on the nature (i.e., envelope) of the sounds produced by the insects, and it reinforces the hypothesis that synchrony ultimately results from the value of exposed chirp endings. The prediction that chorusing format depends on sound type follows from the general conservation of phonatome form and stridulatory movement. For example, all of the *Neoconocephalus* species that have been examined, regardless of whether they are continuous or discontinuous singers, produce phonatomes with more intense sound toward the end via a rapid opening and a slow closing of the forewings. More relevant to synchronous chorusing, however, is that an analogous, long-term increase in sound intensity is obvious toward the ends of multi-phonatome chirps and interrupted buzzes in most discontinuously singing *Neoconocephalus* species (e.g., see Fig. 6 and oscillograms in Walker (1975), Counter & Henke (1977), and Walker & Greenfield (1983)). While no tests have been conducted to determine whether chirp endings are more critical than chirp beginnings in the attraction of females of the "threatening" of other males, the fact that endings are consistently louder is strongly suggestive of this and of consequent selection for mechanisms that would promote following behaviour and the exposure of chirp endings.

A thorough assessment of the above hypothesis could entail formal game-theoretic analysis (see Partridge & Krebs 1978) of the advantages accrued through all possible chirping tactics. Implementation of this analysis, in turn, would rely on a detailed understanding of the pair-formation process. We currently know little about the phonotaxis of *Neoconocephalus* females or their distribution and movement in the field. Information on the latter would be essential for discerning whether chorusing males attract females to their vicinity, or whether females are already present and need only to orient toward particular singers: such possibilities would differentially influence the computation of advantages in any game-theoretic treatment.

Alternation as the Converse of Synchrony

Accepting the lines of reasoning leading to the hypothesis proposed above, another prediction, based on converse logic, would be that alternation of chirps or interrupted buzzes occurs when beginnings, as opposed to endings, bear the most significance. In such species, exposed chirp beginnings would be desirable and could be achieved most reliably via delaying the initiation of one's chirp for a brief interval following the termination of the neighbour's chirp. Again, if two insects are both timing their songs in this manner, alternation would result.

Chorusing by "Unevenly Matched Males"

In outlining the possible evolution of synchrony, it is assumed that the two interacting males were approximately equal in dominance. While this may have been true in the cases of *N. spiza* recorded synchronising in the field and from which the evolutionary model was derived, it should not be suspected in every situation. In particular, the studies of synchrony in *N. nebrascensis* chorusing in the laboratory almost certainly involved "unevenly matched" males. These synchronising pairs always included a consistent leader and follower (Meixner & Shaw 1986). The follower produced interrupted buzzes at a slower rate than the leader during solo singing, retreated from the leader in close encounters, and shortened his buzzes to the extent that they were completely concealed by the leader; i.e. followers in *N. nebrascensis* did not realise exposed buzz endings. Because *N. nebrascensis* followers are most likely subordinate individuals, their abbreviated buzzes may represent a means of singing, albeit rather ineffectively, in the vicinity of a dominant male without the risk of detection by him: following in *N. nebrascensis* may be a variation of the "satellite" tactic known in many other Orthoptera. Shortening of buzzes by followers tends to decrease as inter-male distances (Meixner & Shaw 1986), and it is possible that males separated by long distances in the field, synchronise in an evenly matched fashion lacking a follower/leader distinction.

Unison Bout Singing

The collective singing bouts of continuous singers, such as *N. affinis*, may be due to the same factors I suggested as being contributory to synchrony in *N. spiza* and *N. nebrascensis*. In *N.*

affinis, collective bouts are most conspicuous when the chorusing males are all producing relatively short segments of uninterrupted song (Fig. 3). At such times, the temporal pattern of singing actually resembles the synchrony of chirps or interrupted buzzes in discontinuously singing species, however, the units of sound are longer and not uniform in length. Otherwise, the role of following rotates among the males as in *N. spiza*, with individuals typically initiating their 30 s song segments several seconds after the start on one of their neighbours. *N. affinis* males may effect this long-term analog of synchrony because individuals that follow and terminate collective bouts might accrue a selective advantage. Possibly, the sharp transient from sound to silence at the end of an unobscured song segment is particularly valuable for female orientation.

When some of the individuals in a chorus of *N. affinis* males do not split their singing into a series of short segments, the resemblance to synchrony is less obvious. Nonetheless, certain aspects of the chorus suggest that the essential elements of synchrony are still present. In these choruses, consistent followers and leaders occur (Greenfield 1983). The followers in *N. affinis* call during a relatively small percentage of the nocturnal activity period, sing uninterruptedly during collective bouts of shorter periods than leaders, and do not initiate singing unless one or more neighbours already have (Fig. 7). *N. affinis* followers are most likely subordinate individuals, as indicated by the finding that males who sing infrequently show lower levels of aggression toward playback of recorded song (Brush *et al.* 1985).

Both followers and leaders in *N. affinis* time their songs in a manner conducive to the termination of collective bouts. Followers do not begin singing until at least several seconds have elapsed after the initiation of a neighbour's song, and leaders often resume singing immediately after having ceased if neighbours are calling at that instant (Greenfield 1983). Co-occurrence of following and leading tactics in these choruses suggests that both the termination of collective bouts and the overall duration of singing may be valuable in terms of sexual selection. Individuals who adopt the following tactic may be energetically limited, though, and therefore forced to direct all of their effort toward bout termination.

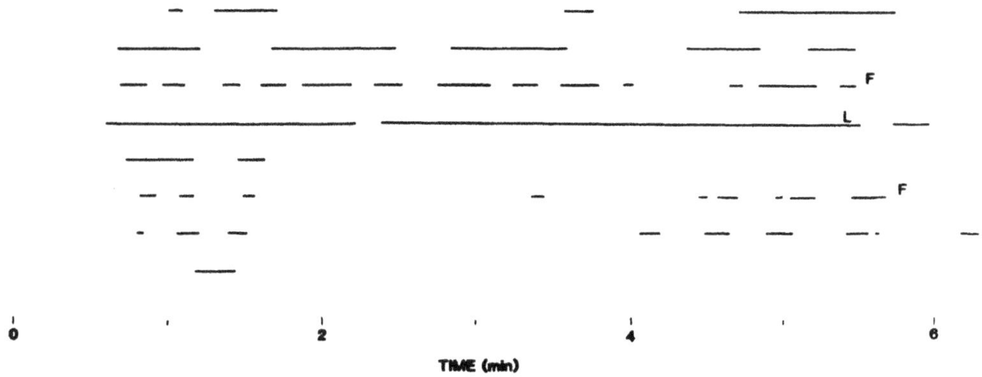

Fig. 7. Temporal pattern of song segments of *N. affinis* males engaged in unison bout singing with "leaders" (L) and "followers" (F) shown (see Greenfield 1983).

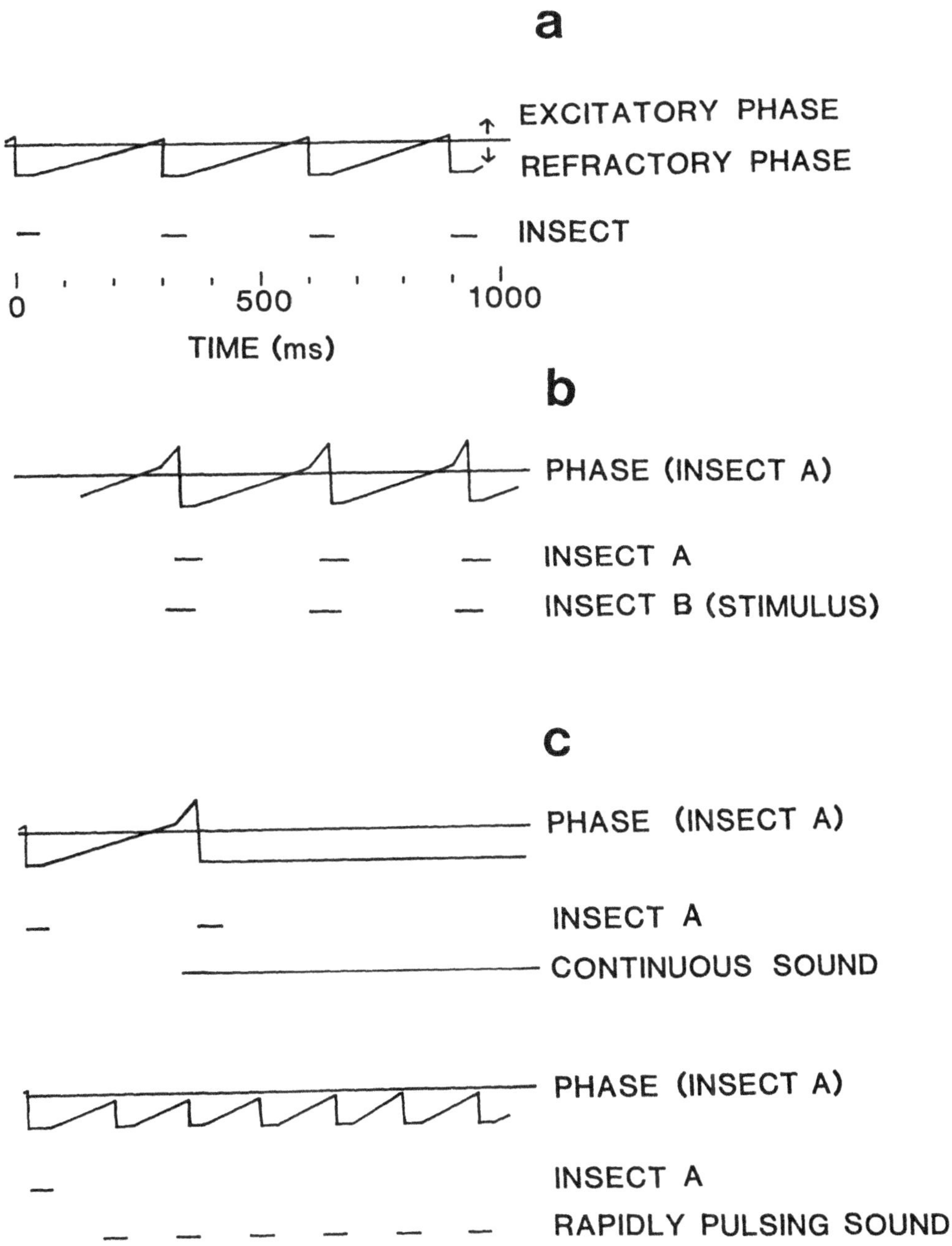

Fig. 8. a. Model depicting alternating excitatory and refractory phases during solo calling of an *N. spiza* male. **b.** Triggering of refractory phases by acoustic stimuli (e.g., chirps of a neighbouring *N. spiza* male) that are "unanswered" immediately after the onset of the stimulus. That is, insect A enters a 200-ms refractory phase 30 ms after the beginning of each stimulus. **c.** Maintenance of an extended refractory phase (= inhibition) by a continuous or rapidly pulsing sound. In these situations, insect A never enters an excitatory phase because gaps of silence 200 ms in length never occur (see Greenfield 1988).

Interspecific Inhibition: An incidental by-product of the "rules" of following

It is noteworthy that the three *Neoconocephalus* species known to experience interspecific inhibition are all discontinuously singing species and that two of these are confirmed synchronisers. This relationship also occurs among decticine katydids (Samways & Broughton 1976; Latimer 1981; Latimer & Broughton 1984), and it may be inferred that chirping species are often inhibited by continuously singing ones. To understand why this unilateral relationship may occur, a more detailed analysis of mechanisms responsible for the rhythm and synchrony of chirps is needed.

In *N. spiza,* a model based on alternating excitatory and refractory phases (Greenfield 1988), can be constructed to account for the timing of chirps. One can postulate that during solo singing a male's propensity to call declines sharply after starting a chirp, and he then enters a refractory phase. After this, the insect's level of excitation increases gradually until production of a chirp is again likely. Normally, an interval of 220 ms would be necessary for this excitatory phase to be reached (Fig. 8a).

To adapt this model to synchronous calling, I suggest that the level of excitation of "insect A" would be elevated by a chirp of "insect B", provided that A is already in, or at least is approaching, an excitatory phase at the chirp's onset. Approximately 30 ms after onset of this stimulus (chirp), A enters a refractory phase even if he has not chirped (Fig. 8b). The model complies with the observation that a 200 ms latency period follows the end of each chirp of B during which time A never initiates a chirp, and vice-versa (Greenfield 1988). After the chirp of B, A gradually approaches an excitatory phase, but his cycle is reset to maximum refractoriness by any acoustic stimulus unanswered by him (Fig. 8b).

The above model predicts an imperfect synchrony of chirps in which neighbours may alternately follow one another. It also indicates that any loud, continuous (or rapidly pulsing) sound with the frequency characteristics of *N. spiza* song would be inhibitory because the insects are perpetually kept in a refractory phase (Fig. 8c). Thus, interspecific inhibition in *N. spiza*, and perhaps other chirping species, may simply be a by-product of conspecific chorusing (see also Latimer & Broughton 1984): the "rules" for following dictate that chirps should not be produced until a certain interval has passed after the ending of a neighbour's chirp, and the buzzes of congeneric species are treated as if they are unusually long *N. spiza* chirps.

Summary

Neoconocephalus katydids comprise a genus of 25 New World species that are noteworthy for the loud and prolonged calling songs of the males. Analyses of the songs show that their principal carrier frequencies are broad-band and range from 7-16 kHz and that they are delivered at wingstroke rates from 3.1-267 s^{-1}. As in many Orthoptera, the species can be divided into two categories based on the amplitude modulation characteristics of their songs: those that produce phonatomes continuously for indefinite periods, described as continuous singers, and that group phonatomes according to various periodic regimens, discontinuous singers. Discontinuous singers may produce "chirps" consisting of one to several phonatomes or "interrupted buzzes" which are trains of 50 or more phonatomes. These song units range from 30 ms chirps delivered at a rate of 14 s^{-1} (*N. ensiger*) to 1300 ms interrupted buzzes separated by 800 ms silent intervals (*N. nebrascensis*). With two exceptions, phonatome rates are roughly constant within a species at a given temperature. However, rates of production of chirps or interrupted buzzes are somewhat more variable in several species.

Based on our knowledge of acoustic communication in *Neoconocephalus*, I formulate a hypothesis that explains the evolution of discontinuous songs and interactions between such songs as the outcome of sexual selection. I argue that chirps and interrupted buzzes arose primarily to provide "acoustic windows" through which singing males could hear and monitor neighbouring conspecific males. Because the insects are unlikely to hear conspecifics during their own stridulation, the silent intervals between chirps or interrupted buzzes allow them to detect intruders and to respond. Responses may take the form of an approach, a retreat, or a call reserved specifically for agonistic encounters. It is expected that occasions during which a male would benefit from making such responses would most frequently occur in dense populations. As predicted, the species of *Neoconocephalus* that most often occur in high densities are discontinuous singers.

Four of the nine discontinuously singing species chorus by synchronising their chirps or interrupted buzzes. In N. spiza a detailed analysis revealed that the synchrony was imperfect and involved "followers" and "leaders" who partially overlapped their chirps. Within a pair of synchronising males the role of follower rotated frequently: this feature suggested that individuals timed their singing in a manner conducive to exposing the endings of their own chirps while at the same time concealing the endings of the neighbour's chirps. Possibly, this tactic evolved because endings are more critical (than beginnings) in female attraction or in the "threatening" of other males. The increased sound intensity toward the endings of chirps in synchronising species supports this notion.

Several chirping species are inhibited from singing by the calls of more continuously singing congeneric species. A model constructed to account for the rhythm of both solo and synchronous chirping suggests that the inhibitory responses are incidental by-products of intraspecific chorusing.

Acknowledgements

Many of the ideas proposed here were originally stimulated during research on *Neoconocephalus* that was supported by the Smithsonian Tropical Research Institute and the Academic Senate of the University of California, Los Angeles. More recently, conversations with Ian Dadour and George Klump helped to mold some of the specific hypotheses. High frequency analysis of songs was assisted by Pat Brown and Joe Benner. I thank Win Bailey, Ted Burk, Ken Shaw, and Todd Shelly for reviewing the manuscript.

References

Alexander R.D. 1956. A comparative study of sound production in insects, with special reference to the singing Orthoptera and Cicadidae of the Eastern United States. PhD. Dissertation, Ohio State Univ., Columus. 529 pp.

Alexander R.D. 1957a. The taxonomy of the field crickets of the eastern United States (Orthoptera: Gryllidae: *Acheta*). Ann. Entomol. Soc. Am. 50: 584-602.

Alexander R.D. 1957b. The song relationships of four species of ground crickets (Orthoptera: Gryllidae: *Nemobius*). Ohio J. Sci. 57: 153-163.

Alexander R.D. 1962. Evolutionary change in cricket acoustical communication. Evolution 16: 443-467.

Alexander R.D. 1975. Natural selection and specialized chorusing behaviour in acoustical insects. In: Insects, Science and Society. Ed. D. Pimentel. Academic Press, New York. pp 35-77.

Alexander R.D. & E.S. Thomas. 1959. Systematic and behavioural studies on the crickets of the *Nemobius fasciatus* group (Orthoptera: Gryllidae: Nemobiinae). Ann. Entomol. Soc. Am. 52: 591-605.

Allard H.A. 1918. Rhythmic synchronism in the chirping of certain crickets and locusts. Am. Nat. 52: 548-552.

Bailey W.J. 1972. The acoustic status of the species *Hormorocoryphus nitidulus nitidulus* Scopoli (Tettigonioidea, Orthoptera) in southern Europe. Zool. Anz. 189: 181-190.

Bailey W.J. 1975. A review of the African species of the genus *Ruspolia* Schulthess (Orthoptera: Tettigonioidea). Bull. Inst. Fondam. Afr. Noire Ser. A 37: 171-226.

Bailey W.J. 1976. Species isolation and song types of the genus *Ruspolia* Schulthess (Orthoptera: Tettigonioidea) in Uganda. J. Nat. Hist. 10: 511-528.

Bailey W.J. 1979. A review of Australian Copiphorini (Orthoptera: Tettigoniidae: Conocephalinae). Aust. J. Zool. 27: 1015-1049.

Bailey W.J. & A.W.R. McCrae. 1978. The general biology and phenology of swarming in the East African tettigoniid *Ruspolia differens* (Serville) (Orthoptera). J. Nat. Hist. 12: 259-288.

Bailey W.J. & D.R. Theile. 1983. Male spacing behaviour in the Tettigoniidae: an experimental approach.In: Orthopteran mating systems: sexual competition in a diverse group of insects. Eds. D.T. Gwynne & G.K. Morris. Westview Press, Boulder, Colorado. pp 163-184.

Belwood J.J. & G.K. Morris. 1987. Bat predation and its influence on calling behaviour in Neotropical katydids. Science 238: 64-67.

Blatchley W.S. 1920. Orthoptera of northeastern America. Nature Publ. Co., Indianapolis. 784 pp.

Brush J.S., V.G. Gian & M.D. Greenfield. 1985. Phonotaxis and aggression in the coneheaded katydid *Neoconocephalus affinis*. Physiol. Entomol. 10: 23-32.

Burk T.E. 1982. Evolutionary significance of predation on sexually signalling males. Fla. Entomol. 65: 90-104.

Busnel M.-C. 1967. Rivalite acoustique et hierarchie chez l'Ephippiger (Insecte, Orthoptère, Tettigoniidae). Z. Vgl. Physiol. 54: 232-245.

Counter S.A. 1977. Bioacoustics and neurobiology of communication in the tettigoniid *Neoconocephalus robustus*. J. Insect Physiol. 23: 993-1008.

Counter S.A. & W. Henke. 1977. Commensal auditory communication in two species of *Neoconocephalus* (Orthoptera). J. Insect Physiol. 23: 817-824.

Dadour I.R. & W.J. Bailey. 1985. Male agonistic behaviour of the bushcricket *Mygalopsis marki* Bailey in response to conspecific song (Orthoptera: Tettigoniidae). Z. Tierpsychol. 70: 320-330.

Doherty J. & R. Hoy. 1985. The auditory behaviour of crickets: some views of genetic coupling, song recognition, and predator detection. Q. Rev. Biol. 60: 457-472.

Ewing A.W. 1984. Acoustic signals in insect sexual behaviour. In: Insect Communication. Ed. T. Lewis. Academic Press, London. pp 223-240.

Fabre J.-H. 1912. Souvenirs entomologiques, Vol. 5, 8th ed. Librairie Ch. Delgrave, Paris, France.

Fretwell S.D. 1972. Populations in a seasonal environment. Princeton University Press, Princeton, New Jersey. 217 pp.

Fulton B.B. 1928. A demonstration of the location of auditory organs in certain Orthoptera. Ann. Entomol. Soc. Am. 21: 445-448.

Fulton B.B. 1934. Rhythm, synchronism and alternation in the stridulation of Orthoptera. J. Elisha Mitchell Sci. Soc. 50: 263-267.

Fulton B.B. 1956. The genus *Anaxipha* in the United States (Opthoptera: Gryllidae). J. Elisha Mitchell Sci. Soc. 72: 222-243.

Gangwere S.K. 1961. A monograph on food selection in Orthoptera. Trans. Am. Entomol. Soc. 87: 67-230.

Greenfield M.D. 1983. Unsynchronized chorusing in the coneheaded katydid *Neoconocephalus affinis* (Beauvois). Anim. Behav. 31: 102-112.

Greenfield M.D. 1988. Interspecific acoustic interactions among katydids (*Neoconocephalus*): inhibition-induced shifts in diel periodicity. Anim. Behav. 36: 684-695.

Greenfield M.D. & K.C. Shaw. 1983. Adaptive significance of chorusing with special reference to the Orthopteran mating systems: sexual competition in a diverse group of insects. Westview Press, Boulder, Colorado.

Greenfield M.D., T.E. Shelly & K.R. Downum. 1987. Variation in host-plant quality: implications for territoriality in a desert grasshopper. Ecology 68: 828-828.

Gwynne D.T. 1977. Mating behaviour of *Neoconocephalus ensiger* (Orthoptera: Tettigoniidae) with notes on the calling song. Can. Entomol. 109: 237-242.

Heath J.E. & R.K. Josephson. 1970. Body temperature and singing in the katydid *Neoconocephalus robustus* (Orthoptera, Tettigoniidae). Biol. Bull. 138: 272-285.

Hedrick A.V. 1986. Female preference for male calling bout duration in a field cricket. Behav. Ecol. Sociobiol. 19: 73-77.

Heller K.-G. & D.v. Helversen. 1986. Acoustic communication in phaneropterid bushcrickets: species-specific delay of female stridulatory response and matching male sensory time window. Behav. Ecol. Sociobiol. 18: 189-198.

Josephson R.K. 1984. Contraction dynamics of flight and stridulatory muscles of tettigoniid insects. J. Exp. Biol. 108: 77-96.

Josephson R.K. 1985. The mechanical power output of a tettigoniid wing muscle during singing and flight. J. Exp. Biol. 117: 357-368.

Latimer W. 1981. Acoustic competition in bush crickets. Ecol. Entomol. 6:35-45.

Latimer W. & W.B. Broughton. 1984. Acoustic interference in bush crickets: a factor in the evolution of singing insects? J. Nat. Hist. 18: 599-616.

Love R.E. & T.J. Walker. 1979. Systematics and acoustic behaviour of scaly crickets (Orthoptera: Gryllidae: Mogoplistinae) of eastern United States. Trans. Am. Entomol. Soc. 105: 1-66.

Meixner A.J. 1976. Acoustical behaviour and spacing in the Nebraska conehead *Neoconocephalus nebrascensis* (Bruner) (Orthoptera: Tettigoniidae). PhD Dissertation, Iowa State University, Ames. 160 pp.

Meixner A.J. & K.C. Shaw. 1979. Spacing and movement of singing *Neoconocephalus nebrascensis* males (Tettigoniidae: Copophorinae). Ann. Entomol. Soc. Am. 72: 602-606.

Meixner A.J. & K.C. Shaw. 1986. Acoustic and associated behavior of the coneheaded katydid, *Neoconocephalus nebrascensis* (Orthoptera: Tettigoniidae). Ann. Entomol. Soc. Am. 79: 554-565.

Morris G.K. 1980. Calling display and mating behaviour of *Copiphora rhinoceros* Pictet (Orthoptera: Tettigoniidae). Anim. Behav. 28: 42-51.

Nutting W.L. 1953. The biology of *Euphasiopteryx brevicornis* (Townsend) (Diptera, Tachinidae), parasitic in the coneheaded grasshoppers (Orthoptera, Copiphorinae). Psyche (Camb.) 60: 69-81.

Otte D. 1977. Communication in Orthoptera. In: How Animals Communicate. Ed. T.A. Sebeok. Indiana University Press, Bloomington. pp 334-361.

Otte D. & J. Loftus-Hills. 1979. Chorusing in *Syrbula* (Orthoptera: Acrididae). Cooperation, interference competition, or concealment? Entomol. News 90: 159-165.

Otte D. & J. Smiley. 1977. Synchrony in Texas fireflies with a consideration of male interaction models. Biol. Behav. 2: 143-158.

Partridge B.L. & J.R. Krebs. 1978. Tree frog choruses: a mixed evolutionary stable strategy? Anim. Behav. 26: 959-960.

Perkins R.C.L. 1899. Orthoptera. Fauna Hawaiiensis 2: 1-30.

Pringle J.W.S. 1954. A physiological analysis of cicada song. J. Exp. Biol. 31: 525-560.

Ragge D.R. 1972. An unusual case of mass migration by flight in *Grillus bimaculatus* Degeer (Orthoptera: Gryllidae). Bull. Inst. Fondam. Afr. Noire Ser. A 34: 869-878.

Redtenbacher J. 1891. Monographie der conocephaliden. Verh. Zool. Bot. Ges. Wien 41: 315-562.

Rehn J.A.G. & M. Hebard. 1915. Studies in American Tettigoniidae (Orthoptera). III. Synopsis of the species of the genus *Neoconocephalus* found in North America north of Mexico. Trans. Am. Entomol. Soc. 40: 365-413.

Robinson D.J., J. Rheinlander & J.C. Hartley. 1986. Temporal parameters of the male-female communication in *Leptophyes punctatissima* Bosc. Physiol. Entomol. 11: 317-323.

Samways M.J. & W.B. Broughton. 1976. Song modification in the Orthoptera. II. Types of acoustic interaction between *Platycleis intermedia* and other species of the genus (Tettigoniidae). Physiol. Entomol. 1: 287-297.

Saussure H. de & A. Pictet. 1898. *Conocephalus*. Biologia Centrali-Americana, Orthoptera 1: 388-394.

Shaw K.C. 1968. An analysis of the phonoresponse of males of the true katydid, *Pterophylla camellifolia* (Fabricius) (Orthoptera: Tettigoniidae). Behaviour 31: 203-260.

Shaw K.C., R.J. Bitzer & R.C. North. 1982. Spacing and movement of *Neoconocephalus ensiger* males (Conocephalinae: Tettigoniidae). J. Kans. Entomol. Soc. 55: 581-592.

Shaw K.C. & P. Galliart. 1987. Acoustic and mating behaviour of a Mexican katydid, *Pterophylla beltrani* (Orthoptera: Tettigoniidae). Fla. Entomol. 70: 354-368.

Thomas E.S. 1933. *Neoconocephalus lyristes* (Rehn and Hebard) in the Middle West. Ann. Entomol. Soc. Am. 26: 303-308.

Thorson J., T. Weber & F. Huber. 1982. Auditory behavior of the cricket. II. Simplicity of calling-song recognition in *Gryllus*, and anomalous phonotaxis at abnormal carrier frequencies. J. Comp. Physiol. 146: 361-378.

Tuttle M.D. & M.J. Ryan. 1982. The role of synchronized calling, ambient light, and ambient noise, in anti-bat-predator behaviour of a treefrog. Behav. Ecol. Sociobiol. 11: 125-131.

Walker T.J. 1957. Specificity in the response of female tree crickets (Orthoptera, Gryllidae, Oecanthinae) to calling songs of the males. Ann. Entomol. Soc. Am. 50: 626-636.

Walker T.J. 1962. The taxonomy and calling songs of United States tree crickets (Orthoptera: Gryllidae: Oecanthinae). I. The genus *Neoxabea* and the *niveus* and *Varicornis* groups of the genus *Oecanthus*. Ann. Entomol. Soc. Am. 55: 303-322.

Walker T.J. 1963. The taxonomy and calling songs of United States tree crickets (Orthoptera: Gryllidae: Oecanthinae). II. The nigricornis group of the genus *Oecanthus*. Ann. Entomol. Soc. Am. 56: 772-789.

Walker T.J. 1969a. Acoustic synchrony: two mechanisms in the snowy tree cricket. Science 166: 891-894.

Walker T.J. 1969b. Systematics and acoustic behaviour of United States crickets of the genus *Cyrtoxipha* (Orthoptera: Gryllidae). Ann. Entomol. Soc. Am. 62: 945-952.

Walker T.J. 1969c. Systematics and acoustic behaviour of United States crickets of the genus *Orocharis* (Orthoptera: Gryllidae). Ann. Entomol. Soc. Am. 62: 752-762.

Walker T.J. 1973. Systematics and acoustic behaviour of United States and Caribbean short-tailed crickets (Orthoptera: Gryllidae: *Anurogryllus*). Ann. Entomol. Soc. Am. 66: 1269-1277.

Walker T.J. 1975. Stridulatory movements in eight species of *Neoconocephalus* (Tettigoniidae). J. Insect Physiol. 21: 595-603.

Walker T.J. 1983. Diel patterns of calling in nocturnal Orthoptera. In: Orthopteran Mating Systems: Sexual Competition in a Diverse group of Insects. Eds. D.T. Gwynne & G.K. Morris. Westview Press, Boulder, Colorado. pp 45-72.

Walker T.J. & D. Dew. 1972. Wing movements of calling katydids: fiddling finesse. Science 178: 174-176.

Walker T.J. & M.D. Greenfield. 1983. Songs and systematics of Caribbean *Neoconocephalus* (Orthoptera: Tettigoniidae). Trans. Am. Entomol. Soc. 109: 357-389.

Walker, T.J. & J.J. Whitesell. 1978a. New species of conehead from Florida Everglades (Orthoptera: Tettigoniidae: *Neoconocephalus*). Entomol. News 89: 27-32.

Walker T.J. & J.J. Whitesell. 1978b. *Neoconocephalus maxillosus*: a Caribbean conehead in south Florida (Orthoptera: Tettigoniidae). Fla. Entomol. 61: 1-3.

Walker T.J. & J.J. Whitesell. 1982. Singing schedules and sites for a tropical burrowing cricket (*Anurogryllus muticus*). Biotropica 14: 220-227.

Walker T.J., J.J. Whitesell & R.D. Alexander. 1973. The robust conehead: two widespread sibling species (Orthoptera: Tettigoniidae: *Neoconocephalus "robustus"*). Ohio J. Sci. 73: 321-330.

Weissman D.B., D.C.F. Rentz, R.D. Alexander & W. Loher. 1981. Field crickets (*Gryllus* and *Acheta*) of California and Baja California, Mexico (Orthoptera: Gryllidae: Gryllinae). Trans. Am. Entomol. Soc. 106: 327-356.

West-Eberhard M.J. 1983. Sexual selection, social competition, and speciation. Q. Rev. Biol. 58: 155-183.

Whitesell, J.J. 1969. Biology of United States coneheaded katydids of the genus *Neoconocephalus* (Orthoptera: Tettigoniidae). M.S. Thesis, University of Florida, Gainesville. 73 pp.

Whitesell J.J. 1974. Geographic variation and dimorphisms in song, development, and color in a katydid: field and laboratory studies (Tettigoniidae, Orthoptera). PhD. Dissertation, University of Florida, Gainsville. 74 pp.

Whitesell J.J. & T.J. Walker 1978. Photoperiodically determined dimorphic calling songs in a katydid. Nature 274: 887-888.

Wolf H. & O.v. Helversen. 1986. "Switching-off" of an auditory interneuron during stridulation in the acridid grasshopper *Chorthippus biguttilus* L. J. Comp. Physiol. 158: 861-871.

Zimmerman E.C. 1948. Subfamily Copiphorinae. Insects of Hawaii. 2: 112-121.

6. THE ACOUSTIC BEHAVIOUR OF MALE *MYGALOPSIS MARKI* (COPIPHORINAE)

Ian R. Dadour
Winston J. Bailey

Introduction

Newly discovered species quickly become new resources for scientific endeavour. One recently described species of copiphorine tettigoniid, *Mygalopsis marki* Bailey (Bailey 1979), has taken on such a fashionable role with studies ranging from its phenology (Lymbery, 1984, 1987), colour polymorphism (Lymbery 1984), male calling behaviour (Sandow, 1980; Thiele, 1981; Thiele & Bailey, 1980; Bailey & Thiele, 1983; Dadour, 1986, 1989; Dadour & Bailey, 1985;), defense (Sandow & Bailey, 1978; Bailey & Sandow, 1983), speciation (Dadour & Johnson, 1983; Barendse, 1986; Barendse, chapter 14 this volume), auditory physiology (Hill & Oldfield, 1981; Oldfield, 1984; Römer & Bailey, 1986; Römer, 1987; Römer *et al.*, 1989; Roessler *et al.*, manuscript) and the biophysics of hearing (Bailey & Stephen, 1978). Its credentials for such a concerted endeavour are that it is the noisiest and one of the most obvious tettigoniids along the coastal regions of Western Australia. The slow moving males are easily captured throughout the year at most sites, and the genus has the added attraction of being endemic to the southwestern part of Australia, showing evidence of recent speciation. There is one detractor however, females are difficult to find and observations of mating, despite hundreds of hours of field work, are very rare! The only opportunity to capture females appears to be while they are feeding during their final larval stages and early imago around the reed margins of ephemeral lakes. They occupy areas of known high male density and are obvious only by their exposed stance while feeding on stems and grass heads. Hence, given that experimental ethologists prefer to study abundant and easily manipulated animals, this chapter concerns the behaviour of adult male insects.

Besides *M. marki* there are three other known species within the genus *Mygalopsis*, *M. pauperculus* Walker, *M. sandowi* Bailey and *M. thielei* Bailey (Plate 18) and a fourth undescribed species appearing as a very close relative to *M. marki* in the southern half of the species' range. The genus was based on *Mygalopsis ferruginae* by Redtenbacher (1891) from the Swan River colony but that species was found to be synonymous with *Pseudorhynchus pauperculus* (Walker, 1869). European taxonomists made the imperial assumption that the Swan River Colony was a single locality, extending throughout southwest Australia. Localities of both specimens were undoubtedly close to the major port of Albany, situated in the middle of the now defined range of *M. pauperculus*. *Mygalopsis* was separated from *Pseudorhynchus*, and the three new *Mygalopsis* species were added (Bailey, 1979). "*Mygalopsis*", derived from two Greek words meaning "the appearance of a field mouse", is a near accurate description of a brown female of *M. marki* with its elongated fastigium and long straight ovipositor. Morphologically similar copiphorines are *Banza* from Hawaii, *Belocephalus* from southern United States and *Ruspolia lemairei* (Griffini) from central Africa

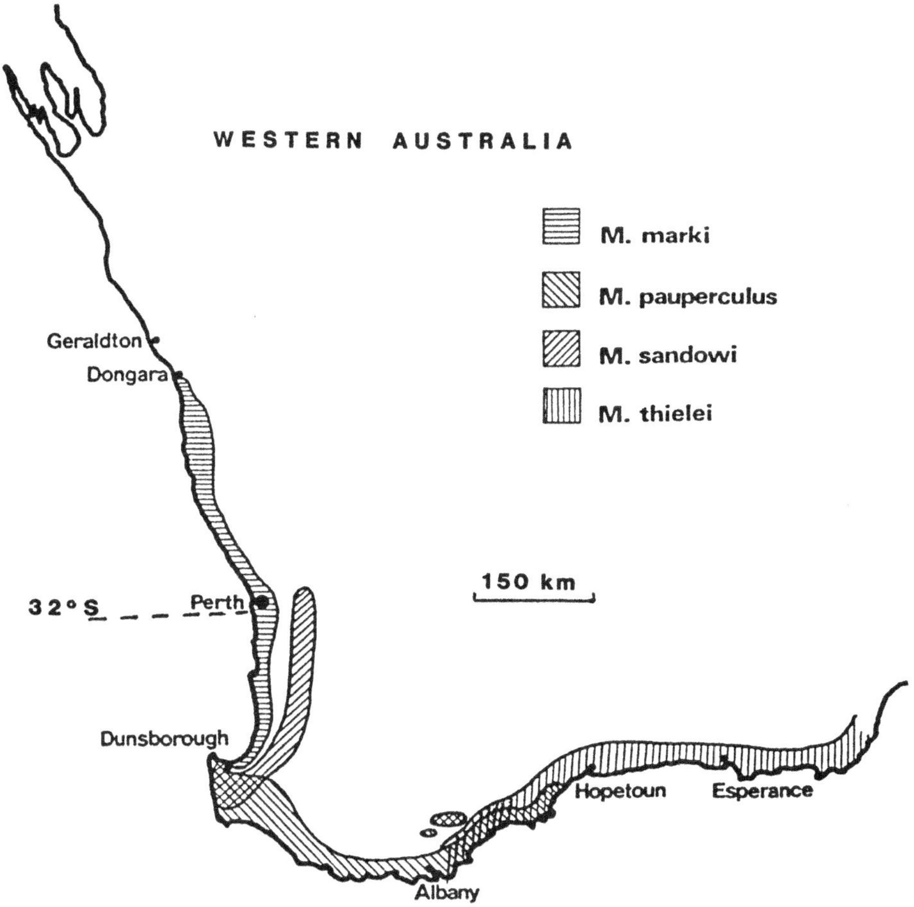

Fig. 1. The distribution of the four species of *Mygalopsis* in Western Australia. The division between the northern and southern species of *Mygalopsis marki* occurs south of Perth close to latitude 32°S.

(Bailey, 1976).

Males are found in aggregations within which movement is extremely restricted. Although site fidelity over the short term (24-48 hours) is a common feature of many male copiphorines (Shaw *et al.* 1981), we have found marked males of *M. marki* to remain on the same bush for periods often in excess of 30 days. The insect's low vagility is reflected in its brachyptery and short hind femur length. Based on measures of gene flow between local populations, Barendse (Chapter 14) estimates that a local population remains discrete over a range of less than 20 m. We have found that *Mygalopsis* is extremely sensitive to fire, and where bush fires have destroyed the habitat, even though the vegetation may regrow within 2-3 years, recolonisation by *Mygalopsis* may take over 5 years, if it occurs at all. Hence, as fire is a very common feature of the coastal heathlands, populations may be separated by substantial barriers to gene flow, and certain genotypes may be removed altogether by a single calamity.

The geographical range of *M. marki*, as described in its initial revision was between Dongara (approx. Lat. 28 00' S) and Dunsborough (approx. Lat. 34 00' S) (See Fig. 1 in Barendse chapter 14). Since then a species south of Lake Richmond (Lat. 32 17' S) has been identified as being genetically distinct from *M. marki* (Dadour & Johnson, 1983). This undescribed species is referred to in this chapter and that of Barendse as *Mygalopsis* sp. B. The genetic divergence of the species occurred after the incursion of the sea over the coastal plain in the late Pleistocene (sometime between 1m. yrs b.p. and 100 000yrs b.p.) (see Barendse, chapter 14 this volume).

Life History

M. marki has a univoltine life history pattern but with two peaks of eclosion in early summer (November/December) and autumn (March/April). There are 8-9 moults from egg to adult, and those adults from the early summer emergence (SE) are significantly larger than their autumn (AE) counterparts from the same habitat (Fig. 1). This size difference is due to the development of the SE population over the mild winter months when food is abundant and lower temperatures extend developmental times. SE and AE populations do not represent successive generations: the SE population hatch in early summer, overwinter as nymphs and emerge as adults early the following summer, whereas the AE population hatch in spring, develop oversummer and emerge as adults in March (Fig. 2) (Lymbery, 1987).

Walker (1964) and Alexander (1968) suggested that allochronic speciation could occur in a number of gryllid and tettigoniid species in the United States where eclosion times were separated. The same possibility exists for *M. marki*, but we have found no evidence of genetic differences between the seasonal morphs. Adult males are long lived (6-9 months) and both SE and AE individuals can be found calling together throughout the autumn and hence mating could occur between them.

Like many copiphorines (Owen, 1969; Bailey & McCrae, 1978), *M. marki* has two major colour morphs, green and brown (Lymbery, 1984). As in the acridids (Rowell, 1971) colour change may occur throughout the developing life of the insect, and Lymbery (1984) found that in *M. marki* colour characters were phenotypically plastic with colour changes restricted to moults: insects can change from green to brown or brown to green at any moult. Changes in colour are associated with water intake and/or water balance. Thus, brown males dominate

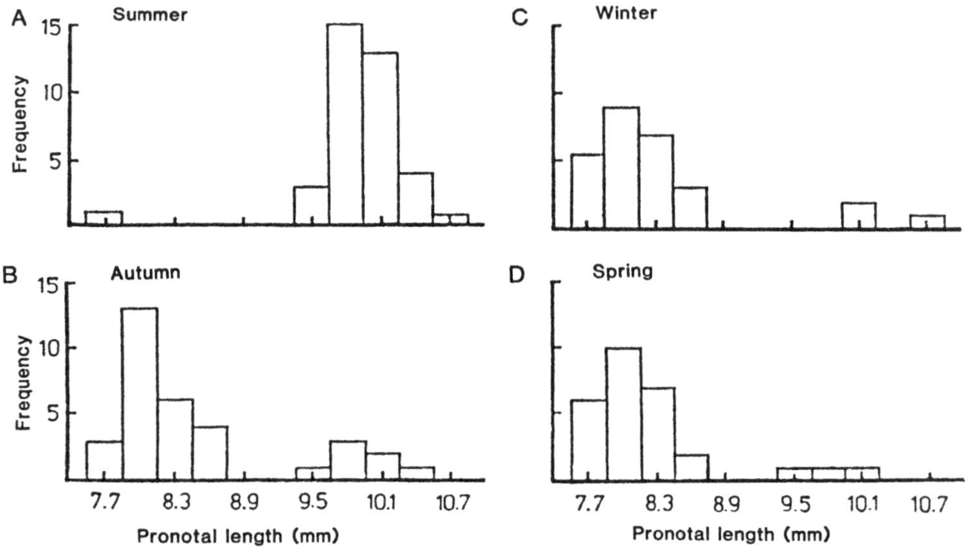

Fig. 2. Distribution of pronotal lengths of adult males sampled from Nicholson Road. A Summer 1980/81 (n = 36); B Autumn 1981 (n = 33); C Winter 1981 (n = 28); D Spring 1981 (n = 28) (After Lymbery 1987).

the drier sclerophyll woodlands of inland areas of the sand plain, whereas green individuals are more common on the wetter coastal heath and around swamps. He also determined that predation induced a level of selective mortality with respect to the insect's colour and the background on which it occurred in nature. Furthermore, adults actively choose background colours to match their own body colour for their diurnal resting sites (Connell, manuscript; Belwood chapter 2).

Sound Production – Defensive Stridulation

Remarkably, *Mygalopsis* has three stridulatory mechanisms, two of which are used in defense and the third, the more usual tegminal stridulation, in social communication. Defensive sounds are made by the larvae and less frequently by the imago, presumably by stridulation of hard parts in the head capsule. The sound, which is a squeak, appears to emanate from behind the labrum which moves with each burst of sound. Similar labral stridulation has been reported for a mecopodine, *Sexava femorata* by Lloyd and Gurney (1975) and a agraecine *Solomona* sp. (Sandow, 1980; Rentz, personal communication). The significant movements of the large head trachea during sound production suggest that they may act as a resonator. As with the hind wing stridulation (Bailey & Sandow, 1983) the noise is only produced when the insect is handled or disturbed by a predator.

Hindwing stridulation involves the downwardly projecting short spines of the highly modified wings engaging a cuticular latticework on the dorsum of the abdomen. The resulting hissing sound is associated with small movements of the abdomen against these spines. Experiments involving a potential predator, the skink *Egernia napoleonis* (Grey), found that stridulation by the hind wings associated with a defensive posture decreased the success of predator attack (Sandow & Bailey 1978). Attacked males attempted to bite the skink, and skinks associated these attacks with the defensive stridulation. As such slight movement is required to produce the sound, insects held within the skinks mouth were still able to produce the defensive hiss.

Social Communication

Social communication is through tegminal stridulation, and follows the typical pattern of most other tettigoniids. The song consists of a more or less continuous series of syllables grouped into chirp sequences with a short inter-chirp interval (Dadour 1989). The song's continuous rattle is the most obvious night call of the coastal heath and woodland, occasionally acting as a masking noise to other calling species (Römer *et al.* 1989). For example the large sagine *Hemisaga denticulata* (White) appears to be inhibited from calling within populations of *M. marki*, confining its calling almost entirely to the day when males of *M. marki* are inactive. In areas where *M. marki* is not present *H. denticulata* also calls throughout the night.

Chirp length and repetition rate are sensitive to temperature and each chirp varies from 2-600 syllables. Chirp rate is also influenced by social interaction with other males, for instance males that occupy the periphery of the aggregation where the density of males is low sing both intermittently and for relatively shorter periods (Dadour, 1989).

There are two major frequency peaks in the song, one of which is audible and close to 11 kHz, while the second is within the ultrasonic range at 22 kHz. The relative intensity of the two frequency peaks varies with the distance from the insect, where the power spectrum of the higher peak shows a greater degree of variance in its intensity than the lower frequency peak. This variance is due to the increased attenuation of the higher frequencies of the song with increasing distance from the calling male (Römer & Bailey, 1986) (Fig. 3).

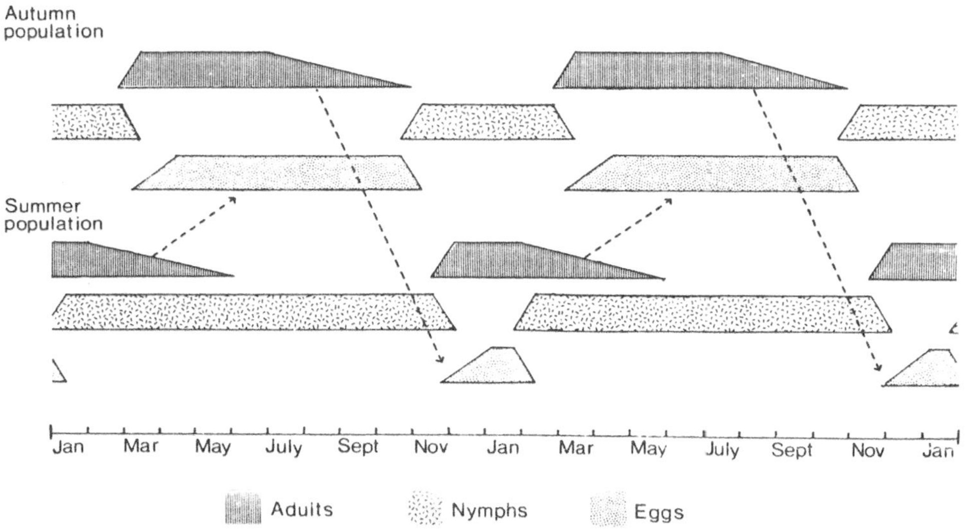

Fig. 3. The seasonal occurrence of the principal developmental stages of *M. marki*. Arrows represent the possibility of gene exchange. (After Lymbery 1987.)

Specificity of Song

The call structure of the song of *M. marki* is distinct from that of the three other species, *M. sandowi*, *M. pauperculus* and *M. thielei* implying a role in species isolation. Southwards along the coastal plain, the geographic range of *M. marki* overlaps with that of *Mygalopsis* sp. B., and within this zone of overlap (Lake Richmond to Peelhurst Beach, 40 km) there is no observable song differences between the two recently separated species, except in their response to temperature changes. Animals south of Lake Richmond have a significantly faster chirp rate, shorter interchirp interval, and a greater number of syllables per chirp than males from the north at preferred temperatures for singing (Dadour, manuscript). However, within the zone of overlap these call parameters are not sufficient for mate recognition (Dadour & Johnson 1983). *M. marki* and *Mygalopsis* sp. B are geographically isolated along their eastern boundary from *M. sandowi* which has a very different song. The southern part of *Mygalopsis* sp.B's range marginally overlaps (approx 1 km) with *M. pauperculus* and here hybridisation is extremely rare. We have found only one male where the genotype matched *Mygalopsis* sp. B, but the song was that of *M. pauperculus*. Sandow (1980) notes that where the range for *M. pauperculus* overlaps with *M. thielei* (approx 50 to 100 km), males of *M. pauperculus* recorded from populations within this zone sang with an increased number of syllables per chirp. This has been confirmed by more recent recordings from the same area (Bailey unpublished). Unfortunately, although we have an understanding of the established geology of the region and the recent events of speciation in this genus, data collection in the two southernmost zones of overlap is at the moment too sparse to provide any real insights into how the respective song types may have differentiated.

Temperature Effects

As indicated above, the temporal aspects of the calling song are affected by temperature. Walker (1975) predicted that in songs where the syllable repetition rate is high, temperature will have a far more profound effect; the slope of repetition rate against temperature will be steeper. This prediction is certainly true when comparing the response to temperature of the rapid syllable repetition rate of *M. sandowi* to that of the 3 other species of *Mygalopsis* (see Fig. 4 for regression equations). However, we know that *M. marki* is able to sing on very cold nights where ambient temperature may fall below 8°C and where syllable repetition rates are as low as 2-3 per chirp. At these low temperatures, the thoracic temperature is 5-6°C above abdominal temperatures due to muscle metabolism, and up to 8°C above ambient temperature (measured by inserting a fine bulb thermometer into the abdomen for 10-15 s and then into the thorax for a further 15 s (Sandow & Bailey, unpublished data)). Insects calling at low temperatures are reluctant to do so once they have been disturbed, implying that when the thoracic temperature falls, stridulation may become energetically too costly.

Unlike males of *Tettigonia viridissima* and *T. cantans* (Nielsen & Dreisig, 1970), which sing from early afternoon to late evening and where their nocturnal stridulatory activity is dependent on the temperature during the period preceding stridulation, males of all 5 species of *Mygalopsis* commence singing at dusk. It is arguable that by calling early, just before sunset, males of *M. marki* may have sufficient body temperature to maintain stridulation as the ambient temperature drops through the night. Hence, continuous calling, as opposed to intermittent chirping would then be an adaptive strategy to maintain muscle temperature and remain broadcasting (Bailey manuscript).

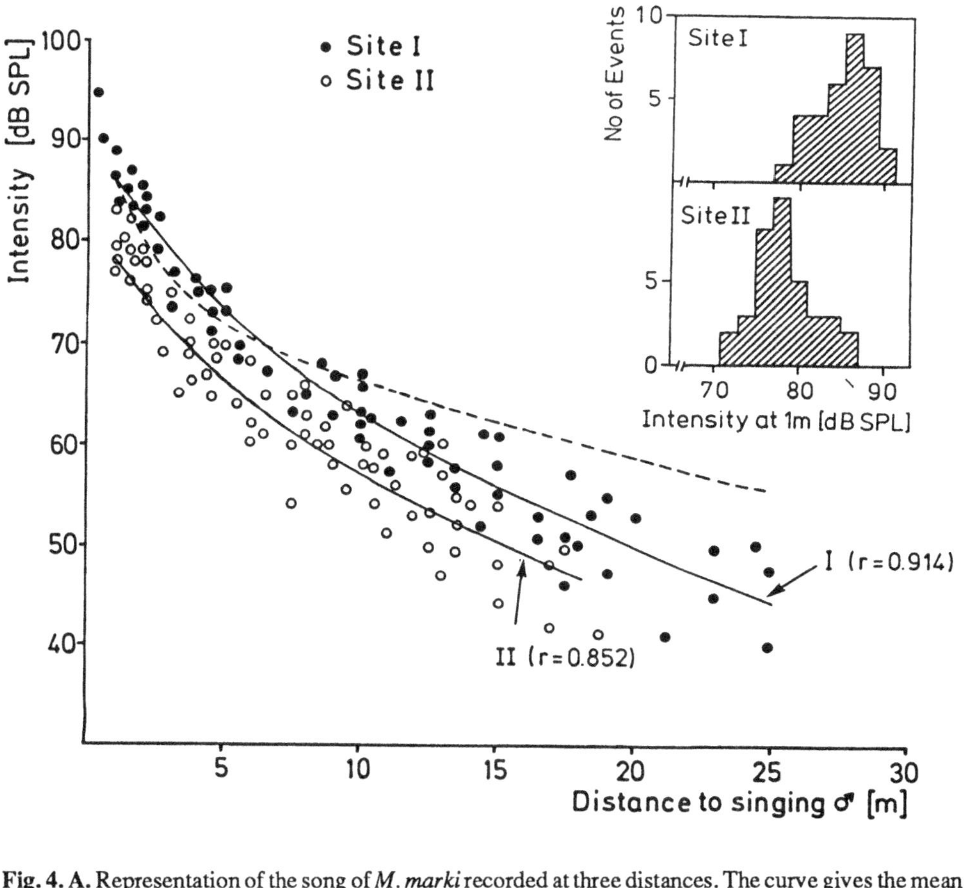

Fig. 4. A. Representation of the song of *M. marki* recorded at three distances. The curve gives the mean rate of attentuation ± SD of the song in one of the populations studied by Römer and Bailey (1986).

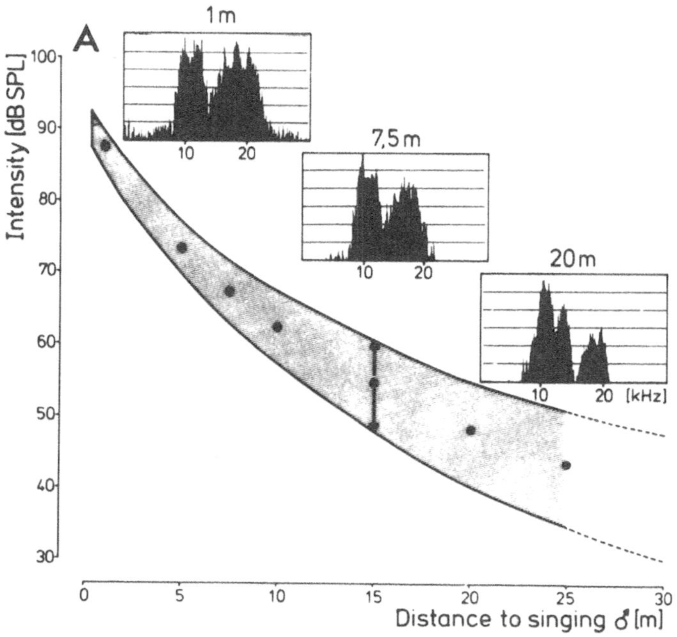

Fig. 4. B. Temporal pattern of the song illustrating the decreasing signal-to-noise ratio with increasing distance from calling insect (Römer 1987).

Body size

Body size and correlates of body size are useful indicators of male quality and have been associated with predicting how successful an individual is in terms of its survival and/or reproductive success (Maynard Smith & Parker 1976; Simmons 1987). Amongst those animals that acoustically communicate, the three components of song, intensity, frequency and temporal pattern have been documented as cues used by females to assess male qualities (anurans, Davies & Halliday 1978; Arak 1983, 1988; Gerhardt 1982, 1987; tettigoniids, Latimer & Sippel 1987; Bailey & Yeoh 1988). Similarly, in male competition these same cues may also indicate the competence of any male to hold a resource or maintain a calling territory. Of the three song characters that provide information in either of these behaviours, variance both in temporal pattern and/or in the frequency of the dominant peak are the most likely to be reliable cues. Temporal pattern in some species is affected by male competition, and the frequency is directly related to the size of the sound producing organ. In the Tettigoniidae, the area of the sound generator on the wing is directly correlated with frequency (e.g. Latimer, 1981) and wing size is correlated with body size, hence larger males will call with lower frequencies and should be preferred by females searching for large males. However, evidence for mate choice based on frequency appears fragile, and in some cases preference operates in the reverse direction, females preferring calls with higher frequencies (Bailey & Yeoh, 1988; Gwynne & Bailey, 1988; Bailey *et al.* in press). Frequency is simply controlled by the size of the sound generator on the tegmen, and although frequency and size are related in all cases where these two parameters have been measured (Morris & Pipher 1972; Latimer 1981; Gwynne & Bailey 1988), and in *Mygalopsis* where the largest species *M. sandowi* sings with a much lower frequency than its smaller relative *M. pauperculus* (Bailey 1979; Sandow 1980), by viewing the relationship at a population level within a species one of the authors (IRD) was able to find no such correlation. A study of 4 northern and 4 southern populations of *M. marki*, where mean size difference is significant there appeared to be no relationship between this and the position of the two major peak frequencies (11.5 kHz peak, $F(6,108) = 1.17$ and the 23.5 kHz peak, $F(6,108) = 2.41$: $p > 0.05$) (Dadour, manuscript).

In both crickets and tettigoniids the intensity of song appears to be the overruling character governing female preference under both laboratory and field conditions (Latimer & Sippel 1987; Bailey & Yeoh 1988; Gwynne & Bailey 1988; Bailey *et al.* in press), and this seems consistent with the behaviour of some frog species (e.g. Arak, 1988). Where males or females respond merely to the most intense signal, the response is one of "passive attraction" (Parker, 1983), and not necessarily the consequence of an active choice process between two conflicting cues. Intermale behaviour, and the selection of mates, if based on call intensity (the "passive choice" of Thornhill & Alcock 1983 p 406), will be compounded by indirect effects unrelated to male size. Thus the density of the surrounding vegetation, the way in which the insect is perched on the plant (sound intensities will be higher towards the dorsum (Thiele & Bailey, 1980)), or even the wind direction will influence not only the total sound intensity but the magnitude of the frequencies within the song. In *M. marki* loudness can influence the size of a males territory, larger males are more widely spaced from their neighbour than smaller males, and call intensity correlates with size (Bailey & Thiele, 1983), therefore despite the complicating effects of habitat attenuation, loudness is a major factor controlling male behaviour. Given this relationship, male-male behaviour has been studied at three levels, (i) the population level as male spacing, (ii) the behaviour of individuals to changes in the call intensity of a neighbour, and (iii) an examination of the way in which the auditory neuropile interprets acoustic information at varying distances from calling males.

Male spacing

Males are both attracted and repelled by the calls of conspecific neighbours, adjusting their final calling position until the aggregation appears as a regularly spaced group of males (Morris, 1967; Thiele & Bailey, 1980; Latimer & Schatral, 1983; Brush *et al.*, 1985). Field experiments with *M. marki* have shown how males move away (centrifugal effect) from other calling males when the perceived intensity of the call is above an acceptable level, and towards (centripetal effect) when the intensity is beneath this level. Further proof of the role of sound intensity in male spacing was that deafened males dispersed to a mean intermale distance of 1.5 m over 7 days whereas undeafened males became regularly spaced with a mean nearest neighbour distance of 11.5 m (Thiele, 1981; Bailey & Thiele, 1983).

Males of *M. marki* from more southerly populations are more closely distributed than northern males. This difference could be due to the influence of the habitat or to the difference in size between males from northern and southern populations. In order to separate the size effect from any habitat effects, Thiele (1981) transplanted a population of males from the south to the north, where after 5 days they resumed their usual southern inter-male distance. Römer and Bailey (1986) were able to establish that males appear to be adjusting their intermale distance to quite precise intensity levels in both populations. The sound pressure level perceived by males from the call of their surrounding neighbours was close to 65 dB.

Based on Thiele's experiments demonstrating centripetal and centrifugal movements by males when released in groups, Römer and Bailey (1986) found that males were not spacing themselves at the auditory threshold of the call. The maximum hearing distance (up to 30m) was 2-3 times that of the nearest neighbour distance. As high frequencies attenuate more rapidly than low frequencies, these workers suggested that males may space themselves on the relative level of the high frequency peak: distance estimation by differential frequency attenuation (Morris *et al.*, 1978). Although such an hypothesis could not be ruled out in these experiments using the auditory neuropile as a "living microphone" (Rheinlaender & Romer, 1986), the variance in the peak power of the 22 kHz peak was far greater than that of the lower peak, and it would be difficult to envisage how such a variable character could be used to maintain a stable spacing system. However, given that high frequency receptors in the ear are associated with quite distinct areas of the auditory neuropile in the prothoracic ganglion, interneurons associated with these centers may have quite different capacities for estimating distance than those more sensitive to lower frequencies. This tonotopically local responses appear to be based on a frequency-dependent synaptic process of excitation and inhibition and may well control distance monitoring behaviour (Römer 1987: Rheinlaender & Römer chapter 13).

Male Strategies

When walking through natural populations of chorusing males of *M. marki*, one frequently observes singing males dropping in and out of the chorus. A closer examination of a single male, shows that it tends to sing from one position in the bush, and then moves within the bush to another perch, or if observed for longer periods may even move to a new bush. Morris (1967,1971) suggested a simple model of male agonistic behaviour, based on 3 species of conocephalines, *Conocephalus nigropleurum*, *Orchelimum gladiator* and *O. vulgare*, which resulted in an even distribution of singing males in an aggregation. When a male calls from within a critical range of another male, the increasing intensity of the song will elicit a response from the resident male. This male may adjust its position or alter its calling pattern. As the intruding male approaches further the interaction may escalate into various forms of aggressive behaviour, eventually resulting in a fight where the loser may withdraw to a distance where the perceived sound intensity will produce no further agonistic behaviour. As a consequence of this males in the aggregation will be spaced with some regularity to each other.

Both *M. marki*'s sedentary nature, and the fact that it occurs in regularly spaced aggregations, makes this an ideal animal to test Morris's predictions. Individuals in a natural aggregation were selected at random and the call of a male, pre-recorded under similar temperature conditions was played at increasing intensities to the focal male. By alternating periods of song with periods of silence, and stepping up the call intensity in 2 dB steps, the experiment simulated an approaching male walking and then calling toward an established calling male. At low intensities (30-65 dB) the resident male would alter its song structure by both increasing the length of its interchirp interval and decreasing the number of syllables within each chirp (Dadour 1989). As the intensity approached 65 dB males began moving from their perch, some beginning this movement by rotating or pivoting around the perch before moving towards or away from the loudspeaker on horizontal branches, or up and down in the plant on vertical stems. As the intensity increased to 65-85 dB the eventual outcome in all experiments was that males moved to the base of the plant and stopped singing. This downward movement was always in stages with a resumption of singing at each stop. At the conclusion of the experiments, 4 males had moved to neighbouring plants and after 1 or 2 hours they could be heard singing from a new perch (Dadour & Bailey 1985).

Games theory predicts that aggressive encounters between individuals should be settled quickly, and that they should only escalate if any asymmetry between contestants is not recognised (Maynard Smith & Parker 1976). In the above experiment there is the possibility of one of at least three asymmetries developing between the intruder and the resident. Firstly, the resident may have a higher 'resource holding power', e.g. one individual may be larger than the other. The second is the 'payoff' asymmetry, whereby either individual may equate different values to a resource and therefore put more energy into a fight either to obtain it or defend it. Thirdly, individuals may use an arbitrary cue, such as prior ownership of a territory or previously established position in a heirarchy, to settle disputes. In the type of experiment we have just described the resident is unable to affect the outcome of the interaction by reciprocal behaviour (Fig 5) in that the intruding male (in our case a simulation) moving closer to the resident becomes increasingly louder, and also the resident has no opportunity to outcompete the intruding male. In most situations involving conflict behaviour the resident normally wins the encounter (Parker 1978), and if one considers the relative stability of males over time in an aggregation (Wright 1975) then we may assume that either male interactions are very rare or if they occur, the resident wins most encounters.

Gradually increasing the intensity of the calling song in a stepwise manner resulted in the male descending the bush in stages and singing at each stop. One explanation of this behaviour is that the resident male may be utilising the structure of the plant to attenuate the intruder's song, and in moving to a new position he may be able to gain some sort of a calling advantage over the intruder. Alternatively, the residènt male, by singing at each stop on his descent, may be bargaining with the intruder over sharing the resource (Maynard Smith & Parker 1976): males may prefer to share space rather than risk a physical contest. This

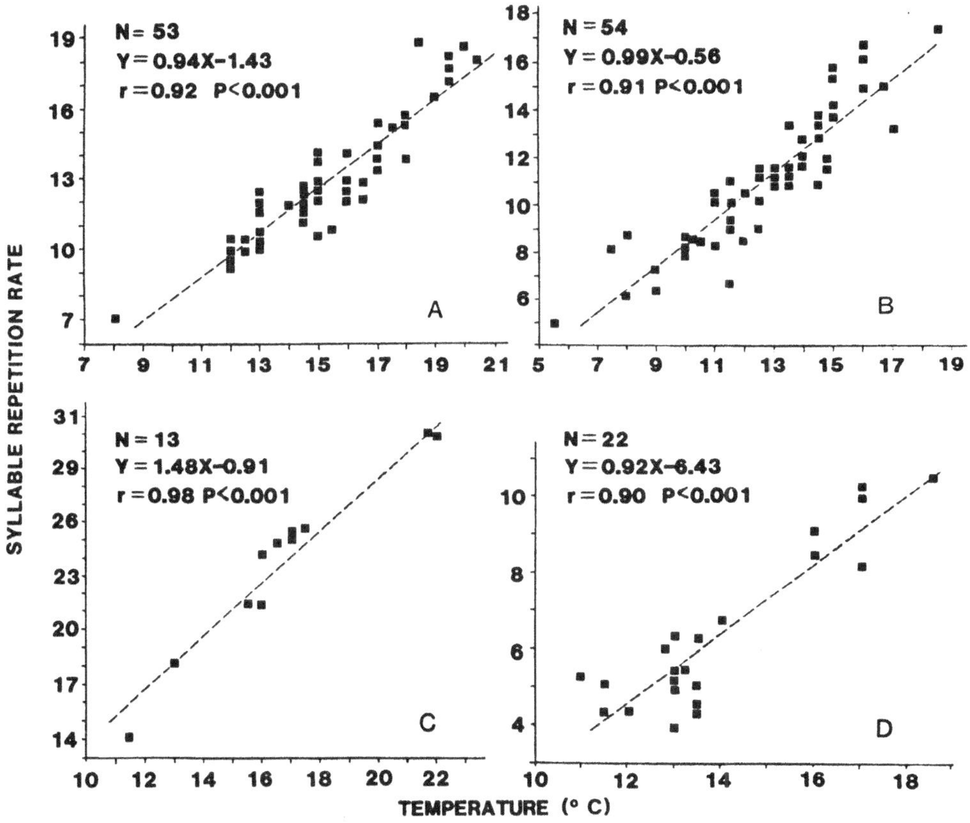

Fig. 5. The relationship between syllable repetition rate and temperature in four species of *Mygalopsis*. A. *M. marki*; B. *M. paupeuculus*; C. *M. sandowi*; D. *M. thielei* (Sandow 1980).

explanation assumes that males are not already on minimum sized acoustic territories. Therefore, the message of intent conveyed by the song will indicate to an intruding male that point where to approach no further. The result as viewed from the population is that males are regularly spaced. In either of the above alternatives, the resident male remains calling and by maintaining the call even from a less exposed position may increase its chance of mating (c.f. *Bufo cognatus* (Sullivan 1983) and *Gryllus bimaculatus* (Simmons 1986)).

This present scenario of males altering their song and then moving away to a new location in response to an intruding male may be considered a "loser strategy" (Barash 1982). The

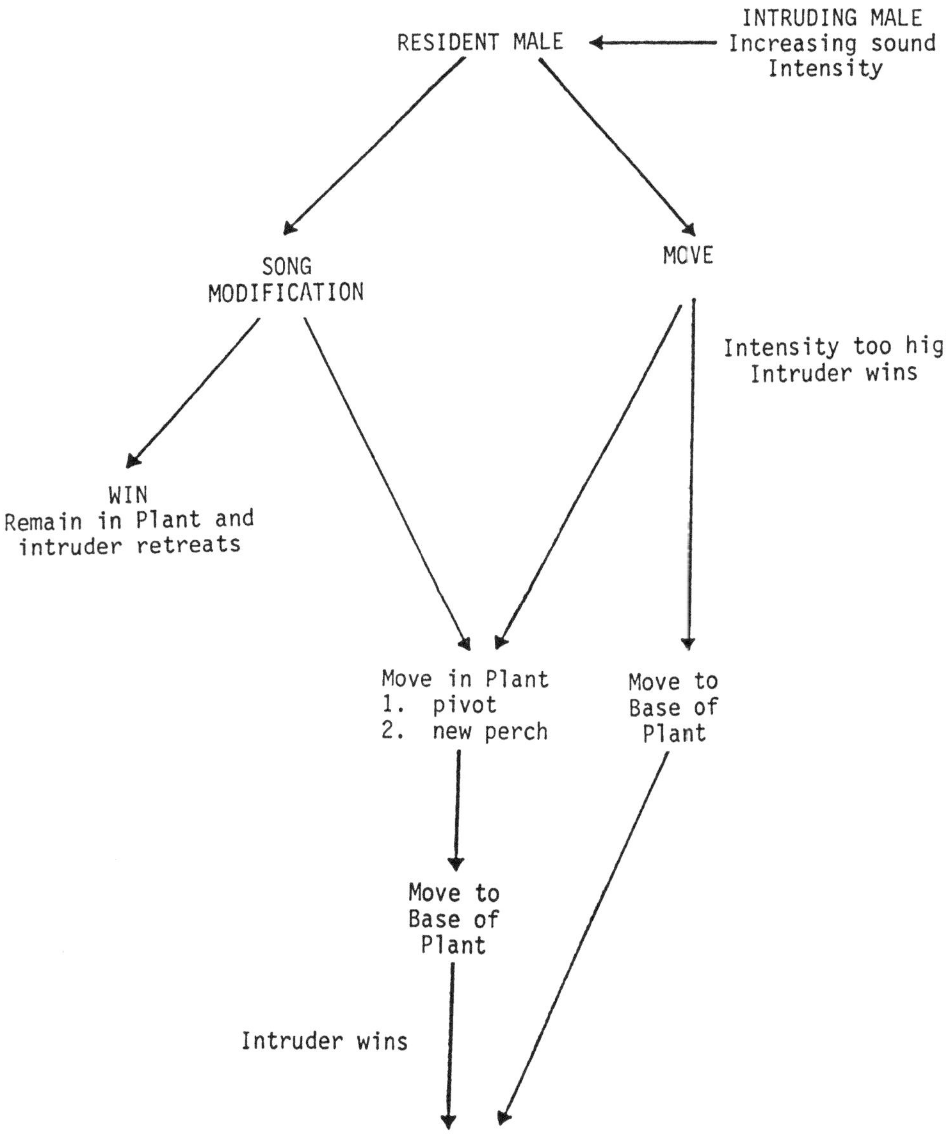

Fig. 6. Schematic representation of the possible outcomes of an interaction between a resident male when confronted with a calling conspecific.

payoff of reducing the agonistic component of their behaviour will be offset by the male being able to continue broadcasting his song. As the perceived signal strength exceeds a certain threshold, the male moves to the base of the plant and finally to a new perch on another plant from where it can again broadcast its song. Avoiding a fight with its inherent risk of injury at least allows the male to mate on another day. Such a strategy may be suitable to an insect that appears to have a very extended calling period (6-9 months (Lymbery 1984)) and few encounters with females. Therefore, at this stage it is difficult to equate such acoustic interactions as being 'loser' behaviours without establishing the relative mating success of each individual. To answer this question one must look to the behaviour of adult females: are females using acoustic criteria to assess the relative status of males?

References

Alexander R.D. 1968. Life cycle origins, speciation, and related phenomena in crickets. Q. Rev. Biol. 43: 1-41.

Arak A. 1983. Sexual selection by male-male competition in natterjack toad choruses. Nature 306: 261-262.

Arak A. 1988. Callers and satellites in the natterjack toad: evolutionary stable decision rules. Anim. Behav. 36: 416-432.

Bailey W.J. 1976. Species isolation and song types of the genus *Ruspolia* Schulthess (Orthoptera: Tettigoniidae) in Uganda. J. Nat. Hist. 10: 511-528.

Bailey W.J. 1979. A review of the Australian Copiphorini (Tettigoniidae: Conocephalinae: Copiphorini: Orthoptera). Aust. J. Zool. 27: 1015-1049.

Bailey W.J. & A.W.R. McCrae. 1978. The general biology and phenology of swarming in the East African tettigoniid *Ruspolia differens* (Serville) (Orthoptera). J. Nat. Hist. 12: 259-288.

Bailey W.J. & J.D. Sandow. 1983. Mechanisms of defensive stridulation in the bushcricket *Mygalopsis marki* Bailey (Copiphorini, Tettigoniidae). Acta Zool. 64: 117-122.

Bailey W.J. & R.O. Stephen. 1978. Directionality and auditory slit-function: A theory of hearing in bushcrickets. Science 201: 633-634.

Bailey W.J. & D.R. Thiele. 1983. Male spacing behavior in the Tettigoniidae: an Experimental Approach. In: Orthopteran Mating Systems: Sexual Competition in a Diverse Group of Insects. Eds. D.T. Gwynne & G.K. Morris. Westview Press, Boulder. pp 163-184.

Bailey W.J. & P. Yeoh. 1988. Female phonotaxis and frequency discrimination in the bushcricket *Requena verticalis*. Ecol. Entomol. 13: 363-372.

Bailey W.J., R.J. Cunningham, L. Lebel, (in press). Song power, spectral distribution and female phonotaxis in the bushcricket *Requena verticalis* (Tettigoniidae: Orthoptera) active female choice or passive attraction. Anim. Behav.

Barash D.P. 1982. Sociobiology and Behavior. 2nd ed. Elsevier New York, Oxford.

Barendse W.J. 1986. Speciation in the genus *Mygalopsis* (Orthoptera: Tettigoniidae) in Southwestern Australia. PhD Thesis, University of Western Australia, Nedlands.

Brush J.S., V.G. Gian & M.D. Greenfield. 1985. Phonotaxis and aggression in the coneheaded katydid *Neoconocephalus affinis*. Physiol. Entomol. 10: 23-32.

Dadour I.R. 1986. Acoustic behaviour of the male bushcricket *Mygalopsis marki* (Orthoptera: Tettigoniidae). PhD Thesis, University of Western Australia, Nedlands.

Dadour I.R. 1989. Temporal pattern changes in the calling song of the katydid *Mygalopsis marki* Bailey in response to conspecific song (Orthoptera: Tettigoniidae). J. Insect Behav. 2: 199-215.

Dadour I.R. & W.J. Bailey. 1985. Male agonistic behaviour of the bushcricket *Mygalopsis marki* Bailey in response to conspecific song (Orthoptera: Tettigoniidae). Z. Tierpsychol. 70: 320-330.

Dadour I.R. & M.S. Johnson. 1983. Genetic differentiation, hybridization and reproductive isolation in *Mygalopsis marki* Bailey (Orthoptera: Tettigoniidae). Aust. J. Zool. 31: 353-360.

Davies N.B. & T.R. Halliday. 1978. Deep croaks and fighting assessment in toads *Bufo bufo*. Nature 274: 683-685.

Gerhardt H.C. 1982. Sound pattern recognition in some North American treefrogs (Anura: Hylidae): implications for mate choice. Am. Zool. 22: 581-595.

Gerhardt H.C. 1987. Evolutionary and neurobiological implications of selective phonotaxis in the green treefrog, *Hyla cinerea*. Anim. Behav. 35: 1479-1489.

Gwynne D.T. & W.J. Bailey. 1988. Mating system, mate choice and ultrasonic calling in a zaprochiline katydid (Orthoptera: Tettigoniidae). Behaviour 105: 202-223.

Hill K. & B.O. Oldfield. 1981. Auditory function in Tettigoniidae (Orthoptera: Ensifera). J. Comp. Physiol. 142: 169-180.

Latimer W. 1981. The acoustic behaviour of *Platycleis albopunctata* (Goeze) (Orthoptera: Tettigoniidae). Behaviour 76: 182-206.

Latimer W. & A. Schatral. 1983. The acoustic behaviour of the bushcricket *Tettigonia cantans*. I. Behavioural responses to sound and vibration. Behav. Proc. 8: 113-124.

Latimer W. & M. Sippel 1987. Acoustic cues for female choice and male competition in *Tettigonia cantans*. Anim. Behav. 35: 887-900.

Lloyd J.F. & A.B. Gurney. 1975. Labral stridulation in a katydid (a coconut- infesting "tree-hopper") (Orthoptera: Tettigoniidae: Mecopodinae). Entomol. News 86: 47-50.

Lymbery A.J. 1984. The Plasticity of coloration in the Bush Cricket *Mygalopsis marki* (Orthoptera: Tettigoniidae). PhD. Thesis, University of Western Australia, Nedlands.

Lymbery A.J. 1987. Seasonal populations in the life cycle of *Mygalopsis marki* Bailey (Orthoptera: Tettigoniidae). J. Aust. Entomol. Soc. 26: 323-330.

Maynard Smith J. & G.A. Parker. 1976. The logic of asymmetric contests. Anim. Behav. 24: 159-175.

Maynard Smith J. & G.R. Price. 1973. The logic of animal conflict. Nature 246: 15-18.

Morris G.K. 1967. Song and aggression in Tettigoniidae. PhD Thesis, Cornell University, Ithaca.

Morris G.K. 1971. Aggression in male conocephaline grasshoppers (Tettigoniidae). Anim. Behav. 19: 132-137.

Morris G.K., G.E. Kerr & J.H. Fullard. 1978. Phonotactic preferences of female meadow katydids (Orthoptera: Tettigoniidae: *Conocephalus nigropleurum*). Can. J. Zool. 56: 1479-1487.

Morris G.K. & Pipher. 1972. The relation of song structure to tegminal movement in *Metrioptera sphagnorum* (Orthoptera: Tettigoniidae). Can. Entomol. 104: 977-985.

Nielsen E.T. & H. Dreisig. 1970. The behaviour of stridulation in Orthoptera Ensifera. Behaviour 37: 205-252.

Oldfield B.O. 1984. Physiology of auditory receptors in two species of Tettigoniidae (Orthoptera: Ensifera). J. Comp. Physiol. 155: 689-696.

Owen D.F. 1969. Green-brown polymorphism in African Tettigoniidae. Rev. Zool. Bot. Afr. 80: 346-351.

Parker G.A. 1978. Evolution of competitive mate searching. A. Rev. Entomol. 23: 173-196.

Parker G.A. 1983. Mate quality and mating decisions. In: Mate Choice. Ed. P. Bateson. Cambridge University Press, Cambridge. pp 141-164.

Redtenbacher J. 1891. Monographie der Conocephaliden. Verh. Zool.-Bot. Ges. Wien 41: 315-562.

Rheinlaender J. & H. Römer. 1986. Insect hearing in the field. I. The use of identified nerve cells as 'biological microphones'. J. Comp. Physiol. 158: 647- 651.

Römer H. 1987. Representation of auditory distance within a central neuropil of the bushcricket *Mygalopsis marki*. J. Comp. Physiol. 161: 33-42.

Römer H. & W.J. Bailey. 1986. Insect hearing in the field. II. Male spacing behaviour and correlated acoustic cues in the bushcricket *Mygalopsis marki*. J. Comp. Physiol. 159: 627-638.

Römer H., W.J. Bailey & I.R. Dadour. 1989. Insect hearing in the field. III. Masking by noise. J. Comp. Physiol. 164:

Rowell C.H.F. 1971. The variable coloration of the acridoid grasshoppers. Adv. Insect Physiol. 8: 146-190.

Sandow J.D. 1980. The morphology, distribution and acoustic repertoire of the genus *Mygalopsis* (Orthoptera: Tettigoniidae). Masters Thesis, University of Western Australia, Nedlands.

Sandow J.D. & W.J. Bailey. 1978. An experimental study of defensive stridulation in *Mygalopsis ferruginea* Redtenbacher (Orthoptera: Tettigoniidae). Anim. Behav. 26: 1004-1011.

Shaw K.C., R.C. North & A.J. Meixner. 1981. Movement and spacing of singing *Amblycorypha parvipennis* males. Ann. Entomol. Soc. Amer. 74: 436-444.

Simmons L.W. 1986. Inter-male competition and mating success in the field cricket, *Gryllus bimaculatus* (De Geer). Anim. Behav. 34: 567-579.

Simmons L.W. 1987. Heritability of a male character chosen by females of the field cricket, *Gryllus bimaculatus*. Behav. Ecol. Sociobiol. 21: 129-133.

Sullivan B.K. 1983. Sexual selection in the great plains toad (*Bufo cognatus*). Behaviour 84: 258-264.

Thiele D.R. 1981. Mechanisms of acoustic spacing within populations of the tettigoniid *Mygalopsis marki*. PhD Thesis, University of Western Australia, Nedlands.

Thiele D.R. & W.J. Bailey. 1980. The function of sound in male spacing behaviour in bush-crickets (Tettigoniidae, Orthoptera). Aust. J. Ecol. 5: 275- 286.

Thornhill R. & J. Alcock. 1983. The Evolution of Insect Mating Systems. Harvard Univ. Press. p 547.

Walker F. 1869. Catalogue of the specimens of Dermaptera Saltatoria in the collection of the British Museum, Vol. 2. pp 225-423.

Walker T.J. 1964. Cryptic species among sound-producing ensiferan Orthoptera (Gryllidae and Tettigoniidae). Q. Rev. Biol. 39: 345-355.

Walker T.J. 1975. Effects of temperature on rates in poikilothermous nervous systems: evidence from the calling songs of meadow katydids (Orthoptera: Tettigoniidae: Orchelimum) and reanalysis of published data. J. Comp. Physiol. 101: 57-69.

Wright A. 1975. Aspects of the acoustic ethology and field ecology of *Mygalopsis marki* (Orthoptera: Tettigoniidae). Hons. Thesis, University of Western Australia, Nedlands.

7. ACOUSTIC COMMUNICATION BETWEEN THE SEXES IN BUSHCRICKETS

David Robinson

Introduction

Acoustic signalling in bushcrickets is predominantly a masculine feature, and in very many species the female has no acoustic role to play other than that of a silent receiver. This would seem perfectly logical if the purpose of the male song was simply to convey information. However, the male calling song is used almost universally by the female to locate the male, and so one could argue that it would be more efficient if the female indicated to the male that she was using his signal for orientation purposes so that, apart from anything else, he didn't stop just before the female located him. If the purpose of acoustic communication in the bushcrickets is to bring male and female together for a successful mating, then the system that we actually find should be the most efficient one to achieve that end. The fact that, at a superficial level, we can find apparent flaws in the design of the system suggests either that the system is indeed flawed or, much more likely, that a more extensive analysis and a deeper understanding is required. This chapter provides that analysis.

Types of acoustic communication

Acoustic signals between the sexes have a number of advantages for bushcrickets over other forms of signal. They are usable round the clock, and many bushcrickets are nocturnal. Bushcrickets are also relatively small, but need to communicate over large distances in complex environments, a feature available from sound, but not vision. The physical parameters of sound can be varied rapidly, so information can be encoded in the sound – potentially a large amount of information though as yet there is no evidence that there is extensive information transfer in bushcrickets. Sound provides reliable cues for location purposes, and can of course be turned on and off at will. The structure of the sound producing organ can impart a directional characteristic to the sound emission, giving a greater range than the same power distributed uniformly. Finally, turning off the signal effectively makes the animal disappear in many situations!

The major disadvantage of sound as a communication medium is that the environment is usually noisy. Other animals produce sound, of course, but wind and the movement of vegetation also generate a lot of noise. So the recipient of the information has to be selective. Sound shares with many communication systems the problems of privacy. It is difficult to ensure that only the intended recipient receives the message. This is particularly true if there is no indication from the recipient that they have received anything.

Amongst the bushcrickets there are several different types of communication system between the sexes. The most common one by far is the one in which the male calls and the female moves towards the sound source without making any signals herself. We have already hinted at some of the apparent problems of this system, but it is important to review its key features before considering systems in which the female has an active signalling role.

The silent female system

Since this is the most common system it has been very well studied. The male's song provides the listening female with at least the minimal information about the identity of the caller and his position. It will also indicate to the female that the male is sexually receptive. The male, of course, only knows if he has been successful in attracting a mate when a female arrives. His investment in calling is therefore of a speculative nature. The longer the male sings, the better his chance of attracting a female, but the more his investment in energetic terms. This last fact may not be of serious consequence since continuous singing is a feature of some species such as *Tettigonia viridissima* and *Ruspolia differens*. Sexually receptive males within hearing distance of each other will be in competition, and male-male interactions are a feature of this system of communication. Males of the copiphorine *Mygalopsis marki* have been observed to space themselves according to their perception of the volume of sound produced by their neighbours (Bailey & Thiele 1983).

Charles Darwin (1871), in formulating his theory of sexual selection in his book "The Descent of Man and selection in relation to sex", concluded that selection involved both competition between males and mate choice by females. Where acoustic signalling is used to bring the sexes together the extent of mate choice that the female has must be limited by the information content of the male's song. This implies that the reproductive fitness of a male can be indicated acoustically. In most tettigoniids the male provides nutrient to the female via the spermatophylax, the non-sperm carrying portion of the spermatophore (see Gwynne chapter 3). In some species the spermatophore may be as much as 40% of the body weight of the male (Gwynne, 1982), representing a significant contribution to the female's diet. It is, therefore, an advantage to the female to select a male who can make the largest possible contribution to the egg-producing effort – but the only source of information is the song. There is evidence that larger males produce louder songs and that females select louder songs (Crankshaw, 1979; Bailey, 1985; Bailey & Yeoh 1989). So, providing male size and fitness are correlated the female is indeed exercising a choice that has evolutionary implications. However, sound is attenuated during its passage through the air, and so the female might perceive a small male singing close to her as louder – and therefore larger and more desirable – than a large male singing some distance away. So, if the female is truly going to choose, she must use additional cues from the song.

In the two other communication systems that are found in bushcrickets the female has an active signalling role, so there is not only the possibility for a refined choice by the female, but the male may be able to exercise a choice as well.

Female song appears to be restricted to only three, unrelated, groups of the Tettigoniidae, occurring commonly amongst phaneropterines and in a few species of ephippigerines. It is possible that it also occurs in some pseudophyllines (Fulton, 1933; Nickle & Carlysle, 1975).

The motile singing female system

This system is characteristic of certain species of ephippigerines. Although the basic behaviour is similar – the receptive male calls and a receptive female moves towards him – the female possesses a well developed stridulatory apparatus comparable to that found on the male. The female uses this to produce song in answer to the male. The male and female stridulatory apparatus in *Steropleurus stali* are shown in Figure 1.

The female file is on the upper surface of the right elytron (modified wing), and so is

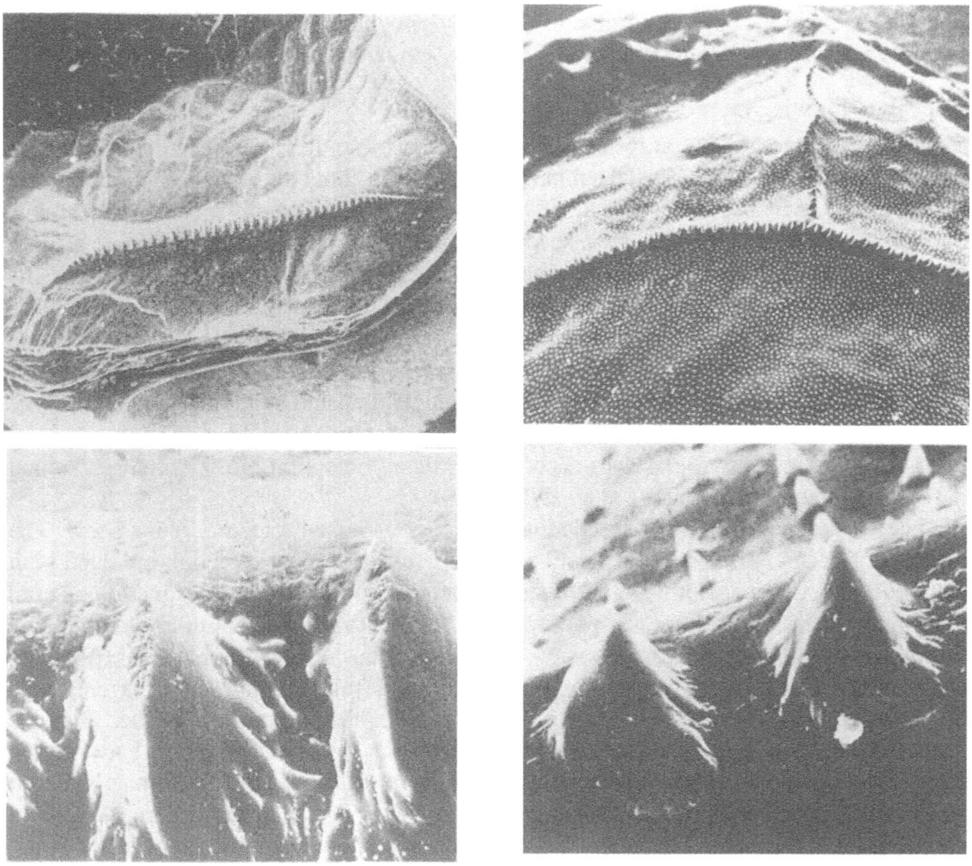

Fig. 1. The stridulatory apparatus of *Steropleurus stali*. (a) The underside of the left elytron of the male showing the file. The articulation is to the left. x 24. (b) Detail of the teeth of the male file. x600. (c) The upperside of the right elytron of the female showing part of the file, and tooth development on the cross vein. x 58. (d) Detail of the teeth of the female file. x1150. (Hartley *et al.* 1974)

morphologically distinct from the male file, which is on the lower surface of the left elytron. I consider the features of this communication system in the next section.

The stationary singing female system

In this system, which appears to be relatively common amongst members of the Phaneropterinae, the female produces a sound in response to the male call, but she remains still whilst the male moves towards her. The female stridulatory apparatus is quite different to that of the male and, as in ephippigerines, is not homologous with it. The structure of the stridulatory apparatus of both a male and a female of *Poecilimon hamatus* is shown in Figure 2. The features of this communication system are discussed below.

Acoustic communication between male and female ephippigerines

In most ephippigerines the female has a well developed stridulatory apparatus. However, only a few species are known in which the female actually produces a song. The usual use is in the production of a disturbance call or reflex cry (Dumortier 1963). This disturbance call in *Ephippiger terrestris* was described by Yersin (1854) as being similar to the male call. The male call in *E. terrestris* is a single chirp so the disturbance call of the female could sound similar. However, disturbance calls are quite variable, and distinct from the normal stridulation of the male.

In those ephippigerine species where the female produces a song, the song pattern is characteristic and is clearly distinct from the disturbance call. A male call and a female response call of *Steropleurus stali* are shown in Figure 3. These calls have been studied by Hartley *et al.* (1974) and have quite a complicated structure. The male call is a single chirp lasting between 252 and 305 ms. There are normally ten diplosyllables within the chirp, and the overall amplitude of the diplosyllables increases throughout the chirp. The female song is also a single chirp and the amplitude of the diplosyllables increases throughout the chirp. There are differences in detail between the two songs, but their overall structure is similar. This is despite the fact that the female stridulatory apparatus is not homologus with that of the male.

The female song is produced under specific conditions. It is always a response song produced after a male call, and the timing of the reply is surprisingly consistent, the reply following the male call within 200 ms (Hartley, *et al.* 1974). Females do not reply to other females. In situations where a number of females are within acoustic range of a calling male it is not necessarily the nearest female that replies. In observations made by Hartley *et al.* (unpublished data), a male would emit several calling chirps over a period of about a minute, to each of which a female would reply. There would then be a period of silence.

In their study of female song in ephippigerines Hartley, *et al.* (1974) recorded a series of acoustic interactions between males and females of *Platystolus obvius*. They found that the

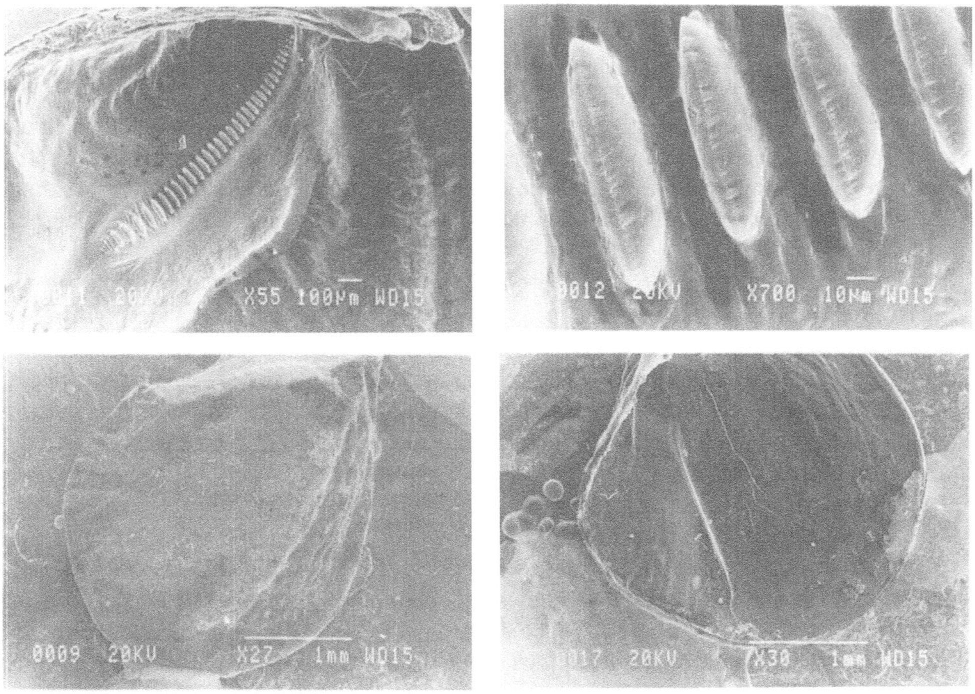

Fig. 2. Stridulatory apparatus of *Poecilimon schmidti.* (a) Underside of the left elytron of the male showing the file. The scale bar represents 100 mm. (b) Detail of the male file. Scale – 10 mm. (c) Upper surface of the right elytron of the female showing the file. Scale – 1 mm. (d) Underside of the left elytron of the female showing the plectrum. Scale – 1 mm. (Courtesy of J. C. Hartley)

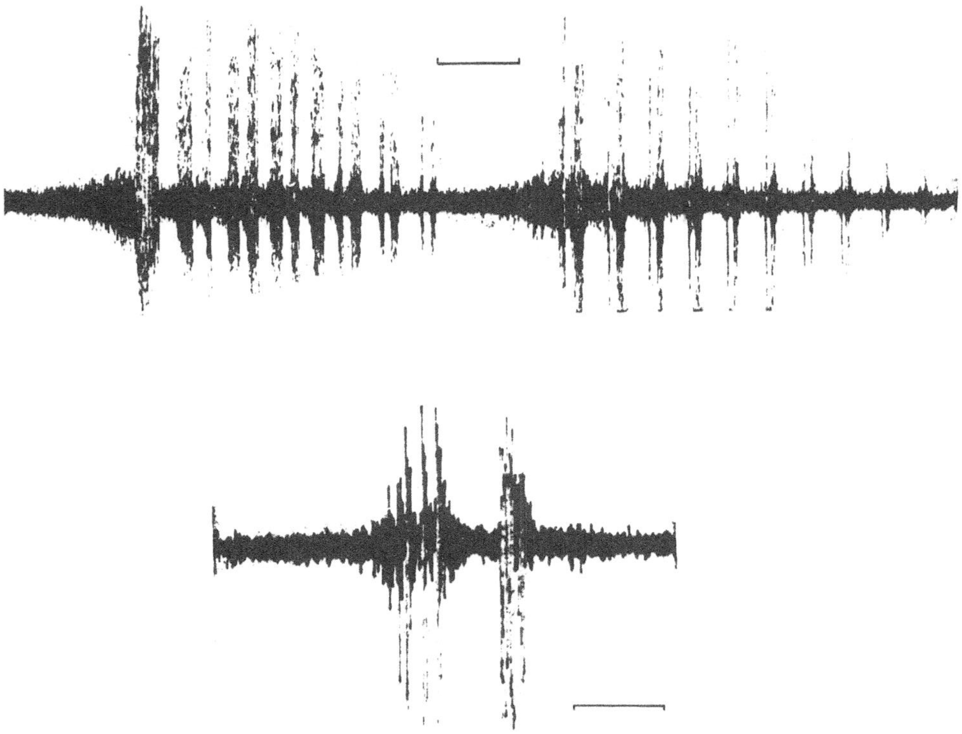

Fig. 3. Oscillogram of the male calling song and the female reply of *Steropleurus stali*. The male call is on the left, and the time marker represents 50 ms. (Hartley *et al.* 1974)

males on their own produced only occasional chirps at about 20 minute intervals. If a female response was received, the male increased his calling rate to around one chirp every minute, a response that continued for some time even after the female stopped responding. Communication between males and females could take place over a distance of at least 10 m.

This example illustrates a system that is very different from the silent female system discussed earlier. Male singing activity is limited, unless there is a female replying. The investment of the male in singing behaviour is greatest when the chance of a successful mating is greatest. We do not know what elements of choice might be present in such a system, but clearly the potential is there for both male and female choice. In *Callicrania monticola* there is a more elaborate interchange between the male and female which may represent a more refined choice mechanism.

A female *Callicrania monticola* will sing only in reply to a male call. Her reply is short, consisting of one to three very short pulses. However, the male call is much longer, and consists of a sequence of syllables of increasing amplitude, with 1 to 4 final syllables of greater length and amplitude. Figure 4 shows three calls produced by the same male, each call having a different number of final syllables. The significance of the variation in number of final syllables is unknown. Observations have shown that there is a range of acoustic interactions in this species (Robinson and Hartley, unpublished data), unlike the other ephippigerines so far described. The calls of two pairs of *C. monticola* were recorded on tape over a period of one hour, and subsequently analyzed. Three typical interactions are shown in Figure 5. and the categories of interaction observed are shown in Table 1.

Table 1. Singing activity in *Callicrania monticola*

	Pair I	Pair II
MALE CALLS	28	21
2 syllable calls	7	8
3 syllable calls	19	13
4 syllable calls	2	0
MALE-MALE CALLS		
3 syllable calls and replies	1	9
MALE-FEMALE CALLS	13	27
Male call with female reply 2 syllable calls	2	4
3 syllable calls	6	7
4 syllable calls	0	1
Male call with female reply and further male call	5	15
Both 2 syllable calls	0	5
Both 3 syllable calls	0	7
First call 2 syllables, second 1	2	0
First call 3 syllables, second 1	2	3
First call 3 syllables, second 2	1	0
TOTAL NUMBER OF CALLS	47	57

In this species a male can get an acoustic reply from either a male or a female. There is, as yet no explanation for the range of interactions observed, nor for their meaning to the individuals taking part. Since female calls are only produced in response to male calls they obviously have a sexual function. There is at least the possibility that the communication has a role in sexual selection.

Those response songs that have been described are a feature of only a few species of ephippigerines yet all the species have essentially the same equipment. This raises the obvious question – why don't females of the other species produce songs? In part the answer may be the traditional one of insufficient study, but species like *Ephippiger ephippiger* are very well known and yet no female song has been observed.

The ability of females to produce sound must have evolved independently of the stridulatory apparatus of the male. The selection pressures on the sexes may, of course, be different but it does seem more likely that parallel evolution of stridulatory mechanisms has taken place, and this may have been in response to the selection pressure provided by predation. All three ephippigerine species whose behaviour has been described above live in low density populations, and they do not have a tendency to aggregate. This is in contrast to *Ephippiger* and *Ephippigerida* species where the males are reiterative singers, often occurring in aggregations. It may well be an advantage for males of species living at a low density to be occasional singers unless there is a female present. The advantages of this strategy are firstly that in a low density population there may well not be a female within hearing range so considerable investment in speculative singing would result in a profitless loss of energy. Secondly, a single insect singing is more likely to attract the attention of a predator to itself than if it were merely one of a large number where its chance of being missed by a predator would be greater.

The ephippigerines are slow, heavy and flightless insects. Alexander (1960) in a survey of American tettigoniids associated disturbance calls with susceptibility to predation because he found four species which produced disturbance chirps and three of these were flightless and slow moving. Apart from the ephippigerines the only other European bushcricket known to produce disturbance chirps is *Polysarcus denticauda* (Faber 1953). It too is large and slow moving like *Ephippiger*.

It is not possible to decide how the female stridulatory apparatus evolved. It may have appeared as a reproductive feature with some species subsequently only producing disturbance chirps. Alternatively, and more likely, it could have evolved as a deterrent to predators with those species that now have female response songs being evolutionarily more recent. Since female stridulatory apparatus has evolved independently in the phaneropterines, we should postpone further discussion of the selection pressures that might produce female song systems until we have considered phaneropterine acoustic behaviour.

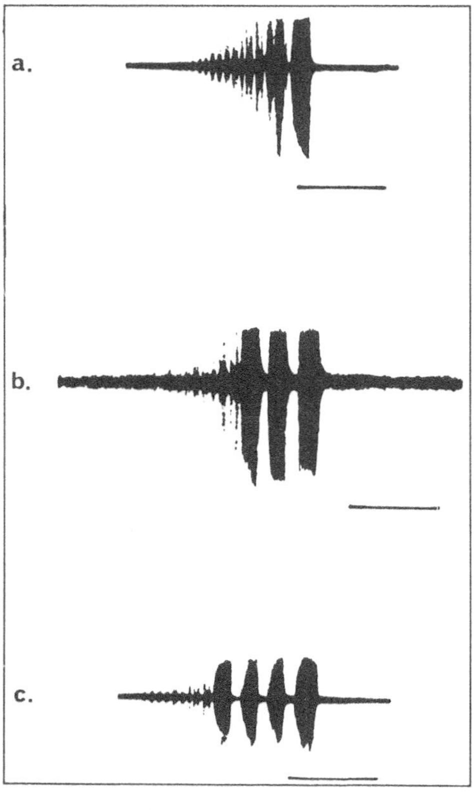

Fig. 4. **(a - c)** Oscillograms of three calls of a male *Callicrania monticola*. The time bar is 1 s.(From Hartley *et al.* 1974)

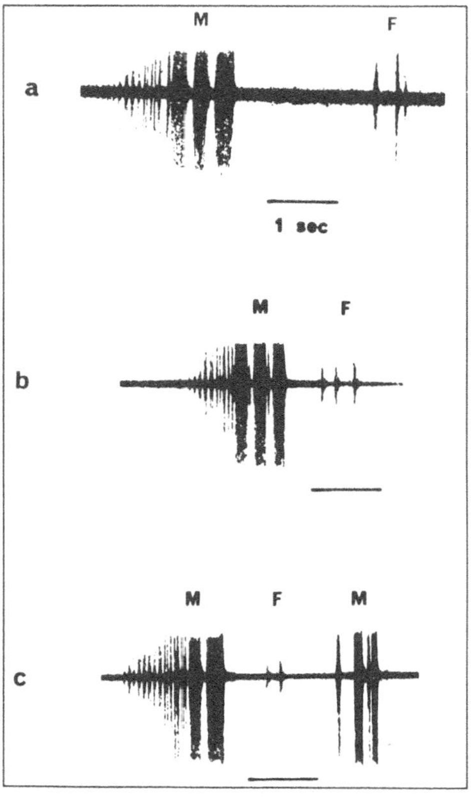

Fig. 5. Oscillograms of three typical song interactions between a male and a female *Callicrania monticola*. The time bar is 1 s.

Acoustic communication between male and female phaneropterines

The 'stationary singing female' system of communication is common amongst members of the Phaneropteridae. The occurrence of female stridulatory organs is widely reported in the literature (*eg Nickle & Carlysle, 1975) and sound-production has been observed directly in some species (*eg Alexander, 1960). Behavioural studies are not so common, and are mainly concerned with North American species (*eg Spooner, 1968). Some palaearctic species have been investigated, notably *Poecilimon* (Heller & von Helversen, 1986; Heller chapter 8), *Isophya* (Zhantiev and Dubrovin, 1977; Heller & von Helversen, 1986) and *Leptophyes punctatissima* (Hartley & Robinson, 1976; Robinson, 1980; Robinson *et al.* 1986; Rheinlaender *et al.* 1986).

Leptophyes punctatissima is a bushcricket that occurs over much of Europe. It is of great interest because of the highly specialised acoustic communication system that it has evolved. Some other members of the genus appear to have similar systems, but less is known about them. The male and female calls are illustrated in Figure 6. The male call (Figure 6a) consists of a chain of five to eight ultrasonic pulses, each of about 1 ms duration. The female song (Figure 6b) is much briefer, being rarely more than one pulse of about 1 ms duration. The complete interchange between male and female lasts less than 100 ms (Hartley & Robinson, 1976; Ahlén & Degn, 1981; Robinson, *et al.* 1986), the duration of the interchange being, at least in part, dependent on temperature. The frequency spectra of the two sexes are similar and for both the peak energy in the spectrum is around a frequency of 40 kHz. There is surprising similarity between the calls, and it is worth emphasizing that the stridulatory apparatus of the female is not homologous with that of the male.

Singing activity in the male

The male calls for periods during the day and the night. In the laboratory a peak of activity in the afternoon has been observed (Figure 7). The activity of the male increases if he receives a response from a female (Figure 8). If the response continues, but the male is prevented from mating there is a gradual decline in his calling rate. The response that the male receives has to be of a specific character, and herein lies the major significant feature of the acoustic behaviour of *Leptophyes*. The female response has to occur within a narrow time window for the male to respond, and in turn his response is to increase the rate of stridulation, and move

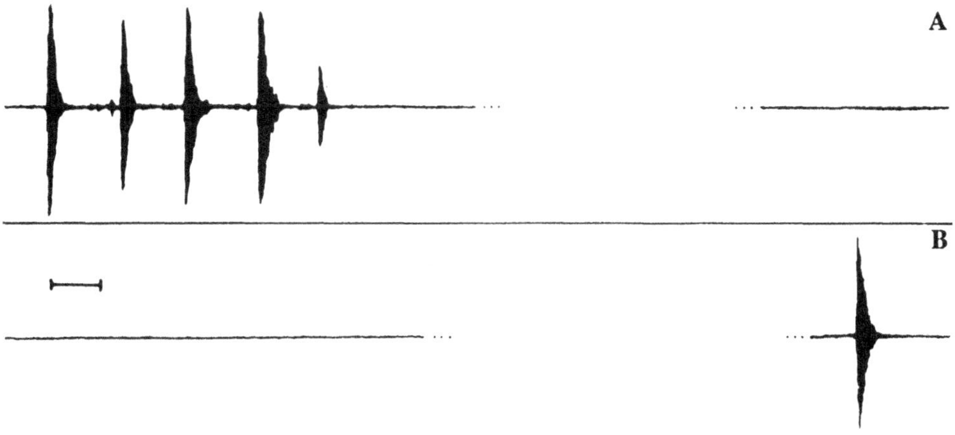

Fig. 6. The male (a) and female (b) songs of *Leptophyes punctatissima*. Songs recorded approximately 10 cm from the singing insects, the sound intensity being of the order of 100 dB SPL for both insects. (Robinson *et al.* 1986)

towards the stationary, responding female. The increase in chirp rate may be quite significant. Using artificial female chirps generated electronically Hartley and Robinson (1976) were able to show that the chirp rate of a single male, which was about 12 per hour in the absence of females, increased to about 450 per hour when a female response was given to each chirp.

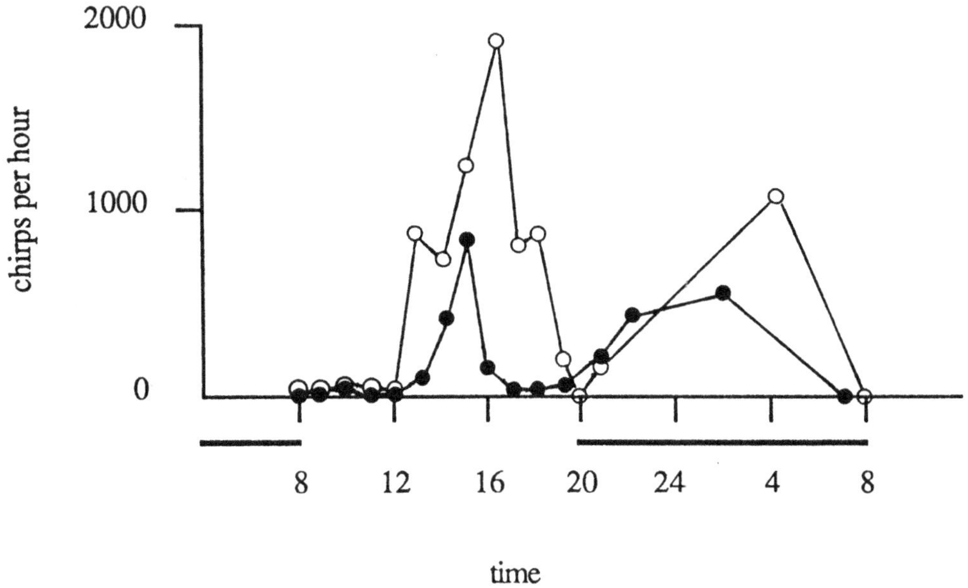

Fig. 7. Singing activity of a male *Leptophyes punctatissima* measured over a 24 hour period (solid circles) and the activity when each call received a synthetic female reply. (Redrawn from Robinson 1980)

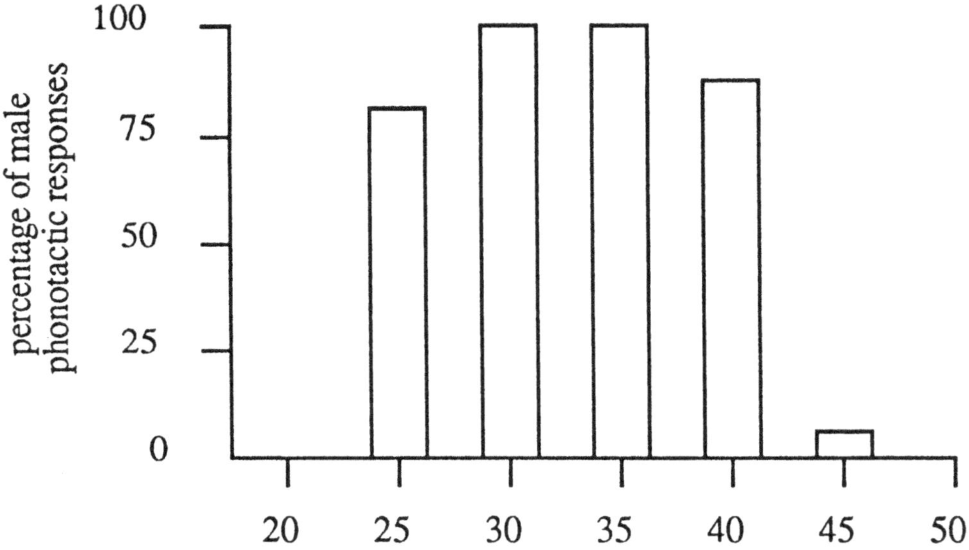

Fig. 8. Percentage of male phonotactic responses of male *Leptophyes punctatissima* to a synthetic female reply at different delay times between the onset of the male song and the onset of the female reply. Sound intensity of the female reply as measured at the male's taking off point was 80 dB SPL. The temperature was 28°C ± 2°. N=25 males. (Hartley *et al.* 1974; Robinson *et al.* (1986))

Female response to male chirps

Females will normally produce a response to any short click that has an ultrasonic component. It was this feature that initially enabled the female song to be identified (Hartley & Robinson, 1976). Subsequently, electronic mimics of the male song were used to investigate the limits of female response. Shorter pulses of ultrasound were more effective at evoking a female response. Pulses of 10 ms duration and 45 kHz produced more than twice as many female responses as pulses of the same frequency, but of 40 ms duration.

The timing of the female response is remarkably consistent. The results of 100 duets (male call and female reply) for each of three insects are shown in Table 2.

This consistency of timing of the female response is not affected by the duration of the male call. Altering the duration of synthetic male calls over the range 2-50 ms does not affect the timing of the reply (Robinson *et al.*, 1986). Ambient temperature does affect the timing, as might be expected, with the female response being faster at higher temperatures.

Table 2. Net delay of the response of female *Leptophyes* to synthetic male calls at a sound intensity well above threshold. The ambient temperature was 28°C ± 2°.

Mean net delay (ms)	Standard Error	N	Range
25.9	1.4	100	24-32 ms
27.5	0.8	100	26-32 ms
32.0	1.4	100	29-37 ms

Phonotaxis

In *Leptophyes* it is the male that performs phonotaxis. The specific stimulus to phonotaxis appears to be the timing of the female reply. In a test of 25 males, Robinson *et al.* (1986) found that males did not respond phonotactically to a synthetic female reply if the reply was more than 50 ms or less than 20 ms after the start of the male call (Figure 8). Within this group range each male had an individual response range of less than 30 ms. This means that the male will only start a phonotactic approach to the female if her reply falls within his 'temporal window'.

The existence of a temporal window during which a reply is expected is not completely novel – it has been observed in duetting acridids where the male 'expects' to hear a reply within 1 s. What is so notable about the temporal window in *L. punctatissima* is the much faster, and much more precise timing. The matching of a male temporal window of 20-50 ms with a female reply whose net delay is 25-32 ms is particularly striking. There are good reasons why the upper limit of the male's temporal window should be greater than the upper limit of the female's net delay. Firstly, the female's delay is affected by temperature, the relationship being inverse as one might expect. Secondly, the duet is so fast that the travel time of the sound through air is significant. Consider a stridulating male three meters from a female. Sound travels at about 0.34 m ms^{-1}, so the female will perceive the start of the male call about 9 ms later (Figure 9). The time taken for the auditory T-fibre in the mesothoracic ganglion to respond is 13 ms in *L. puntatissima*. The minimum net delay measured for females is 23 ms (at 28°C). If we make the reasonable assumption that there is no time for the brain to be involved in the production of the female reply, then the remaining 10 ms is enough time for the generation of flight muscle contraction via at least one motorneuron. The female's reply then has to return to the male, so a further 9 ms travel time is required. Thus the male perceives the female reply 41 ms after he started his call – within his temporal window.

These figures suggest that the maximum distance over which communication leading to

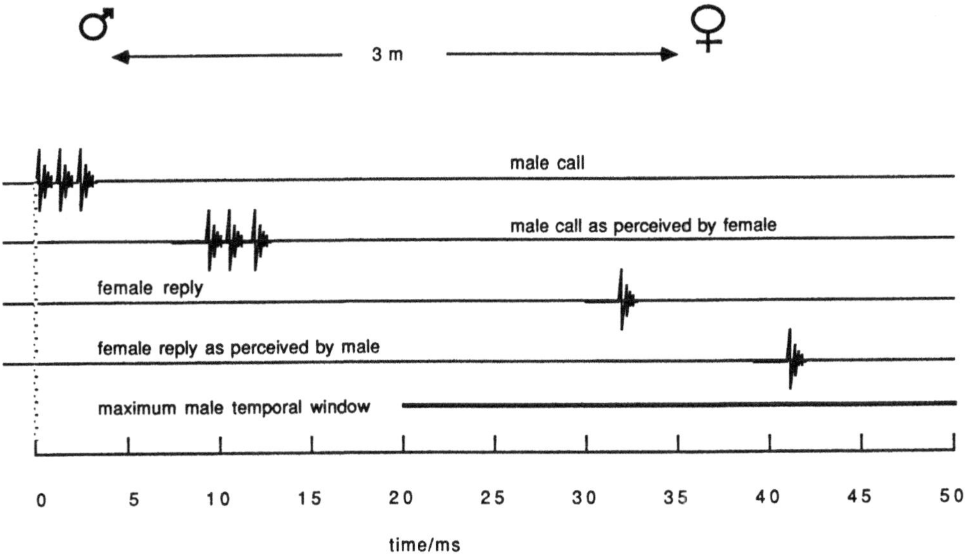

Fig. 9. Timing diagram of a single male *Leptophyes punctatissima* call and the female reply showing the travel time of the sound and the time window of the male.

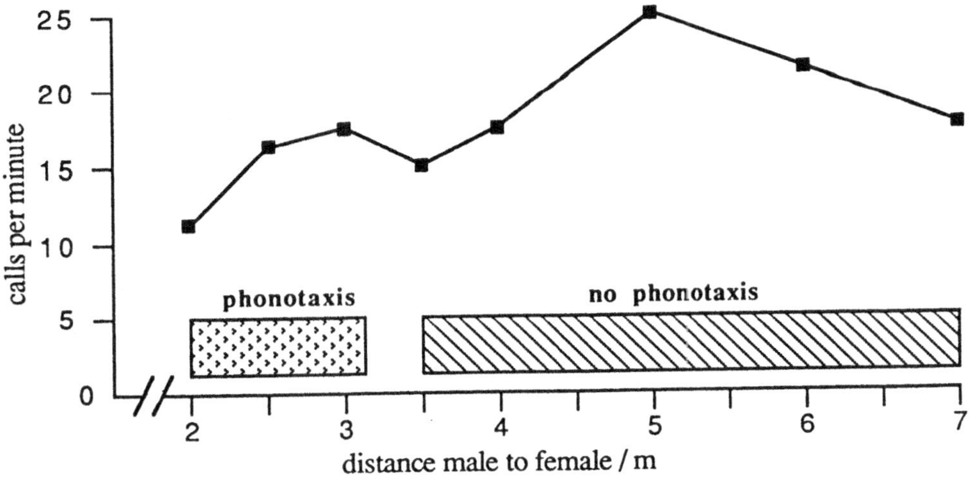

Fig. 10. Acoustic interactions between a male and a female *Leptophyes punctatissima* at different separation distances. The mean rate of singing of the male is plotted against the separation distance. (Redrawn from Zimmerman *et al.* in press)

phonotaxis could take place is 8.5 m at 28°C.

Furthermore, it has been shown by Zimmerman, *et al* (1989) that the distance over which phonotaxis takes place is smaller than the distance over which communication can take place. In Figure 10 the interaction between a male and a female is shown for different separations. Phonotaxis is limited to a distance of just over 3m between the two, although communication occurs over a distance of nearly 7m. There is no apparent decline over the whole 7m in male singing rate. A decline would indicate that he was either not perceiving, or was not receptive

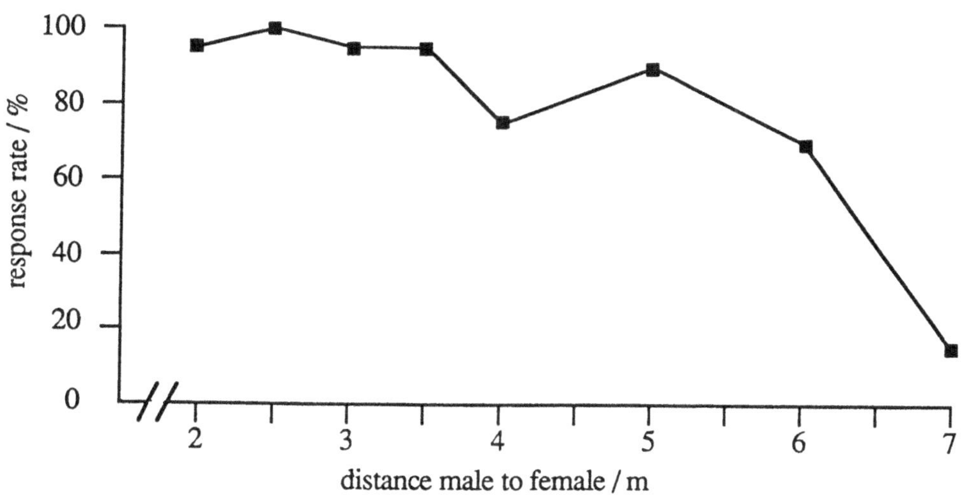

Fig. 11. The response rate of the female as a percent (a), and her overall latency (b), with mean values plotted against distance. The insets in (b) show the distribution of female response latencies at two different distances.(Redrawn from Zimmerman *et al.* in press)

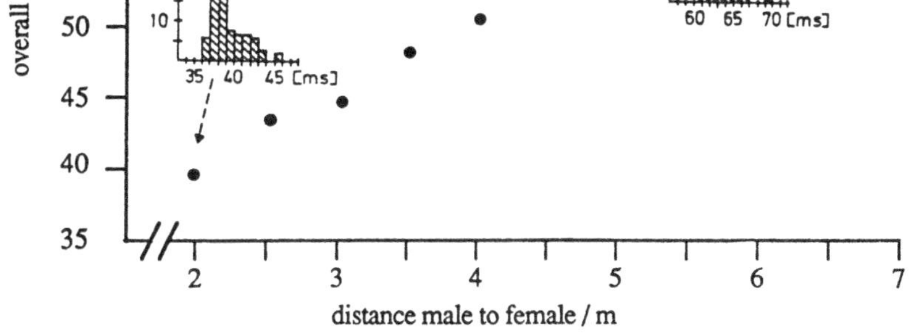

to, the female reply. So there must be some feature of the song that provides a cut-off point. The two most obvious candidates are intensity and delay in the female reply.

For any pair of male and female, the female will have a fixed response time. This is shown in the insets in Figure 11 where the female response is surprisingly consistent. However, as perceived by the male both the intensity of her response, and its perceived delay will vary with separation distance. Perceived delay will depend upon the travelling time of the sound. Since both these relationships can be calculated for a particular set of environmental conditions, it is possible to plot pairs of values of intensity and delay to produce a distance characteristic curve – Figure 12.

This shows that the male could estimate, at least potentially, the distance to the female using either latency or intensity or a combination of the two. Using this curve it is possible to erect a testable hypothesis. If male phonotaxis is elicited by using synthetic female replies in which intensity and delay can be altered independently, using values from the distance characteristic curve should produce phonotaxis up to the point when one or other limiting value is exceeded. This is the key point in the test. If, at this point, intensity is increased while the latency remains at the same value, then a restoration of phonotaxis would indicate that intensity was the limiting value. The three possibilities predicted are shown in Figure 13 (right). In fact, when just such an experiment was carried out by Zimmermann *et al.* (1989) (Figure 13) on 9 males, 3 of them were limited by intensity only, and 6 by a combination of latency and intensity. None were limited by latency alone.

It is not surprising that intensity is a limiting factor. A sound that is of too low an intensity

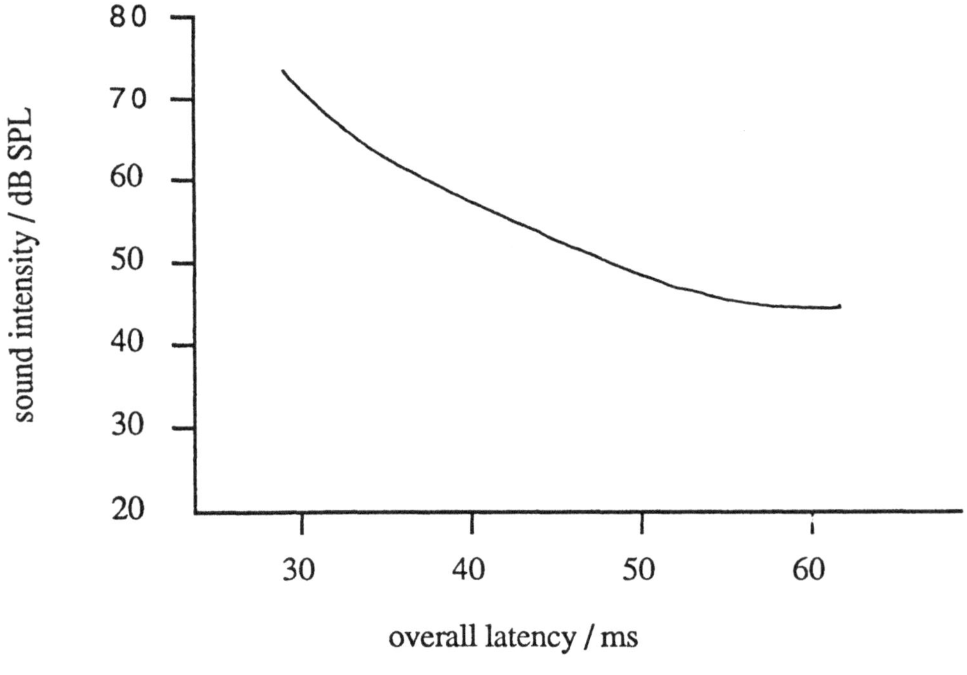

Fig. 12. Distance characteristic curve. The line connects combinations of sound intensity and response latency of an average female at specific distances.

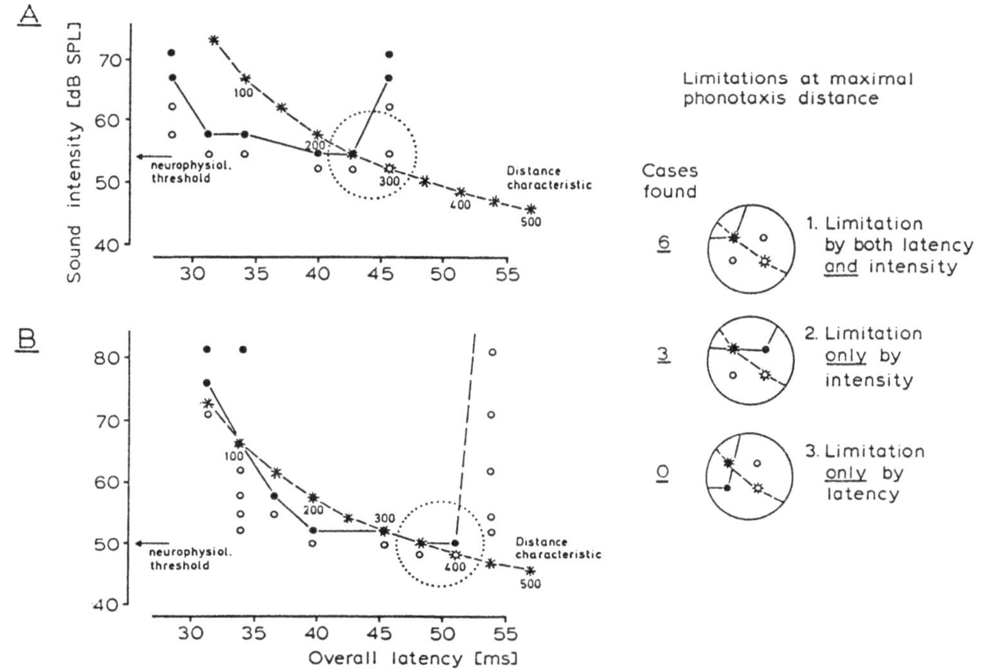

Fig. 13. Threshold curves for two males (A and B), measured with a synthetic female reply. Sound intensity and latency could be varied independently of each other. Closed circles indicate phonotaxis, open circles no phonotaxis. The key section where phonotaxis was abandoned is shown by a dotted circle. The right of the figure shows the three possible cases of limitation and the number of males found in each case. (Redrawn from Zimmerman *et al.* in press)

would not be perceived, so the timing of the reply would be irrelevant. There is no difference in sensitivity of the ear between the male and female (Rheinlaender and Hardt, unpub), but there is a difference in the intensity of sound produced. Male calls are generally around 25 dB louder than the female calls. This means that a pair can apparently be communicating over a large distance, but in fact the male is not perceiving the female's reply. It seems likely that the 'time window' of the male acts as a species isolating mechanism and could operate among related phaneropterine species, and that phonotaxis is elicited by a female reply of sufficient intensity within the time window of the male.

The advantage of this type of mating system is that the time window is independent of the attenuating characteristics of the environment. The amplitude modulation pattern of the sound is largely irrelevant; it is after all, extremely short. The requirements of the system from the male are for a brief pulse(s) of sound at the right frequency, phase-locked to the male. This gives a relatively robust system that is largely immune to interference. Its speed is, of course, an aid to interference immunity.

ARE THERE SELECTIVE ADVANTAGES IN FEMALE SONG?

The male-female communication system in the phaneropterines is very different from that found in the ephippigerines, but both are very different from that generally found in bushcrickets. Undoubtedly the two male-female systems evolved independently, but did similar selection pressures shape them?

Callicrania, *Platystolus* and *Steropleurus* are heavy-bodied, cumbersome insects that do not travel at great speed and are incapable of flight. They also generally live in low density populations, and the males are sporadic singers. All three have a male-female communication system. However, there are plenty of ephippigerines that are similar but do not have such a communication system, for example *Ephippiger*. *Ephippiger* is a reiterative singer often occurring in large aggregations. Indeed, Busnel (1967) reported the occurrence of a local song hierarchy in *E. cruciger*. So, the differences between *Ephippiger* and these three genera may well lie in the population size.

Singing advertises the position of the singer, not only to a predator capable of perceiving the sound but also to a visual predator. The movement of the elytra makes the normally well-camouflaged insect conspicuous. In a large aggregation of singing insects the risk to each individual is probably less than for a male singing on his own.

For a lone male, it would be advantageous to have a short song. Brief, widely spaced signals are not easy for a predator to locate. But if they are not easy for a predator to locate, think of the female who will have similar problems. She also has to make herself conspicuous by moving – walking towards a distant, intermittent beacon.

There appear to be advantages to both sexes in having a female reply. The male gets a positive reinforcement to his singing which encourages him to continue. The female, by singing, elicits song from the male which she can use for accurate location. In economic terms, both sexes are investing effort in singing and phonotaxis only when the probability of a successful mating is high. By contrast, in other bushcrickets the male may well invest heavily in singing when there is no female within range to hear him, and the female may invest time in approaching a male who stops singing before she reaches him, or she may find herself arriving near the male along with several other competing females. By establishing that mating is feasible before phonotaxis commences, *Callicrania*, *Platystolus* and *Steropleurus* reduce the risks from predators and decrease the energetic investment made in each successful mating.

While there are advantages in the female signalling to the male, there may also be disadvantages. The song could attract predators or parasites. Walker (1964) showed that a vertebrate predator – in this case a cat – could locate orthopteran prey using the calling song. Cade (1975, 1979) described the tracking of singing crickets by their song, by females of the parasitic fly *Euphasiopteryx ochracea*.

The stridulatory apparatus of males and females in these three genera are not homologous (See above), which means that stridulation evolved independently in each sex. The development of the acoustic interchanges must have involved parallel evolution. Similarly in the phaneropterines there must also have been parallel evolution of stridulatory apparatus and acoustic behaviour. However, evolution in the phaneropterines has produced an apparently more sophisticated system of communication.

The comments made above about predation and population density can be made also about *Leptophyes punctatissima*. The parallel selection pressure on both males and females of avoidance of predators would tend to produce short songs. As the songs have a very pure carrier frequency – at 40 kHz its well into the ultrasonic range – within the range of frequencies used by bats, there may be a further pressure to keep the song short. *Leptophyes* is very active at night (Figure 7), and there are four bats found in the same areas as *Leptophyes* in the UK that pick insects from vegetation (Blackmore, 1964). The calls of all of these overlap the frequency range of *Leptophyes*, so both predator and prey could detect each other. The possibility that shrews, with sound emissions in the frequency range of 30-60 kHz are also potential predators should not be overlooked.

There is good evidence that neotropical bushcrickets suffer predation by bats that pick insects off foliage (see Belwood chapter 2). Belwood and Morris (1987) baited mist nets with singing bushcrickets and caught bats. Control nets with silent bushcrickets did not trap nets. Analysis of prey remains showed that of the insect remains found in roosts of one species of bat 40.1% were remains of bushcrickets, though at leasts 50% of these came from females that are not known to sing. Belwood and Morris also examined the songs of the bushcrickets that lived sympatrically with the bats. The calls have sinusoidal frequencies in the high audio or ultrasonic range, are infrequent and of short duration. They found that bats took longer to locate infrequent singers, and did not locate silent insects. They suggest that the high number of females that are eaten is a result of normal echolocation of prey by the bats, but that the phonotaxis performed by the females means that they spend more time moving in the open and they are, of course, often close to a singing male. Evidence that the male song may be a cue used by predators to take females comes from a study by Sakaluk and Belwood (1984). They observed geckos locating singing male crickets. The male would be immune from predation in his burrow, but females attracted to his song were caught by the geckos in the open.

Predators using ultrasound are not only a threat to *Leptophyes* . They are also a source of acoustic interference. The cluttered environment which *Leptophyes* occupies is noisy, particularly in the ultrasonic range and, with ultrasound having short wavelengths, there are considerable echoes. As the calls of both sexes are brief, they are also more difficult to locate. Additionally, the information that can be imparted by a very short song is sparse. The disadvantages of these short high frequency signals appear to be offset by the very tight coupling between male call and female reply, and this coupling may also provide a means of species recognition.

The observations made on *Leptophyes* so far, suggest that the male has a sensory 'window', the only time during which a female reply will be effective. This window would provide a marked improvement on the signal-to-noise ratio of the female signal, and reduce confusion with other short signals in the environment. The female response is relatively unspecific, so there are only two parts of the acoustic interchange that could provide the vital species- specific isolation: (i) the frequency of the female song as perceived by the male, and (ii) the time window of the male.

The amplitude modulation pattern of the song is not a likely candidate since the songs are so short. The song of the female has a peak frequency close to 41 kHz (Figure 5). The song is a pure tone produced by resonance (R.O. Stephen, unpubl.). However, to act as a species isolating mechanism, the male would have to be able to distinguish this frequency from others, or receive it preferentially. The slit input to the tympanic organ has been shown to be tuned to the fundamental call frequency in certain bushcrickets (Bailey & Stephen, 1978; Stephen & Bailey, 1982) but there is no evidence yet of this in *Leptophyes.*

In *Poecilimon ornatus* a synthetic female signal at 94 dB was always effective at eliciting phonotaxis at frequencies of 10, 20, 30 and 40 kHz, suggesting that, in this species at least, frequency has no appreciable role to play in species recognition (Heller & von Helverson, 1986).

The time window of the male, if it varied from species to species, would provide an isolating mechanism. Such variation has been observed in a group of barbitistine species from the genera *Andreiniimon*, *Isophya*, and *Poecilimon* (Heller & von Helversen, 1986; see Heller chapter 8). The range of mean delays for the six species, which were all measured at the same temperature, was 16 ms to 430 ms. Clearly the time window varies substantially between species. A time window during which a response is expected has been observed in other animals. For example, the calls of tick-tock cicadas are answered by the female with a remarkably uniform time delay of 70-85 ms (Gwynne, 1987). However, the precise timing that is observed in *Leptophyes punctatissima* , and the sharply tuned window of the male are outstanding examples by comparison.

There are two problems with the phenomenon of time windows that remain unresolved at present:

(i) the neural basis for the time window is completely unknown.

(ii) the observed effect of temperature on the time window in *Leptophyes* (close to 1 ms increase per degree centigrade fall in temperature) is greater than would be expected, even when both the temperature effects on the sound itself and the known temperature coefficients for neurons are taken into account.

The selection pressures that have shaped the male-female communication system in the phaneropterines appear to have been predation and signal to noise ratio. Selection must have been effective on both sexes to match the time window of the male with the response delay of the female. There has to be a trade-off between time window and maximum communication distance, because the travel time of sound is a significant component of the overall delay perceived by the male. The time window is also influenced by the intensity of the call. A low intensity call allows a very narrow time window since the window only has to function for a small range of separation distances between male and female.

The matching must be the result of parallel evolution, and it may be this fact, more than any other that explains why such systems have not appeared throughout the Tettigoniidae.

References

Alexander R.D. 1960. Sound communication in Orthoptera and Cicadidae. In: Animal Sounds and Communication. Eds. W.E. Lanyon & W.N. Tavolga. American Institute of Biological Sciences. pp 38-92.

Ahlén I. & H.J. Degn. 1981. Lovvårtbitarens *Leptophyes punctatissima* sang. Fauna och Flora (Stockh.) 75: 265-268.

Bailey W.J. 1985. Acoustic cues for female choice in bushcrickets (Tettigoniidae). In: Acoustic and Vibrational Communication in Insects. Eds. K. Kalmring & N. Elsner. Paul Parey, Berlin, Hamburg. pp 101-110.

Bailey W.J. & R.O. Stephen. 1978. Directionality and auditory slit function: A theory of hearing in bushcrickets. Science 201: 633-634.

Blackmore M. 1964. Chiroptera. In: Handbook of British Mammals. Ed. H.N. Southern. Blackwell Scientific Publications, Oxford.

Bailey W.J. & P. Yeoh. 1988. Female phonotaxis and frequency discrimination in the bushcricket *Requena verticalis*. Ecol. Entomol. 13:363-372.

Belwood J.J. & G.K. Morris. 1987. Bat predation and its influence on calling behaviour in neotropical katydids. Science 238, 64-67.

Busnel M.C. 1967. Rivalité acoustique et hiérarchie chez l'Ephigger (Insecte: Orthoptère: Tettigoniidae). Z. vergl. Physiol. 54, 232-245.

Cade W.H. 1975. Acoustically orienting parasitoids: fly phonotaxis to cricket song. Science 190, 1312-1313.

Cade W.H. 1979. The evolution of alternative male reproductive strategies in field crickets. In: Sexual Selection and Reproductive Competition in Insects (Ed. Blum, M.S. and Blum, N.A.). New York: Academic Press. pp 343-379.

Crankshaw O.S. 1979. Female choice in relation to calling and courtship songs in *Acheta domesticus*. Anim. Behav. 27: 1274-1275.

Dumortier B. 1963. Ethological and physiological study of sound emissions in Arthropoda.

In: Acoustic Behaviour in Animals. Ed. R.-G. Busnel. Elsevier, London. pp 277-345.

Faber A. 1953. Laut-und Gebärdensprache bei Insekten: Orthoptera. I. Stuttgart.

Fulton B.B. 1933. Stridulating organs of female Tettigoniidae (Orthoptera). Entomol. News 44: 270-275.

Gwynne D.T. 1982. Mate selection by female katydids (Orthoptera: Tettigoniidae, *Conocephalus nigropleurum*). Anim. Behav. 30: 734-738.

Gwynne D.T. 1987. Sex-biased predation and the risky mate-locating behaviour of male tick-tock cicadas (Homoptera: Cicadidae). Anim. Behav. 35,571-576.

Gwynne D.T. 1983. Male nutritional investment and the evolution of sexual differences in Tettigoniidae and other Orthoptera. In: Orthopteran mating systems: Sexual competition in a diverse group of insects (Ed. Gwynne, D.T. and Morris, G.K.). Colorado: Westview Press. pp 337-366.

Hartley J.C., D.J. Robinson & A.C. Warne. 1974. Female response song in the Ephippigerines *Steropleurus stali* and *Platystolus obvius* (Orthoptera, Tettigoniidae). Anim. Behav. 22: 382-389.

Hartley J.C. & D.J. Robinson. 1976. Acoustic behaviour of both species of the speckled bush cricket *Leptophyes punctatissima*. Physiol. Entomol. 1: 21-25.

Heller K. G. & D. von Helversen. 1986. Acoustic communication in phaneropterid bushcrickets: species-specific delay of female stridulatory response and matching male sensory time window. Behav. Ecol. Sociobiol. 18: 189-198.

Nickle A.D. & T.C. Carlysle. 1975. Morphology and function of female sound- producing structures in ensiferan Orthoptera, with special emphasis on the Phaneropterinae. Int. J. Insect Morphol. Embryol. 4: 159-168.

Rheinlaender J., M. Hardt & D. Robinson. 1986. The directional sensitivity of a bush cricket ear: a behavioural and neurophysiological study of *Leptophyes punctatissima*. Physiol. Entomol. 11: 309-316.

Robinson D.J. 1980. Acoustic communication between the sexes of the bush cricket *Leptophyes punctatissima*. Physiol. Entomol. 5: 183-189.

Robinson D., J. Rheinlaender & J.C. Hartley. 1986. Temporal parameters of male- female sound communication in *Leptophyes punctatissima*. Physiol. Entomol. 11: 317-323.

Sakaluk S.k. & Belwood J.J. 1984. Gecko phonotaxis to cricket calling song: a case of satellite predation. Anim. Behav. 32,659-662.

Spooner J.D. 1968. Pair-forming acoustic systems of phaneropterine katydids (Orthoptera: Tettigoniidae). Anim. Behav. 15: 197-212.

Stephen R.O. & W.J. Bailey. 1982. Bioacoustics of the ear of the bushcricket *Hemisaga* (Saginae). J. Acoust. Soc. Am. 72: 13-25.

Walker T.J. 1964. Experimental demonstration of a cat locating orthopteran prey by the prey's calling song. Florida Entomologist 47, 163-165.

Yersin M. 1854. Memoire sur quelques faits relatifs a la stridulation des orthoptères et leur distribution geographique en Europe. Bull. Soc. Vaudoise Sci. Nat. 4: 108-128.

Zhantiev R.D. & N.N. Dubrovin. 1977. Sound communication in the genus *Isophya* (Orthoptera, Tettigoniidae). Zool. Zh. 56: 40-51.

Zimmerman U., Rheinlaender J. & Robinson D. 1989. Cues for male phonotaxis in the duetting bushcricket *Leptophyes punctatissima*. J. Comp. Physiol. A. 164:621-628.

8. EVOLUTION OF SONG PATTERN IN EAST MEDITERRANEAN PHANEROPTERINAE: Constraints by the Communication System

K-G. Heller

Introduction

Communication in bushcrickets is usually mediated by species' specific songs, and the diversity of song pattern within the group is the result of differing selective pressures (similarly discussed in Ryan (1985) for acoustically communicating amphibia). On the one hand are internal factors, the physiological constraints of the animal and on the other, the physical limitations of the habitat. For instance the production of the syllable rate by the forewings of the male is determined by its muscle physiology and can only be exceeded by special "tricks" such as raising the body's temperature (Josephson, 1984; Heller, 1986). Environmental constraints also set limitations on the song, and these affect the song's transmission. Here, not only are there the exaggerated effects of excess attenuation on high frequencies, but also interference from biotic and abiotic noise. These factors combine to disrupt the song's clarity (Michelsen, 1985; Römer *et al.* 1988, Schatral chapter 9, Rheinlaender & Römer chapter 13 this volume). Finally, the song's structure itself may be a result of a co-evolution process between its attractiveness to a mate and as a cue for predators and parasites (Belwood & Morris, 1987; Bailey chapter 12 this volume).

Producing a signal is not all, for the demands of the receiver are equally important at both the physiological and behavioural level. There are, however, few studies that adequately illustrate the influence of selective forces on the receiver, with their reciprocal effect on the sender. In fact most studies concentrate on the behaviour of the receptor system to differing signal types. In the present paper I intend to show how the interaction of both sender and receiver, in the rather specialised communication system of the Phaneropterinae, influences the evolution of the temporal structure of the male calling song.

The Nature of Duetting

Typically a sexually responsive female bushcricket silently approaches a calling male. However, in several bushcricket species the female also emits an acoustic signal in response to the male's call (see Robinson, chapter 7), and although this behaviour occurs sporadically amongst the subfamilies of the Tettigoniidae, it appears to be most common among the Phaneropterinae, a group representing about 2000 of 5000 known Tettigoniidae species (Kevan, 1982). Indeed the production of a "responding" signal by the female appears as one of the characteristics of this subfamily (for a review see Heller & Helversen, 1986; see also Robinson *et al.*, 1986; Zhantiev & Korsunovskaya, 1986; Zimmermann *et al.* 1989).

It is obvious that the stridulatory organs of the female have evolved quite independently to those of the male (see Pfau & Schroeter, 1988), and indeed quite differently from similar organs in other subfamilies. Hence, this behaviour may support the hypothesis that the Phaneropterinae represents a sister group to all other Tettigoniidae.

When the two sexes are involved in a "duet" the female responds with a very short signal, which in most cases is acoustically distinct from the male song (Fig. 1 A). This compares to the situation in the Ephippigerinae where the female in some species can respond with a song structure quite similar to that of the male (Hartley *et al.*, 1974; Fig. 1 B). The meaning of the different female signals, and also of the male's response (Pfau & Schroeter, 1988), is unclear. Phaneropterine females emit their short and simple structured signal within a brief and limited time period immediately after the male has produced its song. This response time maybe as short as 10-20 ms (Heller & Helversen, 1986). If her signal falls within this prescribed time window, the male then responds by walking to the female. In order to compensate for the time the sound waves require to propagate from the caller, the time window of the male is slightly wider than that of the female's response (Robinson *et al.*, 1986; Heller & Helversen, 1986).

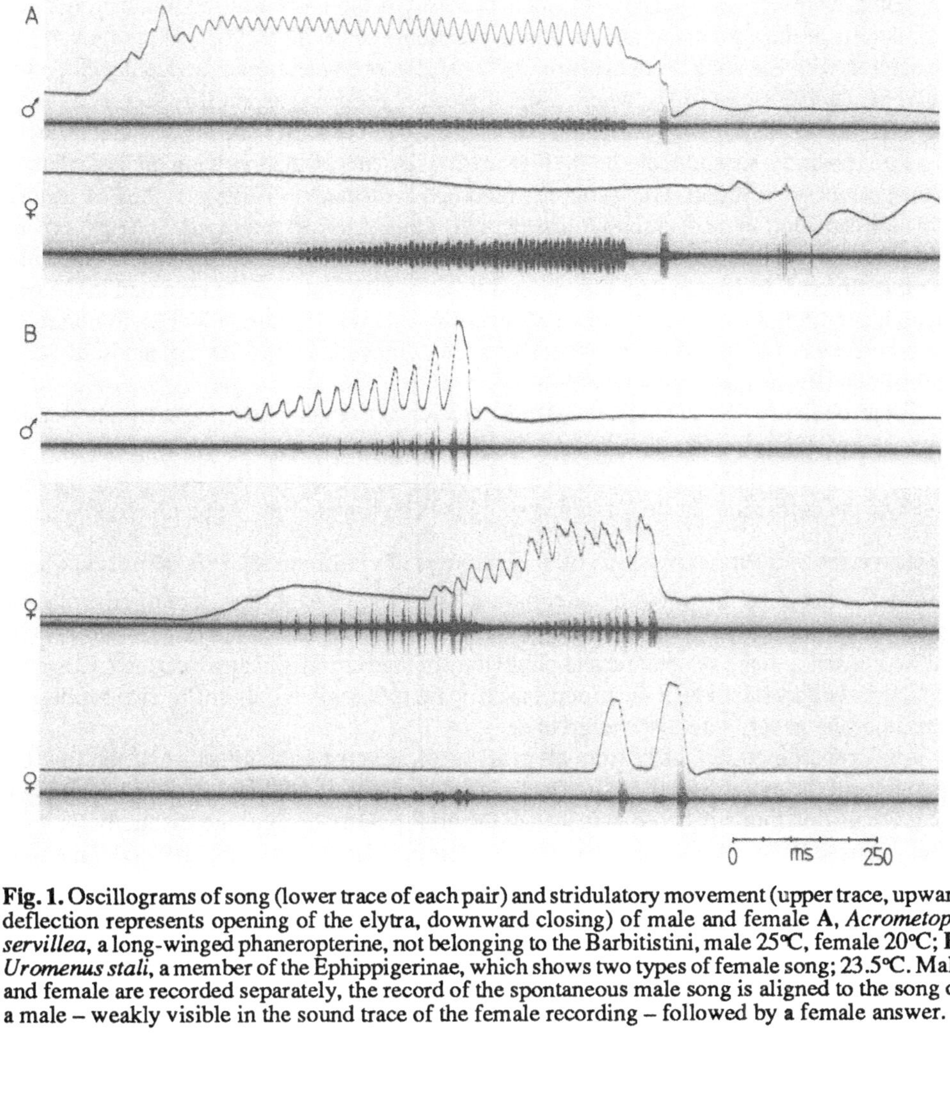

Fig. 1. Oscillograms of song (lower trace of each pair) and stridulatory movement (upper trace, upward deflection represents opening of the elytra, downward closing) of male and female **A**, *Acrometopa servillea*, a long-winged phaneropterine, not belonging to the Barbitistini, male 25°C, female 20°C; **B**, *Uromenus stali*, a member of the Ephippigerinae, which shows two types of female song; 23.5°C. Male and female are recorded separately, the record of the spontaneous male song is aligned to the song of a male – weakly visible in the sound trace of the female recording – followed by a female answer.

This form of antiphony between male and female requires a well structured song pattern in the male. The song must show one (or several) character(s) that can be easily identified by the female, and so use these as a trigger for her response. Furthermore, an interval should immediately follow the signal in order for the male to hear the response of the female. The importance of these two requirements on the song's structure becomes clear when the song, and behaviour of a number of different species are compared. Such a comparative approach becomes even more powerful in a monophyletic group, since the evolutionary trends and the process of evolution may be reconstructed from a theoretical archetype.

The species of the tribe Barbitistini provide such an opportunity. The Barbitistini, comprised of some 200 species, occur in the eastern part of the Mediterranean region. By using the song patterns of 90 species of this monophyletic group, it can be shown how (1), the requirements of the communication system are fulfilled, and (2) the song of the male has undergone changes in those cases where the female has lost her acoustic response behaviour.

Methods

The songs were recorded and the movement of the elytra monitored during sound production. (For details of the equipment and recording procedure see Heller, 1988.) Wing movements associated with the song were analyzed by using the opto-electronic device developed by Helversen and Elsner (1977), where a small piece of reflective tape (3M, Scotchlite 7610) was attached to the forewing and illuminated by a system of lenses and mirrors that reflected to a light position-sensitive diode. In this way movements of the forewing elicited changes in the current of the diode, which could be recorded as changes in voltage on the FM-channel of a tape recorder. In most cases a Racal store 4DS tape recorder was used. Song pattern and wing movements produced by the two species of the genus *Phonochorion* and of *Poecilimon zonatus* Bolivar, *P. kutahiensis* Werner and *P. serratus* Karabag were recorded in the field on an Uher 4200 IC using a separate FM recording adaptor (Vetter, model 2). For analysis, the recordings of song and wing movement were displayed on the screen of an oscilloscope and filmed (Recordine Tonnies K-831).

The stridulatory files of the males of most species were examined using scanning electron microscopy. Descriptions of the song, and illustrations of the stridulatory files of those European species that are not depicted in this paper can be found elsewhere (Heller, 1988). A list of the collecting sites used in this study is to be found in the Appendix to this paper.

Systematics and Phylogeny of the Barbitistini (Tettigoniidae, Phaneropterinae)

In order to describe the radiation of certain characteristics over time, it is of advantage to know the group's phylogeny. This should be based on as many characters as possible. In the following account the systematic and phylogeny of the Barbitistini are described. However, our current understanding of the group, based on morphology, is only sufficient to subdivide certain of the genera into discrete groups.

The Phaneropterinae differ from all other bushcrickets by the structure of the first two segments of the tarsus, and these are not grooved laterally. In addition, in these species with long wings, the hindwings are longer than the elytra. Also the eggs are distinctly flattened (Bei-Bienko, 1954). While the last two characters are clearly independent developments of this subfamily (autapomorphic features), the morphology of the tarsal segments is still rather primitive. Since the tarsi are laterally grooved in all other tettigoniids, this provides additional evidence to suggest that the Phaneropterinae are probably a sister group of all other subfamilies (grooving might extend the basal attachment surfaces, possibly as an adaption to a higher mobility of the predatory species).

This huge subfamily has received scant treatment since it was reviewed in Brunner von Wattenwyl's work (1878, 1891), and subdivides into several groups that have been regarded as subfamilies or tribes by later authors, but this is only a partial reflection of the evolutionary developments within the subfamily. Brunner von Wattenwyl attributed all micropterous species of the Phaneropterinae to the Barbitistini, and called the whole group the Odonturae. The reduction of the wings obviously occurred several times independently, however, the genus *Odontura* appears to be closely related to the long-winged genus *Ducetia* (Ragge, 1980). Thus, the possession of long or short wings is not a sufficient criterion on which to define the Barbitistini as a monophyletic group.

As a second important characteristic I have considered is the sabre-like ovipositor, which according to Bei-Bienko (1954), in one of the two groups within the Barbitistini bears a short, finely toothed ovipositor, while the other shows a long, roughly toothed ovipositor. The short ovipositor is similar in structure to that of many long-winged Phaneropterinae. Therefore, it is not clear at present if these forms evolved independently from the long-winged species, or have a common ancestor with those species that have a long ovipositor. All genera with a short ovipositor have relatively long legs (*Andreiniimon*, *Orthocercodes*), and in this way are similar to the long-winged species. This observation provides support for the hypothesis that they may have separated at a very early stage from the other genera, or have become micropterous quite independently.

Since the long, roughly toothed ovipositor is uncommon among the Phaneropterinae, it is quite probable that it has been newly developed by the ancestor of the respective genera as synapomorphic feature. It presumably developed as an adaptation to oviposition in the ground, which is typical of all species of this group, but is not common among the Phaneropterinae. The genera that have a long, roughly toothed ovipositor are undoubtedly a monophyletic group, and in the following text will be called the Barbitistini.

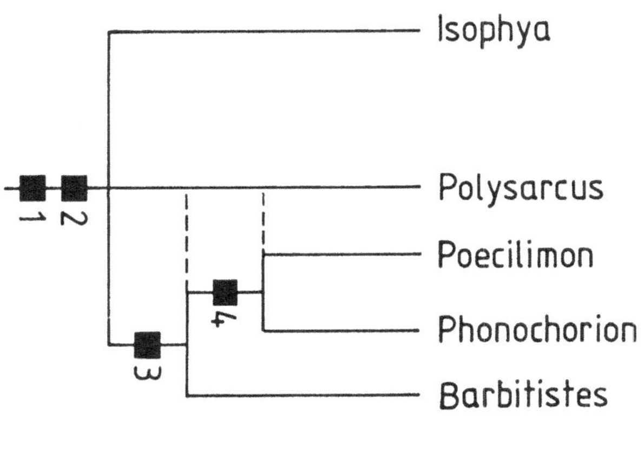

Fig. 2. Diagram of the phylogenetic relationships of the Barbitistini (considered only genera with more than one species, other genera see text). Apomorphic characteristics: 1 micropterous, 2 ovipositor long, roughly toothed, 3 ventral edge of ovipositor straight, 4 pronotum extends over the base of the forewings (for *Polysarcus* see text).

THE GENERA OF THE BARBITISTINI

Within the Barbitistini are several genera that form different groups based on their morphological characteristics (Fig. 2). The exact relationships between the genera within each group are not known. The groups are listed and in brackets are the number of known species for each genus. This number is followed by the number of species of which the acoustic behaviour has been investigated.

Group 1: Genus *Isophya* (75/27)
2: *Polysarcus* (4/2)
3: *Barbitistes* (7/6), *Metaplastes* (4/3), *Ancistrura* (1/1)
4: *Poecilimon* (105/44), *Parapoecilimon* (1/1), *Poecilimonella* (1/0), *Micrimon* (1/0), *Artvinia* (1/1), *Isoimon* (1/0)
5: *Phonochorion* (3/2)

1. Genus *Isophya*: this genus is characterized by plesiomorphic features. The ovipositor is ventrally curved, and the pronotum does not overlap the forewings in either sex. The genus occurs from the Pyrenees to the Altai mountains in Mongolia. This wide distribution (larger than in all other genera, apart from the parthenogenetic species *Poecilimon intermedius* (Fieber)) may indicate the antiquity of the genus.

2. Genus *Polysarcus*: The genus shows the following apomorphic characters. The insects have short forelegs, the pronotum (strangely shaped, and therefore considered as convergence to *Poecilimon*) partly overlaps the forewings, and the fastigium is extremely wide. However, the shape of the pronotum might also indicate that the genus *Polysarcus* is related to the genus *Poecilimon*. In addition, the ovipositor is more similar to the ovipositor of those genera described below than to that of *Isophya*. The animals occur from northern Spain throughout Europe and Asia Minor to northern Iran. They occupy a much larger area than any of the following genera. Hence, the genus may have a long history of development of its own.

The genera of the other three groups have one common and presumably derived (apomorphic) characteristic; the ventral edge of the ovipositor is straight in all species and hence very different to the ancestral species. A comparison should be made between *Phaneroptera* and other long-winged species. In addition the transverse groove of the pronotum is in front of the mid-line, perhaps indicating a trend to protect the attachment of the forewing by the pronotum (see *Poecilimon*).

3. Genera *Barbitistes*, *Metaplastes*, *Ancistrura*: The three genera do not have any common features apart from those mentioned above, however, the males have complex cerci and are of a different shape (*Barbitistes*, *Metaplastes*), and in addition, the form of their subgenital plates differs (*Metaplastes*, *Ancistrura*). The geographic range of these genera is from northern Spain to the north west of the USSR (but not including Asia Minor and the Caucasus area).

4. Genera *Poecilimon* etc.: In the genera of this group the apomorphic characteristic is the extension of the base of the forewings. The geographic distribution is from the northern Balkans and Italy to Asia Minor to the south of the USSR and the Caucasus. Only the parthenogenetic species *Poecilimon intermedius* is distributed over a much broader area.

5. Genus *Phonochorion*: Because of the extremely specialised shape of the pronotum and the subgenital plate of the male, this genus forms a unique group. Nevertheless, similar developmental trends occur in several species of the genus *Poecilimon*; *Phonochorion* may have *Poecilimon*-like ancestors. The three species are confined to an area in the north east of Turkey and the neighbouring USSR.

The Ancestral Song Pattern of the Barbitistini

To get some idea of the evolution of the song pattern, a knowledge of the primitive song structure of this group is of primary importance. For several reasons I assume that the ancestor emitted a song consisting of a sequence of single, and relatively drawn-out syllables that were loosely arranged and gradually increased in amplitude (crescending syllables). The syllables were separated by distinct intervals (see also Heller, 1984). During sound production the forewings were closed slowly at first but then increasingly faster. In favour of the existence of this ancestral type of song pattern are the observations that, (a) syllables of the described structure occur in four of the five groups mentioned above, and (b) these syllables are produced by those species of the genera *Poecilimon* and *Phonochorion*. These have the simplest calling patterns of all, and their stridulatory files are without the specialisations found in other species of the genus *Poecilimon*. In the genus *Isophya*, syllables with this presumably primitive structure can be located in both simple and complex song patterns. Finally (c), the structure of all complex songs can be traced back to this primitive song pattern.

Description of the Song Pattern

1. Group: Genus *Isophya*

The species of the genus *Isophya* differ only slightly in their morphology, and therefore they cannot be easily identified. It is not clear at the moment just how the species are related. Various song types are produced, and the difference between the types depends on amplitude modulation and complexity (see below). However, at present I am unable to decide as to whether similar song patterns occur in phylogenetically related groups, or if they developed under convergence.

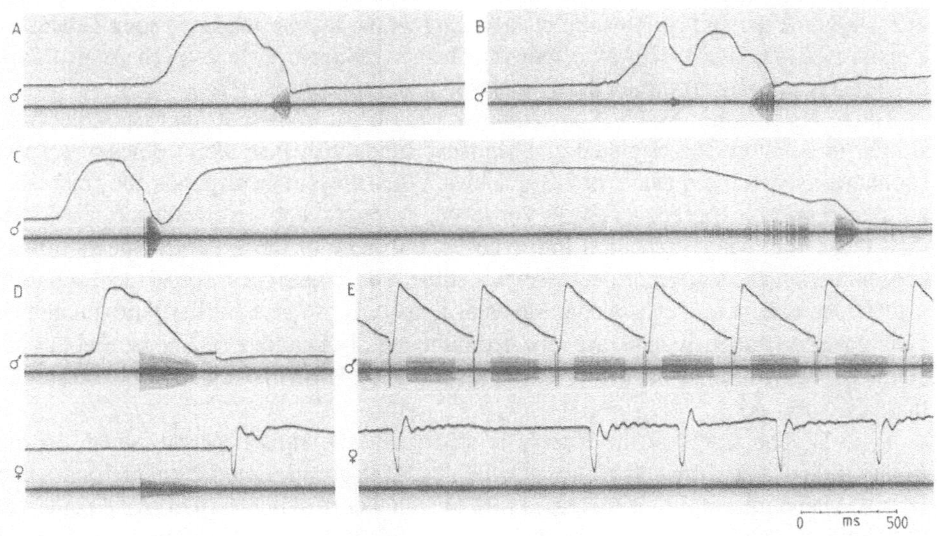

Fig. 3. Oscillograms of song (lower trace of each pair) and stridulatory movement (upper trace; elytra opening upward, closing downward) of **A, B** *Isophya savignyi* (male, two different types of syllables; 27.5°C), **C** *I. speciosa* (male; 26°C), **D** *I. modestior* (male and female; 26°C), **E** *I. kraussii* (male and female; 23.5°C/22°C; all recordings as in fig. 1).

Type A. The song consists of single syllables with increasing amplitude (crescending syllables), which is very similar to the ancestral song pattern of the Barbitistini. Single pulses may occur that are distinctly separated from the previous dense pulse sequence: *I. stysi* Cechan (with certain reservations), I. spec. (both Zhantiev & Dubrovin, 1977; Zhantiev & Korsunovskaya, 1986), *I. savignyi* Brunner (Fig. 3 A, see also Type D). The response behaviour of the females is unknown. Presumably they use the end of the syllable as a marker, in the same manner as that observed for *I. stepposa* Bey-Bienko (see below).

Type B. Songs of this type consist of single syllables that show a distinct beginning and gradually decrease in their amplitude (decrescending syllables): *I. taurica* Brunner (Zhantiev & Dubrovin, 1977; Zhantiev & Korsunoskaya, 1986), *I. pyrenaea* (Serville), *I. leonorae* Kaltenbach, *I. tosevski* Pavicevic, *I. modestior* Brunner (Fig. 3 D), *I. thoracica* Karabag, *I. zernovi* Miram, *I. ilkazi* Ramme (with a tendency to produce several syllables similar to type C; data on the two last species unpublished).

In many species (*taurica, leonorae, tosevski, modestior*) the response behaviour of the females has been studied in some detail (Zhantiev & Korsunoskaya, 1986; Heller & Helversen, 1986; Fig. 3 D; unpublished). In comparison to other species, all females in this group respond relatively late after the male's song (compared to *Poecilimon* species, Heller & Helversen, 1986). They presumably use the beginning of the song as a trigger for their own response. A long latency helps to avoid an overlap between their signal with the end of the male's song. It is remarkable that the males produce one or several pulses just when the response of the female is to be expected. For *I. thoracica* and *I. ilkazi* this pulse is emitted much earlier, after the beginning of the song (syllable length shorter), and therefore a shorter response latency of the female is likely.

Type C. The song consists of a fast sequence of rather long (*I. stepposa* (Zhantiev & Korsunovskaya, 1986), *I. kraussii* Brunner) or short (*I. nervosa* Ramme unpubl.) single syllables, which often end with distinctly isolated pulses.

The female responds when these pulses are produced (*I. stepposa, I. kraussii* Fig. 3 E), as in *I. stepposa* the end of the song clearly triggers the female response song (Zhantiev & Korsunovskaya, 1986). This song type may be derived from Type A (perhaps also Type B in some species) by arranging the syllables in a fast sequence.

The songs of *Isophya* spec. A and *Isophya* spec. B, also consist of single syllables that are similar in structure and arranged in a sequence (unpublished results). Males of *I.* spec. A spontaneously emit sequences of 2-3 syllables. When the female responds, they increase the number of syllables per sequence to six or seven. The first syllable is always softer and shorter than those following (transition into type E), but none of the syllables shows a distinct beginning, and the amplitude decreases with time. The female can respond to all syllables of a sequence except to the first one. Males of *I.* spec. B produce similarly modulated short syllables spontaneously, and are repeated quite rapidly in a long sequence of about 20-30 elements. The response behaviour of the female is not known, but is likely to be similar to that of *I.* spec. A.

Type D. The songs of the closely related species *I. staneki* Maran and *I. hospodar* (Saussure) consist of several groups of syllables. Each group comprises numerous syllables (Fig. 4 A) and the amplitude increases with the number of syllables. This song pattern does not show any obvious trigger for the response of the female. The female (*I. hospodar*) apparently responds to the first, very soft sound of the syllable group (see Discussion for the problems arising from this situation).

Type E. The song consists of several different types of syllables: *I. rossica* Bey-Bienko

(Zhantiev & Dubrovin, 1977; Zhantiev & Korsunovskaya, 1986), *I. amplipennis* Brunner, *I. speciosa* (Privaidsky) (Fig. 3 C). In the songs of these three species the different syllables are emitted with an interval of several seconds. The female of *I. rossica* responds briefly after the beginning of the second syllable (Zhantiev & Korsunovskaya, 1986). In this second syllable the pulses are emitted with a distinctly larger interval than in the first syllable. These songs can be traced back to a one-syllable pattern (type B), on the assumption that the isolated pulse(s) produced at the time when the female responds (at the end of the syllable) is (are) extended into a syllable. When the time interval is large between the song of the male and the response of the female, indicative of many species of the type B song, there is sufficient time for the male to open its wings again and to produce a second syllable. Presumably in a derived state, the beginning of this sound triggers the response of the female.

Different types of syllables are also emitted by *I. brunneri* Retowski, *I. gracilis* Miram (Zhantiev & Dubrovin, 1977; Zhantiev & Korsunovskaya, 1986), *I. straubei paucidens* Heller, *I. reticulata* Ramme (unpublished), *I. major* Brunner (Fig. 4 B) and occasionally *I. savignyi* (Fig. 3 B). The intervals are short between syllables. The response behaviour of the females is known only for *I. major*, where the female apparently responds to the first isolated pulse of the second syllable type (between one and two syllables of type I are followed by several syllables of type II and the female is able to respond to them all). This response may be similar to that in *I. straubei* and *I. brunneri*. For *I. savignyi* and *I. straubei* the first syllable has developed as an incomplete and autonomous beginning of a syllable (in a similar way as for *I. major*). In *I. reticulata* a single, or two short syllables are followed by a loud syllable which increases slowly in amplitude (crescending). It is likely that the female responds to this syllable.

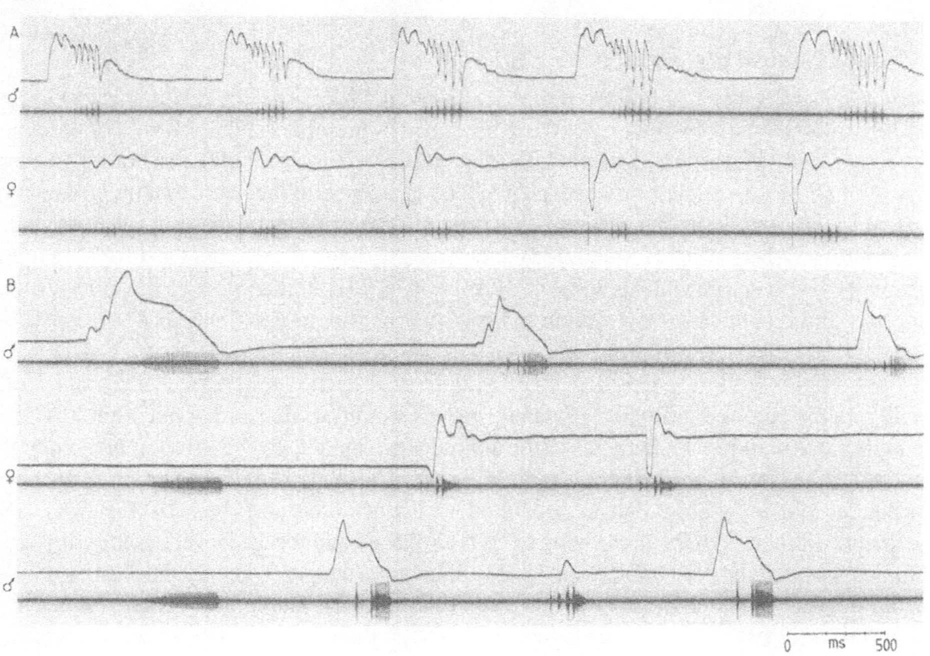

Fig. 4. Oscillograms of song (lower trace of each pair) and stridulatory movement (upper trace; elytra opening upward, closing downward) of **A** *I. hospodar* (male and female: 25°C) **B** *I. major* (male, female, and male responding to another male; 25°C; all recordings as in fig. 1).

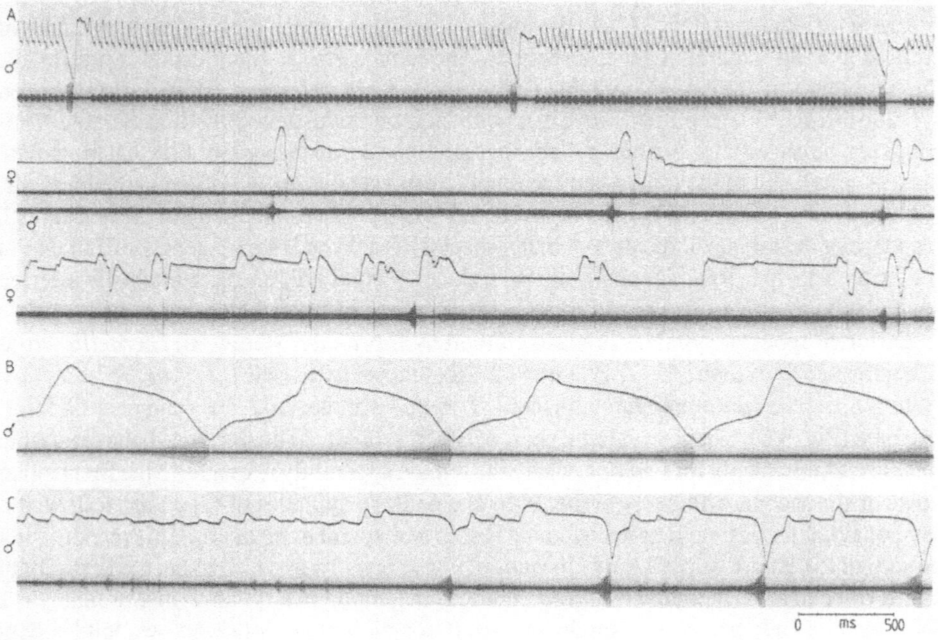

Fig. 5. Oscillograms of song (lower trace of each pair) and stridulatory movement (upper trace; elytra opening upward, closing downward) of **A** *Polysarcus denticaudus* (male and female, see text; 25°C/30°C; recording as in fig. 1), **B** *Phonochorion uvarovi* (male; 19.5°C), **C** *Phonochorion satunini* (male; 27°C).

2. Group: Genus *Polysarcus*

The calling songs produced by the males of the genus *Polysarcus* are distinct from the calls of almost all other Barbitistini except those produced by members of the genus *Phonochorion*. The songs are continuous and emitted for long periods of time (minutes and hours) on sunny days. Not all of the species, however, have been examined to the same extent, however, *P. denticaudus* (Charpentier) has been studied in detail (descriptions of the song and behavioural observations Faber, 1953; Heller, 1988), whereas only song recordings exist for *P. scutatus* (Brunner) (Heller, 1988). The song of *P. zacharovi* (Shchelkanovtsev) has not yet been analyzed, but it sounds to my own ear to be similar in structure to the calls of the other two species.

The complex calling song of *P. denticaudus* is described by Faber (1953) and Heller (1988). At the beginning (Phase I) the male moves around whilst calling but after 20 s-2 min the male stops walking – mostly when the animal is on top of the vegetation. It then increases its wingstroke rate and the syllable intervals become shorter (Phase II). After several seconds it produces louder syllables that are separated by distinct intervals (Phase III; Fig. 5 A). After a sequence of these syllables, the song turns into phase I and the animal starts moving again. Females that are ready to mate use the loud syllables as a trigger, and respond within the short silent interval after each syllable (Fig. 5 A; female above). The female remains motionless when calling. When a male hears the response of the female he shortens phase I considerably, and more often emits phase III or else he extends phase III by producing a longer series of loud syllables. The male can be observed orienting towards the female after having emitted a loud syllable. Some females, perhaps those that are highly motivated, start responding during the fast calling period (Fig. 5 A, female below; A behaviour observed not only in the laboratory but also in the field). However, the singing male obviously cannot use this response for his orientation.

The behaviour of *P. scutatus* is unknown. The song structure resembles that of *P. denticaudus*, and indicates that the female responds in the intervals after the longer and louder "marking" syllables. The significance of the separately occurring "buzzing" call is not known.

3. Group: Genera *Barbitistes, Metaplastes, Ancistrura*

Each of the three genera represents a monophyletical group, however, there is insufficient evidence to show how the groups and the species within each group are related. In contrast to *Polysarcus* these species inhabit the upper vegetation layer of bushes and trees. The song consists of homogeneous short syllables that are emitted in a specific temporal sequence by those species in which complex song patterns are typical. The song patterns fall into two groups:

Type A. The call consists of several syllable-groups separated by intervals that are larger than the syllable intervals themselves. Within each group the syllable interval is not variable as shown by *Barbitistes serricauda* (Fabricius) and *Metaplastes ornatus* (Ramme) (Fig. 6 A).

The female of *Metaplastes* can respond to any syllable of each group (Fig. 6 A), and this is probably true for *B. serricauda* and may also be the case for *B. fischeri* (Yersin). In the latter species the males emit long sequences of syllables with variable amplitude.

Type B. The song consists of several syllable-groups. The interval between the last and the previous syllable of each group is much longer than for the normal syllable interval. Therefore, this last syllable may be close to the first of the next group, e.g. *B. obtusus* Targioni-Tozetti, *B. yersini* Brunner, *B. ocskayi* (Chapentier) (Fig. 6 B), *M. pulchripennis* (Costa), *M. oertzeni* (Brunner)(Fig. 6 C), *A. nigrovittata* (Brunner).

In those species in which the response behaviour of the females has been studied (*A. nigrovittata* Heller & Helversen, 1986; *B. ocskayi* (Fig. 6 B); *B. yersini*, unpublished) the females respond immediately after the last syllable of each group.

4. Group: Genus *Poecilimon* (and related genera)

This group contains the most species and is also the most variable in its acoustic behaviour. The song patterns range from the simple ancestral type to those of great complexity showing different syllable types. The most striking observation, however, is that in some species the females have lost their responsive behaviour. Such losses may have occurred at least twice and quite independently, first in *P. pergamicus* Brunner, *P. kutahiensis* Werner, and second in all other species. The forewings of the female are reduced to tiny non-overlapping scales, or are missing altogether. This change in the communicating ability of the female has some distinct effects on the calling pattern of the male (see below).

Heller (1984) investigated the relationship of many European species of the genus *Poecilimon*, but many of the song patterns shown below are produced by species from Asia minor, where nothing is known about their phylogeny.

The songs of the more than 40 species investigated can be subdivided into eight types. There are, however, several transitional forms that may provide important information as to just how the song patterns evolved.

Type A. The calling song consists of single syllables that can be rather long. The amplitude increases towards the end of the syllable (crescending), but the magnitude of this increase varies between species. The increase is lowest in a group of closely related species in which the females are not able to respond to the male song (*P. orbelicus* Pancic, *P. similis* Retowski (Fig. 7 A), *P. bosphoricus* Brunner, *P. scythicus* Shchelkanovtsev, *P. boldyrevi* Miram, *P. beibienkoi* Tarbinski (data of the last three species in Zhantiev & Korsunovskaya, 1986)). The ending of the syllable is much more pronounced in those species that belong to the *P. affinis* (Frivaldski) group (*P. affinis* (Fig. 7 D), *P. pindos* Willemse, *P. hoelzeli* (Harz), *P. obesus* (Brunner), *P. artedentatus* (Heller), *P. nobilis* Brunner, *P. brunneri* (Frivaldski) and relatives (*P. macedonicus* Ramme, *P. elegans* Brunner), and *P. gracilis* (Fieber). In some species (*P. nobilis*, *P. gracilis* etc.) the first and longer part of the syllable consists of several isolated

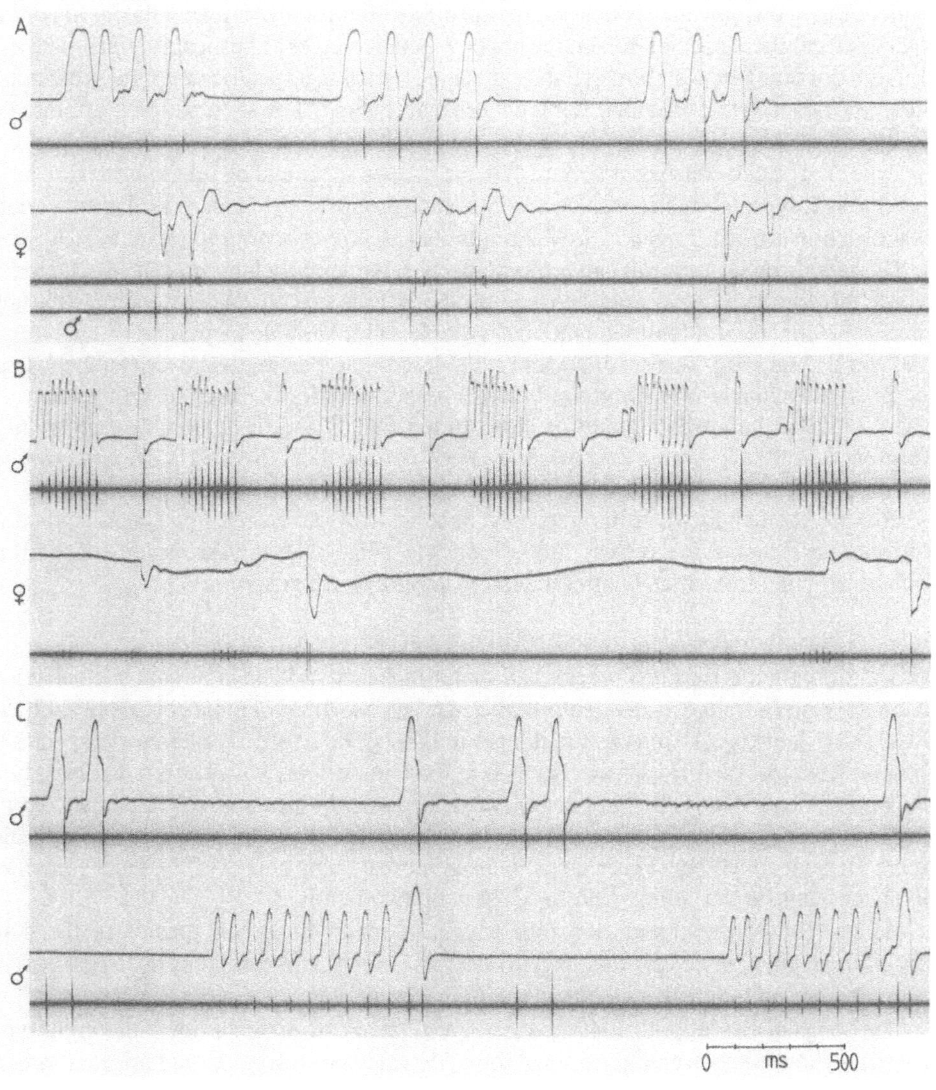

Fig. 6. Oscillograms of song (lower trace of each pair) and stridulatory movement (upper trace; elytra opening upward, closing downward) of **A** *Metaplastes ornatus* (male and female; 29°C), **B** *Barbitistes ocskayi* (male and female; 23.5°C), **C** *Metaplastes oerzteni* (two males, the second (lower) male produces a rival song; 27°C/28°C; all recordings as in fig. 1).

pulses, and only the ending is clearly audible as a result of its dense pulse sequence (tendency to type B). The ending of the syllable is clearly marked in *P. sanctipauli* Brunner (Fig. 7 B) in a similar way to that in *Artvinia tschorochensis* (Adelung). In both species the forewing movement stops before the call's loud ending, so that the first part of the syllable is separated by an interval. In *Artvinia*, two syllables of this type are arranged as a sequence, and the first of the two syllables shows a very soft sequence of pulses as a first part.

The female responds immediately after the syllable of the male (*P. sanctipauli*, Heller &

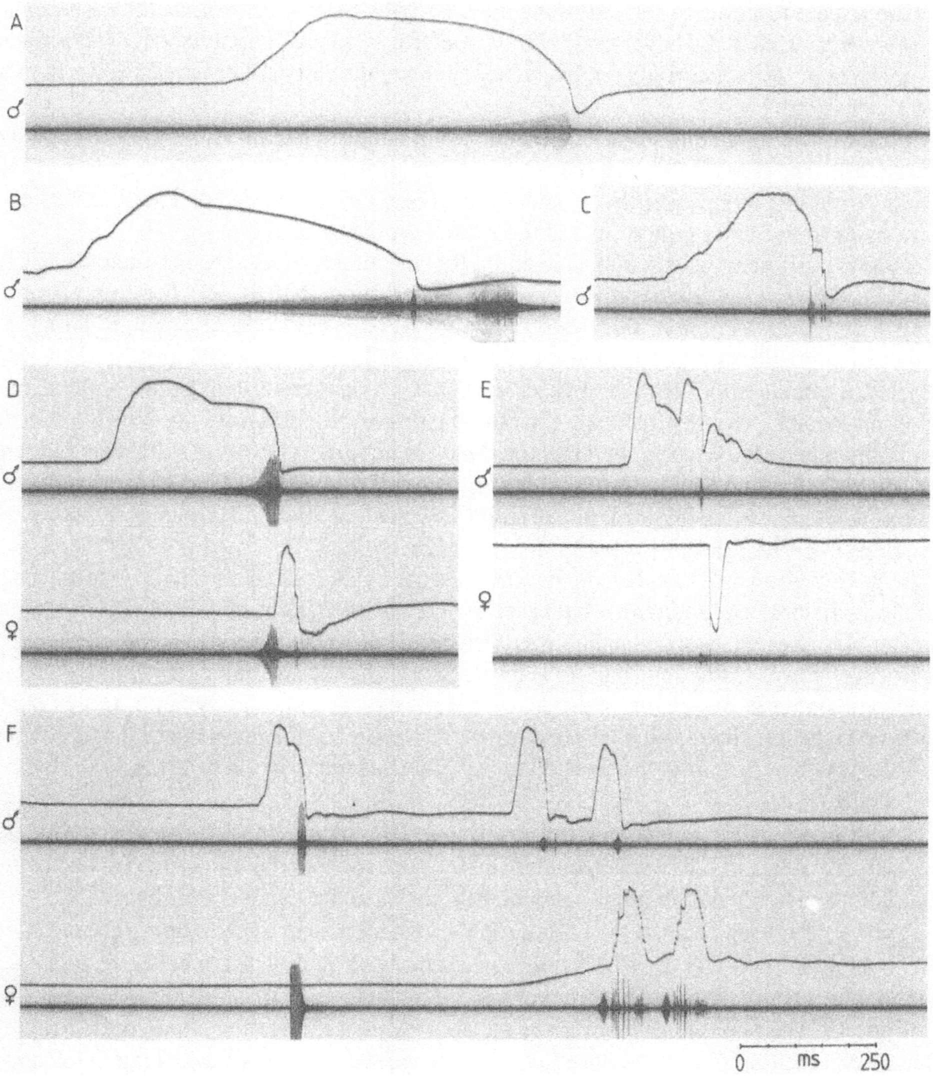

Fig. 7. Oscillograms of song (lower trace of each pair) and stridulatory movement (upper trace; elytra opening upward, closing downward) of **A** *Poecilimon similis* (male; 27°C), **B** *Poecilimon sanctipauli* (male; the end of the syllable is masked by the song of a male of *P. obsesus*; 23°C), **C** *Poecilimon zonatus* (male; 19°C), **D** *Poecilimon affinis* (male and female; 26.5°C), **E** *Poecilimon hamatus* male and female; 24°C), **F** *Parapoecilimon antalyaensis* (male and female; 22°C; all recordings as in fig. 1).

Helversen, 1986; *P. affinis*, Fig. 7 D; *P hoelzeli*, unpublished).

Type B. The song consists of very short syllables, and the syllable interval may last for several seconds: *P. ornatus* (Schmidt), *P. jonicus* (Fieber) and relatives, *P. zwicki* Ramme, *P. laevissimus* (Fischer), *P. thoracicus* (Fieber), *P. schmidtii* (Fieber) (Fig. 8 A), *P. zonatus* Bolivar (Fig. 7 C). This song has obviously developed from type A, and such a hypothesis is supported by the structure of the stridulatory file. In many of the species mentioned above, a finely structured part of the file exists which is not used during stridulation. This part is much longer in the species of type A, and is in this case responsible for the first part of the syllable. In *P. schmidtii* the teeth over part of the file are completely reduced, yet the normal stridulatory movement is still maintained (Fig. 8 A, detailed description see Heller 1984).

The female responds shortly after the male's single syllable (*P. ornatus*, *P. laevissimus*, *P. thoracicus* Heller & Helversen, 1986; *P. jonicus*, *P. zwicki*, unpublished). The acoustic response of the female is triggered by the beginning of the call in *P. ornatus* (Heller, in prep; see Discussion)

Type C. The song consists of a fast sequence of single syllables: *P. pergamicus*. Like *P. kutahiensis* (see type G) this species is related to *P. jonicus* and the response abilities of the female were lost independently of other species (see type A, F, G, H). The song has been developed from that in type B, in that the elements are repeated quickly.

Type D. Different types of single syllables (or very short syllable groups) characterize the song: *P. erimanthos* Willemse & Heller, *P. cretensis* Werner (Fig. 8 C, D), *P. inflatus* Brunner (unpubl.), *Parapoecilimon antalyaensis* Karabag(Fig 7 F). Loud and simple structured syllables alternate with fast sequences of complex syllables or syllable groups in the three species of *Poecilimon*. Weak tendencies to such a song pattern also exist in *P. jonicus*. For all three species it is not yet known what these types of syllables mean, or how the females communicate with the males. In *Parapoecilimon antalyaensis* one loud syllable is followed by between two and four softer though irregular syllables (Fig. 7 F). The song pattern is therefore similar to the song structure of *Isophya major* (see above). The latency of the female response to each of the soft syllables is very short.

Type E. A sequence of syllables that show a steadily increasing amplitude is followed by one loud syllable emitted with a larger interval: *P. bilgeri* Karabag (Fig. 8 B). The female presumable uses this isolated syllable as a trigger for her response (see *Barbitistes* type B). A first step towards this song pattern is the song of *P. hamatus* Brunner (Fig. 7 E), in which only one single soft syllable is produced before the second loud syllable. Occasionally isolated and short single syllables are emitted (movement of the forewing as in type B). The female usually responds to the second loud syllable and rarely to the soft (Fig. 7 E; Heller & Helversen, 1986).

The following three song types are only produced by species in which the female does not acoustically respond. They are characterized by a fast repetition of the syllables, and neither the time pattern nor the amplitude modulation shows any distinct markings.

Type F. The song consists of single syllables, where neither the beginning nor the end of the syllable is pronounced: *P. thessalicus* Brunner and the related species, *P. anatolicus* Ramme (occasionally with slightly pronounced syllable ending).

Type G. The song consists of short syllable sequences that are often emitted rapidly. Because of their amplitude modulation, many of the song patterns that belong to this group can be derived from type A or F. For instance, song structure in *P. zimmeri* Ramme is clearly derived from type F. Certain populations still emit a song consisting of single syllables. *P. mariannae* Williamse & Heller also belongs to this group. The song patterns of *P. ledereri* Ramme (Fig. 8 G) and *P. tuncayi* Karabag (unpublished) can be derived from type A. In both

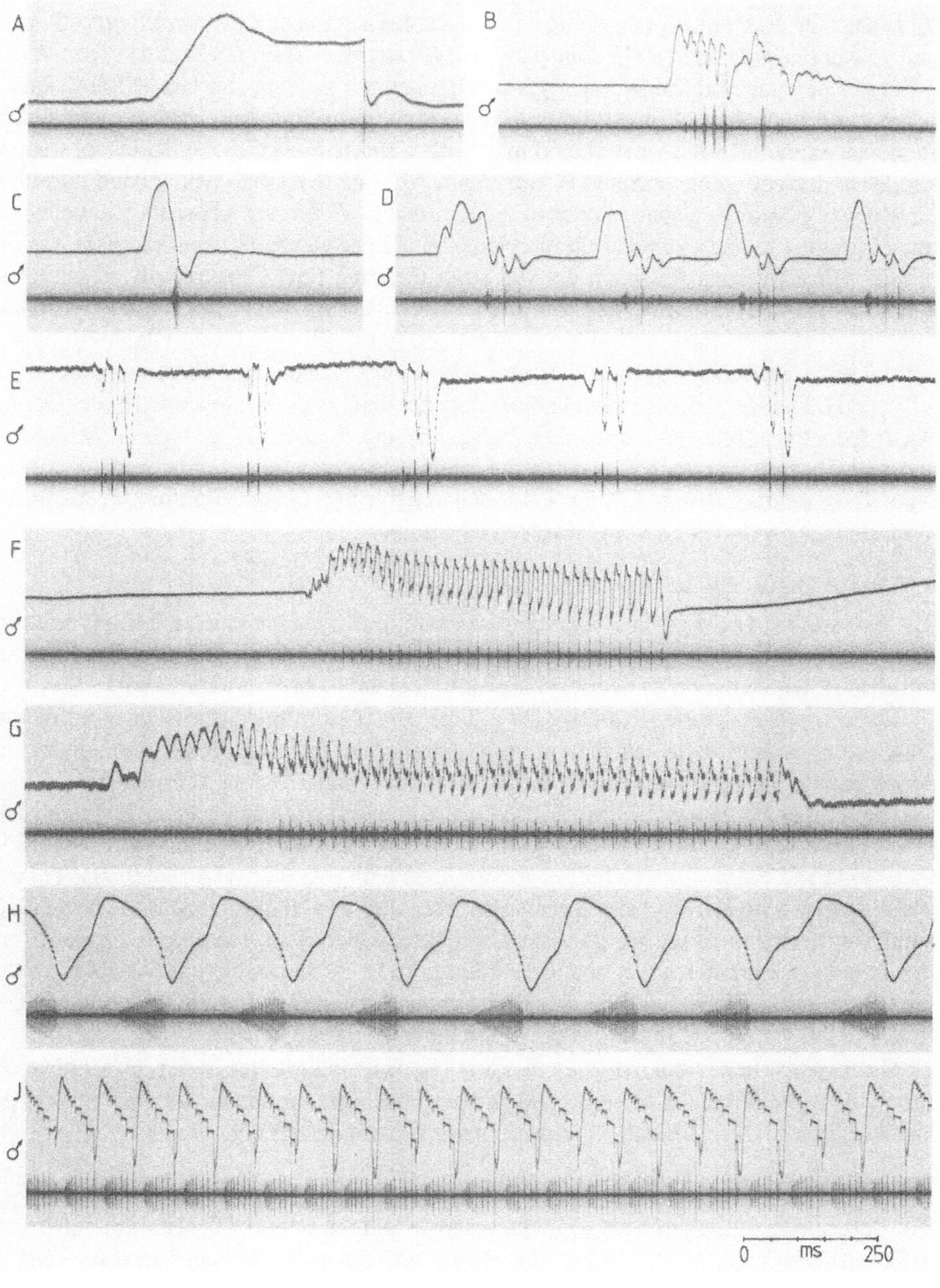

Fig. 8. Oscillograms of song (lower trace of each pair) and stridulatory movement (upper trace; elytra opening upward, closing downward) of **A** *Poecilimon schmidtii* (male: 22.5°C), **B** *Poecilimon bilgeri* (male; 23°C), **C**, **D** *Poecilimon cretensis* (male, two different types of syllables; 26°C), **E** *Poecilimon serratus* (male, 15°C), **F** *Poecilimon kutahiensis* (male, 10.5°C), **G** *Poecilimon* spec. A (male; 26.5°C), **H** *Poecilimon ledereri* (male; 24.5°C), **I** *Poecilimon ampliatus* (male; 23°C).

species the length of the syllable sequence is obviously variable, and very excited animals produce a song similar to type H. The song patterns of *P. chopardi* Ramme and *P. propinquus* Brunner are presumably also derived from type A, although they appear to be somewhat different. The crescending beginning of the syllables is reduced (in *P. propinquus* there are only one or two pulses left). The song patterns of *P. kutahiensis* (Fig. 8 E) and *P.* spec. A (Fig. 8 F) have developed through convergence, despite their striking similarities. Both species belong to groups that differ distinctly in regard to their morphology. These rapid syllable sequences may have been derived from a further shortening of the syllables in a pattern similar to that of *P. chopardi* and *P. mariannae* (*P.* spec. A). On the other hand they could have developed from a song similar to that emitted by *P. bilgeri* where they have lost the marking syllable and extended the previous syllable sequence (*P. kutahiensis*). It is also possible that the song has been derived from the song type C by rapidly repeating the syllables. The very different song pattern of *P. serratus* (Fig. 8 E) still lacks any reasonable interpretation. Here, the male holds its wings open and rapidly produces syllable-groups that consist of between one and three elements with a very unusual amplitude modulation.

Type H. The song consists of syllable sequences that last from a few up to many seconds. Apart from the species *P. propinquus*, *P. ledereri* and *P. tuncayi* (see type G), which only sporadically emit such long sequences, the following species consistently produce this song type: *P. ampliatus* Brunner (Fig. 8 H) and relatives (*P. ebneri* Ramme, *P. klisurensis* Williamse) as well as *P. harveyi* Karabag (unpublished).

5. Group: Genus *Phonochorion*

The two species of the genus *Phonochorion* examined here emit extended songs similar to those observed for the genus *Polysarcus*. The insects inhabit dense and high meadows with tall shrubs, where they may occur syntopically (personal observations).

The song of *P. uvarovi* Karabag (Fig. 5 B) consists of condensed sequences of long, crescending single syllables that are very similar in their amplitude modulation and accompanying wing movements to the ancestral song pattern of the Barbitistini.

The song of *P. satunini* Uvarov (Fig. 5 C) also contains similarly structured syllables, but is more complex, showing some similarities to that of *Polysarcus denticauda*. Five to ten of these loud and crescending syllables are arranged with relatively short intervals. They are separated by between two and three softer and shorter syllables produced by relatively smaller movements of the wings. After this phase, which is very similar to phase III in *P. denticauda*, an uninterrupted sequence (duration 15-20 s) of the soft syllables follows (similar to phase I in *P. denticauda*). Then the song is continued again, with a sequence of loud syllables (Fig. 5 C shows the transition between the two phases).

The response behaviour of the females is not known for any of the two species, but I would predict that the female is likely to respond after the long crescending syllables (compare *Isophya kraussii*, Fig 3 E and *Polysarcus denticauda*, Fig. 5 A).

Discussion

The description of the different song types shows that the songs of the Barbitistini are very different from the calling patterns produced by other bushcricket subfamilies, particularly the Tettigoniinae and the Conocephalinae (for a comparison see for instance: Alexander, 1956; Dubrovin & Zhantiev, 1970; Walker, 1975; Bailey, 1976; Rentz, 1985; Heller, 1988). The songs of many of the species I have examined are short and produced with long intercall intervals. The calls of the male appear to have distinct markers on which the female may cue her response. These markers enable the female to time her response accurately. The species of the tribe Barbitistini use three different methods to produce such markings (see below: a – c; Fig. 9), and in those species in which the female does not respond acoustically, the male song is changed in quite a different way (see below: d; Fig. 9).

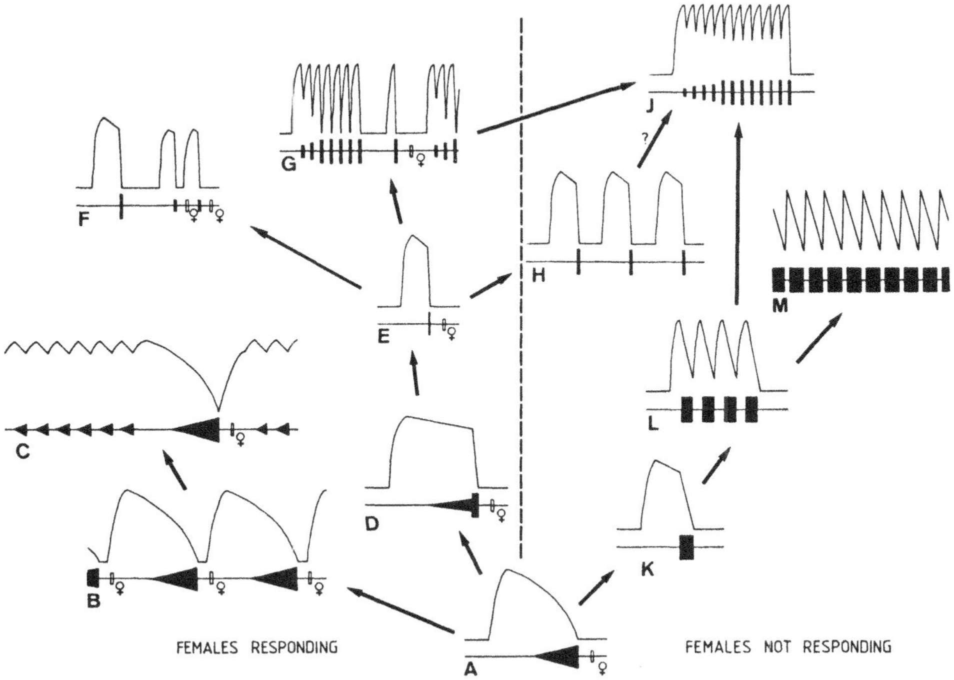

Fig. 9. Evolution of the song pattern within the genera of the Barbitistini (without genus *Isophya*). The upper line represents the stridulatory movement (upward deflection opening of the elytra, downward closing), the lower line represents the song (schematized oscillogram). **A** Ancestral song pattern (in species with and without responding females), **B** *Phonochorion satunini*, **C** *Phonochorion uvarovi*, *Polysarcus denticauda*, **D** *Poecilimon* type A with pronounced syllable ending, **E** *Poecilimon* type B, **F** *Poecilimon* type D, **G** *Barbitistes* type B, *Poecilimon* type E (by way of *Barbitistes* type A, not shown), **H** *Poecilimon* type C, **J** *Poecilimon* type G/H, **K** *Poecilimon* type F, **L** *Poecilimon* type G, **M** *Poecilimon* type H (details see text).

(a) Amplitude modulation within a syllable in songs showing one syllable type.

Presumably the primitive song pattern of the Barbitistini showed syllables with a pronounced ending (Fig. 9 A; female response existed). The ending of the syllable becomes more pronounced in the genus *Poecilimon* (Fig. 9 D), and may be on the one hand a result of changes within the neuromuscular system. By a short stop of the stridulatory movement, an interval is inserted before the loud ending of the syllable (*P. sanctipauli*). On the other hand, it may be due to changes in the structure of the stridulatory file (*P. brunneri* group), enabling the insect to insert the final part of its song abruptly. In other species (*P. affinis* group) combinations of these two methods may occur. As evolution has proceeded further the quieter part of the syllable is completely lost and only the loud, short ending is maintained (for instance *P. schmidtii*, *P. ornatus*; Fig. 9 E). Here the "duets" between the two sexes can take place during a very short time period. In several species of the genus *Isophya*, the beginning of the song is strongly pronounced; the amplitude modulation of the syllable is decrescending. In most of the examined species (see *Isophya*, type B) the female responds relatively late, but faster duets may also occur.

(b) Amplitude modulation of the song by using different syllable types.

Amplitude modulation using different syllable types has been realized most clearly in the genera *Phonochorion* and *Polysarcus*, where there is the excellent opportunity to follow the development from simple to complex song pattern.

Phonochorion uvarovi produces a song that is most similar to the primitive call of the Barbitistini (Fig. 9 B). The song of the second species, *P. satunini*, has apparently increased in complexity since a second softer syllable type is now inserted. It is arranged between the loud syllables in a special pattern. Finally, the song of *Polysarcus denticauda* has developed smoothly from this type of calling pattern: The complexity of the song is enhanced by the insertion of another syllable type (Fig. 9 C).

In the species of the genera *Isophya* and *Poecilimon*, with calling songs showing several syllable types, the females only respond at a very specific time. In the few species that have been examined, the female does not appear to respond to the loudest part of the song, but rather to syllables (even distinctly softer ones: *Parapoecilimon*; Fig. 9 F) that are temporarily related to the loudest syllable of the call.

(c) Song of homogeneous syllables emitted in a fixed temporal sequence.

Importance is attached to the temporal sequence of the syllables, and this is illustrated within the genera *Metaplastes*, *Barbitistes* and *Ancistrura*. However, in contrast to these species, male *Isophya* and *Poecilimon* only produce one syllable type. The relatively simple song of *Metaplastes ornatus* can be derived from a sequence of shortened syllables, similar to that of *Poecilimon laevissimus*. The more complex temporal patterns may have developed from this song structure, where the female responds only to the marking syllable, which is separated by a longer interval from the previous syllable group (Fig. 9 G).

(d) The emission of song patterns of several or many syllables

In those species on the genus *Poecilimon* in which the female does not respond to the male calling song, the males produce song patterns that are completely different from those described above. In general the duty cycle (the time in which a sound signal is emitted) increases, and the song patterns are remarkably similar to the "normal" calling song structure of most other bushcrickets. The ancestral song type of the Barbitistini is altered, and consists of single syllables in which the highest amplitude is neither at the beginning nor at the end. This variation occurs within one population: in one animal the beginning is louder, and in another it is the end song type F). However, many species produce a song consisting of short sequences of homogenous syllables that result from rapid movements of the forewings (song type G; Fig. 9 L). In some species the variation in the length of this syllable sequence is low (in *P. chopardi* sequences of 3-5 syllables), while in others it is quite large. Perhaps the extreme case is represented by the song of *P. ledereri*, in which a sequence may range from a single syllable to syllable series which last (rarely) for several seconds. This latter form of song is similar to the pattern emitted by species that always produce long songs (song type H; Fig. 9 M). *P. pergamicus* has ancestors that produce very short single syllables, and the females have independently lost the ability to respond acoustically to the male's song. In this species the increase in the duty cycle has developed as a convergence by emitting a rapid sequence of short syllables.

Because the songs show such differences in their complexity, and present many transitional stages, the evolution of the calling song patterns can often be traced with ease. It is much more difficult, however, to find the reasons for the observed changes. In the following discussion I will provide some hypotheses to explain these developments. However, as long as the behaviour and ecology of many of the species remains undescribed, any scenario as to why a song pattern has changed in one species but not in another will remain speculative.

One reason for a change in the song structure could be that certain call patterns are susceptible to acoustic rivalry between males of the same species. For the *Isophya* and *Poecilimon* species, that produce long single syllables, it can be frequently observed that a second animal chimes into the syllable of the first male. In species of *Poecilimon* at least, the end of the syllable may be obscured (Fig. 7 B) and the response of the female may not always be audible: alternatively the female may respond to the song of the second male. In species of *Isophya* the song can also be masked by the call of the rival. The beginning of a second syllable may affect the choice of the female, confuse her or even delay her response. In the *Poecilimon* species that emit much shorter syllables, the complete duet is so fast that a rival cannot interfere.

The presence of these short syllables leads to a secondary problem in that the brevity of the signal may be too short to convey any species' specific information. In those cases in which the beginning of the male's signal triggers the response of the female (for instance in *Poecilimon ornatus*, Heller, in prep., or in *Leptophyes albovittata* (Kollar), which is a phaneropterine that does not belong to the Barbitistini *sensu strictu*, Zhantiev & Korsunovskaya, 1986), it is unclear as to whether the amplitude modulation of the syllable conveys species' specific information. Thus, the female may respond to any noise she hears. However, the sensitivity of the receptor cells, which are mostly tuned to the species' specific frequencies, reduces the range of possible stimuli. A female may be also affected by the song she has previously heard, or she may evaluate the rhythm of the male song (see the different syllable intervals of the otherwise very similar songs of the *Leptophes* species; Faber, 1953).

Many of the more complex songs are probably derived from such a system. Short repeated syllables are arranged with different intervals. More recent studies (Dobler, unpublished data) would appear to favour the idea that the first syllable group contains species' specific information, whilst the following, isolated element, is the trigger for the female's response. However, these patterns could be exploited and disturbed, or even interrupted by rivals. A second male could attract the female's response by producing a single syllable shortly before the trigger syllable of the first male (Fig. 6 C; note that the meaning of the complex rivalry song of *Metaplastes oertzeni* is not fully explained by this). The songs of the other species (*Barbitistes*, type B) may be less susceptible to disturbances. In these songs the interval before the trigger syllable is much shorter. For example, a male *Isophya major* is able to respond quite rapidly to the first syllable (type I) of the first animal, responding with a type II syllable which is more usually the trigger for the female response (Fig. 4 B).

Male-male interactions are not the only factor to influence the changes of song pattern. The habitat is also very important, and those species that produce homogeneous syllables with different intervals are more inclined to live in bushes and trees, where these patterns may be less disturbed by the vegetation compared to more complex amplitude modulated syllables. The same may be true for vibratory signals, which may add conspecific information, and this communication channel could be of greater significance for those species living in bushes. This idea is supported by the presence of purely vibratory signals in the males of *Ancistrura nigrovittata* (Heller & Helversen, 1986) and *Barbitistes* species (Heller, 1988).

Not much is known about the acoustic properties of habitats that are capable of affecting the development of the long and permanent songs of the *Phonochorion* and *Polysarcus* species. *Polysarcus denticauda* shows a song structure and behaviour that is very similar to that of the tettigoniine *Decticus verrucivorus* (Linnaeus), an insect which lives in a similar habitat (Schatral *et al.*, 1985). The males in both the species move around while stridulating. By comparison, *Psorodonotus illyricus* Ebner , another tettigoniine species with a similar behaviour, shows a totally different song pattern (Keuper at al, 1986).

Depending on the species and the syllable type, various reasons are possible as to why the complexity of the song has increased in species producing several different syllable types. In *Parapoecilimon antalyaensis* the female responds to several soft syllables that are emitted after a single loud syllable. Quite obviously, louder syllables can be perceived over far greater distances than the soft syllables. Thus, the female may respond only to males that are relatively close, or when she is in a position to hear the softer syllables. This has the advantage that the female responds only at the distance when the male is able to hear her response. The song of the female is often softer than the song of the male. Thus, in species in which the male produces only one syllable type, the female may respond without being heard by the male. In many populations, however, due to the animals being at high density, these problems may not be so relevant. Here, the response of the female might travel to more than one calling male (in large species as *Poecilimon affinis* the response of the female is audible to a male at a distance of 8 to 12 m). In other species that live in populations at low densities, and in which the response of the female is particularly conspicuous, or acoustically hunting predators or parasitoids occur, a separation of the male song in a close and a far range signal may be of advantage both for female attraction and male competition.

Females have been described approaching the male in the genus *Isophya* (Zhantiev & Korsunovskaya, 1986), and animals which have a very low threshold show positive phonotaxis under laboratory conditions. However, the frequency and biological significance of this behaviour has yet to be explained in field studies. The acoustic reaction of a phonotactically responding female seems to be only meaningful if it leads to an increase in the singing activity of the male and therefore facilitates the search of the female for the male (possibly an important function of the female response in the ephippigerine *Platystolus obvious*, Hartley *et al.* 1974). In *Barbitistes* and *Metaplastes* both males and females move towards each other, at least in the laboratory, and if this occurs in the field the risk of becoming the victim of predators cueing on movement should be the same for both sexes.

A curious unexplained phenomenon occurs in the genus *Isophya*, particularly *I. rossica* but also in species such as *I. kraussii* in which the syllables end with isolated pulses. Females respond shortly after the beginning of the second syllable of the male (Zhantiev & Korsunovskaya, 1986), and in other species the isolated pulses are produced exactly at the time when the male expects to hear the response of the female. These conflicting clicks may mask the calls of searching females. This behaviour also occurs in many other genera such as *Poecilimon*, *Barbitistes*, *Metaplastes* and *Polysarcus*. Here males may even produce single pulses or groups of pulses and it is unlear if these sounds affect the behaviour of the calling male. Female *I. hospodar* also responds at the beginning of the syllable group at a time when the male is acoustically active.

To summarize, within the Barbitistini the evolution of song pattern in species with responding females follows three models: (1) amplitude modulation of single syllables, (2) use of different syllable types, and (3) use of homogeneous syllables in a fixed temporal sequence. In species in which the females have lost the response behaviour song patterns have evolved similar to those of other bushcrickets, in particular the tettigoniines and conocephalines. It would be very interesting to know if other groups of the Phaneropterinae showed similar behaviour patterns of female phonoresponse, and if the songs of both species have diverged along the same lines and under the same circumstances as I have described for the Barbitistini.

Acknowledgements

I would like to thank Winston Bailey and Otto von Helversen for the critical review of this manuscript. I am also indebted to all who helped collect animals, in particular E. Blum, M. Gebhardt, C. Hartley, D. & O. v.Helversen, M. Volleth, R. Weid and H. Wolf. Special thanks go to A. Schatral for translation.

Appendix

Localities (international code of the country, province, locality) and date of capture of the species figured in the text:

Acrometopa servillea (Brulle), male: GR, Kavalla, Pangaion mountains, 2.8.82; female: GR, Ilia, Pirgos, 1984;
Uromenus stali (Bolivar), male: E, Madrid, Puerto de Somosierra, 17.8.84; female A: E, Avila, P. de Pena Negro, 14.8.84; female B: E, Avila, P. de Serranillos, 14.8.84;
Isophya savignyi: IL, Ha Galil, 17 km northnortheast Zefat, 20.4.86;
I. modestior: YU, Serbia, Grdelica, 1986;
I. kraussii, male: D, Bavaria, Wrzburg, 1981; female: D, Bavaria, Altheim, 8.5.84;
I. speciosa: YU, Serbia, Grdelica, 19.5.86;
I. major : TR, Mersin, Anamur, 18.4.87;
I. hospodar: TR, Edirne, Ipsala, 28.4.87;
Polysarcus denticauda, male: GR, Kavalla, Pangaion mountains 1.8.82, female: YU, Dalmatia, Rijeka 1986;
Metaplastes ornatus: GR, Fthiothis, Makrakomi, 1987;
M. oertzeni: GR, Akhaia, Kalentzi, 4.7.86;
Barbitistes ocskayi: YU, Makedonia, 10 km northwest. Miravci, 1984, 1985;
Phonochorion satunini, Ph. uvarovi: TR, Rize, pass south of Ikizdere, 4.8.83;
Poecilimon similis: TR, Giresun, 20 km south of Dereli, 1.8.83;
P. sanctipauli: GR, Rhodos, Lindos, 19.4.83;
P. zonatus: TR, Erzurum, pass northwest. Ispir, 4.8.83;
P. serratus: TR, Bursa, Ulu Dagh, 23.7.83;
P. affinis: GR, Pieria, Pieria mountains, 1986;
P. hamatus: GR, Rhodos, Lindos, 19.4.83;
P. schmidti: YU, Makedonia, west. Tetovo, 8.8.81;
P. bilgeri: TR, Antalya, Termessos, 8.6.85;
P. kutahiensis: TR, Bursa, 5 km northwest. Inegöl, 24.7.83;
P. cretensis: GR, Kreta, 1986;
P. ledereri: TR, Aydin, Nyssa, 5.5.85;
P. spec. A: TR, Aydin, Priene, 29.4.85;
P. ampliatus: YU, Istra, Slavnik, 1983;
Parapoecilimon antalyaensis: TR, Antalya, Termessos, 8.6.85.

References

Alexander R.D. 1956. A comparative study of sound production in insects with special reference to the singing Orthoptera and Cicadidae of the Eastern United States. Dissertation. Ohio State University. pp 1-529.

Bailey W.J. 1976. Species isolation and song types in the genus *Ruspolia* Schulthess (Orthoptera, Tettigonioidea) in Uganda. J. Nat. Hist. 10: 511-528.

Bailey W.J. 1990. The ear of the bushcricket. In: The Tettigoniidae: Biology Systematics Evolution. Eds. W.J. Bailey & D.C.F. Rentz. Crawford House Press.

Belwood J.J. & G.K. Morris. 1987. Bat predation and its influence on calling behaviour in neotropical katydids. Science 238: 64-67.

Bei-Bienko G.Y. 1954. Tettigonioidea. Phaneropterinae. In: Fauna of the USSR, Orthoptera, Vol II/2. Moskau, Leningrad. (Translated, Jerusalem 1965).

Brunner von Wattenwyl C. 1878. Monographie der Phaneropteriden. Wien. pp 1-401.

Brunner von Wattenwyl C. 1891. Additamenta zur Monographie der Phaneropteriden. Verh. Zool. Bot. Ges. Wien. 41: 1-196.

Dubrovin N.N. & R.D. Zhantiev. 1970. Acoustic signals of katydids (Orthoptera, Tettigoniidae). Zool. Zh. 49: 1001-1014.

Faber A. 1953. Laut- und Gebärdensprache bei Insekten. Orthoptera (Geradflügler). Staatl. Museum für Naturkunde. Stuttgart.

Hartley J.C., D.J. Robinson & A.C. Warne. 1974. Female response song in the ephippigerines *Steropleurus stali* and *Platystolus obvius* (Orthoptera, Tettigoniidae). Anim. Behav. 22: 382-389.

Heller K.-G. 1984. Zur Bioakustik und Phylogenie der Gattung *Poecilimon* (Orthoptera, Tettigoniidae, Phaneropterinae). Zool. Jb. Syst. 111: 69-117.

Heller K.-G. 1986. Warm-up and stridulation in the bushcricket, *Hexacentrus unicolor* Serville (Orthoptera, Conocephalidae, Listroscelidinae). J. Exp. Biol. 126: 97-109.

Heller K.-G. 1988. Bioakustik der europäischen Laubheuschrecken. Josef Margraf. Verlag. In press.

Heller K.-G. & D.v. Helversen. 1986. Acoustic communication in phaneropterid bushcrickets: species-specific delay of female stridulatory response and matching male sensory time window. Behav. Ecol. Sociobiol. 18: 189-198.

Helversen O.v. & N. Elsner. 1977. The stridulatory movements of acridid grasshoppers recorded with an opto electronic device. J. Comp. Physiol. 122: 53-64.

Josephson R.K. 1984. Contraction dynamics of flight and stridulatory muscles of tettigoniid insects. J. Exp. Biol. 108: 77-96.

Keuper A., K. Kalmring, A. Schatral, W. Latimer & K. Kaiser. 1986. Behavioural adaptations of ground living bushcrickets to the properties of sound propagation in low grassland. Oecologia 70: 414-422.

Kevan D.K.M. 1982. Orthoptera. In: Synopsis and Classification of Living Organisms, Vol. II. Ed. S.P. Parker. McGraw-Hill Book Company, New York *et al.* pp 352-361.

Michelsen A. 1985. Environmental aspects of sound communication in insects. In: Acoustic and Vibrational Communication in Insects. Eds. K. Kalmring & N. Elsner. Parey Verlag, Berlin, Hamburg. pp 1-10.

Pfau H.K. & B. Schroeter. 1988. Die akustische Kommunikation von *Platystolus martinezi* (Bolivar) – ein schnelles Antwort-Rückantwort-System (Orthoptera, Tettigoniidae, Ephippigerinae). Bonn. Zool. Beitr. 39: 29-41.

Ragge D.R. 1980. A review of the African Phaneropterinae with open tympana (Orthoptera: Tettigoniidae). Bull. Br. Mus. Nat. Hist. (Entomol.) 40: 67-192.

Rentz D.C.F. 1985. The Tettigoniinae. A monograph of the Tettigoniidae of Australia, Vol 1. CSIRO, Australia. pp 1-384.

Rheinlaender J. & H. Römer 1990. Acoustic cues for sound localisation and spacing in orthopteran insects. In: The Tettigoniidae: Biology Systematics Evolution. Eds. W.J. Bailey & D.C.F. Rentz. Crawford House Press. pp XX – XX

Robinson D.J. 1990. Acoustic communication between the sexes in bushcrickets. In: The

Tettigoniidae: Biology Systematics Evolution. Eds. W.J. Bailey & D.C.F. Rentz. Crawford House Press.

Robinson D.J., J. Rheinlaender & J.C. Hartley. 1986. Temporal parameters of male-female sound communication in *Leptophyes punctatissima*. Physiol. Entomol. 11: 317-323.

Römer H., W.J. Bailey & I. Dadour. 1988. Insect hearing in the field. III. Masking by noise. J. Comp. Physiol. 164: 609-628.

Ryan M.J. 1985. The Tungara frog. A study in sexual selection and communication. University of Chicago Press, Chicago, London. pp 1-230.

Schatral A., W. Latimer & K. Kalmring. 1985. The role of the song for spatial dispersion and agonistic contacts in male bushcrickets. In: Acoustic and Vibrational Communication in Insects. Eds. K. Kalmring & N. Elsner. Parey Verlag, Berlin, Hamburg. pp 111-116.

Walker T.J. 1975. Stridulatory wing movements in eight species of *Neoconocephalus*. J. Insect Physiol. 21: 595-603.

Zhantiev R.D. & N.N. Dubrovin. 1977. Sound communication in the genus *Isophya* (Orthoptera, Tettigoniidae). Zool. Zh. 56: 38-51.

Zhantiev R.D. & O.S. Korsunovskaya. 1986. Sound communication in bushcrickets (Tettigoniidae, Phaneropterinae) of the European part of the USSR. Zool. Zh. 65: 1151-1163.

Zimmermann U., J. Rheinlaender & D.J. Robinson. 1988. Cues for male phonotaxis in the duetting bushcricket *Leptophyes punctatissima*. J. Comp. Physiol. 164: 621-628

9. INTERSPECIFIC ACOUSTIC INTERACTIONS IN BUSHCRICKETS

Andrea Schatral

Introduction

In nature an acoustically signalling animal is exposed to the sounds of both its own species and to those of others. The sounds of conspecifics have a recognized influence on the caller, as well known for birds (e.g. Helversen & Wickler, 1971; Thorpe, 1963), frogs (Wells, 1977; Schwartz & Wells, 1985) and many orthopteran insects (e.g. Busnel, 1964; Jones, 1966; Latimer, 1981b; Greenfield & Shaw, 1983). Such song interactions are a common feature of aggregations where selection may have acted on this particular behaviour, either through male competition or indeed through female choice (Shaw & Greenfield 1983). Interspecific responses have also been described in these major acoustical groups, for example among birds by Wassermann, (1977), Cody and Brown, (1969), Popp *et al.* (1985), Wickler, 1972 and Ficken *et al.*(1974), frogs by Schwartz (1987), Littlejohn and Martin (1969) and Littlejohn *et al.* (1985), and the Orthoptera by Samways (1976; 1977 a), Latimer (1981 a,b), Latimer and Broughton (1984) and Greenfield (1988). However, the biological significance of these interactions is not clear. As the temporal pattern of the conspecific's call is to function in species' recognition (e.g frogs: Sullivan & Leek (1986) and orthopterans (Hill *et al.*, 1972; Paul, 1976; Popov & Shuvalov, 1977), any changes in this pattern may have consequences for individuals attracting a mate.

In this chapter I am concerned only with interspecific acoustic interactions of bushcrickets, and early observations in both field and laboratory have indicated that males may respond to the song of a different species (Dumortier, 1963; Broughton, 1965; Busnel *et al.*, 1956). Up to now, interspecific acoustic interactions have been recorded for several genera of the Tettigoniidae, indicating that the phenomenon may be widespread in this family. Moreover, it has been shown that bushcrickets are able to perceive a wide range of song frequencies (Kalmring *et al.*, 1985) and show similar auditory thresholds (Römer *et al.*, 1989) across species. Finally, interspecific acoustic interference may be of biological relevance for bushcrickets, since several species frequently occur in sympatric communities and show similar daily patterns of calling activity (Bailey, 1976; Samways, 1977 b; Latimer & Broughton, 1984).

Thus, orthopteran insects seem to be ideal to discuss the following questions: (1) Is the response to conspecific song the same as the response to heterospecific song? (2) Is there evidence for heterospecific recognition (Morris & Fullard, 1983; Gwynne & Morris, 1986), or rather for the "supra-normal (conspecific) stimulus" hypothesis (Latimer, 1981a)? (3) Is the frequency at which the song is broadcast the principal cause for song interference as suggested by Latimer and Broughton (1984)? (4) Have strategies evolved to prevent interspecific acoustic interactions?

To tackle these problems in this chapter I review results of recently published work on European tettigoniids and present my own research on some Australian species.

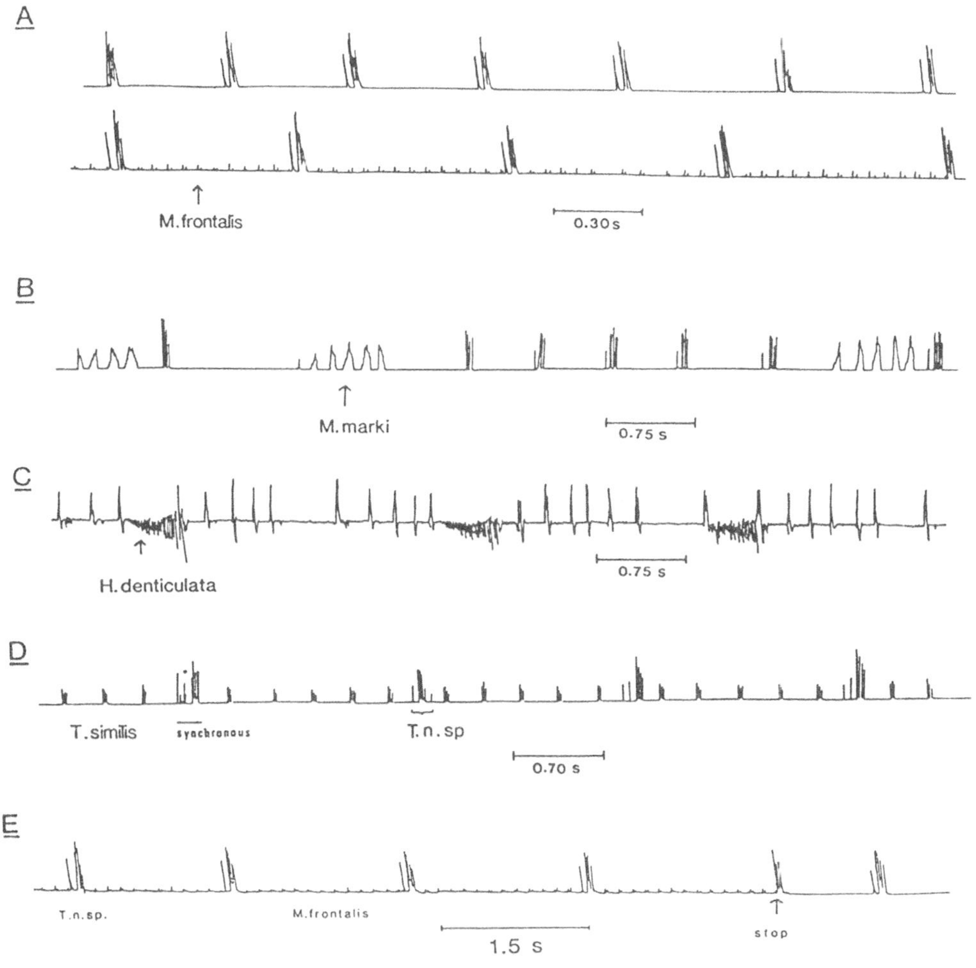

Fig. 1. Acoustic response of *T. similis* to the songs of (A) *M. frontalis* (Upper line: Solo song of *T. similis*, lower line: duet between *T. similis* and *M. frontalis*). (B) *M. marki*, (C) *H. denticulata*, (D) *T.*n.sp. (E) Acoustic response of *T.*n.sp. to the song of *M. frontalis*.

Effects of allospecific calls on song structure

In Fig. 1 and 2 I have schematized various acoustic interactions between bushcrickets based principally on studies carried out on European species such as the decticine genera *Platycleis* and *Metrioptera* (Samways & Broughton, 1976; Samways, 1977 a, b; Latimer, 1981a; b; Latimer & Broughton, 1984), the new world copiphorine *Neoconocephalus* (Greenfield, 1988) and my own research on the acoustic interactions between two *Tympanophora* (Tympanophorinae) species, *T.*n.sp. 1 (an undescribed species, D. Rentz, pers. comm.) and *T. similis*, and the copiphorine *Mygalopsis marki* Bailey, the tettigoniine *Metaballus frontalis* (Walker), and the sagines, *Pachysaga* n.sp., and *Hemisaga denticulata* (White) (Schatral &

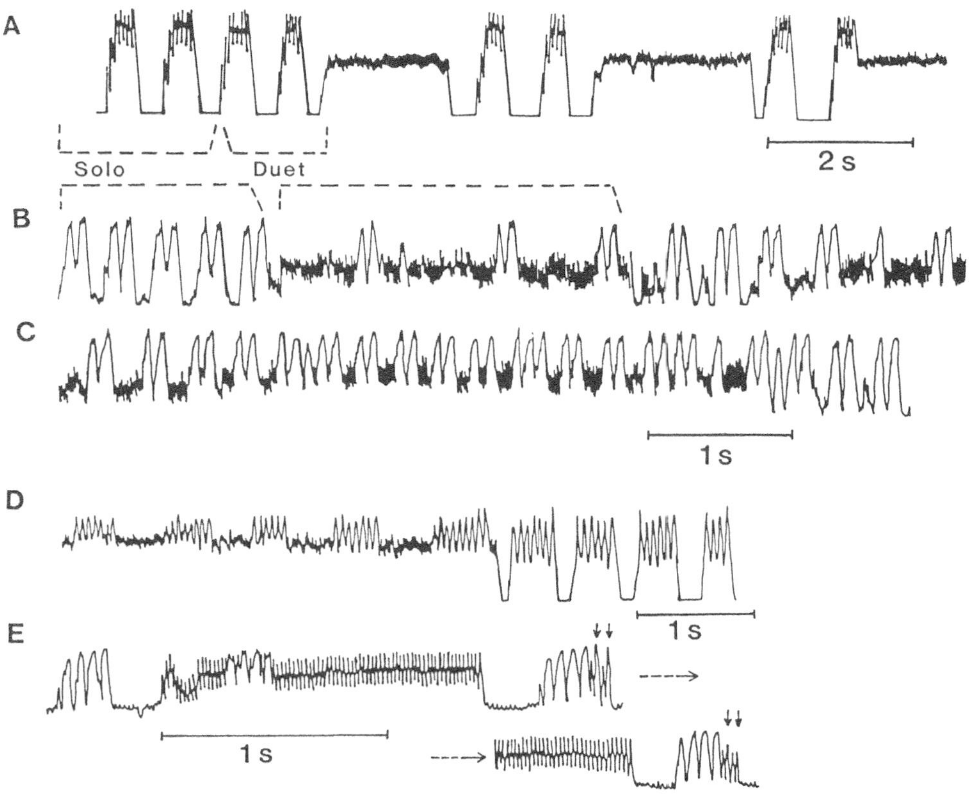

Fig. 2. Acoustic responses of *P. sabulosa* to the song of (A) *M. roeselii* (inhibition), (B) *P. intermedia* (reduction in chirp rate) and (C) *M. roeselii* (increase in chirp duration). Acoustic responses of *P. albopunctata* to the song of *M. roeselii*. (D) Increase in chirp duration, (E) production of microsyllables. (from: Latimer & Broughton, 1984).

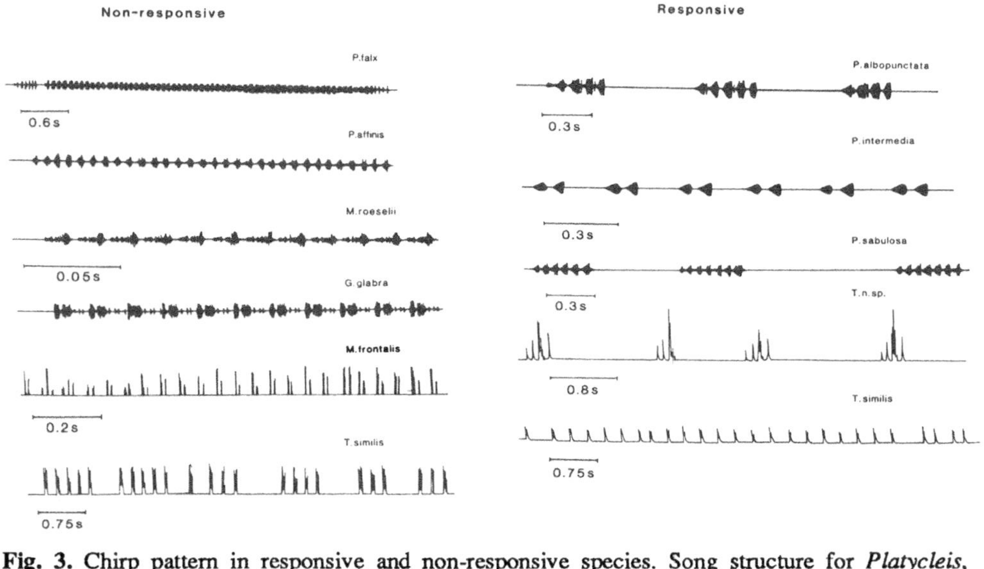

Fig. 3. Chirp pattern in responsive and non-responsive species. Song structure for *Platycleis*, *Metrioptera* and *Gampsocleis* from: Samways, 1976; Latimer, 1981 and Latimer & Broughton, 1984.

Yeoh, in press). In addition, I refer to casual observations in both the field and laboratory of interactions between species of *Ephippiger* (Busnel *et al.*, 1956), between *P. intermedia* and *Pholidoptera femorata*, and between *P. intermedia* and *Tylopsis lilifolia* (Samways, 1977 a). However, these observations have not been included in the discussion in this chapter due to the paucity of data.

Terminology

In this chapter the following terms will be used:

Duet: Two males are calling during the same time period.

Acoustic interaction: The song of at least one of the duetting males is influenced by the call of its duetting partner. The various forms of acoustic interactions are synchrony, alternation and unison singing. During alternation a male is calling after its singing partner has stopped. During synchrony the singing units of each male are emitted at the same time maintaining the species'specific syllable pattern. Unison singing occurs when the singing units are produced simultaneously but without true synchrony (see Latimer, 1981b for terminology).

Chirp: An acoustic unit which is audible to the human ear.

Syllable: Sound produced by one movement of opening and closure of the sound producing apparatus (see also Latimer 1981b and Broughton, 1963). A chirp comprises several syllables.

General responses

During heterospecific acoustic interactions the following responses have been observed:

(i) (Complete) inhibition: a male stops singing during the chirp of its singing partner. In some cases the inhibition lasts longer than 1 min after the end of the stimulus.

(ii) Initial inhibition, that is the singing activity of a male is only inhibited during the beginning of the chirp of the second male, and then the first male increases its chirp rate.

(iii) Change in chirp rate relatively to the chirp rate of an acoustically isolated (solo) singer; the chirp rate either drops or increases.

(iv) Change in chirp length relatively to the chirp length produced by solo males.

(v) Qualitative changes in syllable length; the emission of microsyllables (these are syllables that are shorter and of lower amplitude than normal syllables). The duration of these elements is distinctly smaller and their intensity lower than in normal syllables.

Although all five modifications of singing behaviour have been described in different species, seldom more than two have been reported as occurring in the same interaction. A reduction in chirp rate, inhibition (including initial inhibition), and an increase in chirp length are the most common responses. The song may also be affected some time after the interaction (post-interaction or post-duet), where complete inhibition as well as rebound effects (increase in chirp rate) and changes in chirp length may occur (Latimer & Broughton, 1984). Species that respond to the song of a conspecific by adjusting their own song output so that the emissions of the two singing partners are synchronised or alternated are more likely to change song parameters in response to heterospecific song. These are called "reactive" species (Latimer, 1981b). In general, reactive species emit discrete chirps (discontinuous singers), and the chirp interval lasts at least as long as the chirp itself. These species are represented by *Platycleis intermedia* (Serville), *P. sabulosa* Azam, *P. albopunctata* (Goeze), *Metrioptera brachyptera* (L.) (Samways, 1976 a, b; Latimer & Broughton, 1984), *N. spiza* Walker & Greenfield (Greenfield, 1988) and *Tympanophora* n.sp.1 (Schatral & Yeoh, submitted). Species that do not appear to adapt their song pattern to the structure of the conspecific song but affect heterospecific callers are *Platycleis affinis* Fieber, *P. falx* (Fabricius), *Metrioptera roeselii* (Hagenbach) and *Gampsocleis glabra* (Herbst). In general the conspecific call has only a stimulatory effect (see Latimer, 1981b; Samways, 1976). These species will be called "non-reactive" species in the following account (see also Latimer, 1981b). The chirps are either significantly longer than the chirp intervals (*M. marki* Bailey, *M. roeselii*, *G. glabra*;) or chirp interval and chirp are very short (high chirp rate) (*T. similis* Riek, *M. frontalis*) (Fig. 3). Thus "non-reactive" species tend to produce fast and incessant songs. A decrease in chirp rate is the only acoustic response to the conspecific song in *M. marki* (Dadour, 1989). However, the difference between reactive and non-reactive species is not always clear, and males of the Australian species *T. similis* may alternate (typical of a reactive species) but also call in unison (typical of a non-reactive species) with singing conspecifics.

Many of the heterospecific interactions are one-way interactions (for terminology see Samways & Broughton, 1976). During an one-way interaction only the song of one of the two duetting males is affected. Whether a one-way or two-way interaction occurs appears to be related to the responsiveness of each of the two males involved in the duet: One way interactions characterize the interactions between a reactive and a non-reactive species, two-way interactions the duet between two reactive species. But Greenfield (pers. comm.) has noted an exception to this one-way interaction between the reactive species *N. spiza* and *N. affinis* (Beauvois). In the following discussion the song and acoustic responses of reactive and non-reactive species to conspecific and heterospecific songs are described in more detail (Table I).

Table I: Responses to conspecific and heterospecific song (cs and hs respectively) for six reactive species (0 = no change, x = change reported).

Response	*P.intermedia*		*P.sabulosa*		*P.albopunctata*		*M.brachyptera*		*N. spiza*		*T.n.sp.1*	
	cs	hs	cs	hs	cs	hs	cs	hs	cs	hs	cs	hs
Complete inhibition		x		x		x		x		x		
Chirp interval												
increases	0	x		x				x			x	x
decreases	0	x		0			x			x		
Chirp duration												
increases	0	x	0	x		x		x				
decreases	0	x	0									
Microsyllables						x		x				
Unison singing		x		x	x	x						x
Alternation		x	x	x					x		x	x
Synchrony	x	0							x			

The calling behaviour of reactive species

(1) *Platycleis intermedia* emits two syllables/chirp when acoustically isolated (solo song), and also when involved in conspecific duets (Samways, 1976 a). During a conspecific duet males appear to synchronise their chirps, however, during heterospecific interactions with species such as *P. sabulosa*, *P. affinis*, *P. falx*, and *P. albopunctata*, males sing either in unison or alternate their chirps. Chirp rate does not differ between solo song and conspecific duet in *P. intermedia*, but drops considerably in interactions with some heterospecifics (Samways & Broughton, 1976; Latimer & Broughton, 1984) and increases in a duet with *M. roeselii*. During a heterospecific duet, chirp length increases, producing polysyllabic chirps (chirps consisting of many syllables), or decreases with the occasional inclusion of microsyllabic

chirps (Samways & Broughton, 1976; Latimer & Broughton, 1984). Further, after an interaction the chirp rate may increase as illustrated by *P. affinis*, *P. albopunctata* or *M. roeselii* (rebound effect) (Samways & Broughton, 1976; Latimer & Broughton, 1984). An increase in chirp length has been found for *P. intermedia* during the post-duet song between this species and *P. affinis*.

(2) *Platycleis sabulosa* produces mostly 6-8 syllables/chirp during solo songs and apparently maintains this number during conspecific duets (Samways, 1976a). Males mostly alternate their chirps when singing with a conspecific (Samways, 1976a), but when interacting with *M. roeselii*, males may either alternate or sing in unison (Latimer, 1981b; Latimer & Broughton, 1984). Samways (1976a) made no mention as to whether chirp rate dropped during a conspecific duet compared to the chirp rate produced in a solo song.

During an interaction with *M. roeselii* (Latimer & Broughton, 1984), *P. intermedia* or *P. albopunctata* (Samways & Broughton, 1976) the reduction in chirp rate is considerable. Chirp length does not seem to change during conspecific duets, but does increase during duets with *M. roeselii* and during the post-interaction period with this species, chirp length also increases (Latimer & Broughton, 1984). After the interaction with *P. intermedia* a rebound in chirp rate can be observed (Samways & Broughton, 1976).

(3) *Platycleis albopunctata* emits 4-5 syllables/ chirp and usually alternates chirps with a conspecific during the first phase of the duet. Then one male sings continuously at the normal solo-chirp rate, whilst the other male sings sporadically. During alternation, each insect produces chirps at half the normal rate (Latimer, 1981b), and alternation also characterizes the duet with *M. roeselii*, *Gampsocleis glabra* (Latimer & Broughton, 1984), and *P. intermedia* (Samways & Broughton, 1976). Although chirp length usually increases when a male stridulates simultaneously with *M. roeselii* (Samways & Broughton, 1976; Latimer and Broughton, 1984), sometimes chirp length decreases, and microsyllables are produced during the song of *M. roeselii* (Latimer & Broughton, 1984). Chirp length may also increase during a duet with *P. intermedia*. After the duet with *M. roeselii*, chirp rate is often higher than during solo-song (rebound effect) (Latimer & Broughton, 1984).

(4) *Metrioptera brachyptera* produces three syllables/chirp when acoustically isolated (Latimer & Broughton, 1984). During the duet with *M. roeselii* the number of syllables/chirp increases and microsyllables are produced. Chirp rate appears to decrease compared to the chirp rate found during duets with conspecifics (Latimer & Broughton, 1984).

(5) *Neoconocephalus spiza* produces discrete chirps (Duration: 50-60 ms). Chirp interval can be variable (up to 3.5 chirps/s). During conspecific duets, chirps are alternated or sung in synchrony, whereas during stimulation with the song of *N. affinis* calling is inhibited (Greenfield, 1988).

(6) *Tympanophora* n.sp.1 emits discrete chirps that are alternated in conspecific duets. During alternation chirp interval increases significantly. Alternation of chirps can be also observed when T. n.sp. 1 is duetting with calling *Pachysaga* n.sp. Alternation, but mostly unison singing characterizes the interactions with *T. similis*, and unison singing occurs when duetting with *Metaballus frontalis*. Chirp interval increases during a duet with *M. frontalis* and with *T. similis*. Effects on chirp structure or length have not been found (Schatral & Yeoh, submitted).

In summary, for all species tested the responses to conspecific song and those to heterospecific calls differed. If this is due to the ability of the animals to discriminate between heterospecific and conspecific song, or just the result of different excitatory effects forms part of the following discussion.

Calling behaviour of non-reactive species

The calls of conspecifics may stimulate or inhibit the singing activity of non-reactive species (see above), whilst in general heterospecific calls do not affect calling (see Latimer & Broughton, 1984). However, *M. marki* appears to increase its chirp interval when stimulated with the discontinuous song of *H. denticulata* (Bailey, pers. comm), and males of *T. similis* increase their chirp interval when duetting with *M. frontalis* or *M. marki*.

Mechanisms of Interspecific Acoustic Interactions

Three important prerequisites for interspecific acoustic interactions have been suggested. First, song frequencies should be over a similar range (Samways & Broughton, 1976; Samways, 1976b; Latimer, 1981b; Latimer & Broughton, 1984). Second, males should be able to respond to the conspecific song by alternating or synchronising chirps (Latimer & Broughton, 1984), and third, there should be similarities in the syllable structure of the songs (Samways, 1977a).

The emission of similar song frequencies

Broad-band tuning of the receptors enables males to perceive a wide range of signals (see Römer, 1987; Römer *et al.*, 1989; Kalmring *et al.*, chapter 11). The importance of the high frequency peaks of the song for the behaviour of tettigoniids has been demonstrated for *Tettigonia cantans* (Fssly) by Latimer and Sippel (1987), for *Requena verticalis* Walker by Bailey and Yeoh (1988), and for an undescribed species of a zaprochiline genus by Gwynne and Bailey (1988). Latimer and Broughton (1984) assumed that interacting bushcricket species that were broadcasting similar song frequencies would more easily influence the song of a second species (Latimer & Broughton 1984). This was also suggested for acoustic interference in frogs. In all frog species that have been reported to be involved in heterospecific acoustic interactions the (narrow) song spectra overlap (Schwartz & Wells, 1985; Littlejohn & Martin, 1969; Littlejohn *et al.*, 1985).

Latimer and Broughton (1984) used the song of the tettigoniid *Gampsocleis glabra* as a control stimulus. The bushcricket emits a much lower carrier frequency (peak at about 12 kHz) than all the other species tested (frequency range: 25-35 kHz) but the chirp rate is similar to that of *M. roeselii*. Although the authors conclude that *G. glabra* does not affect the song of *P. albopunctata*, because of this dissimilarity in the carrier frequency, the results show – in contrast to their statement – that chirp rate is significantly reduced in *P. albopunctata* when *G. glabra* is singing (Latimer & Broughton, 1984). Moreover, the song spectra of these bushcrickets spectra do overlap, although the frequency peaks differ. Again a difference in song frequency has been suggested as the reason why *P. affinis* and *P. albopunctata* are not interacting acoustically, but Latimer and Broughton (1984) presented no data to substantiate this claim.

The response of a first order neuron in the ventral cord (Omega neuron) in *H. denticulata* is totally suppressed in the presence of the song of *M. marki*. This could be due to both songs having similar frequency spectra (the peaks in the song spectra of the two species overlap at 12 kHz), or to the incessant song pattern of *M. marki* (Römer & Bailey, pers. comm.).

These observations raise the important question as to whether it is an overlap in the song's frequency range that affects the way in which species interact, and the degree of modification on just how much the frequency spectra do overlap. The results I have obtained from a study of the acoustic interactions between the two *Tympanophora* species with other heterospecifics support the following hypotheses.

(1) The degree of overlap of both frequency as well as the temporal pattern affects the

degree of interference. Greenfield (1988) found that song structure was important in that the song pattern of *N. spiza* was affected by the inter-pulse interval of heterospecific songs produced by play-back within the laboratory.

(2) A high similarity in the song's frequency spectrum does not necessarily lead to a strong interference.

During duets between the Australian *Tympanophora* species, chirp interval and often song structure were affected, but singing was only inhibited during the initial part of the interaction. Complete inhibition, similar to that found during the duets between other species such as *Platycleis* and *N. spiza*, did not occur. The observed effects may be related to the high frequency elements in the songs of *T. similis* and *T.* n.sp. 1, which overlap with those of their heterospecific neighbours. However, the dominant frequency, as well as the shape of the whole spectrum, clearly differ between the *Tympanophora* species and other sympatric heterospecifics. The songs of *H. denticulata*, *M. frontalis*, *M. marki* and *Pachysaga* n.sp. show a dominant frequency which is between 4 and 8 kHz higher than the calling frequency produced by males of the genus *Tympanophora* (see Schatral & Yeoh, in preparation). The two *Tympanophora* species show three frequency peaks, the songs of *H. denticulata* and *M. frontalis* are broad-banded with several peaks, and the song of *Pachysaga* sp. nov. shows just a single narrow band at 16 kHz.

Finally, it is very interesting to note that the song of *T. similis* inhibits the song of *T.* n.sp.1 only during the initial part of the duet although both the species emit calls which are very similar in syllable structure and frequency spectrum.

Responsiveness

The previous discussion suggests that the ability of a male to respond to the conspecific song is one indicator of the male's behaviour towards heterospecific calls. Reactive species alternate and/or synchronise chirps during duets between conspecifics. These species are characterized by a very high response plasticity, in that they react to incessant as well as to chirping songs. The behaviour of *T. similis*, which is rather a "non-reactive" species, is particularly interesting because its song affects the chirping call of *T.* n.sp.1 and is also affected by the continuous songs of *M. marki* and *M. frontalis*. However, the call is not affected by the discrete chirps of *T.*n.sp.1 or *H. denticulata*. Thus, males of *T. similis* may be able to adapt to the time characteristics of the stimulus. This behavioural possibility is also reflected in the high variability of the chirp interval of the solo singers.

In general, the reason why non-reactive species are not affected by the calls of reactive species may be that short acoustic signals are not be "powerful" enough to affect the song of non-reactive species. Within the genus *Platycleis*, similarities in song pattern at the syllable level have been regarded as promoting reactivity of *P. intermedia*, but dissimilarities at the chirp level are responsible for song modification itself (Samways & Broughton, 1976). This may lead to a strong reaction, like the production of microsyllables, or complete inhibition, which this species shows in the presence of heterospecific songs. As the data demonstrate, inhibitory songs may be composed of long or of very short chirps as long as chirp interval is sufficiently short. Therefore the inhibitory effect they exert seems to depend on the continuity of the signal rather than on a certain fine structure of the song.

Excitation and inhibition have been suggested as the mechanisms involved in interspecific interactions (Samways & Broughton, 1976; Latimer & Broughton, 1984). Inhibition operates when a male stops singing during a duet, or slows down its chirp rate: the observed increase in chirp interval results from at least one male being inhibited: mutual inhibition leads to both males increasing their chirp interval. In addition, excitation is responsible for an increase in chirp rate and chirp length during duets and post-duets. The rebound in singing rate may result from inhibition related excitation (Latimer & Broughton, 1984; Heiligenberg, 1966, 1969): the level of excitation is increased with the length of inhibition. Excitatory and inhibition phenomena have been also observed to lead to intraspecific male-male interactions in a number of genera (Jones, 1966; Heiligenberg, 1969; Jones & Dambach, 1973; Shaw, 1968). Therefore response patterns that were observed in heterospecific interactions may have evolved in conspecific interactions (Samways & Broughton, 1976).

Heterospecific Acoustic Interactions in the Field

Species that have been studied in the laboratory have also been examined in the field. For example, *Platycleis* has been investigated by Samways (1976b) and Latimer and Broughton (1984), and *N. spiza* by Greenfield (1988). *T. similis* and *T.* n. sp.1 have been the subjects of my own research.

In France *P. intermedia*, *P. affinis*, *P. falx*, *P. sabulosa* and *P. albopunctata* sing at the same time of the year, but are not syntopic in any of the areas investigated by Samways (Samways, 1976b). *P. affinis* and *P. falx*, the two non-reactive and continuous singers, stridulate readily throughout the day and night, whereas *P. intermedia*, *P. sabulosa* and *P. albopunctata* are primarily nocturnal. Samways (1977b) suggested agricultural disturbance as the reason why he was able to record acoustic interactions for *P. intermedia* and *P. affinis*, as well as for *P. falx* and *P. sabulosa* in the field. Response patterns in the field were the same as those in the laboratory.

Latimer and Broughton (1984) surveyed several bushcricket communities in southern France and Spain where they found that in France *P. albopunctata* occurred in the same vegetation with *M. roeselii* or *G. glabra*. In Spain – unlike in France – *P. albopunctata* is associated with *P. affinis*, whereas in other areas *P. albopunctata* sings close to *Pycnogaster jugicola*. *P. albopunctata* was observed duetting with *M. roeselii*; and on four occasions *P. albopunctata* appeared to be inhibited by its song (Latimer, 1981 a,b). Both species are usually allopatric. In habitats in north Germany *M. roeselii* was interacting acoustically with *M. bicolor* (Philippi) (McHugh, 1972).

Finally, in Panama, *N. spiza* shares its habitat with three other *Neoconocephalus* species but acoustic interactions have not been observed (Greenfield, 1988).

In Western Australia, in those habitats that have been investigated so far, *T.* n.sp.1 is sympatric with *T. similis*; both are found singing from the same type of vegetation. Furthermore, both species are also sympatric with *H. denticulata*, *M. frontalis* and *Pachysaga* n.sp., and in some study sites they are sympatric with these four species as well as with *Mygalopsis marki*. Although in general, the daily singing patterns of all species overlap, *T.* n.sp. 1 starts singing three weeks earlier in the year than *T. similis* and when *T. similis* begins to sing *T.* n.sp.1 shifts its calling to a much earlier time. However, I was unable to observe any effects of the call of *T. similis* on the chirp pattern of *T.* n.sp.1..

In summary field studies indicate that interspecific interactions between species do occur in the field. However, the allopatric occurrence of the five *Platycleis* species in France indicate that (1) duets between species are not frequent, and (2) one mechanism to avoid heterospecific acoustic interactions is spatial separation. Many authors suggested that spatial separation occurs between two species and results in allopatry, or the selection of different micro-habitats (Alexander, 1967; Walker, 1974; Samways, 1976a, 1977a; Latimer, 1981b; Latimer & Broughton, 1984).

How spatial separation may be achieved has been demonstrated in the laboratory. Males of *P. intermedia* move away from singing *P. falx* a species which exerts a strong inhibitive effect on the singing activity of *P. intermedia* (Samways & Broughton, 1976). McHugh (1972) suggested that interspecific acoustic interactions between male *M. bicolor* and *M. roeselii* could lead to their spatial separation, and finally negative phonotaxis has been also shown for males of *P. albopunctata* in response to the song of *M. roeselii* (Latimer, 1981 b).

How to Avoid Heterospecific Interactions

Habitat preferences may have evolved to prevent dysgenic acoustic interactions between species with similar song parameters (Latimer, 1981b). Latimer and Broughton (1984) base their argument on the observation that sympatric species have dissimilar frequency spectra. The authors conclude from a survey on a bushcricket community in Spain that sympatric species usually differ in their song frequency spectra. However, Römer *et al.* (1989) found six sympatric tettigoniids on the coastal sand plain of Western Australia with substantially overlapping song frequencies. Further, male *Neoconocephalus spiza* are often syntopic with other coneheads that produce songs over the same frequency range (Greenfield, 1988), and Bailey (1976; 1979) found similarly overlapping ranges of closely related copiphorines in Uganda (*Ruspolia differens* (Serville), *R. fuscopunctata* (Karny) and *R. flavovirens* (Karny)), northern Australia (*Pseudorhynchus lessoni* Serville and *P. selonis* Bailey) and southwest Australia (*Mygalopsis pauperculus* (Walker), *M. sandowi* Bailey and *M. thielei* Bailey). It is also intriguing that a still unidentified morphological type of *Tympanophora*, perhaps a separate species from *T. similis*, which emits a song very similar to the call of *T. similis* has not been observed sharing the same habitat with *T. similis*.

A second mechanism preventing heterospecific interactions is temporal isolation (Alexander, 1960; Walker, 1974; Samways, 1976a; Latimer & Broughton, 1984). Seasonal differences between interacting species have been found for species in the genus *Tympanophora*. Thus males of *T.* n.sp.1. start singing earlier in the year than males of *T. similis*. (An explanation for this phenomenon could be that males of *T.* n.sp.1 develop under cooler conditions than *T. similis* and/or show different food preferences.) I have observed that at the beginning of its seasonal activity, males of *T.* n.sp.1. start singing one hour after sunset. When male *T. similis* developed into calling adults, the activity periods of both species overlapped for several days, and during this time *T.* n.sp.1 would begin to stridulate about one hour earlier in the evening. Also the number of calling males decreases when *T. similis* begins to call. It is therefore tempting to speculate that the shift in the daily singing activity observed for *T.* n.sp.1 is due primarily to the increasing number of singing *T. similis*. Some evidence for this hypothesis results from a laboratory experiment in which the number of calling males declined during stimulation with the call of *T. similis* (unpubl.observations). Greenfield (1988) reports on a reversal in diel periodicity of singing in *Neoconocephalus*. *N. spiza* sings during the day when syntopic with other species of the same genus. Males of these species emit songs that appear to inhibit the singing activity of *N. spiza*, however, the males of *N. spiza* call at night time when they can be considered allochronic to other species within the habitat.

Finally, an improvement in directional hearing may lead to an increase in the intensity of the conspecific signal at the receptor level – a third factor reducing interspecific acoustic interactions (Latimer & Broughton, 1984). Although theoretically plausible, there are no experimental data supporting this hypothesis. A suitable experiment should test whether hearing is more directional in species which are exposed to heterospecific song than in insects which are never share the habitat with heterospecifics.

In summary, the distribution patterns and singing cycles observed for bushcrickets in the field may be not only be the result of adaptation to the habitat and competition between conspecifics, but in certain cases they also may have evolved in response to calling heterospecific males, avoiding acoustic interference by these species.

Heterospecific Interactions, Female song Recognition: the course for future research

It has been shown that the temporal pattern of the song is important for species' specific recognition in crickets and frogs. Unfortunately only few data exist on bushcrickets (Morris & Fullard, 1983) (see Elsner & Popov, 1978 for a review). Changes in song structure during heterospecific acoustic interactions may affect the quality of the conspecific song. Here, only one study has focused on this problem. Female *Conocephalus nigropleurum* (Bruner) were confused by vertically projected calls of the congeneric species *C. brevipennis* in a laboratory trial (Bailey & Morris, 1986). Further experiments should test in greater detail how females of reactive species respond to the song of their own species recorded in the presence of various singing heterospecifics.

The problem has been studied in frogs, where females have been presented with masked conspecific signals (Narins, 1982; Ehret & Gerhardt, 1980). Many of the interspecific song interactions are regarded as having been derived from the normal plasticity of behaviour within the species (Broughton, 1965), and the response pattern are thought to have evolved for interactions between conspecifics (Samways & Broughton, 1976). It has been argued (Latimer & Broughton, 1984) that any increase in chirp duration during heterospecific interactions, similar to conspecific duets, may be the result of rebound excitation following a period of inhibition. Latimer (1981a) suggested that continuous songs may have the effect of a "supra-normal conspecific stimulus", and may therefore elicit stronger responses than normal (such as the production of microsyllables). However, even if the mechanisms underlying heterospecific interactions are identical to those leading to conspecific interactions, this does not necessarily imply that a male simply mistakes a heterospecific's call for that of a conspecific male. Nevertheless, there is no evidence to suggest that the responses reported during heterospecific interactions have anything to do with heterospecific recognition, as suggested for females of *Conocephalus nigropleurum* by Gwynne and Morris (1986). Table I indicates that the responses during heterospecific and conspecific songs are different in all conspecific interactions compared to heterospecific duets. Latimer's (1981a) "supra-normal stimulus hypothesis" (which explains heterospecific interactions as the result of stronger-than-normal reactions to the conspecific song, see above) seems doubtful, since there is a qualitative and not only quantitative change in the responses during heterospecific duets.

Finally, there is evidence that neither frogs nor birds just mistake the heterospecific song (Wickler, 1972; Schwartz & Wells, 1983). Schwartz and Wells (1983) conclude from their studies on frogs that conspecific calls may be simply masked by the heterospecific song. Just how important the frequency content of the song is for acoustically interacting males is not clear. A first attempt to elucidate this problem has been made by Greenfield (1988) where he stimulated males of *Neoconocephalus spiza* with tone pulses of different frequency and duration, and found the strongest effect (complete inhibition) at stimulus frequencies similar to the frequency peaks of the conspecific song. Field work should be extended, to find out (1) how common heterospecific interactions are, (2) if responsive species follow a certain strategy to avoid the close proximity of non-reactive species. This may take the form of negative phonotaxis, or shifts in the time used for calling. We also need to establish if (3) species singing close to a heterospecific have fewer chances of attracting a female, and (4) if they are more mobile than individuals stridulating near by conspecifics.

ACKNOWLEDGEMENTS

My own research was generously supported by grants from the German Research community and the Australian Research Grant Scheme (ARC). I am grateful to M.D.Greenfield and W.J.Bailey for critical comments on this manuscript.

REFERENCES

Alexander R.D. 1960. Sound communication in Orthoptera and Cicadidae. In: Animal Sounds and communication. Eds. W.E.Lanyon & W.N.Tavolga. Symp.Amer.Inst.biol.Sci. 7: 38-92.

Alexander R.D. 1967. Acoustical communication in arthropods. Ann.Rev.Entomol. 12: 485-526.

Bailey W.J. 1976. Species isolation and song types of the genus *Ruspolia* Schulthess (Orthoptera: Tettigoniidea). J.nat. Hist. 10: 511-528.

Bailey W.J. 1976. A review of Australian Copiphorini (Orthoptera: Tettigoniidae: Conocephalinae). Aust.J.Zool: 27, 1015-49.

Bailey W.J. & G.K. Morris. 1986. Confusion of phonotaxis by masking sounds in the bushcricket *Conocephalus brevipennis* (Tettigoniidae: Conocephalinae). Ethology 73: 19-28.

Bailey W.J. & P.B.Yeoh. 1988. Female phonotaxis and frequency discrimination in the bushcricket *Requena verticalis* (Tettigoniidae: Listroscelidinae). Physiol.Entomol. 13: 363-372.

Broughton W.B. 1963. Method in Bioacoustic terminology. In: Acoustic behaviour of animals. Ed. R.G.Busnel, Elsevier Amsterdam. pp. 3-24.

Broughton W.B. 1965. Song learning among grasshoppers? New Sci. 27: 338 – 341.

Busnel M.C. 1967. Rivalité acoustique et hierarchie chez l'Ephippiger (Insecte, Orthoptere, Tettigoniidea). Z.vergl.Physiol. 54: 232-245.

Busnel R.G., M.C. Busnel & B.Dumortier 1956. Relations acoustiques interspecifiques chez les Ephippigeres (Orthopteres, Tettigoniidae). Ann.des Epiphyt. 7: 451-469.

Cody M.L. & J.H.Brown 1969. Song asynchrony in neighboring bird species. Nature 222: 778-780.

Dadour J.R. 1989. Temporal pattern changes in the calling song of the bushcricket *Mygalopsis marki* (Bailey) in response to conspecific song. Journal of Insect Behaviour 2: 199-215.

Dumortier B. 1963. Ethological and physiological study and sound emissions in Arthropoda. In: Acoustic Behaviour of Animals. Ed. R.G.Busnel, Elsevier, Amsterdam, pp. 583-654.

Ehret G. & H.C.Gerhardt. 1980. Auditory masking and effects of noise on responses of the green tree frog (*Hyla cinera*) to synthetic mating calls. J.comp.Physiol. 141: 13-18.

Elsner N. & A.V. Popov. 1978. Neuroethology of acoustic communication. Adv.Insect Physiol. 13: 229-356.

Ficken R.W., M.S.Ficken & J.P. Hailman. 1974. Temporal pattern shifts to avoid acoustic interference in singing birds Science N.Y.. 183: 762-763.

Greenfield M.D. 1988. Interspecific acoustic interactions among katydids (*Neoconocephalus*): inhibition-induced shifts in circadian rhythms. Animal Behav. 36: 684-695.

Greenfield M.D. & K.C.Shaw 1983. Adaptive significance of chorusing with special reference to the Orthoptera. – In: Orthopteran mating systems. Eds. D.T.Gwynne and G.K.Morris. Westview, Boulder, Colorado, pp. 1-27.

Gwynne D.T. & W.J.Bailey. 1988. Mating System, mate choice and ultrasound calling in a zaprochiline katydid (Orthoptera: Tettigoniidae). Behaviour: 105, 202-223.

Gwynne D.T. & G.Morris. 1986. Heterospecific recognition and the origin of behavioural isolation in acoustic orthopterans (Insecta). Evolutionary Theory 8: 33-38.

Helversen D.v. & W.Wickler. 1971. Über den Duettgesang des afrikanischen Drongo *Dicurus adsimilis* Bechstein. Z.Tierpsychol. 29: 301-321.

Heiligenberg W. 1966. The stimulation of territorial singing in house crickets (*Acheta domesticus*). Z.vergl.Physiol. 53: 114-129.

Heiligenberg W. 1969: The effect of stimulus chirps on a cricket's chirping (*Acheta domesticus*). Z.vergl.Physiol. 65: 70-97.

Hill K.G. 1974. Carrier frequency as a factor in phonotactic behaviour of female crickets (*Teleogryllus commodus*). J.comp.Physiol. 93: 7-18.

Jones M.D.R. 1966. The acoustic behaviour of the bush cricket *Pholidoptera griseoaptera*. I. Alternation, synchronism and rivalry between males. J.exp.Biol.: 45, 15-30.

Jones M.D.R. & M. Dambach. 1973. Response to sound in crickets without tympanal organs (*Gryllus campestris* L.). J.comp.Physiol. 87: 89-98.

Kalmring K, W.Kaiser, C.Otto & R.Kühne. 1985. Co-processing of vibratory and auditory information in the CNS of different tettigoniids and locusts. In: Acoustic and vibrational communication in Insects. Eds. K.Kalmring and N.Elsner, Paul Parey, Berlin & Hamburg, pp. 193-202.

Kalmring K., A.Keuper & W.Kaiser (in press). Aspects of acoustic and vibratory communication in seven European bushcrickets. In: The Tettigoniids. W. J.Bailey.

Latimer W. 1981 a: Acoustic competition in bush crickets. Ecol. Entomol. 6: 35-45.

Latimer W. 1981 b. The acoustic behaviour of *Platycleis albopunctata* (Goeze) (Orthoptera, Tettigoniidae). Behaviour 76: 182-206.

Latimer W. & W.B. Broughton. 1984. Acoustic interference in bushcrickets. J.Nat.Hist. 18: 599-616.

Latimer W. & M.Sippel 1987. Acoustic cues for female choice and male competition in *Tettigonia cantans*. Anim.Behav. 35: 887-900.

Littlejohn M.J. & A.A.Martin. 1969. Acoustic interaction between two species of Leptodactylid frogs. Anim.Behav. 17: 785-791.

Littlejohn M.J., P.A.Harrison & R.C.Mac Nally. 1985. Interspecific acoustic interactions in sympatric populations of *Ranidella signifera* and *R. parinsiguifera* (Anura: Leptodactylidae). In: The Biology of Australasian frogs and reptiles. Eds. G.Grigg, R.Shine & H.Ehrmann. Royal Society of New South Wales; pp. 287-296.

McHugh R. 1972. Aspects of acoustic interactions in the bushcricket genus *Metrioptera* (Orth., Tettigonioidea), PhD thesis, London University.

Morris G.K. & J.H.Fullard. 1983. Random Noise and Congeneric discrimination in *Conocephalus*. In: Orthopteran Mating System. Eds. D.T.Gwynne & G.K.Morris. Colorado, Boulder Press. pp 73-96.

Narins P.M. 1982. Effects of masking noise on evoked calling in the Puerto Rican coqui (Anura: Leptodactylidae). J.comp.Physiol. 147: 439-446.

Paul R.C. 1976. Species specificity in the phonotaxis of female ground crickets (Orthoptera: Gryllidae: Nemobiinae). Soc.Amer. 69: 1007-1010.

Popov A.V. & V.F.Shuvalov 1977. Phonotactic behaviour of crickets. J.comp.Physiol. 119: 111-126.

Popp J.W., R.W.Ficken & J.A.Reinartz 1985. Short-term temporal avoidance of interspecific acoustic interference among forest birds. Auk 102: 744-748.

Römer H. 1987. Representation of auditory distance within a central neuropil of the bushcricket *Mygalopsis marki*. J.comp.Physiol. 161: 33-42.

Römer H., W.J.Bailey & J.Dadour. 1989. Insect hearing in the field. III. Masking by noise. J.comp.Physiol. 164: 609-620.

Samways M.J. 1976 a. Song modification in the Orthoptera. I. Proclamation songs of *Platycleis* spp. (Tettigoniidae). Physiol.Entomol. 1: 131-149.

Samways M.J. 1976 b. Song modification in the Orthoptera. III. Microsyllabic echemes in *Platycleis* spp. (Tettigoniidae). Physiol.Entomol. 1: 299-303.

Samways M.J. 1977 a. Song modification in the Orthoptera. IV. The *Platycleis intermedia/ P. affinis* interaction quantified. Physiol.Entomol. 2: 301-315.

Samways M.J. 1977 b. Bushcricket interspecific interactions in the field (Orthoptera, Tettigoniidae). J.Nat.Hist. 11: 155-168.

Samways M.J. & W.B.Broughton 1976. Song modification in the Orthoptera. II. Types of acoustic interactions between *Platycleis intermedia* and other species of the genus (Tettigoniidae). Physiol.Entomol. 1: 287-297.

Schatral A. & P.B.Yeoh. in press. Spatial distribution, calling and interspecific acoustic interactions in two species of the Australian tettigoniid genus *Tympanophora* (Orthoptera, Tettigoniidae). J. Zool.

Schwartz J.J. & K.D.Wells 1985. Intra- and interspecific vocal behaviour of the neotropical treefrog *Hyla microcephala*. Copeia 1985: 27-38.

Schwartz J.J. 1987. The importance of spectral and temporal properties in species and call recognition in a neotropical treefrog with a complex vocal repertoire. Anim.Behav. 35: 340-347.

Shaw K.C. 1968. An analysis of the phonoresponse of males of the true katydid *Pterophylla camellifolia* (Fabricius) (Orthoptera: Tettigoniidae). Behaviour 31: 203-260.

Sullivan B.K. & Leek M.R. 1986. Acoustic communication in the Woodhouse's toad *Bufo woodhousei*). I. Response to calling males to variation in spectral and temporal components of advertisement calls. Behaviour 98: 305-319.

Thorpe W.H. 1963. Antiphonal singing in birds as evidence for avian auditory reaction time. Nature 197: 774-776.

Walker T.J. 1974. Character displacement and acoustic insects. Am.Zool. 14: 1137-1150.

Wassermann F.E. 1977. Intraspecific acoustical interference in the White-throated sparrow (*Zonotrichia albicollis*). Anim.Behav. 25: 949-952.

Wells K.D. 1977. The social behaviour of anuran amphibians. Anim. Behav. 25: 666-693.

Wickler W. 1972. Duetttieren zwischen artverschiedenen Vogeln im Freiland, Z.Tierpsychol.31: 98-103.

10. NEUROANATOMY OF TETTIGONIIDS

Reinhard Lakes
Thomas Schikorski

Introduction

The highly developed auditory and vibratory communication system of tettigoniids has attracted the attention of biologists since the early studies of Burmeister (1832). The tympanic organ, situated beneath the knee of the foreleg, has been studied at the level of the single cell, both at the periphery and within the central nervous system (e.g. Kalmring *et al.*, 1978; Oldfield, 1984; Römer *et al.*, 1988). Its complex acoustic structure is reviewed by Bailey (chapter 12). A comparative study on the analagous ear of crickets and the ear of acridids, situated in the first abdominal segment, has been reported earlier (Ander, 1939). However, despite our knowledge of the anatomy and physiology of this system our understanding lags far behind that for other orthopterans, such as the ensiferan gryllids, and caeliferan acridids.

In this chapter we review data on the two major sub-families, and extend them with our own results on the Tettigoniidae. We commence the review by taking a general perspective of the disposition of mechanoreceptors in bushcrickets on the legs as well as those on the mouthparts. Furthermore, the gross anatomy of the thoracic ganglia is discussed, allowing us to assess the relationship of single cells both to each other, and to more central landmarks such as those commissures or neuropile areas within the insect's central nervous system.

The Peripheral Nervous System and its Sense Organs

The nerve pattern and sense organs in the different body parts are considered here. We concentrate on the mechanoreceptors whose cell bodies lie in the periphery, sending their axons to the central nervous system where they end with a specific central projection.

The innervation of the wings

The wings of the bushcrickets are not only used for flight but also for stridulation. In many tettigoniid species the wings are short and used only in sound communication. The wings bear campaniform sensilla as well as different types of hair receptors, with most receptor hairs innervated by more than one sensory cell (Erhardt, 1916; Fudalewicz-Niemcyk, 1956, 1958). Left to right dimorphism in terms of the arrangement of the sensilla is common between the forewings, and less surprisingly the organisation of receptor groups differs from fore to hind wings. However, the innervation pattern has been used only as a phylogenetic criterion within the orthopteroidea (Fudalewicz-Niemcyk, 1956). Although wing morphology has been examined in relation to sound production in a range of tettigoniid species (Keuper *et al.*, 1988), no attempt has been made to show the role of the sensilla in the different contribution of each wing during stridulation. A feedback input from the campaniform sensilla of the wing to the neural control of stridulation has been shown in crickets (Schäffner & Koch, 1987), and in locusts wing sensilla modulate the motor pattern of the flight system (Wolf & Pearson, 1987). It is even possible that the same wing sensilla are involved in both functions - stridulation and flight.

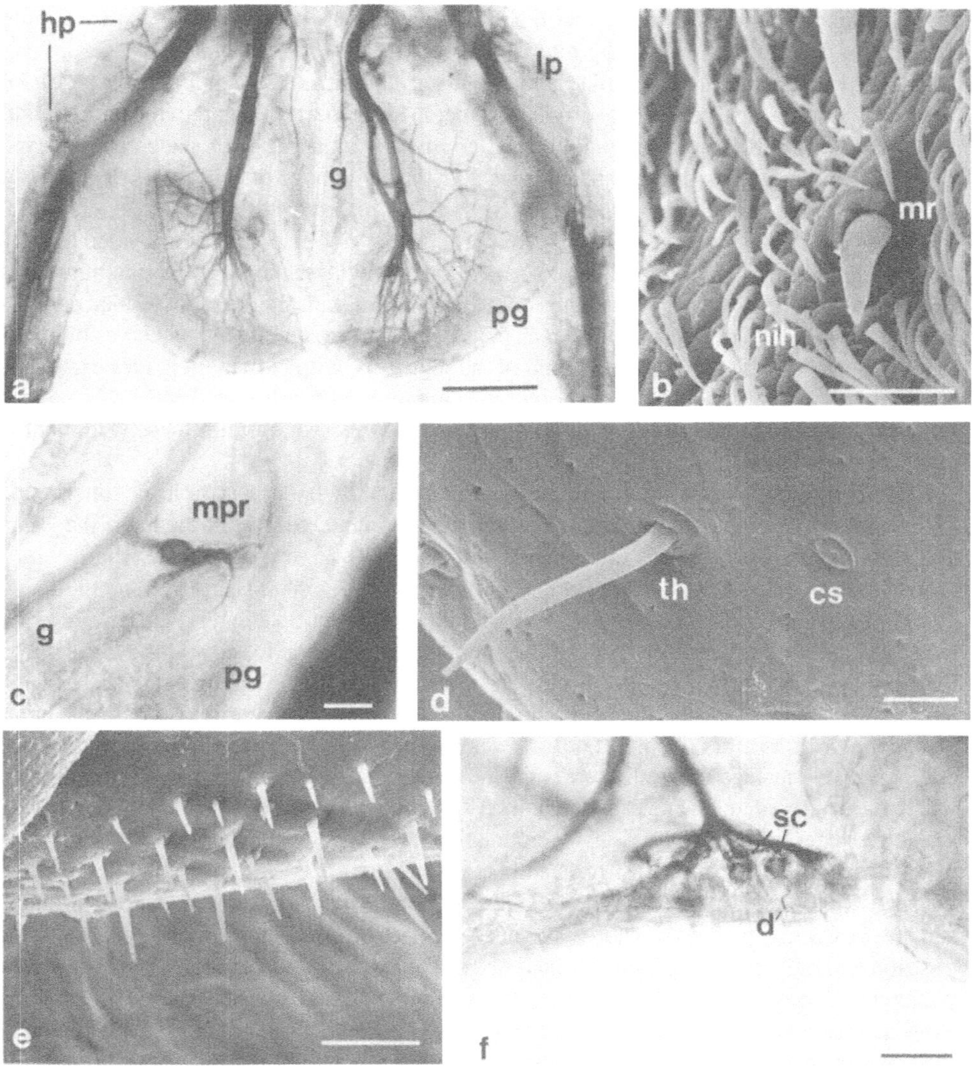

Fig. 1. Sensilla and their innervation of the mouth parts of European tettigoniids, showed with Cobalt backfillings and scanning electron microscopy. a) Labium of *Tettigonia cantans* with Cobalt stained nerves. Branches of the nerve enter the palp (lp), glossa (g) and paraglossa (pg), where they arborize and innervate the cuticle sensilla (scale: 500 fm). b) Detail of the epipharyngeal surface of the labrum of *Decticus verrucivorus*, showing innervated mechanorecetor hairs (mr) and non-innervated hairs (nih). Scale: 20 *fm*. c) Multipolar receptor (mpr) at the joint of glossa and paraglossa of the labium of *Tettigonia cantans*. Scale: 50 *fm*. d) Trichoid mechanoreceptive hair (tr) and campaniform sensilla at the maxilla of *Decticus albifrons*. Scale: 20 *fm*. e) Hairplate at the base of the maxillary palp of *Decticus albifrons*. Scale: 50 *fm*. f) Innervation of a hairplate at the labial palp of *Tettigonia cantans*. Each hair is innervated by one sensory cell (sc) whose dendrite (d) is attached to the base of the hair. Scale: 100 *fm*.

The innervation of the mouth parts

The mouth parts are involved in food recognition and food handling and bear many chemo- and mechanoreceptive sensilla to perform these tasks. Whereas the anatomy of these receptors has received some attention in locusts (e.g. Blaney & Chapman, 1969a,b) and crickets (e.g. Charline *et al.*, 1984; Klein, 1981), the tettigoniids appear as an underworked area and hence, we report some of our data from the European bushcrickets with scanning electron microscopy (SEM) studies and Cobalt backfillings.

The labrum, mandibles, maxillae and the fused labium is each supplied by a pair of bilaterally symmetrical nerves that originate either in the tritocerebral part of the brain (labral nerve) or in the suboesophagal ganglion. After entering an appendage, each of the nerves containing motor and sensory axons bifurcates. In the labium one branch enters the palp and the other the separated glossa and paraglossa (Fig. 1a), and both nerves divide again, supplying the anterior and posterior parts of the appendage. The glossa is innervated by only one nerve covering the inner and outer surface. The paraglossal nerve splits into numerous branches which often form complex anastomoses, and contains the axons of bipolar sensilla at the cuticle (Fig. 1a). The sensilla of each palpal segment are innervated by nerves of minor order which originate at the posterior end of the segments, with both main nerves extending to the tip of the palp where they arborize into numerous branches to the densely packed sensilla at the tip (Rath, 1888). This basic scheme of innervation within each segment is repeated in the legs.

The chemoreceptors of the mouth parts are identified by both their multiple innervation and the morphology of the hair shaft. Scanning electron microscopy of the tip of the labial and maxillary palps shows trichoid sensilla with short (10-25 μm) curved hair shafts with a pore at the tip. These sensilla are multiply innervated (Rath, 1888; Hsü, 1938; unpubl. res.) in contrast to many of the hairs more proximal of the palp. Multiple innervated hairs are also found at the margins of the interior appendages such as the paraglossa or at the inner rim of the epipharyngeal surface of the labrum. The outer surface of the labrum bears trichoid and basicon sensilla with mechanoreceptive function, as do most of the proximal palp segments and the interior mouth appendages. But not all of the hairs are innervated. At the median epipharyngeal surface of the labrum, non-innervated hairs form a dense pattern amongst which are scattered mechanoreceptive hairs (Fig. 1b). The presumed mechanical function of the single hairs is based on the long smooth hair shaft (Fig. 1d) and the innervation by only one sensory cell. The campaniform sensilla are either organized in fields of 8-15 cells (as at the segmental border between the cardo and stipes of the maxilla), or are localized as single sensilla at various places over the appendage (Fig. 1d). Hair plates which register the leg position and movement are found at the bases of the palpal segments (Fig. 1e,f). These hair plates consist of 20-30 short trichoid sensilla each with one sensory cell (Fig. 1f). Besides the external mechanoreceptors, internal sensory cells also monitor the position or movement of the appendage. One example is the multipolar receptor located at the joint of the glossa and paraglossa (Fig. 1c). These receptor cells are found at most of the joints, where their dendrites stretch into both segments. Chordotonal organs are at the first segments of the palps and attached to apodemes fixed at the next segment border.

The innervation of the legs

The sense organs and muscles of the legs are supplied by two nerves, N3 and N5 (after Campbell, 1961). Nerve N3 innervates two hair plates at the proximal coxa in addition to the muscles and sense organs of the pleurite. Nerve N5 bifurcates before entering the coxa into an anterior (N5B1) and a posterior branch (N5B2) (Fig. 2a). In some orthopteroid species the latter branch can be missing (Friedrich, 1930), but in crickets and grasshoppers the basic pattern is almost the same (Grosch *et al.*, 1985; Mücke & Lakes, 1988). The leg nerve (N5B2) lies ventro-posteriorly in the femur and innervates the muscles and posterior sensilla with small branches within this segment. Just below the genual joint the "subgenual nerve" branches off and the main trunk divides into the "tibial nerve" and the "tarsal nerve" (Schumacher, 1973). The subgenual nerve contains not only axons of the posterior scolopidial sensilla of the subgenual organ, but also those from a large tibial field (9-13 sensilla) of campaniform sensilla located dorsally to the subgenual organ, and from hair sensilla of the posterior proximal tibia. The "tarsal nerve" extends to the tarsi and supplies ventral external sensilla of the tibia (e.g. the campaniform sensilla at the base of the distal spines). By contrast, the anterior proximal spines are sometimes supplied by the "tympanic nerve" (N5B1) (Fig. 5a). At the base of the spines a campaniform sensillum is found, and in a few cases (e.g. the midleg of *Saga pedo*) there are multiple innervated trichoid hairs (Fig. 3, 5a). The "tibial nerve", the second main trunk of the leg nerve in the tibia, bifurcates again and one branch extends dorsally and anteriorly to the tip of the tarsus. The axons of the 10-13 receptor cells of the tibio-tarsal chordotonal organ (e.g. in *Ephippiger ephippiger*) join this nerve trunk. The "tibial branch" runs near the middle axis of the tibia and supplies the muscles and dorsal lateral sensilla.

The "tympanal nerve" (N5B1) innervates the anterior trochantal and coxal hair plates as well as fields of campaniform sensilla, and another side branch bifurcates and supplies the proximal and distal scoloparium of the femoral chordotonal organ (Lakes & Mücke, 1989).

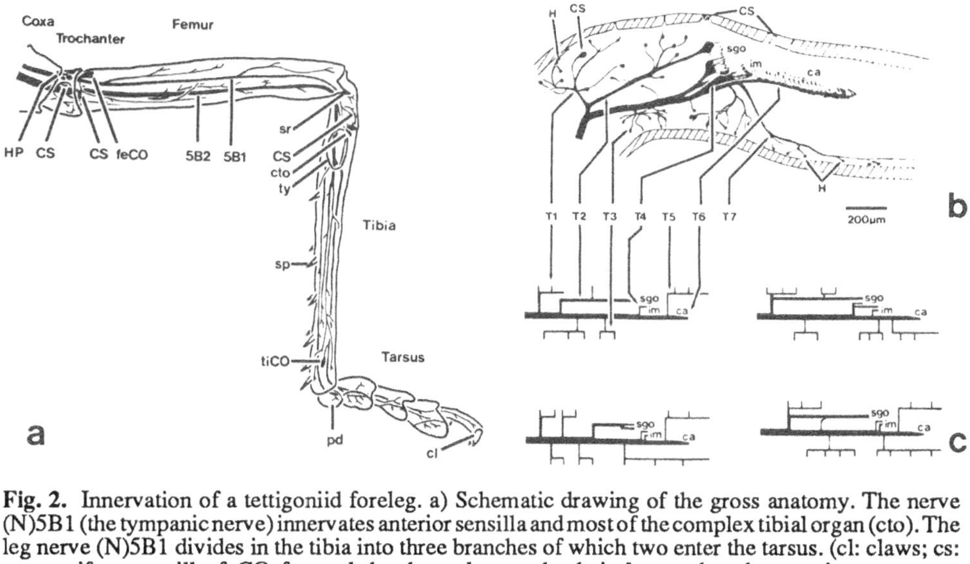

Fig. 2. Innervation of a tettigoniid foreleg. a) Schematic drawing of the gross anatomy. The nerve (N)5B1 (the tympanic nerve) innervates anterior sensilla and most of the complex tibial organ (cto). The leg nerve (N)5B1 divides in the tibia into three branches of which two enter the tarsus. (cl: claws; cs: campaniform sensilla; feCO: femoral chordotonal organ; hp: hairplates; pd: pads; sp: spines; sr: spurs; tiCO: tibial chordotonal organ; ty: tympanum. b) Semischematic drawing of the branching of the tympanic nerve in the tibia of *Ephippiger ephippiger*. The main branches are indicated with T1 - T7. T1, T3, T5 and T7 innervate the cuticle sensillae and T2, T4 and T6 the subgenual organ (sgo), intermediate organ (imo) and the *crista acustica* (ca), respectively. h: hair sensilla. c) Schematic innervation pattern of the tympanic nerve in four individual of *Ephippiger ephippiger*. For indication of the nerve branches see upper left drawing. Note the differences in the branching points and the number of nerves innervating the organs. (after Mücke & Lakes, in press).

In contrast to the locust, this organ is located in the proximal femur of the hindleg of tettigoniids, and not in the distal part of the femur (Usherwood *et al.*, 1968; Theophilidis, 1986). The branching pattern of the tympanal nerve within the tibia consists of seven branches (T1-T7), innervating the anterior surface and the scolopidial organs (Fig. 2b; Lakes & Mücke, 1989; Schikorski & Lakes, 1988). Although the basic pattern, which is probably laid down in a sequence of pathfinding processes during embryogenesis (Keshishian &

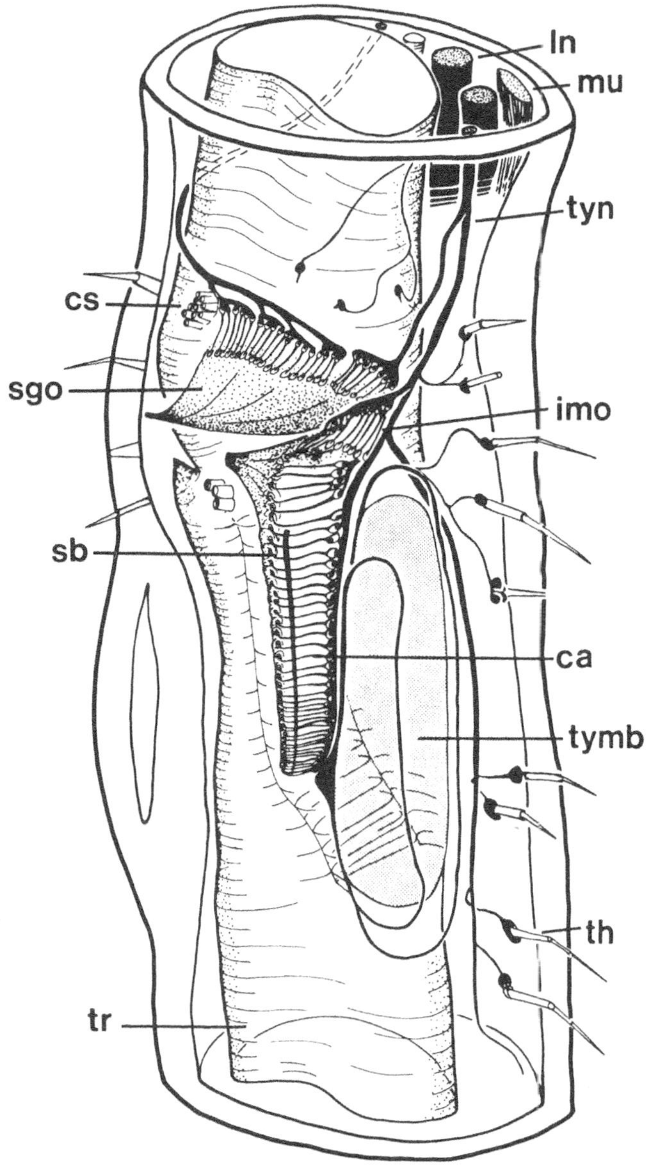

Fig. 3. The proximal tibia of a foreleg with the orientation of the receptor cells. ca: *crista acustica*; cs: campaniform sensilla; imo: intermediate organ; ln: leg nerve; mu: muscle; sb: supporting band; sgo: subgenual organ; th: trichoid sensilla; tr: trachea; tymb: tympanal membrane (grey) and tyn: tympanic nerve.

Bentley, 1983 for the grasshoppers), does not vary in those species we have investigated, the minor branches do show some inter-individual differences (Fig. 2c). Axons could join the main nerves at different locations, or even run in different nerves. Thus, subgenual axons could run within one branch of the tympanic nerve and leave it joining the "subgenual nerve" – a branch of the leg nerve. It appears that single axons have the possibility of "choosing" their own pathways. However, unlike in crickets where in regeneration processes the original innervation pattern could be restored (Huber, 1987), the nerves in regenerated legs of the tettigoniids are completely changed (Lakes & Mücke, 1989).

The complex tibial organ

A pattern of four scolopidial organs, on the femur, the proximal and distal tibia, and the tarsus is common among insects (Eggers, 1928; Debaisieux, 1938; Ander, 1939). In tettigoniids, the scolopidial organs in the proximal tibia of each leg are the main sensory organs transducing airborne and substrate sound (Kalmring *et al.*, this volume). These receptors form the complex tympanal organ which was originally described by Burmeister (1832), and then followed by the studies of v. Siebold (1844), Hensen (1866), Graber (1876), and v. Adelung (1892). An excellent description of the organs was given by Schwabe (1906) whose figures are still used for illustration of the structures. Friedrich (1927, 1928) extended the investigations to the mid-and hindlegs of tettigoniids. More recently electronmicroscopic studies were done by Schumacher (1975, 1979) and Burakov & Zhantiev (1984). However, in spite of the detailed data on the structures of the organ the transduction process is still unclear.

The tympanal organ consists of three main parts (i) the subgenual organ, (ii) the intermediate organ and (iii) the *crista acustica* (Fig. 3). The tympanal membranes at the anterior and posterior side are often cupped by a cuticle wall (Bailey, this volume), and consist of a dorsal fibrillar or membranous part with a thicker part on the ventral side covered by a hypodermal cell layer (Schumacher, 1973). The anterior membrane in most species is smaller than the posterior, but not to the same extent as in crickets (Knetsch, 1939). The membranes are closely connected with tracheal walls from the "acoustic trachea". The trachea has a permanent opening, the acoustic spiracle in the prothorax, and extends without branching from this opening to the tibia. Its structure provides resonances that tune the ear to airborne sound (Lewis, 1974; Chukanov & Zhantiev, 1978; Hill & Oldfield, 1981; Bailey chapter 12). The trachea splits into two branches that are joined together with a chitinous septum at the tympanal organ. The main trachea of the mid- and hindleg have no acoustic spiracle. They branch below the knee and are connected by a broad strand of tissue. The tibial cuticle of these legs show no sign of any tympanal differentiation (Schikorski & Lakes, 1988), however, we were able to find all three receptor organs in each leg pair.

The subgenual organ lies in the dorsal haemolymph canal and is highly developed in orthopterans compared to other insect orders (Autrum & Schneider, 1948). The sensory cells are supplied by two nerves and arranged in a halfcircle at the dorsal anterior part of the leg. They are inserted into cap cells, which are joined together and form a fan-shaped "membrane" that is fixed at one point to the dorsal posterior side of the leg hypodermis, just beneath a field of campaniform sensilla. The "membrane" is additionally covered by a layer of accessory cells. The scolopales of the receptor cells are smaller than those of the *crista acustica* (Schumacher, 1979). The subgenual organ of the three leg pairs are similar in anatomy and the cell numbers are almost equal (Friedrich, 1927; Schumacher, 1979). In *Psorodonotus illyricus* 40 receptor cells in each leg (unpubl. data) form two scoloparia of 20 cells each. The organ apparently perceives the acceleration of substrate borne sound within a frequency range of 1000 to 5000 Hz (Autrum, 1941; Autrum & Schneider, 1948; Schnorbus, 1971) and high intensity airborne sound up to 3 kHz (Kalmring *et al.*, 1978).

The least known organ in respect to function is the intermediate organ (v. Adelung, 1892), located between the subgenual organ and the *crista acustica*. It is divided into two parts (Friedrichs, 1927) and normally innervated by two nerves, although many variations of this configuration can be found (see above). The receptor cells form a continuous row which is extended to the *crista acustica* (Fig. 5b,c). The scolopidial cells resemble those of the subgenual organ although the accessory cells are missing (Schumacher, 1979). The cap cells are joined with tight junctions and their fibrils extend to the tectorial membrane (Schumacher, 1979). These *membrana tectoria* are fixed at two points at the dorsal cuticle (Fig. 4,5c) near

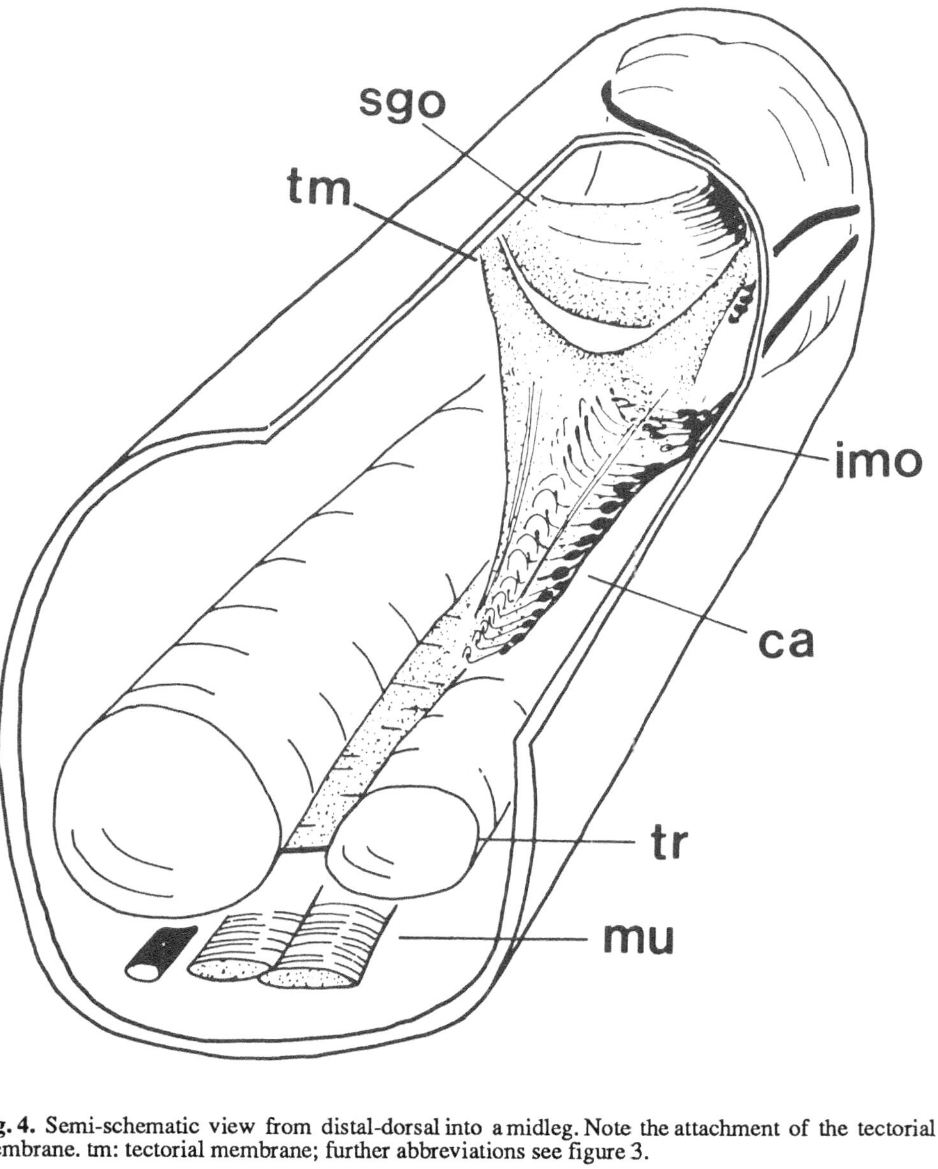

Fig. 4. Semi-schematic view from distal-dorsal into a midleg. Note the attachment of the tectorial membrane. tm: tectorial membrane; further abbreviations see figure 3.

the proximal end of the tympanal organ and covers the *crista acustica*. The number of scolopidial cells is species specific and decreases from the foreleg to the hindleg. For example, in *Tettigonia viridissima* 17 receptor cells were found in the foreleg, 12 in the midleg and 11 in the hindleg (Schumacher, 1979).

As already stated by Knetsch (1939), the continuous row of receptors from the intermediate organ to the *crista acustica* makes the differentiation of the organs difficult. In most of the preparations we have examined, one nerve innervates the sensilla in a transition zone between both organs (Fig. 5b,c). The proximal cells belong to the intermediate organ, and the distal cells belong to the *crista acustica*. Thus, intermediate forms occur e.g. with a small scolopale (characteristic of the intermediate organ) and with a cap cell attached to the tectorial membrane.

This near continuous sequence with the *crista acustica* is also reflected in the organ's physiology. Recordings of responses of single receptor cells to acoustic stimuli reveal more cells with a characteristic type of threshold curve than there are sensilla in the *crista acustica* (Kalmring *et al.*, in prep.). Thus, the additional sensilla are probably located in the intermediate organ, which have a response to airborne sound from 2-14 kHz with a 5-7 kHz optimum (Zhantiev & Korsunvoskaja, 1978; Kalmring *et al.*, 1978). The organ is present in the Haglidae, which are considered as the base group of tettigoniids and gryllids, and the receptor cells form one row with the *crista acustica* cells (Mason, pers. comm.). The gryllids do not possess this organ (Michel, 1974) although some authors subdivide the tracheal organ into three closely related parts (Eibl, 1978). In the cockroach *Periplaneta americana*, a "distal organ" is described which reacts to vibration frequencies of 30-700 Hz. This organ is also thought to be an organ for measuring the haemolymph pressure (Schnorbus, 1971).

The *crista acustica* is a row of scolopidial cells that lies on the anterior side of the tympanic trachea and ends distally before the trachea join again. The receptor cells are formed in the distal part as an alternating row, whereas the cap cells lie in one continuous row (unpubl. res.; Schumacher, 1979). However, the dendrites of two cells can cross: a more proximal receptor cell may insert in a more distal cap cell than its neighbouring cell and vice versa (Kalmring *et al.*, in prep). Each scolopidium consists of four cells: the bipolar receptor cell, an enveloping cell, the scolopale cell and a cap cell. The dendrites stretch at the dorsal surface of the trachea to the median axis of the leg. It terminates with a ciliary segment of a 9+2 filament structure which proceeds to the cap cell. The cilium is anchored at two basal bodies and is dilated just before entering the scolopale cap, and the scolopale (a differentiation of the scolopale cell) is a thickened wall with 8 ribs (Schumacher, 1975, 1979). The cap cells are closely covered by the tectorial membrane (Hensen, 1866). Two supporting bands (Schwabe, 1906; Schumacher, 1975) beneath the cap cells stretch the tectorial membrane to the cuticle insertion. The supporting bands have a fibrillous surface and increase in thickness to the proximal end from 7 to 20 μm in diameter.

The *crista acustica* has both a species specific length and cell number, but these are not correlated with each other (Schumacher, 1973, 1979). Closely related species have similar cell numbers, for example *Ephippiger perforata* has about 27 receptor cells, *E. ephippiger* 29 cells, and the related *Uromenus rugosicollis* 30. In those species that have investigated the number varies between 15 in *Meconema* (Schumacher, 1979) and 59-61 in *Psorodonotus denticauda* (Kalmring *et al.*, in prep.). However, the variation in cell number in one species is rarely more than two cells, which is similar to the crickets (Young & Ball, 1974a). Species with an uncovered tympanum seem to have generally higher cell numbers than those with the membranes cupped behind slits.

The *crista acustica* is the main sensory organ for transduction of airborne sound. Several authors showed the responses of individual cells to airborne sound from 5 to 50 kHz with recordings from this organ in different species (e.g. Zhantiev & Korsunovskaja, 1978; Oldfield, 1984). The anatomical order is reflected in a tonotopical pattern; distal receptors transduce airborne sound of higher frequencies than more proximal cells. But the relevant structures for the transduction of the sound are still in doubt. The stretching of the cilium in its long axis is probably the way in which the airborne signal is translated into electrical changes of the dendrite (Michel, 1974). Thus the structure of the sensillum itself, determined by its somatotopic position could be important in reflecting the tuning characteristic of the receptor cell ("intrinsic factor", see Oldfield, 1988). On the other hand the attached structures to the dendrite, the tectorial membrane and the acoustic trachea may be important for the stretching of the dendrite.

The *crista acustica* of the mid-and hindlegs resembles those of the foreleg in their overall morphology (Fig. 4, 5b,c; Friedrich, 1927, 1928; Schumacher & Houtermanns, 1975; Schumacher, 1979), a feature also observed in crickets (Young & Ball, 1974b). In tettigoniids the scolopidia of the atympanate legs are longer and smaller than in the foreleg (Houtermans & Schumacher, 1974) and are attached to the leg structures differently. The trachea is divided into two branches that are separated by a band of connective tissue (Fig. 4). The receptor cells lie at the anterior trachea and extend their dendrites to the middle of the leg (Fig. 5b). The dendrites insert in cap cells which are enlongated (Fig. 5c) and are covered by a tectorial membrane including the cap cells of the intermediate organ as in the foreleg. In contrast to the reported absence of the supporting bands in the mid- and hindlegs (Houtermans & Schumacher, 1974), our SEM investigations show fibrillous thickenings of the tectorial membrane in a 10 μm wide band, running towards the attachment points at the cuticle as in the forelegs (unpublished ms.).

The conditions in the atympanate legs resemble to a certain degree, the situation in the foreleg early in the postembryonic development. In the second larval instar the trachea of the foreleg are separated and join each other in the later instars, while the tympanic membranes are progressively differentiated (Graber, 1876). The morphology of the foreleg in the completely atympanate tettigoniid *Phasmodes ranatriformis* is similar to that of the mid- and hindlegs of other adult tettigoniids (Lakes *et al.*, in manuscript). The number of sensilla of the *crista acustica* seemed to increase in all three leg pairs until the later instars as revealed with Cobalt stainings. However, at least the majority of the receptor cells are already formed during embryogenesis (Graber, 1876; Meier & Reichert in prep.). The receptor cells of the *crista acustica* in all three pairs of legs differentiate probably from early anlagen, whereas a differentiated intermediate organ is not visible in the embryo (Meier, pers. comm.).

Commonly the number of receptor cells decreases from the foreleg to the hindleg, and although the rate of decrease is not constant within the different species, a reduction of one third to the next caudal leg pair is observed frequently in European tettigoniids. One example is *Psorodonotus illyricus* where there are 29-30 cells in the foreleg (Kalmring *et al.*, in prep.) 21 cells in the midleg and 11 cells in the hindleg. Other species have a similar cell number in the rear legs. *Polysarcus denticaudus* has about 12-13 cells in the atympanate legs, compared to about 60 cells in the foreleg. In the hindleg the cell number can be as low as 4 cells (*Meconema*, Schumacher, 1979).

"Standard" plan of a pregenital abdominal segment

To give a "standard" plan of peripheral innervation within the heteronomous segmentation of the insect body, we have considered the part muscles and sense organs have played in influencing the branching pattern of segmental nerves. The unspecialized pregenital abdominal

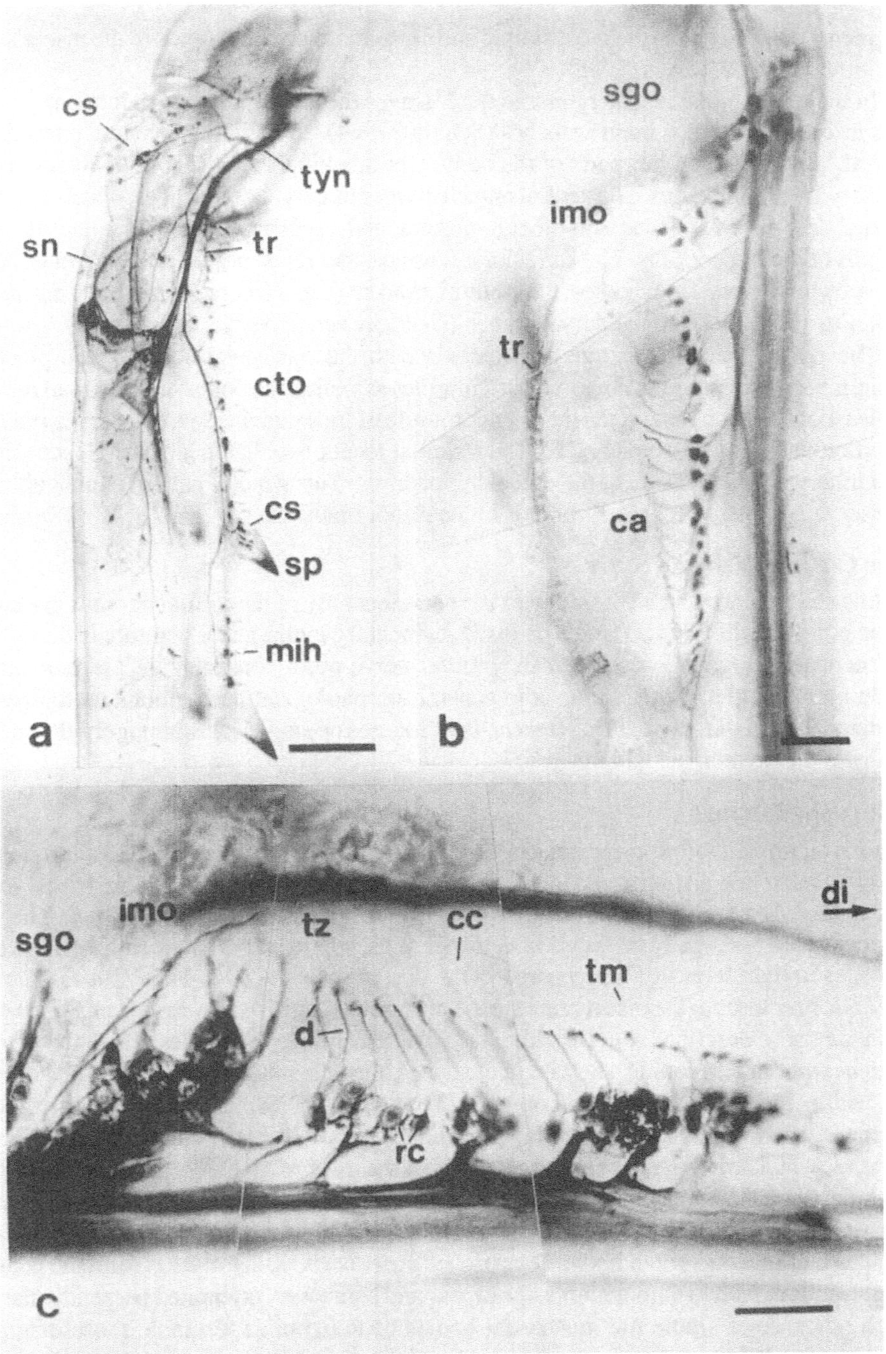

Fig. 5. Proximal tibia and *crista acustica* of the midleg of *Saga pedo*. a) Lateral view with stained tympanal and subgenual nerve. cs: campaniform sensilla; cto: complex tibial organ; mih: multiple innervated hair; sn: subgenual nerve; sp: spine; tr: trachae and tyn: tympanal nerve. b) Ventral view on the *crista acustica* (ca) and intermediate organ (imo). Note the continuous row from the *crista acustica* to the intermediate organ. c) Lateral view of a *crista acustica* and an intermediate organ. The dendrites (d) are stretched to the cap cells (cc) which are covered by the tectorial membrane (tm). di: distal; rc: receptor cell and tz: transition zone.

segments probably have the most simple and uniform innervation fields in the insect's body (Schmitt, 1964; Weevers, 1985).

In the tettigoniids one lateral nerve trunk leaves the corresponding abdominal ganglion and bifurcates into two main branches (Schmitt, 1964), supplying the ventral (sternal) and dorsal (pleural and tergal) parts of the body. Our labelling shows that both nerves contain sensory and motor axons. The ventral branch divides into smaller branches, which supply the sternal, tergo-sternal and sterno-pleural muscles, and a sensory branch innervates the sense organs of the hypodermis. The spiraclular muscles, and sense organs are innervated from a nerve which rises from either the ventral branch (e.g. *Neoconocephalus exilscanoris*, Schmitt, 1964), or from the dorsal branch (*Decticus verrucivorus*, unpubl. res.).

The dorsal branch forms an anastomosis with the transverse nerve of the preceding ganglion, and supplies the dorso-ventrad muscles as well as the dorsal-longitudinal muscles. At least one branch innervates the opener muscle of the spiracle. Sensory nerves run to the chordotonal organ of the pleural fold, at a similar location to that in locusts (Hustert, 1978), and innervate hair sensilla at the dorsal hypodermis. This general pattern is undoubtedly a conservative feature within variations in the minor branches.

The Central Nervous System

In this second part we give a survey of the neuroanatomy of the central nervous system. As in other insects the CNS is formed by fused segmental ganglia, and is commonly divided into the cephalic nervous system and the ventral nerve cord, consisting of the thoracic and abdominal ganglia. Apart from some general morphological descriptions on the nervous system (Steinmann, 1961, 1962), recent work focuses on single cell stainings of the auditory pathway in the thoracic ganglia.

The cephalic ganglia

Main structures of the protocerebrum are the optic lobes, the mushroom bodies and the central body which is considered as the highest center for integration of information. In tettigoniids the optic lobes are well developed and highly differentiated (Nesbitt, 1941). The optic pathways are practically the same as in the acridids, where, for example, information about changes in light intensity is processed in the ocellar pathway (Goodman, 1981). The most obvious structures of the insect brain are the mushroom bodies (*corpora pedunculata*) which were already described by Dujardin (1850). Hanström (1940) described the *corpora pedunculata* in tettigoniids showing the calyces, the peduncle and the lobes. The cortex is formed by the somata of the Kenyon-cells (Kenyon, 1896), which are the intrinsic neurons of the *corpora pedunculata* forming the connections to other brain structures. The two calyces, which are the input structures, connect the mushroom body via the tractus optico-globularis with the optic lobes, a junction that is not found in the acridiid *Locusta migratoria* (Ernst *et al.*, 1977). The *tractus olfactorio-globularis* links the antennal lobes with this input structure of the mushroom body. The parallel fibers of the Kenyon-cells form the peduncle which ends at its two output points, the alpha- and beta lobes, orientated perpendicularily to each other. As a whole the mushroom bodies of tettigoniids resemble those of gryllids (Jawlowski, 1954).

The antennal nerve is the main nerve of the deutocerebrum and ends within the antennal lobes, which are divided into a series of spherical neuropil compartments called glomeruli. It is thought that each glomerulus represents the convergence of inputs with some similarity of response. Gross morphological observations show that the antennal lobe in bushcrickets is divided into a ventral and a dorsal part (Hanström, 1940), and the glomeruli are larger but fewer in number than in *Periplaneta americana* (Boeckh & Ernst, 1987). The more ventrally located tritocerebral glomeruli, which are formed by the terminals of the labral gustatory receptors, are also highly developed (Jawlowski, 1954).

Morphology of the ventral nerve cord

The pro- and mesothoracic ganglia are equal in size, and each consists of one neuromere (Nesbitt, 1941). The metathoracic ganglion of tettigoniids is fused with the first abdominal neuromere (Steinmann, 1962). The six abdominal ganglia are located in the segments 1, 3-6 and 8. The first abdominal ganglion consists of two neuromers and supplies the second and third segments. The terminal ganglion consists of four neuromeres and innervates the genital tract and the numerous sensilla on the cerci.

In recent years special focus has been given to the neuroanatomy of the afferent auditory pathway of bushcrickets (Rehbein, 1973; Kalmring *et al.*, 1978; Zhantiev & Korsunovskaja, 1978; Oldfield, 1983, 1988; Romer, 1983, 1985, 1987). However, as the descriptions were almost entirely restricted to the shapes of single cells within the prothoracic ganglion, nothing can be said about the functional connections of neuronal networks based on morphological data. The anatomy of the meso- and prothoracic ganglia is described below and is based on stained sections (after Wiggelsworth, 1957) of different European tettigoniids.

The gross plan of the mesothoracic ganglion in tettigoniids corresponds to that of the locusts, allowing us to use the nomenclature of Tyrer and Gregory (1982). Thus nine longitudinal tracts can be readily distinguished, of which the larger tracts are the lateral dorsal tract (LDT),the median dorsal tract (MDT), the dorsal intermediate tract (DIT) and the ventral intermediate tract (VIT). And by comparison, the dorsal median tract (DMT), the ventral lateral tract (VLT), the ventral median tract (VMT), the median ventral tract (MVT) and the lateral ventral tract (LVT) (Fig. 6a) are smaller in diameter. All ten commissures described in locusts can also be found in bushcrickets. Six of them are dorsal commissures (DC I-VI), and four lie more ventrally (VC I, ventral commissural loop (VCL II), the supra median commissure (SMC) and the posterior ventral commissure (PVC) (Fig. 6b). In addition a small median commissure (MC) runs through the aRT (anterior ring tract), which could be more easily seen in the prothoracic ganglion. Similar to the locust auditory neuropile an anterior ventral commissure (AVC) can be found in tettigoniids at a comparable place, but it is not known whether it is homologous. However, in tettigoniids the longitudinal and transverse tracts appear to be less compact compared to the locusts. Further obvious structures are the t-shaped tract (TT), the c-shaped tract (CT) and the ring tract (RT). The RT is a collection of fibers of different origin which form a horizontal ring. The dorsal part of the VCL II forms the anterior and the SMC forms the posterior margin of the RT, respectively. Within the anterior part of the RT a fine fibrous neuropil lies close to the midline and is called the anterior part of the ring tract (aRT) (Fig. 6b, g). Three parts are recognized in the insect brain: the proto-, deutero-, and tritocerebrum.

Two further finely textured neuropils exist within the mesothoracic ganglion, one is the anterior ventral association center (aVAC) which is located ventrally and anterior to the aRT, and the second is the ventralmost center (vVAC) which lies posteriorly and ventrally to the aVAC (Fig. 6b, g).

All structures described for the mesothoracic ganglion are also found within the prothoracic ganglion of tettigoniids, however their spatial relationships to each other are different, due to the large arborization area of the auditory fibers within the aRT (Fig. 6d, h). A similar enlarged neuropile has not be found in crickets (Wohlers & Huber, 1985) but it has been described in the metathoracic neuromer of locusts (Tyrer & Gregory, 1982; Halex *et al.*,

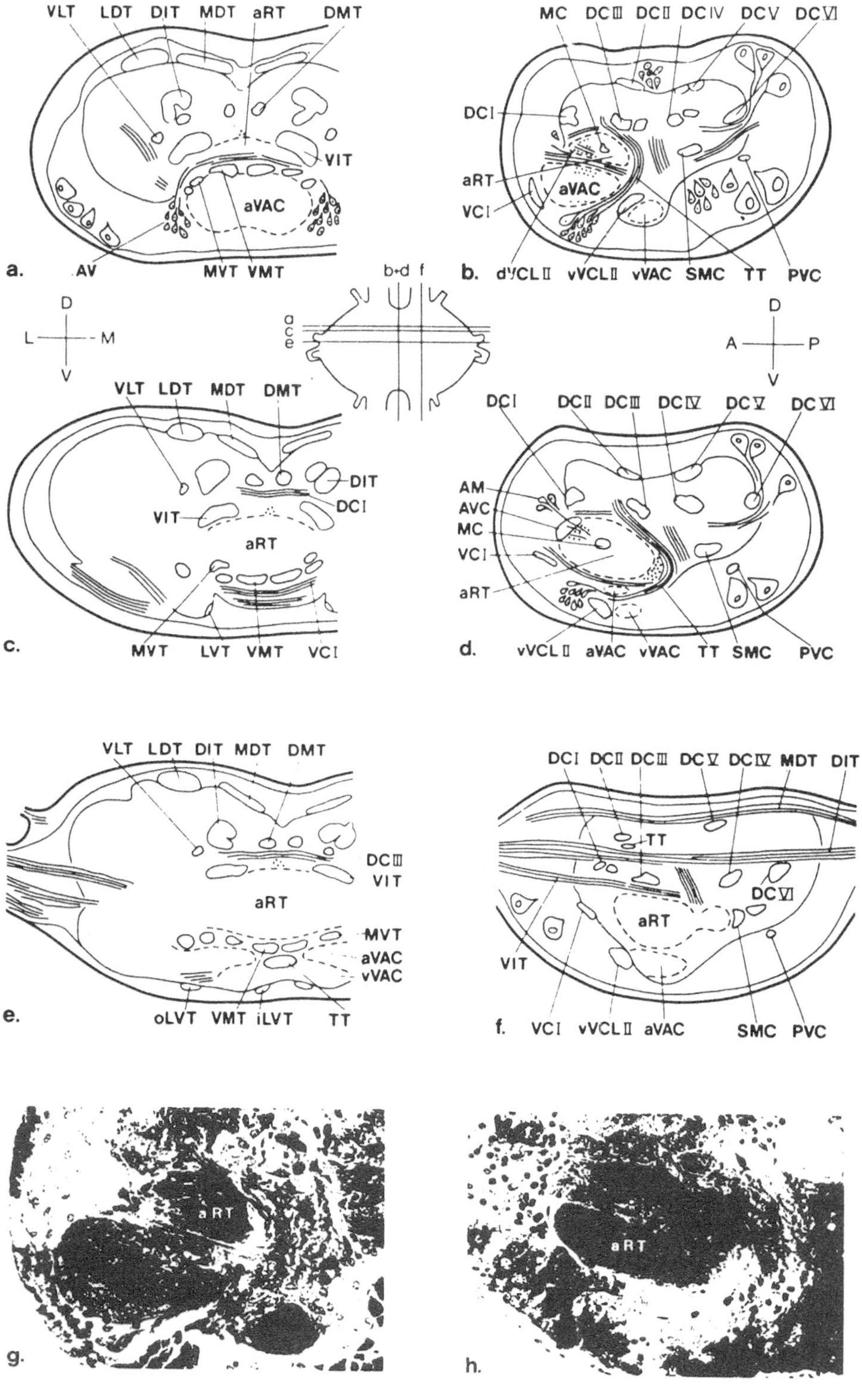
VLT LDT DIT MDT aRT DMT
VIT
aVAC
a. AV MVT VMT
MC DCIII DCII DCIV DCV DCVI
DCI
aRT
aVAC
VCI
b. dVCLII vVCLII vVAC SMC TT PVC
b+d f
a
c
e
D
L M
V
D
A P
V
VLT LDT MDT DMT
DIT
DCI
VIT
aRT
c. MVT LVT VMT VCI
DCI DCII DCIII DCIV DCV DCVI
AM
AVC
MC
VCI
aRT
d. vVCLII aVAC vVAC TT SMC PVC
VLT LDT DIT MDT DMT
DCIII
VIT
aRT
MVT
aVAC
vVAC
e. oLVT VMT iLVT TT
DCI DCII DCIII DCV DCIV MDT DIT
TT
DCVI
aRT
VIT
f. VCI vVCLII aVAC SMC PVC
aRT
g.
aRT
h.

1988). This auditory neuropil (presumably homologous to the aRT) of tettigoniids is extended in the dorsoventral and the rostocaudal axis, compared to the situation found in the mesothoracic ganglion. Fibers of different neuron clusters are closely associated with the aRT; one cluster consists of small cells located anteriorly between both connectives, and is named the anterior median cell body group (AM) in cockroaches (Gregory, 1984). Sagittal sections show that the fibers of these neurons run as a small bundle along the midline through the aRT (Fig. 6d). These morphological data suggest that these neurons could be involved in the processing of auditory information. The same may be true for an anterior ventral cell body group (AV), whose fibers surround the lateral and dorsal margin of the aRT and run in the AVC. Furthermore, fibers from the larger tracts run into the aRT, for example, fibers of the VIT are involved in the MC and fibers of the SMC run rostral into the aRT.

The enlargement of the aRT causes a series of changes in the position of the tracts and neuropile areas. The aVAC is shifted ventrally to the vVAC, lying between the aRT and the vVAC. Both VAC's are slightly reduced in size and extend more laterally than in the mesothoracic ganglion (Fig. 6a, e). The TT arises from a horizontal plane between aVAC and vVAC and turns vertically to the posterior margin of the aRT. The VC I and vVCL II are not affected with the VC I now running adjacent to the auditory neuropil, and the vVCL II lying at the anterior margin of the aVAC. The nerve root III however, together with the aVAC is shifted more posteriorly.

The auditory-vibratory system of the thoracic ganglia

The central projections of mechanoreceptors of proximal leg parts

Single cell recordings at the entry of the leg nerve into the ganglion reveal fibers with a phase-locked response pattern to sinoidal vibratory stimuli of 100 Hz and lower. Simultaneous stainings of the fibers with nickel chloride (after Delcomyn, 1981) exhibit at least three different shapes (Fig. 7a-c). The stain does not mark the peripheral cell body, and the origin of the projection has to be determined from comparative backfills in tettigoniids and other species. The structure in Fig. 7b is similar to the projections of campaniform sensilla described in crickets (Hustert, 1978) and locusts (Hustert *et al.*, 1981). Immediately after entering the ganglion the axon bifurcates into an anterior and a posterior branch, both running towards the midlineforming finer and finer arborizations. This morphology, including the branches of higher order neurons, is very similar to the projection of campaniform sensilla of a trochanter field (Tr1; Hustert *et al.*, 1981). Figure 7a shows a fiber with a similar projection to that of the trochanteral hair-plate receptors in locusts (Pflüger *et al.*, 1981). Although the central projections are close to those of the campaniform sensilla, some differences in the first bifurcation point and the extent of the collateral branching in the ganglion can be used for discrimination of both types of fibers.

The cell type shown in figure 7c has not been described in locusts or in crickets. The axon projects without any collaterals to the center of the hemiganglion where it bifurcates equally

Fig. 6. Gross morphology of the pro- (c-f) and mesothoracic (a,b) ganglia of *Tettigonia viridissima*. Orientation of the sections is given by the scheme in the center. a) Transverse section of the mesothoracic ganglion near the anterior margin of the aRT. The anterior ventral cells (AV) are involved in the dVCL II. b) Sagittal section of the mesothoracic ganglion near the midline. c) Transverse section of the prothoracic ganglion near the anterior margin of the aRT. d) Sagittal section through the prothoracic ganglion. Note the enlarged aRT and the reduced VAC's compared to the mesothoracic ganglion. e) Transverse section of prothoracic ganglion. All three neuropiles are visible. f) Sagittal section of the prothoracic ganglion. The posterior lobes of the aRT are visible. g) h) Photomicrographs of sagittal sections of the meso- and prothoracic ganglion, respectively. Further abbreviations see text.

with some small branches at the end of the projection. In most preparations these branches run to the midline and end ventrally to the auditory neuropile. These fibers probably originate from the femoral chordotonal organ because Cobalt backfillings of whole nerves in locusts display a similar morphology (Mucke pers. comm.). This is further confirmed by the location of the axons within the leg nerve and the physiology of these receptors. The phase-locked response pattern is found in the femoral chordotonal organ of the locust hindleg (Usherwood *et al.*, 1968). This selective response to the velocity and or displacement of the vibrational stimuli is also described for femoral receptors of *Decticus albifrons* (Theophilidis, 1986). Another receptor which could be found in single cell recordings of tettigoniids has a relatively large axon diameter, and the threshold curve follows the acceleration of the stimulus in the range of 100 to 600 Hz. The TT forms together with the CT and fibers of the anterior ventral median cell group a prominent fascicle that connects the dorsal and ventral part of the ganglion. Similar acceleration receptors are found in the stick insect *Cuniculina impigra* (Hofmann & Koch, 1985), and in the tettigoniid *Decticus albifrons* (Theophilidis, 1986). The morphology is similar to the projection pattern from the second part of the femoral chordotonal organ of locusts (Mücke pers. comm.) and the axon ends posterior and dorsal to the auditory vibratory neuropile (Fig. 7d). Two main branches project anterior to the axon and smaller branches occur on the posterior side.

Comparing the different projections (Fig. 7a-d), these mechanoreceptors are clearly distinguishable, but arborize almost in the same area of the ganglia. Furthermore, different mechanoreceptors have similar projections even after ectopic transplantation (Murphey *et al.*, 1985), or in other insect groups like the flies (Pollack *et al.*, 1988), indicating a common plan across a number of species.

The projection of receptors of the complex tibial organ in atympanate legs

Fibers from the subgenual organ, intermediate organ and the *crista acustica* have a common projection area within the aRT. Römer (1985) showed the projection of two different types of vibrating fibers in the prothoracic ganglion at the posterior and anterior margin of the aRT. Thus the endings of the vibratory receptors surround the central projection of the auditory fibers.

In the mesothoracic ganglion fibers with a similar morphology to those in the prothoracic ganglion can be found. Receptors with a best frequency (BF) of 300 to 500 Hz have branches along their axons as they cross the ganglion forming a dense group of terminals in the posterior lobe of the aRT. Receptors with a higher BF that also respond to low frequency airborne sound, display a second collateral projecting to the anterior margin of the aRT (Fig. 7f; unpubl. res.). Both collaterals are more adjacent to each other than in the prothoracic ganglion as might be expected from the smaller size of the aRT. The reaction to airborne sound is an unspecific response to high intensity stimuli within the frequency range of 2-3 kHz. A further receptor, with a projection that misses the posterior branch (Fig. 7e), has a BF to vibrational stimuli up to 3000 Hz but the same reaction to airborne sound as the former receptor (Schikorski & Lakes, 1988). The projection is restricted to the aRT, and is similar to the anterior branch of the fibers described above. It was presumed that fibers with only one posterior termination, and those with both anterior and posterior branches have their origins in the subgenual organ. Those with only one anterior branch probably originate in the intermediate organ and/or in the *crista acustica* of the midlegs. This is supported by the homologous positions of their axons to those of high frequency auditory axons within the leg nerve of the foreleg, and the similarity of their projections to the low frequency receptors. In contrast to the prothoracic ganglion, where the projection of the complex tibial organ is spread over the whole aRT in the mesothoracic ganglion, the fiber endings are divided into two parts. Fibers with relatively high BF's arborize in the anterior part, whereas receptors with low BF's project to the posterior lobe of the aRT. It is not known which fibers project into the center of the aRT.

The projection of the primary auditory neurons

In the last decade the projection of receptors of the *crista acustica* has been studied to some detail (Rehbein, 1973; Kalmring *et al.*, 1978 and this volume; Oldfield, 1983 and reviewed 1988; Römer, 1983, 1985, 1987 and this volume). The fibers of the tympanal organ extend to the prothoracic ganglion, where they end within the ipsilateral half of the ganglion. They form the auditory neuropile, which on its anterior and posterior margins is surrounded by the endings of vibratory fibers of the subgenual organ. The auditory-vibratory neuropile spreads into the whole aRT (Romer, 1985) except for a small dorsal anterior region, the AVC.

The central projections of single cells display a tonotopic organisation within the aRT (Oldfield, 1983; Römer 1983, 1985; Kalmring this volume). The receptors tuned to low frequencies project to the ventral anterior region of the auditory neuropile, whereas the

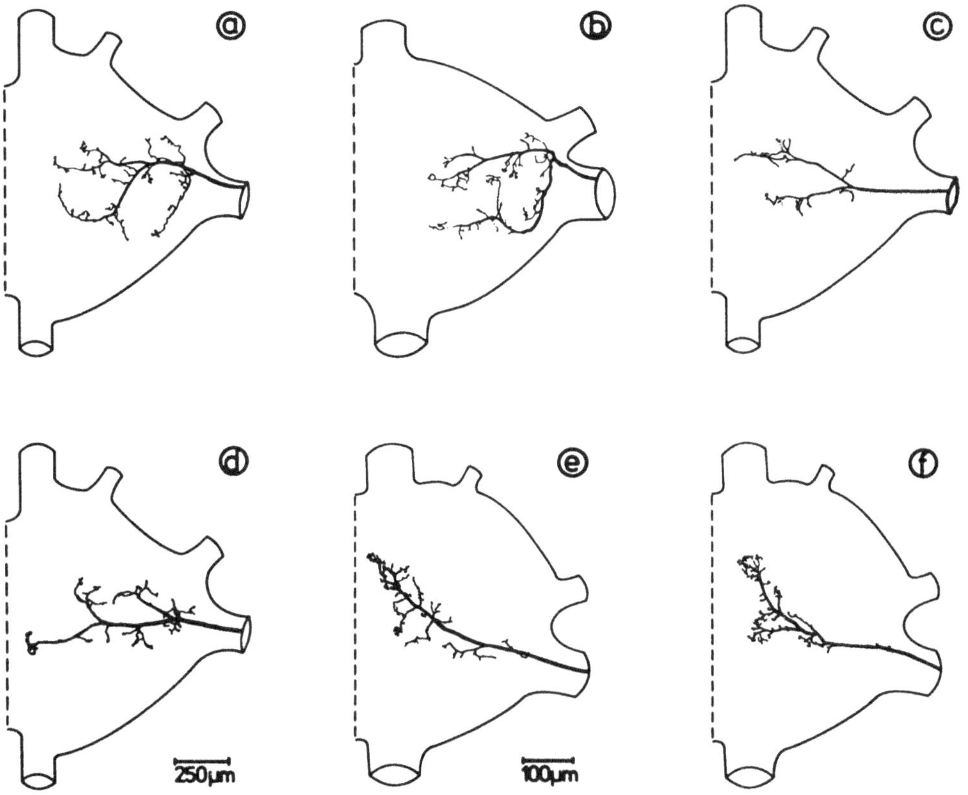

Fig. 7. Drawings of whole mount preparations of vibroreceptors of the midleg of *Ephippiger ephippiger* (a-d) and of *Decticus albifrons* (e,f). a-c) Vibroreceptors with a phase-locked response pattern. d) Acceleration receptor of the complex tibial organ. Anterior is toward the top of the figure. For further details see text.

receptors tuned to higher frequencies form axon terminals in the ventral posterior part of the neuropile (Fig. 8). The ultrasonic receptors, however, curve to the dorsal part of the aRT with those most sensitive to high frequencies close to the anterior margin. They have a less dense arborisation and their dorsal-ventral extension is smaller than for those projecting to the ventral part. The tonotopic organization starts at the ventral anterior edge and extends to the posterior margin of the neuropile finally reaching the dorsal anterior region. The most sensitive fibers exhibit the largest arborisation field and are suggested to synapse onto a larger number of interneurons, or have a stronger effect than other primary auditory neurons (for further details see Kalmring *et al.* this volume).

Auditory and vibratory interneurons

The auditory interneurons

The anatomy of one of these interneurones was first predicted from its physiology (Suga, 1963). The similar response in ascending and descending connectives suggested a large t-

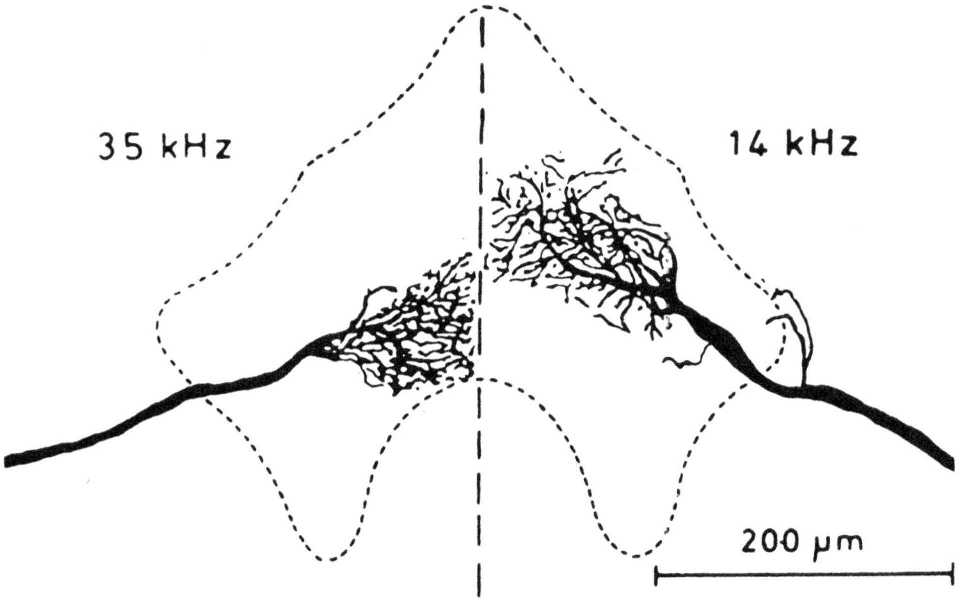

Fig. 8. The projection of two auditory receptors within the auditory neuropile. The receptor tuned to 14 kHz projects anteriorly and the projection with a best frequency of 35 kHz projects posteriorly in the aRT. The dashed line indicates the margin of the auditory neuropile. (after Römer, 1983).

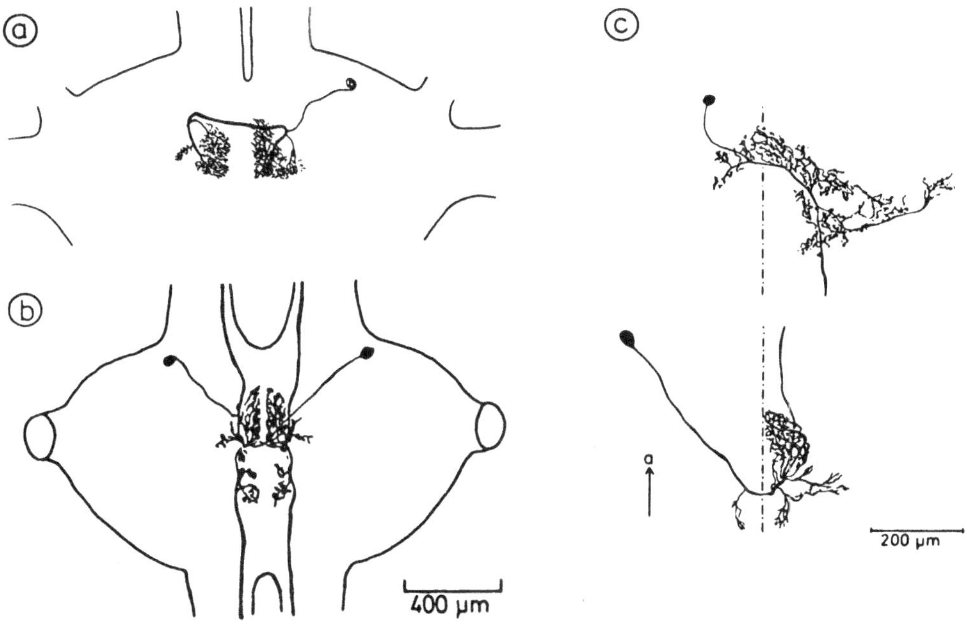

Fig. 9. a) The shape of the omega neuron within the prothoracic ganglion of *Tettigonia cantans* (after Zhantiev & Korsunovskaja, 1983). b) The mirror image paired large T-fiber within the prothoracic ganglion of *Tettigonia viridissima* (redrawn after Rheinlaender & Römer, 1986). c) Two examples of auditory interneurons within the auditory neuropile of *Tettigonia viridissima* (after Römer, 1987).

shaped fiber with an input region in the prothoracic ganglion. Staining of the neuron confirmed this hypothesis and showed the anterior-lateral location of the cell body in the ganglion (Kalmring *et al.*, 1979; Zhantiev & Korsunovskaja, 1983; Oldfield & Hill, 1983 and Römer *et al.*, 1988; Fig. 9b). The dendrites are restricted to one half of the ganglion and are contralateral to the soma. The axon extends posteriorwards to the abdominal ganglia and anteriorly to the brain where it ends in the lateral region of the protocerebrum (Rheinlaender *et al.*, 1972; Kalmring *et al.*, 1979). Branches arising from the axon in the different ganglia run median in the anterior and posterior ganglion halves, and in the thoracic ganglia a lateral dorsal branch can be found. A connection of analogus branches to motoneurons is discussed by Kalmring *et al.* (1974) in locusts and similarly in crickets by Atkins and Pollack (1987). This neuron has monosynaptical connections to the auditory receptor fibers (Römer *et al.*, 1988) and is inhibited via campaniform sensilla and other interneurons such as the omega neuron (McKay, 1969; Kalmring *et al.*, 1979; Römer *et al.*, 1988). The former synaptic connection is thought to be an important factor causing the growth of contralateral dendritic sprouts after unilateral deafferentiation of the auditory input (Lakes *et al.*, in press).

The omega neuron (Fig. 9a) is a local interneuron that has been described by Zhantiev and Korsunovskaja (1983) and by Römer (1985) within the prothoracic ganglion of *Tettigonia cantans* and *T. viridissima*. The cell body lies in the anterior ventro-lateral margin of the ganglion. The arborisations ipsilateral to the soma have a smooth shape which is typical for dendrites, and they reach almost the midline of the ganglion whereas the contralateral branches have a bouton-like structure. The arborisations ipsilateral to the soma have been confirmed as dendrites monosynaptically connected to the auditory receptors (Römer *et al.*,

1988). Sections show that both arborise within the corresponding aRT. The dendrites overlap with the auditory receptors except in the anterior region, where low frequency receptors arborize. Consequently the threshold of the omega-neuron follows approximately the threshold of the whole tympanic nerve up to 120 kHz with an exception at low frequencies (Zhantiev & Korsunovskaja, 1983; Römer, 1985).

Recently Romer and co-workers (1987, 1988) described the morphology of further auditory interneurons within the prothoracic ganglion of *Tettigonia viridissima* (Fig. 9c). All these interneurons have dendritic fields within the aRT but with a different overlap to the auditory primary fiber endings. The degree of overlap is displayed in the physiology of the interneurons which receive monosynaptic input from the auditory receptors. However, the dense network of fibers within the aRT, and the occurrence of dendrites of auditory interneurones outside of the aRT suggest a more complex processing of auditory information.

Vibratory interneurons

Although the physiology of the vibratory neurons is well established (Kalmring & Kühne, 1980) the morphology of these neurons is almost unknown in bushcrickets. Two vibratory interneurons with cell bodies in the prothoracic ganglion are described by Römer (1985) (Fig. 10a,b). Firstly, a descending neuron which arborises at the anterior and the posterior margin

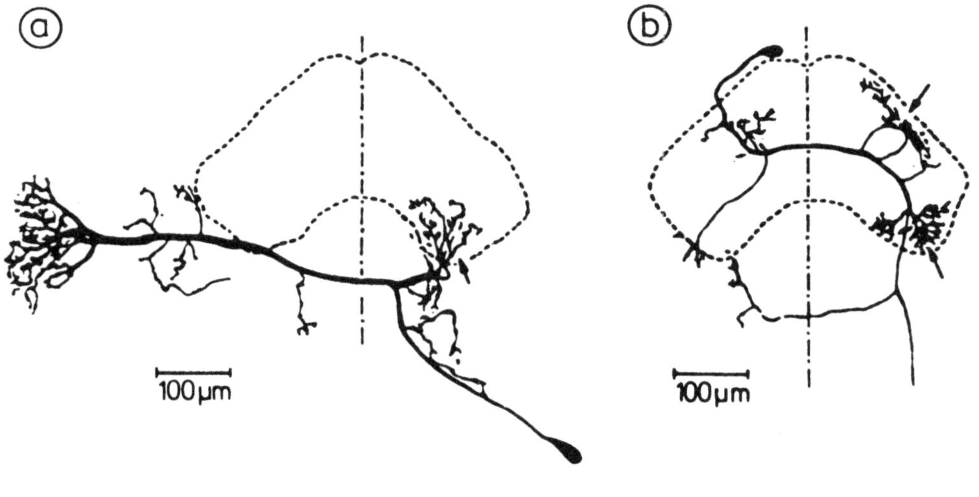

Fig. 10. Thoracic interneurons which respond to vibrational stimuli. a) Local vibratory interneuron within the prothoracic ganglion of *Tettigonia viridissima*. The dashed line indicated the margin of the auditory neuropile (after Römer, 1985). b) Descending vibratory interneuron within the prothoracic ganglion of *Tettigonia viridissima* (after Römer, 1985).

of the aRT and overlaps with the projection of the primary vibratory fibers (Fig. 10b). Its physiology again correlates with its morphology because the interneuron responds to a substrate-borne sound above 800 Hz and to low frequency air-borne sound stimuli with relatively high intensity. Second, the dendrites of a local interneuron arborizes almost exclusively in the posterior lobe of the aRT and the axon crosses the midline projecting with a dense branching pattern at a lateral region of the contralateral hemiganglion (Römer, 1985; Fig. 10a). As predicted by its morphology the response of this interneuron is limited to low frequency vibration and no discharge can be elicited with airborne sound stimuli.

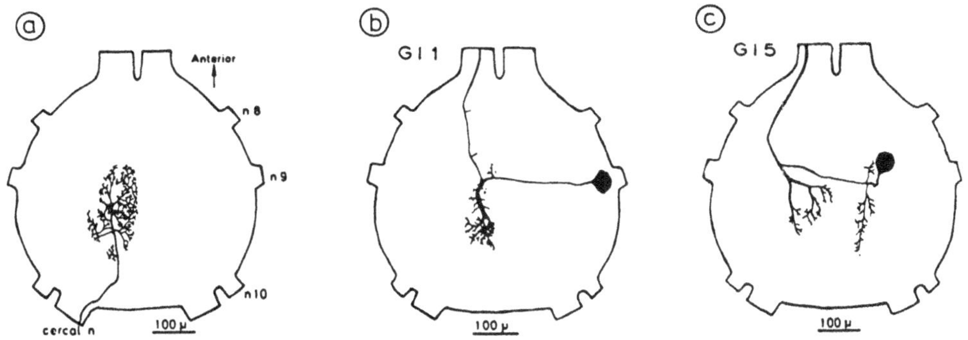

Fig. 11. Some examples of neurons within the cercal mechanoreceptive system. a) The projection of a discrete fiber ending of a cercal hair receptors within the terminal ganglion. b) c) Two cercal giant interneurons with dendritic fields within the terminal ganglion (after Shen, 1983).

The cercal mechanoreceptive system

The abdominal cercus to giant interneuron system was analyzed in *Tettigonia cantans* by Shen (1983). Cobalt-backfillings of the cercal nerve showed the central projections of the cercal hair receptors within the terminal ganglion. The projections are restricted entirely to the ipsilateral half of the cercal ganglion (Fig. 11a), where they end with bouton like structures.

Seven pairs of giant interneurons (GI) with a large axon diameter of 5-15 µm have their cell bodies and dendrites in the terminal ganglion; a similar situation to that in cockroaches (Westin *et al.*, 1977; Fig. 11b,c). The cell bodies form a dorsal and a ventral group, in which each GI has a unique morphological structure, as defined by its dendritic fields and projections into the brain. The dendritic fields of the GI's display a considerable overlap with the projections of the cercal receptors. The extent of overlap is sometimes limited to the contralateral side (with respect to the soma), and sometimes dendritic fields can be found on each side of the ganglion. The high degree of overlap of cercal projection and the dendritic field of the GI's within the cercal ganglion suggests monosynaptic transmission between the cells as has shown to be the case in some GI's of the crickets and the locusts (Shepherd *et al.*, 1988; Boyan, 1988). In the prothoracic and metathoracic ganglion there are axon collaterals that project medially in the anterior and posterior half of these ganglia, and seldom cross the midline. The axons run through the tritocerebrum and deuterocerebrum and end in the mediolateral region of the posterior protocerebrum. The GI's respond to air puffs and low frequency airborne sounds with a constant directional sensitivity independent from the position of the cercus (Shen, 1983, Rozhkova *et al.*, 1984).

Acknowlegdements

These results have been achieved at the Philipps-University, AG Prof. Dr. K. Kalmring. We wish to thank Dr. K. Kalmring, Dr. G. Pollack and A. Mücke for their support and helpful discussion. We are grateful to Mrs. E. Schäfer and S. Völk for technical assistance. The work was supported by DFG (Ka 498/5-1 and Ka 498/6-1).

References

Adelung v.N. 1892. Beitrage zur Kenntnis des tibialen Gehorapparates der Locustiden. Z. Wiss. Zool. 54: 316-349.

Ander K. 1939. Vergleichende anatomische und phylogenetische Studien über die Ensiferen. Opusc. Entomol. Suppl. 2: 1-306.

Atkins G. & G.S. Pollack. 1987. Correlation between structure, topographic arrangement, and spectral sensitivity of sound sensitive interneurons in crickets. J. Comp. Neurol. 266: 398-412.

Autrum H.J. 1941. Uber den Gehör-und Erschütterungssinn bei Locustiden. Z. Vgl. Physiol. 28: 580-637.

Autrum H.J. & W. Schneider. 1948. Vergleichende Untersuchungen über den Erschütterungssinn der Insekten. Z. Vgl. Physiol. 31: 77-88.

Blaney W.M. & R.F. Chapman. 1969a. The anatomy and histology of the maxillary palp of *Schistocerca gregaria* (Orthoptera, Acrididae). J. Zool. 157: 509-535.

Blaney W.M. & R.F. Chapman. 1969b. The fine structure of terminal sensilla on the palp of *Schistocerca gregaria* (Forskal) (Orthoptera, Acrididae). Z. Zellforsch. 99: 74-77.

Boeckh J. & K.D. Ernst. 1987. Contribution of single unit analysis in insects to an understanding of olfactory function. J. Comp. Physiol. 161: 549-565.

Boyan G.S. 1988. Presynaptic inhibition of identified wind-sensitive afferents in the cercal system of the locust. J. Neurosci. 8: 278-2757.

Burakov V.V. & R.D. Zhantiev. 1984. Ultrastructural organization of the tympanal nerve in the bush cricket *Tettigonia cantans* F. (Orthoptera, Tettigoniidae). Entomol. Obozr. 63: 24-28.

Burmeister H. 1832. Handbuch der Entomologie, Vol. I & II. Berlin 1832

Campbell J.I. 1961. The anatomy of the nervous system of the mesothorax of *Locusta migratoria* (R&F). Proc. Zool. Soc. Lond. 167: 403-432.

Charline T., K. Kubra, V.K. Brown & R. Beck. 1984. Comparison of the distribution and nervous innervation of the sensilla on the labrum of *Gryllus bimaculatus* (deGeer) and *Acheta domesticus* (L.) (Orthoptera: Gryllidae) and an account of their development in *A. domesticus*. Int. J. Insect Morphol. Embryol. 13: 81-103.

Chukanov V.S. & R.D. Zhantiev. 1978. The effect of the acoustic trachea on the functional

characteristics of the hearing organs in *Tettigonia cantans* F. (Orthoptera, Tettigoniidae). Vestn. Mosk. Univ. Ser. XVI Biol. 4: 43-47.
Debaisieux P. 1938. Organes scolopidiaux des pattes d'insectes. II. La Cellule 47: 77-202.
Delcomyn F. 1981. Nickel chloride for intracellular staining of neurons in insects. J. Neurobiol. 12: 623-627.
Dujardin F. 1850. Mémoire sur le systeme nerveux des insects. Ann. Sci. Nat. B 14: 195-206.
Eggers F. 1928. Die stiftführenden Sinnesorgane. Zool. Bausteine 2: 1-353.
Eibl E. 1978. Morphology of the sense organs in the proximal parts of the tibia of *Gryllus campestris* and *Gryllus bimaculatus* (Insecta, Ensifera). Zoomorphologie 89: 185-205.
Erhardt E. 1916. Zur Kenntnis der Innervierung und der Sinnesorgane der Flügel von Insekten. Zool. Jahrb. Anat. Ontog. 39: 293-334.
Ernst K.D., J. Boeckh & V. Boeckh. 1977. A neuroanatomical study on the organization of the central antennal pathway in insects. II. Deutocerebral connections in *Locusta migratoria* and *Periplaneta americana*. Cell Tissue Res. 176: 285-308.
Friedrich H. 1927. Untersuchungen über die tibialen Sinnesapparate in den mittleren und hinteren Extremitäten von Locustiden. I. Zool. Anz. 73: 42-48.
Friedrich H. 1928. Untersuchungen über die tibialen Sinnesapparate in den mittleren und hinteren Extremitäten von Locustiden. II. Zool. Anz. 75: 86-94.
Friedrich H. 1930. Weitere vergleichende Untersuchungen der tibialen Skolopidialorgane der Orthopteren. Z. Wiss. Zool. 137: 30-55.
Fudalewicz-Niemcyk W. 1956. The innervation and sense organs in the wings of the grasshopper *Locusta cantans* (F.). Bull. Entomol. Pologne 25: 127-160.
Fudalewicz-Niemcyk W. 1958. Analysis of the wing venation on the background of their innervation in *Stauroderus biguttulus* (L.) and *Tettigonia cantans* (F.) and *Phyllodromia germanica* (L.). Bull. Entomol. Pologne 28: 59-89.
Goodman L.J. 1981. Organization and physiology of the insect dorsal ocellar system. In: The Handbook of Sensory Physiology. Comparative Physiology and Evolution of Vision in Invertebrates, Vol. VII/6c. Ed. H.J. Autrum. pp 201-286.
Graber V. 1876. Die tympanalen Sinnesapparate der Orthopteren. Denkschrift der Akademie, Wien, Math.-Natws. Klasse 36: 1-140.
Gregory G.E. 1984. Neuroanatomy of the mesothoracic ganglion of the cockroach *Periplaneta americana* (L.). Median neuron cell body groups. Philos. Trans. R. Soc. Lond. B 306: 191-218.
Grosch A., F. Callender, M. Petersen, A. Cokl & K. Kalmring. 1985. Vibration receptors of larvae and of imagines in locusts: location on the legs, central projections and physiology. In: Acoustic and vibrational communication in insects. Eds. K. Kalmring & N. Elsner. Parey Verlag, Hamburg. pp 151-161.
Halex H., W. Kaiser & K. Kalmring. 1988. Projection areas and branching patterns of the tympanal receptor cells in migratory locusts, *Locusta migratoria* and *Schistocerca gregaria*. Cell Tissue Res. 253: 517-528.
Hanström B. 1940. Inkretorische Organe, Sinnesorgane und Nervensystem des Kopfes einiger niederer Insektenordnungen. Svenska. Vet. Akad. Handl. 18: 1-265.
Hensen V. 1866. Über das Gehörorgan von Locusta. Z. Wiss. Zool. 16: 190-207.
Hill K.G. & B.P. Oldfield. 1981. Auditory function in Tettigoniidae (Orthoptera, Ensifera). J. Comp. Physiol. 142: 169-180.
Hofmann T. & U.T. Koch. 1985. Acceleration receptors in the femoral chordotonal organ of the stick insect, *Cuniculina impigra*. J. Exp. Physiol. 114: 225-237.
Houtermans B. & R. Schumacher. 1974. Zur Morphologie der atympanalen tibialen

Scolopalorgane von *Tettigona viridissima* L. Z. Morphol. Tiere 78: 281-297.
Hsü F. 1938. Les sensilla des Insectes. La Cellule 47: 5-60.
Huber F. 1987. Plasticity in the auditory system of crickets: phonotaxis with one ear and neuronal reorganization within the auditory pathway. J. Comp. Physiol. 161: 583-604.
Hustert R. 1978. Segmental and interganglionic projection from primary fibres of insect mechanoreceptors. Cell Tissue Res. 194: 337-351.
Hustert R., J.H. Pfluger & P. Bräunig. 1981. Distribution and specific central projections of mechanoreceptors in the thorax and proximal leg joints of locusts. III. The external mechanoreceptors: the campaniform sensilla. Cell Tissue Res. 216: 97-111.
Jawlowski H. 1954. Uber die Struktur des Gehirns bei Saltatoria. Ann. Univ. Mariae Curie C 8: 403-444.
Kalmring K., B. Lewis & A. Eichendorff. 1978. The physiological characteristics of primary sensory neurons of the complex tibial organ of *Decticus verrucivorus* (L.) (Orthoptea, Tettigoniidae). J. Comp. Physiol. 127: 109-121.
Kalmring K. & R. Kühne. 1980. The coding of airborne sound and vibration signals in bimodal ventral-cord neurons of the grasshopper *Tettigonia cantans*. J. Comp. Physiol. 139: 267-275.
Kalmring K., H. Rehbein & R. Kühne. 1979. An auditory giant neuron in the ventral cord of *Decticus verrucivorus* (Tettigoniidae). J. Comp. Physiol. 132: 225-234.
Kalmring K., H. Römer & H. Rehbein. 1974. Connections of acoustic neurons within the CNS of grasshoppers. Naturwiss. 61: 454-455.
Kenyon F.C. 1896. The brain of the bee. A preliminary contribution to the morphology of the nervous system of the arthropoda. J. Comp. Neurol. 6: 133-210.
Keshishian H. & D. Bentley. 1983. Embryogenesis of peripheral nerve pathways in grasshopper legs. II. The major nerve routes. Dev. Biol. 96: 103-115.
Keuper A., S. Weidemann, K. Kalmring & D. Kaminski. 1988. Sound production and sound emission in seven species of european tettigoniids. I. The different parameters of the song; their relation to the morphology of the bushcricket. Bioacoustics 1: 31-48.
Klein U. 1981. Sensilla of the cricket palp: Fine structure and spatial organisation. Cell Tissue Res. 219: 229-252.
Knetsch H. 1939. Die Korrelation in der Ausbildung der Tympanalorgane, der Flügel, der Stridulationsapparate and anderer Organsysteme bei den Orthopteren. Archiv. Natur. Berlin 8: 1-69.
Lakes R. & A. Mücke. 1989. Regeneration of the foreleg tibia and tarsi of *Ephippiger ephippiger* (Orthoptea: Tettigoniidae). J. Exp. Zool. 250:176-187.
Lewis D.B. 1974. The physiology of the tettigoniid ear. I. The implications of the anatomy of the ear to its function in sound reception. J. Exp. Biol. 60: 821-837.
McKay J.M. 1969. The auditory system of Homorocoryphus (Tettigoniidae, Orthoptera). J. exp. Biol. 51:787-802.
Michel K. 1974. Das Tympanalorgan von *Gryllus bimaculatus* DeGeer (Saltatoria, Gryllidae). Z. Morphol. Tiere 77: 285-315.
Mücke A. & R. Lakes. 1988. Sensory nervous system of intact and regenerated legs of locusts. Proc 16th Gottingen Neurobiology Conference. Eds. N. Elsner & F.G. Barth. Thieme Verlag, Stuttgart, New York. 34.
Murphey R.K., J.P. Bacon & S.E. Johnson. 1985. Ectopic neurons and the organization of insect sensory systems. J. Comp. Physiol. 156: 381-389.
Nesbitt H.H.J. 1941. A comparative morphological study of the nervous system of the Orthoptera and related orders. Ann. Entomol. Soc. Am. 34: 51-81.

Oldfield B.P. 1983. Central projections of primary auditory fibers in tettigoniidae (Orthoptera: Ensifera). J. Comp. Physiol. 151: 389-395.

Oldfield B.P. 1984. Physiology of auditory receptors in two species of tettigoniidae (Orthoptera, Ensifera). Alternative tonotopic organizations of auditory organ. J. Comp. Physiol. 155: 689-696.

Oldfield B.P. 1988. Tonotopic organization of the insect auditory pathway. TINS 11: 267-270.

Oldfield B.P. & K.G. Hill. 1983. The physiology of ascending auditory interneurons in the tettigoniid *Caedicia simplex* (Orthoptera; Ensifera): response properties and a model of integration in the afferent auditory pathway. J. Comp. Physiol. 152: 495-508.

Pflüger J.H., P. Bräunig & R. Hustert. 1981. Distribution and specific central projections of mechanoreceptors in the thorax and proximal leg joint of locusts. II. The external mechanoreceptors: Hair plates and tactile hairs. Cell Tissue Res. 216: 79-96.

Pollack G.S., D. Possidente & R.K. Murphey. 1988. Modality specific afferent projection in the flies Phormia and Drosophila. Abstr. Soc. Neurosci. 14: 869.

Rath O.v. 1888. Über die Hautsinnesorgane der Insekten. Z. Wiss. Zool. 46: 413-454.

Rehbein H.G. 1973. Experimentell-anatomische Untersuchungen über den Verlauf der Tympanalfasern im Bauchmark von Feldheuschrecken, Laubheuschrecken und Grillen. Verh. Dtsch. Zool. Ges. 66: 184-189.

Rheinlaender J., K. Kalmring & H. Römer. 1972. Akustische Neuronen mit T- Struktur im Bauchmark von Tettigoniiden. J. Comp. Physiol. 77: 208-224.

Rozhkova G.I., H.I. Rodinova & A.V. Popov. 1984. Two types of information processing in cercal systems of insects: directional sensitivity of giant interneurons. J. Comp. Physiol. 154: 805-815.

Römer H. 1983. Tonotopic organization of the auditory neuropile in the bushcricket *Tettigonia viridissima*. Nature 306: 60-62.

Römer H. 1985. Anatomical representation of frequency and intensity in the auditory system of Orthoptera. In: Acoustic and Vibrational Communication in Insects. Eds. K. Kalmring & N. Elsner. Parey Verlag, Hamburg. pp 25-32.

Römer H. 1987. Representation of auditory distance within a central neuropil of the bushcricket *Mygalopsis markii*. J. Comp. Physiol. 161: 33-42.

Römer H., V. Marquart & M. Hardt. 1988. Organization of a sensory neuropile in the auditory pathway of two groups of orthopterans. J. Comp. Neurol. 275: 201-215.

Schäffner K.H. & U.T. Koch. 1987. A new field of wing campaniform sensilla essential for the production of the attractive calling song in crickets. J. Exp. Biol. 129: 1-23.

Schikorski T.H. & R. Lakes. 1988. The atympanate scolopidial organ of the tettigoniids. A morphological and physiological approach. Proc 16th Göttingen Neurobiology Conference. Eds. N. Elsner & F.G. Barth. Thieme Verlag, Stuttgart, New York. 150.

Schmitt J.B. 1964. Nervous system of pregenital abdominal segments of two orthoptera. J. Entomol. Soc. New York 72: 202-209.

Schnorbus H. 1971. Die subgenualen Sinnesorgane von *Periplaneta americana*: Histologie und Vibrationsschwellen. Z. Vgl. Physiol. 71: 14-47.

Schumacher R. 1973. Morphologische Untersuchunge der tibialen Tympanalorgane von neun einheimischen Laubheuschrecken. Z. Morphol. Tiere 75: 267-282.

Schumacher R. 1975. Die *crista acustica* der Tettigoniidea (Orthoptera). Z. Naturforsch. 30: 552-553.

Schumacher R. 1979. Zur funktionellen Morphologie des auditiven Systems der Laubheuschrecken. Entomologia Gen. 5: 321-356.

Schumacher R. & B. Houtermans. 1975. Vergleich des primären Rezeptorbereiches der tympanalen und atympanalen tibialen Skolopalorgane von 14 mitteleuropäischen Laubheuschrecken-Arten (Orthoptera, Tettigoniidae). Entomologia Gen. 1: 97-104.

Schwabe J. 1906. Beiträge zur Morphologie und Histologie der tympanalen Sinnesapparate der Orthopteren. Zoologica 50: 1-54.

Shen J.X. 1983. The cercus-to-giant interneurons system in the bushcricket *Tettigonia cantans*: Morphology and response to low-frequency sound. J. Comp. Physiol. 151: 449-459.

Shepherd D., G. Kämper & R.K. Murphey. 1988. The synaptic origins of receptive field properties in the cricket cercal sensory system. J. Comp. Physiol. 162: 1-11.

Siebold T.H.v. 1844. Uber das Stimm- und Gehororgan der Orthopteren. Wiegmann's Arch. Naturgesch. 10: 52-81.

Steinmann H. 1961. Beiträge zur Kenntnis des Zentralnervensystems von *Ephippiger ephippiger*. Folia Entomol. Hung. 14: 111-118.

Steinmann H. 1962. On the central nervous system of the meso- and metathorax of orthopterous insects. Ann. Hist. Nat. Mus. Hung. 54: 237-254.

Suga N. 1963. Central mechanism of hearing and sound localization in insects. J. Insect Physiol. 9: 867-873.

Theophilidis G. 1986. The femoral chordotonal organs of *Decticus albifrons* (Orthoptera: Tettigoniidae). I & II. Comp. Biochem. Physiol. B 84: 529-543.

Tyrer N.M. & G.E. Gregory. 1982. A guide to the neuroanatomy of locust suboesophageal and thoracic ganglia. Philos. Trans. R. Soc. Lond. B 297: 91-123.

Usherwood P.N.R., H.I. Runion & J.I. Campbell. 1968. Structure and physiology of a chordotonal organ in the locust leg. J. Exp. Biol. 48: 305-323.

Weevers R. de G. 1985. The insect ganglia. In: Comprehensive Insect Physiology, Biochemistry and Pharmacology, Vol. 5. Eds. G.A. Kerkut & C.I. Gilbert. Pergamon Press, Oxford. pp 215-297.

Westin J., J.J. Langberg & J.M. Camhi. 1977. Responses of giant interneurones of the cockroach *Periplaneta americana* to windpuffs of different directions and velocities. J. Comp. Physiol. 121: 307-324.

Wigglesworth V.B. 1957. The use of osmium in the fixation and staining of tissue. Proc. R. Soc. Lond. B 147: 185-199.

Wolf H. & K.G. Pearson. 1987. Flight motor patterns recorded in surgically isolated sections of the ventral nerve cord of *Locusta migratoria*. J. Comp. Physiol. 161: 103-114.

Young D. & E. Ball. 1974a. Structure and development of the auditory system in the prothoracic leg of the cricket *Teleogryllus commmodus*. II. Adult structure. Z. Zellforsch. 147: 293-312.

Young D. & E. Ball. 1974b. Structure and development of the tracheal organ in the mesothoracic leg of the cricket *Teleogryllus commodus*. Z. Zellforsch. 147: 325-334.

Zhantiev R.D. & O.S. Korsunovskaja. 1978. Morphofunctional organization of tympanal organs in *Tettigonia cantans* F. (Orthoptera, Tettigoniidae). Zool. J. 57: 1012-1016.

Zhantiev R.D. & O.S. Korsunovskaja. 1983. Structure and function of two auditory neuron in the bushcricket *Tettigonia cantans* F. (Orthoptera, Tettigoniidae). Entomol. Obozr. 62: 462-468.

11. ASPECTS OF ACOUSTIC AND VIBRATORY COMMUNICATION IN SEVEN EUROPEAN BUSHCRICKETS

Klaus Kalmring, Armin Keuper and Werner Kaiser

Introduction

Bushcrickets communicate by sound. The vibratory energy transmitted by the calling insect can be divided into two components as perceived by the receptor system; airborne sound and substrate vibration. An essential feature of this communication process is the transmission of a temporal image distinctive for each species. In most bushcrickets this patterned sound is transmitted through the air by frequencies often extending into ultrasound. However, additional information may be conveyed over a shorter range by the substrate on which the animal sits. These low frequencies can range from 1 Hz to 5000 Hz.

This chapter examines the problem of acoustic transmission in seven European bushcrickets. We attempt to relate the morphology of the calling male and the structure of its call, to both its habitat and its mode of sound reception (Table 1).

Problems and Solutions of Airborne Sound and Vibration Transmission

Signal Loss

The factors that attenuate and distort airborne-sound signals fall into two groups: (i) atmospheric absorption and (ii) excess attenuation by both the vegetation and the ground (reviews: Michelsen 1978, Wiley & Richards 1978). The main components of atmospheric absorption are spherical divergence and molecular absorption which is dependant on temperature and humidity.

The effects of attenuation due to vegetation and the ground over which the signal is transmitted are much greater than those produced by atmospheric absorption. Because this excess attenuation is the result of a number of different phenomena (reflection, diffraction, multiple scattering and reverberation), Haupt (1973) classed all these as a single factor. These phenomena become particularly important when the wavelength of the sound corresponds to the size of the intervening structures (e.g. leaves, branches and grass stalks).

Two Solutions in Nature

The two decticines *Decticus verrucivorus* and *Psorodonotus illyricus* live in low grass. Measurements on the transmission of sound in this habitat show that the acoustic conditions are more complex than those in dense bushes (Michelsen & Larsen 1983; Keuper *et al.* 1986). The ground and low vegetation tends to act as a low-pass filter reducing high and ultrasonic frequencies. Acoustic communication near the ground is further complicated by the effects of temperature and wind. On sunny days, negative temperature gradients are often found in the late morning and in the afternoon where they have a profound effect on the propagating sound waves that are bent upwards. This creates shadow zones quite close to the stridulating male.

Thus, in the morning, in the late afternoon and at night, temperature gradients produce the minimum distortion of the signal (e.g. channeling effects) (Michelsen 1978). The behaviour of ground living species such as *D. verrucivorus* and *P. illyricus*, may have evolved as an adaptation to these problems of sound transmission in that they are most active in the morning, thus probably avoiding the effects of shadow zones. We have also observed that the males of both species stridulate in the late afternoon. A second strategy circumventing the problems of shadow zones due to stationary singing near the ground is that males tend to move around whilst stridulating. Despite this constant movement, both species exhibit regular spacing within quite closely knit communities. Moving in this manner increases the probability of hearing and interacting with other males, and if females are also as mobile, they will encounter males. Such a behaviour obviates the problem of communicating with high frequencies in such complex vegetation (Keuper *et al.* 1986).

The acoustic properties of the habitats of *Tettigonia cantans*, *T. viridissima*, *D. albifrons*, *Ephippiger ephippiger* and *E. discoidalis* are very similar. Within these habitats, the dominant plants are bushes and tall dense grass, and the dimensions of these plant structures

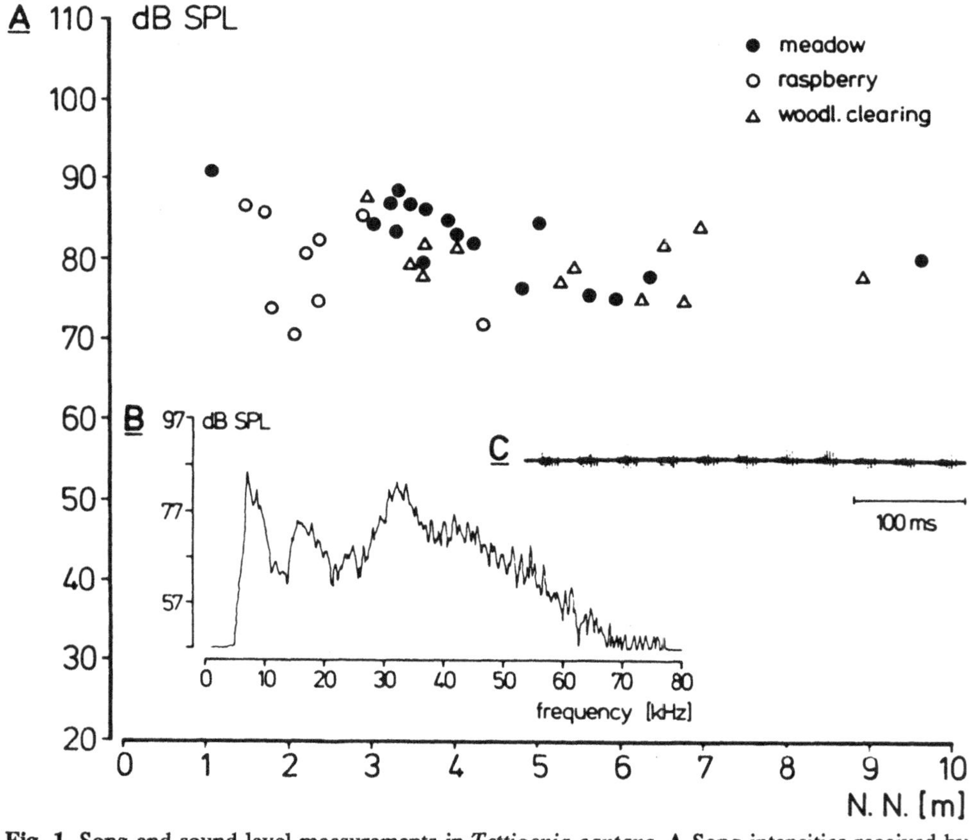

Fig. 1. Song and sound level measurements in *Tettigonia cantans*. **A** Song intensities received by nearest neighbours plotted against nearest neighbour distances (N.M./m). Sound pressure levels measured with an 8 kHz filter. Intensities measured at 32 kHz are more attenuated at distances over 3 m attaining values of 69 dB SPL. Data have been collected from three sites. **B** Power spectrum, and **C** Oscillogram of the conspecific song. (Changed after Schatral *et al.* 1984.)

range from a few mm (grass stalks) to some cm (leaves). For this reason sounds above 10 kHz will be affected by excess attenuation due to vegetation (Keuper & Kühne 1983); the vegetation again acts as a complex low-pass filter. Hence, the effective transmission of the songs of these tettigoniids depends on the density of the vegetation as well as general excess attenuation due to the production of high frequencies; high frequency components of the song are attenuated more rapidly than low frequencies. The extent of damping will depend on both the height of the sender's position as well as that of the height of the receiver (Keuper & Kühne 1983).

Field observations on the acoustic behaviour of *T. cantans* show that males stridulate near the top of high plants, and in addition these animals tend to occupy stable song perches (Schatral *et al.* 1984). Singing males are found to be regularly spaced, with the spacing being dependent on the sound pressure level perceived from the nearest singing neighbour (Fig. 1, Schatral *et al.* 1984). The perceived sound pressure of the low frequency component of the song (at about 8 kHz, measured with a sound level meter and an octave filter at 8 kHz-centre frequency) is about 82 dB SPL whereas the mean value for the 32 kHz-component the second main frequency band of the song, is about 69 dB SPL (Schatral *et al.* 1984).

Since these bushcrickets call high in the plants, they are able to transmit relatively high frequencies. We have noticed that females often approach a stridulating male near the top of plants that are some meters away from the male. This strategy may enable them to receive – and to recognize – the complete frequency spectrum of the song, unaffected by distortions resulting from frequency-dependent attenuation by the vegetation. However, this frequency-dependent excess attenuation may be used as a tool in male location. It has been shown that the relative intensities of the low and high frequency components of the song change with increasing distance (Römer & Bailey 1986). Fig. 2a shows the spectra of *T. cantans* measured at different distances from the sound source in addition to the threshold curve of a central vibratory-auditory neuron. At greater distances only the first dominant peak at around 8 kHz can be detected, and with decreasing distance the relative increase in the perceived intensity of the 16 kHz and 32 kHz components are more than the amplitude of the 8 kHz peak (Keuper & Kühne 1983).

As a part of the male location, information may be gained on distance by a comparison between the intensities of these three frequency components (see also: Morris *et al.* 1978). Approaching the sound source would not only lead to the perception of a high overall intensity, but also of smaller differences between the intensities of the different frequency peaks. This phenomenon is in accordance with the strategy theoretically predicted by Wiley and Richards (1978).

Substrate Vibration

When an approaching conspecific female reaches the plant or bush from which a male is stridulating phonotaxis may be enhanced by the perception of vibratory signals. The sensitivity of the vibratory receptors is such that the central neurons can process the vibratory information within 1.5-2 m of the calling male (Fig. 2b) (Kalmring *et al.* 1985; Kalmring & Kühne 1980). Although vibrations that are produced during stridulation seem to differ from other environmental vibrations in both their frequency spectrum and in the time-amplitude structure, the time pattern of the signal may be masked by complex secondary vibrations produced by the signal itself. Thus, a travelling wave passing along a stem or branch will be reflected from the end of the branch as well as from the base of the stem itself. In other words there will be a number of interfering waves travelling along the substrate. However, in combination with the airborne-sound signal, some characteristic features of the vibratory

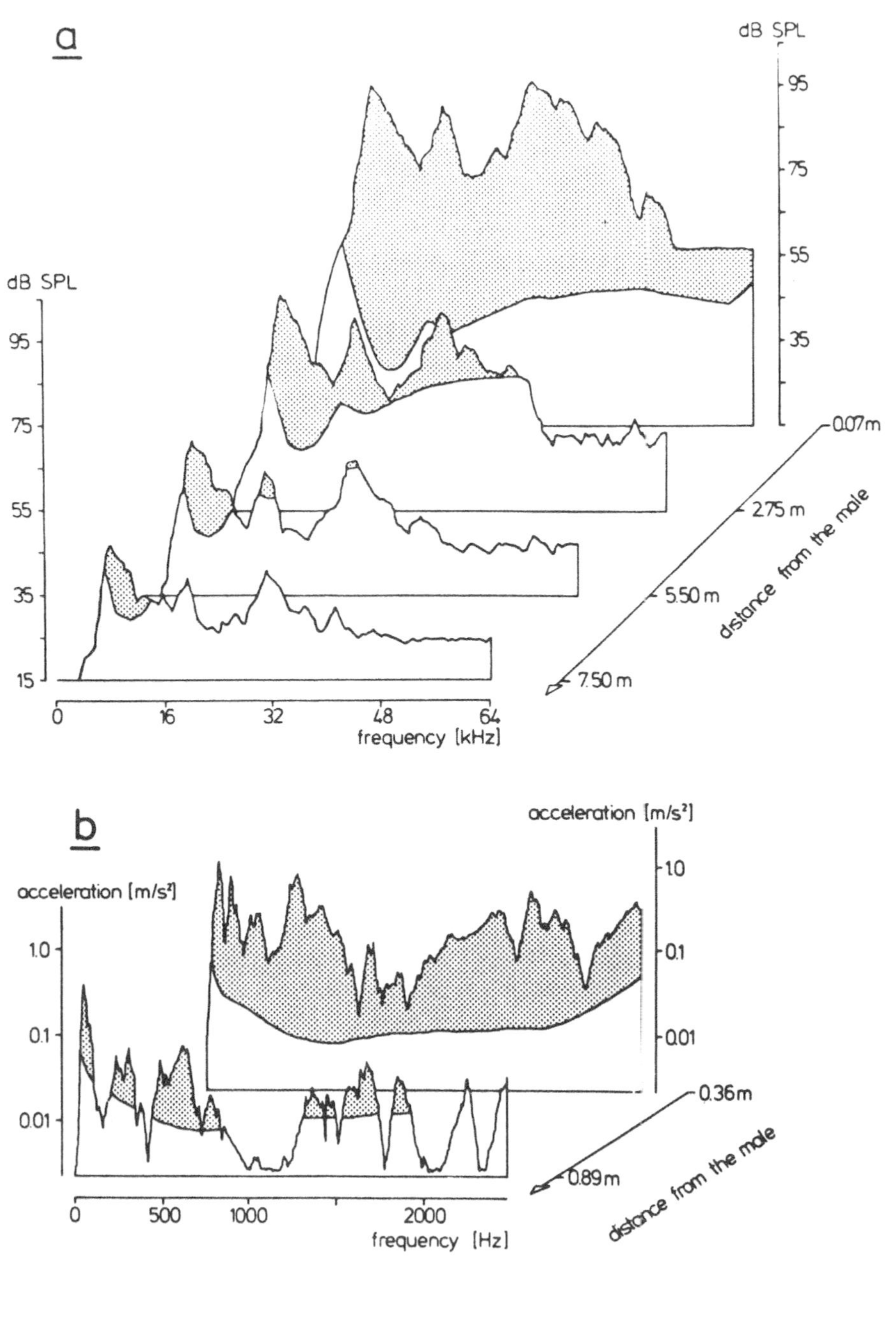

Fig. 2. A Comparison of the spectra of the stridulatory song, emitted via a loudspeaker 1 m above ground and measured at different distances in the biotope, with the threshold curve of a central vibratory-auditory mauron in *Tettigonia cantans* (VS4 neuron) – **B** Comparison of stridulation-induced vibrations in a grass stalk, recorded at different distances from the stridulating male, with the vibratory threshold curve of the VS4 neuron. The threshold curves of the VS4 neuron are redrawn after Silver *et al.* (1980). (Changed after Keuper and Kuhne 1983.)

signals (frequency dispersion and propagation velocity) can be used to estimate the distance to the sound source (Markl 1983).

The stridulation of a bushcricket causes a vibration on the whole animal's body. These vibrations are transmitted via the legs and perhaps the abdomen to the substrate. The main frequency components of the vibration signals induced by stridulation are below 3000 Hz (Keuper & Kühne 1983). The structure of the sound waves consists mainly of bending waves when branches are set into vibration (Michelsen *et al.* 1982), or are surface waves when the ground transmits the signal (Brownell 1977). Often these wave forms occur simultaneously (Cremer *et al.* 1973). Both bending and surface waves propagate with a frequency-dependent dispersion; some frequencies will travel faster than others. Hence, it is possible that the varying characteristics of the signal transmitted in these waves can be used by the orienting insect. However, the waves are rapidly dampened in the plant, and this damping factor will depend on the plant's tissue. In soft and fleshy plants the damping factor is comparatively large (about 40-50 dB/m), and the propagation velocity of a 30 Hz wave is about 15 m/s, whereas a wave of 2000 Hz is transmitted with a propagation velocity of about 130 m/s. By comparison in stiff and woody plants the damping factor is only about 20 dB/m or even less, and the propagation velocity for a 30 Hz wave is up to 30-40 m/s and that for 2000 Hz is close to 300 m/s.

The intensity of vibrations transduced through a plant by a stridulating male can reach acceleration amplitudes of about 1 m/s^2 close to the singing animal (Keuper & Kühne 1983). Since the threshold of insect's vibration receptors is about 0.01 m/s^2 (Kalmring *et al.* 1978) the effective range of the vibratory component can be estimated to be from a few cm to 1.5 m depending on the damping factor.

Directional information on bending waves can be obtained from the frequency dispersion and/or propagation velocity of the vibration. The frequency dispersion of the vibratory part of the stridulatory signal will cause an extension of the "syllable", and thereby its time/amplitude pattern as the distance increases from the singer. Moving closer to a calling male will lead to an improved correspondence between the time patterns of the airborne sound and those of substrate vibration. Hence, another way insects may obtain directional information is from the time difference perceived by the mechanoreceptors of the different legs. In locusts it has been shown that delay times of about 0.5 ms between different leg pairs are processed in central vibratory neurones (Cokl *et al.* 1985). Such a time difference would be adequate for directional location in tettigoniids.

Substrate Vibration and Acuity

In combination with the auditory pathway, the vibratory sense of tettigoniids may add information to auditory acuity. Laboratory experiments with female *T. cantans* show that they cannot discriminate the small angles between adjacent branches of a bush when stimulated with airborne species-specific song alone; the measured acuity was greater than 8° of the midline (Fig. 3). This compares with 5° as reported for *Pachysaga (=Pachysagella) australis* by Bailey and Stephen (1984) to airborne sound. However, when simultaneous vibratory stimulation is provided to *T. cantans* the angle of acuity is less the 8°. Further the female is able to select that branch along which the vibrational information is transmitted (Fig. 3) (Latimer & Schatral 1983).

Recently obtained results with *E. ephippiger* by O. Stiedl (personal communication) on a multi-stemmed branch of broom show quite clearly that the choice between adjacent branches is improved when the stem is vibrated with the species-specific temporal pattern near the airborne sound source. Ephippigerines, like some other tettigoniid species have a

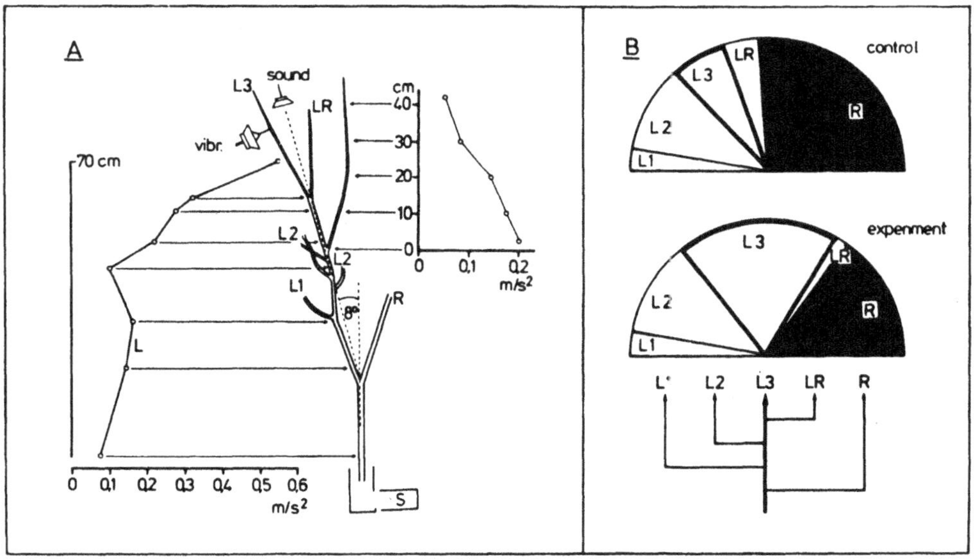

Fig. 3. A Diagram of a *Salix cinerea* bush showing major choices available to *Tettigonia cantans* (males and females) and the positions of the loudspeakers. Choices have been labelled R (the right branch of the first fork), L1, L2 (for closely grouped branches) and at the final fork LR and L3 (the vibrated branch). Curve L shows the vibration intensities recorded at the points indicated. Acceleration values increase towards the source (set at 1.5 m/s2 at the proba). – **B** Results of control (above, without vibration) and experimental trials (bottom) showing the percentage number of insects choosing the named branches (see (A); L3 was vibrated in experimental trials). N tested = 52, N reacting = 35 for control trials; for the experiments N tested = 87, N reacting = 51. (After Latimer and Schatral 1983.)

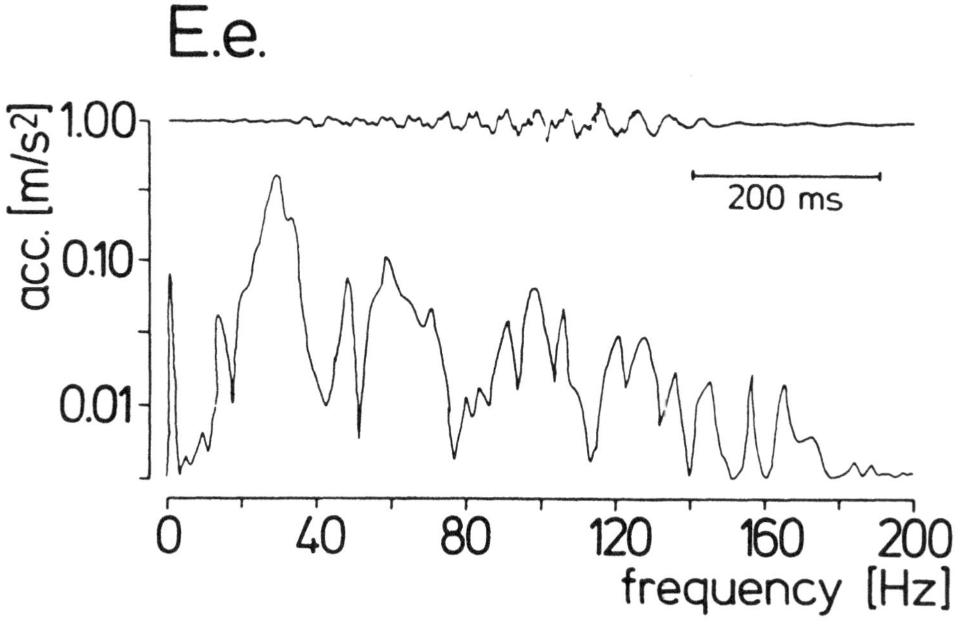

Fig. 4. Oscillogram and power spectrum of a tremulation signal emitted by *Ephippiger ephippiger* (E.e.), the signal was recorded on a plant in a distance of about 15 cm from the tremulating animal. The amplitudes in the power spectrum are given as the acceleration values (acc.) of the vibration. (After Keuper *et al.* 1985.)

special form of vibratory communication, where a singing male stops singing once it detects an approaching female and produces pure vibratory signals by a tremulation of the whole body. These tremulations have already been described in this species by Busnel *et al.* (1956). The acceleration of the signal reaches values of 5-10 m/s^2 measured at the point of the tremulating animal, the duration is about 500 ms and the main frequency lies between 20 and 30 Hz (Fig. 4) (Keuper *et al.* 1985). This form of communication plays an important role in the courtship behaviour of this species. Approaching females show an improved localization of tremulating males when they are within some 30 cm (O. Stiedl, personal communication).

Table 1. The Seven Experimental Animals; their Morphology, their Habitats and their Songs

(Values given as mean ± SD)

a) *Tettigonia viridissima* (T.v.) (Tettigoniinae)

Morphology: a dainty and elegant bushcricket (Fig. 5a).

Body length (head to cerci)	39.7 ± 1.3 mm
width (abdomen)	9.0 ± 0.4 mm
Length forelegs	23.5 ± 1.2 mm
hindlegs	54.7 ± 2.9 mm
Tegmina length	46.1 ± 1.3 mm
width	10.3 ± 0.4 mm

Both sexes fly readily. Habitat: found throughout Europe. In Central Europe it is found up to heights of 300 m above sea level, but in Mediterranean regions as e.g. in Yugoslavia,it is found up to 1000 m above sea level, it prefers either high and dense grass or bushes. At night singing males often climb trees. Song: starts singing in the early afternoon (1400-1500h) and it doesn't stop before midnight. It produces a continuous disyllabic song (Fig. 6).

Syllable duration	(1st syllable)	13.9 ± 1.3 ms
	(2nd syllable)	16.2 ± 2.0 ms
Syllable repetition rate at 24°C		30 syllables/s
Call intensity (20 cm)		99.4 ± 4.8 dB SPL
Main frequency components		10.3 ± 0.4 kHz
		30-55 kHz (Fig. 7)

b) *Tettigonia cantans* (T.c.) (Tettigoniinae)

Morphology: a dainty and elegant bushcricket.

Body length	35.1 ± 1.5 mm
width	8.9 ± 0.4 mm
Length forelegs	23.0 ± 1.0 mm
hindlegs	49.0 ± 2.3 mm
Tegmina just cover the abdomen	
length	25.6 ± 1.0 mm
width	10.3 ± 0.6 mm

Able to fly. Habitat: found throughout Europe. In Central Europe it prefers heights of 300 m up to 1000 m above sea level. In Northern Germany and in Scandinavia it is also found in plains, prefers dense vegetation, either high and dense grass or bushes. At night singing males often climb trees. Song: starts singing in the early afternoon (1400-1500h) and it doesn't stop before midnight. It produces a continuous monosyllabic song (Fig. 6).

Syllable duration	15.0 ± 1.1 ms
Repetition rate at 24°C	30 syllables/s
Intensity (20 cm)	90.0 ± 2.6 dB SPL
Main frequency components	8.6 ± 0.7 kHz
	30-55 kHz (Fig. 7)

Table 1. The Seven Experimental Animals; their Morphology, their Habitats and their Songs (cont.)

(Values given as mean ± SD)

c) *Decticus albifrons* (D.a.) (Decticinae)

Morphology: has a stout and strong body

Body length	46.5 ± 1.8 mm
width	11.2 ± 0.7 mm
Length hindlegs	84.8 ± 3.7 mm
forelegs	24.7 ± 1.2 mm
Wings project over the abdomen	
length	47.6 ± 2.7 mm
width	8.9 ± 0.6 mm

Can fly well. Habitat: lives in the Mediterranean regions and prefers coastal habitats. It prefers high and dense grass or bushes. Song: normally sings in the morning till noon and then starts again in the late afternoon, but stops before darkness (Busnel 1955). It produces a monosyllabic song.

Syllable duration (24°C)	8.1 ± 1.9 ms) (Fig. 6)
Repetition rate (max)	7-10 syllables/s
Intensity (20 cm)	96.1 ± 4.2 dB SPL
Main frequency components	8.4 ± 0.6 kHz
	30-55 kHz (Fig. 7)

d) *Decticus verrucivorus* (D.v.) (Decticinae)

Morphology: has a stout and strong body (Fig. 5b)

Body length	37.8 ± 1.6 mm
width	9.6 ± 0.7 mm
Length hindlegs	63.6 ± 3.4 mm
forelegs	18.6 0.8 mm
Wings just cover abdomen (length)	29.9 ± 3.1 mm
width	7.2 ± 0.4 mm

Normally doesn't fly. Habitat: widespread in Southern and Central Europe. It prefers the mountain regions up to heights of about 1000 m above sea level, a ground dweller living in dry or moist grassland. Song: normally sings in the morning till noon and then starts again in the late afternoon, but stops before darkness (Busnel 1955). It produces chirps of about six syllables.

Duration of main syllables	12.3 ± 1.5 ms
Chirp repetition rate (24°C) (max).	4-7 chirps/s (Fig. 6)
duration	93.4 ± 11.0 ms
Inter-chirp interval	112.9 ± 11.4 ms
Intensity (20 cm)	91.2 ± 2.5 dB SPL
Main frequency components	12.4 ± 0.8 kHz
	30-55 kHz (Fig. 7)

Table 1. The Seven Experimental Animals; their Morphology, their Habitats and their Songs (cont.)

(Values given as mean ± SD)

e) *Psorodonotus illyricus* (P.i.) (Decticinae)

Morphology: has a stout and strong body

Body length	47.4 ± 1.8 mm
width	12.2 ± 0.8 mm
Length hindlegs	53.4 ± 1.5 mm
forelegs	20.2 ± 0.5 mm
Brachypterous – tegmen length	11.0 ± 0.5 mm
width	9.2 ± 0.6 mm

Unable to fly. Habitat: we have found it only in Northern Yugoslavia isolated in the mountains east of Rijeka. A ground dweller living in dry grassland. Song: normally sings in the morning till noon and starts again in the late afternoon, but stops before darkness. It produces a monosyllabic song.

Syllable duration (24°C)	124.1 ± 28.4 ms (Fig. 6)
Repetition rate (max)	0.4 – 0.6 syllable/s
Inter-syllable interval	2.43 ± 0.39 s
Intensity (20 cm)	78.3 ± 2.1 dB SPL
Main frequency components	6.2 ± 0.4 kHz
	30-55 kHz (Fig. 7)

f) *Ephippiger ephippiger* (E.e.) (Ephippigerinae)

Morphology: has a comparatively clumsy body (Fig. 5c)

Body length	36.3 ± 1.8 mm
width	12.1 ± 0.8 mm
Length hindlegs	37.3 ± 3.2 mm
forelegs	21.2 ± 1.9 mm
Brachypterous – tegmen length	5.4 ± 0.2 mm
width	5.4 ± 0.3 mm

Unable to fly. Habitat: lives in Mediterranean regions (e.g. in Southern France) and is found mostly in the mountain regions up to 1000 m above sea level. It prefers either high and dense grass or bushes. Song: sings in the late hours of the night and in the morning till noon (Dumortier 1963). Chirps of 1-3 double syllables (Fig. 6).

Duration of small 1st syllable	15.9 ± 4.0 ms
main 2nd syllable	86.5 ± 9.7 ms
Chirp repetition rate (24°C)	1 chirp/s
duration of 2 double syllables	300 ± 18 ms
Inter-chirp interval	979 ± 124 ms
Intensity (20 cm)	93.2 ± 2.7 dB SPL
Main frequency components	12-40 kHz
(Dominant)	15.0 ± 0.7 kHz (Fig. 7)

Table 1. The Seven Experimental Animals; their Morphology, their Habitats and their Songs (cont.)

(Values given as mean ± SD)

g) *Ephippiger discoidalis* (E.d.) (Ephippigerinae)

Morphology: a dainty and elegant bushcricket

Body length	28.3 ± 1.8 mm
width	8.1 ± 0.7 mm
Length hindlegs	41.6 ± 3.3 mm
forelegs	21.0 ± 1.2 mm
Brachypterous – tegmen length	5.4 ± 0.2 mm
width	5.4 ± 0.3 mm

Unable to fly. Habitat: lives in Mediterranean regions. We have collected it in the mountains of Northern Yugoslavia. It prefers either high and dense grass or bushes. Song: sings in the late hours of the night and in the morning till noon (Hartley, personal communication). It produces chirps which consist of 4-7 double syllables (Fig. 6).

Duration of small 1st syllable	23.6 ± 2.3 ms
main 2nd syllable	86.5 ± 9.7 ms
Syllable repetition rate (24°C)	0.5 chirp/s
Syllables/chirp	4/6 ± 0.9
Chirp duration	663 ± 130 ms
Inter chirp interval	1.74 ± 0.77 s
Intensity (20 cm)	97.9 ± 2.9 dB SPL
Main frequency components	12-40 kHz
	15.0 ± 0.7 kHz (Fig. 7)

Note: The values in this section are taken from Keuper *et al.* (1987 a,b), where further details on the sound production and sound emission of bushcrickets can be found.

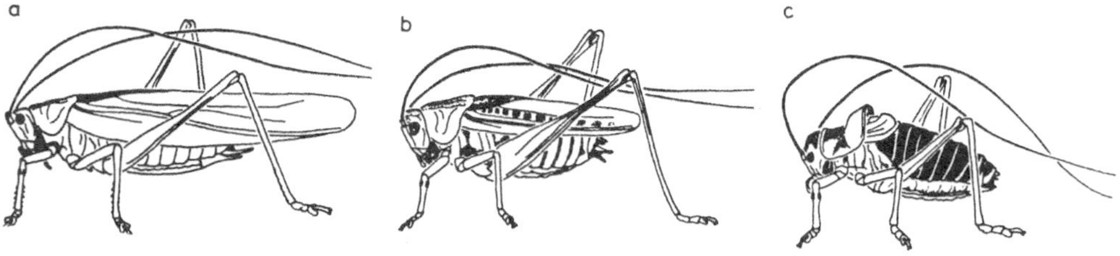

Fig. 5. Drawings of three of the examined species of bushcrickets a) *Tettigonia viridissima*, a long winged species; b) *Decticus verrucivorus*, a species with medium-sized wings; c) *Ephippiger ephippiger*, a brachypterous species. Note the different morphological characteristics. The scale in a) is also valid for b) and c). Drawings by A. Mücke. (After Keuper *et al.* 1987a.)

Fig. 6. Oscillograms showing the time/amplitude pattern of the songs of the seven examined species: *Psorodonotus illyricus* (P.i.), *Ephippiger phippiger* (E.e.), *Ephippiger discoidalis* (E.d.), *Decticus albifrons* (D.a.), *Decticus verrucivorous* (D.v.), *Tettigonia viridissima* (T.v.), and *Tettigonia cantans* (T.c.). Note the existence of the "small" syllables before the "main" syllable (with the exception of P.i.). The time mark is valid for all the oscillograms. (After Keuper *et al.* 1987a.)

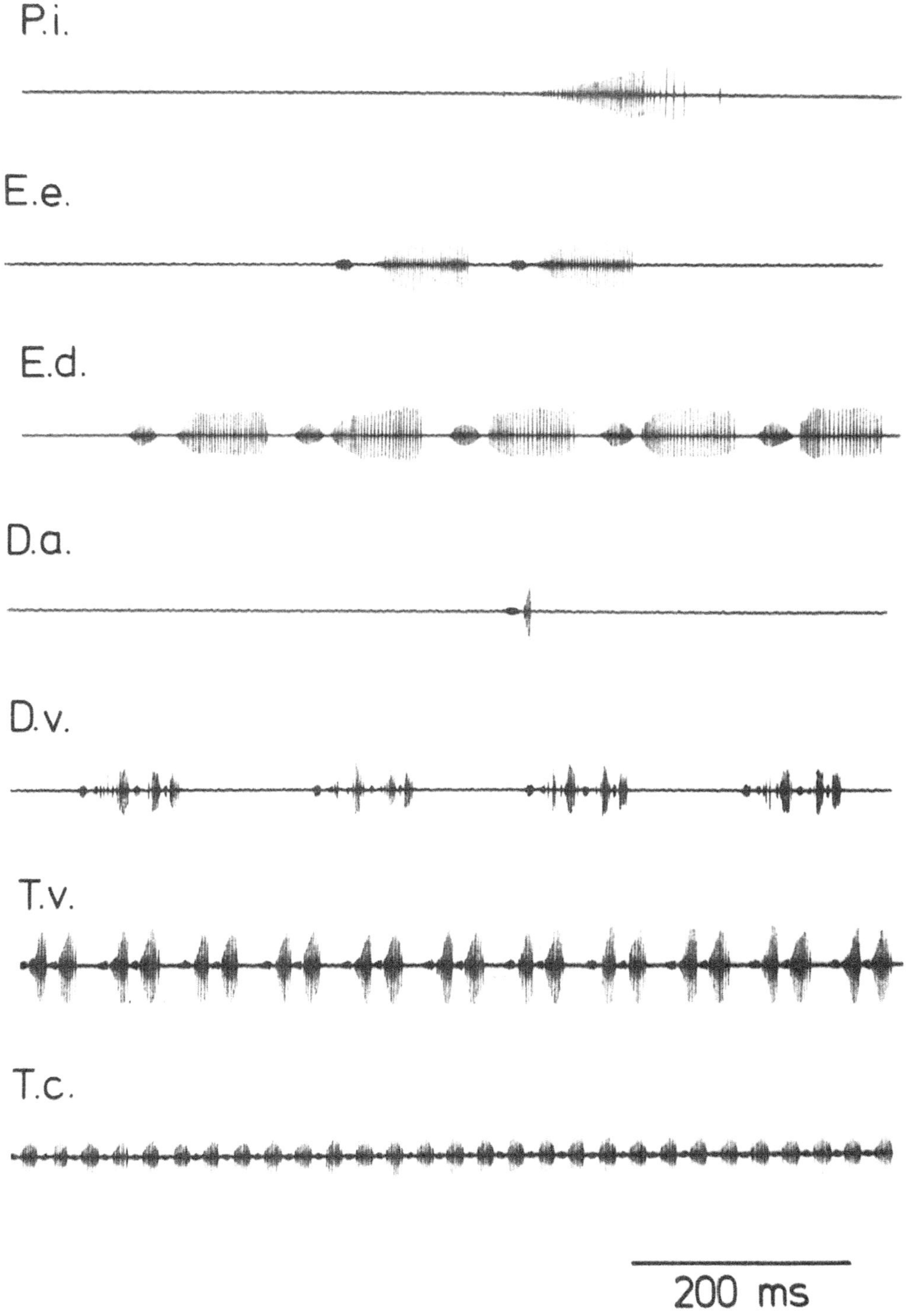
P.i.
E.e.
E.d.
D.a.
D.v.
T.v.
T.c.
200 ms

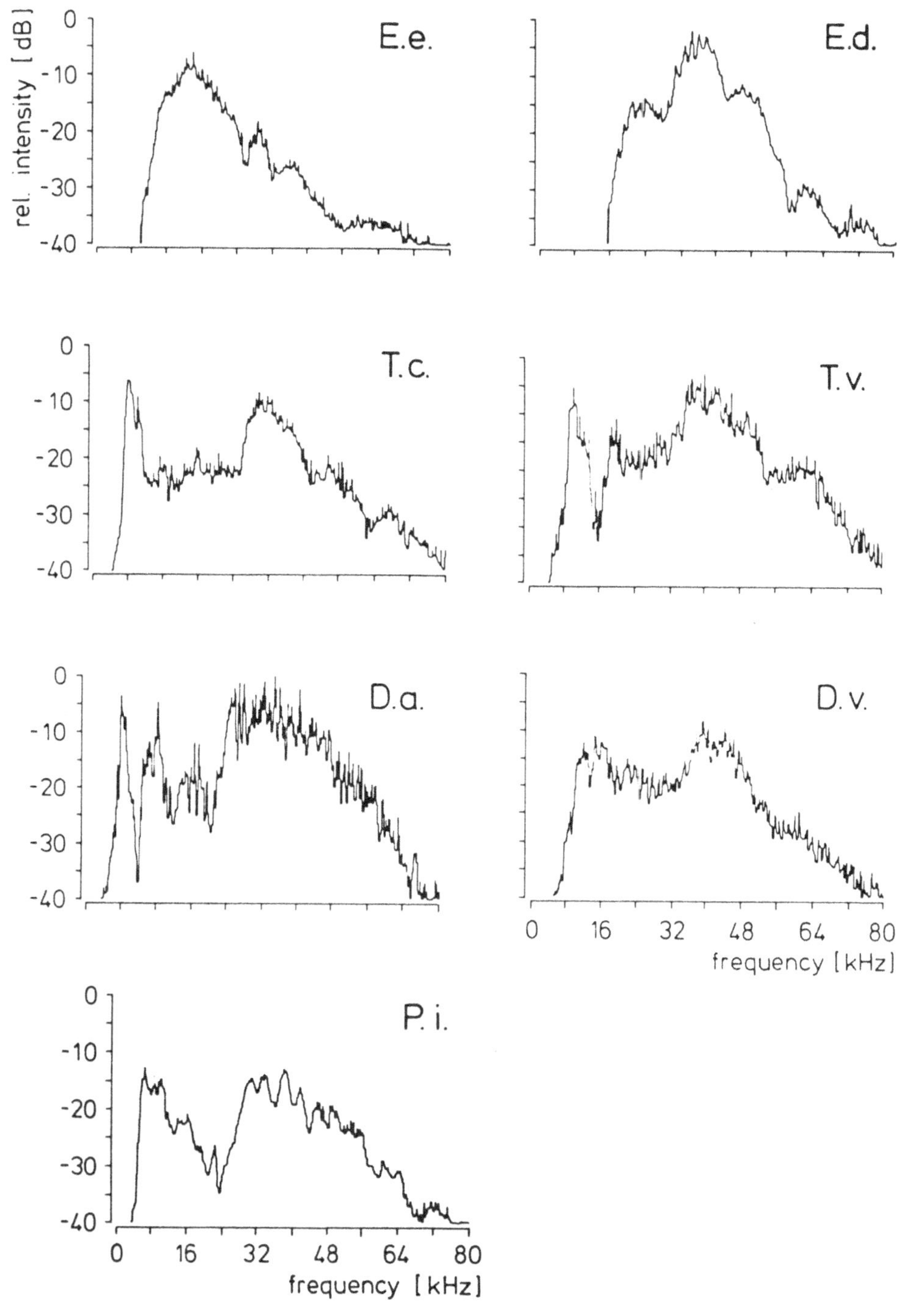

Fig. 7. Power spectra of the stridulatory songs dorsally emitted of the seven examined species: *Ephippiger ephippiger* (E.e.), *Ephippiger discoidalis* (E.d.), *Tettigonia cantans* (T.c.), *Tettigonia viridissima* (T.v.), *Decticus albifrons* (D.a.), *Decticus verrucivorus* (D.v.), and *Psorodonotus illyricus* (P.i.). (After Keuper et al. 1987a.)

THE ACOUSTIC RECEIVER SYSTEM OF THE BUSHCRICKETS

From the previous behavioural data it is evident that bushcrickets possess a combined auditory-vibratory pathway within their central nervous system. The following discussion examines the ways by which the information is perceived by the receptors, and integrated in the central nervous system.

Structure and Function of the Different Receptor Organs

The receptor organs that are most important for the perception of sound and vibration are the complex tibial organs of the forelegs. These tibial organs consist of the tympanal organ, the intermediate organ and the subgenual organ. They exist in a smaller pattern in the mid and the hindlegs. However, the essential structures detecting airborne sound can only be found in the forelegs (i.e. the tympana and the associated trachea). The morphology of the receptor organs, and the arrangement of the corresponding receptor cells are well known (Schwabe 1906; Schumacher 1973, see also Lakes & Schikorsky, chapter 10), and their central projections have been described by Oldfield (1983), Römer (1985, 1987), Grosch *et al.* (1985). Apart from the complex tibial organs, the numerous campaniform sensilla and probably chordotonal organs that are also situated in the six legs function as receptors for the combined auditory-vibratory pathway.

a) The tympanal organs are situated in the proximal parts of the tibia of the forelegs. The stimulus conducting structures are not only built up by the two tympana (in the examined species these tympana are protected by cuticular covers leaving small slits), but also by the acoustic trachea which open into a large spiracle (see Bailey chapter 12).

The *crista acustica* of those species examined in this study are built up by a varying number of auditory cells: the two ephippigerine species possess only 24 (*E. discoidalis*), and 27-28 (*E. ephippiger*) receptor cells, whereas *D. verrucivorus* has 33, *T. cantans* 35, *T. viridissima* 36 and *D. albifrons* 36 receptor cells (see Lakes & Schikorsky chapter 10). The length of the tympanal trachea, and the width of the opening of the spiracle, also differ between the seven species (length of the tympanal trachea: *D. albifrons*: 22.8 ± 1.2 mm; *D. verrucivorus*: 16.8 ± 1.2 mm). Despite these morphological differences, the acoustic threshold curves as measured in the whole tympanic nerve are very similar. Thus the receptor cells do not appear to differ between all the species investigated. Although the threshold curves cover a similar frequency range, the sensitivity differs between the species. The two *Decticus* species are the most sensitive, whereas the two ephippigerines are least sensitive differing up to 10-15 dB (Fig. 8)

Single cell recording from the axons of the receptor cells in the seven species again show a similarity between the threshold curves of identified single neurons in the different species. We frequently found threshold curves with a similar shape (Fig. 9 & 10). Most of the receptor cells of the seven species can be divided into distinct functional types which are present in all the species. This would indicate that despite the variation in the external morphology of the ears of the different species, the mechanical transduction of the stimulus at the level of the receptor organ shows a remarkable parallel between species.

We frequently recorded more receptor cells, with a characteristic frequency close to frequencies corresponding to those of the conspecific songs. Additionally these cells have a more extensive arborisation in the auditory neuropile than the other cells. Compare for example the projections of *D.v.* 77 and *D.a.* 36 in Fig. 10 both having a characteristic frequency of 9 kHz, which is the dominant frequency of *D. albifrons*. This may mean that these cells with these particular frequency characteristics are more numerous in the tympanal organs, or it may be that these cells are recorded preferentially by the experimenter in the

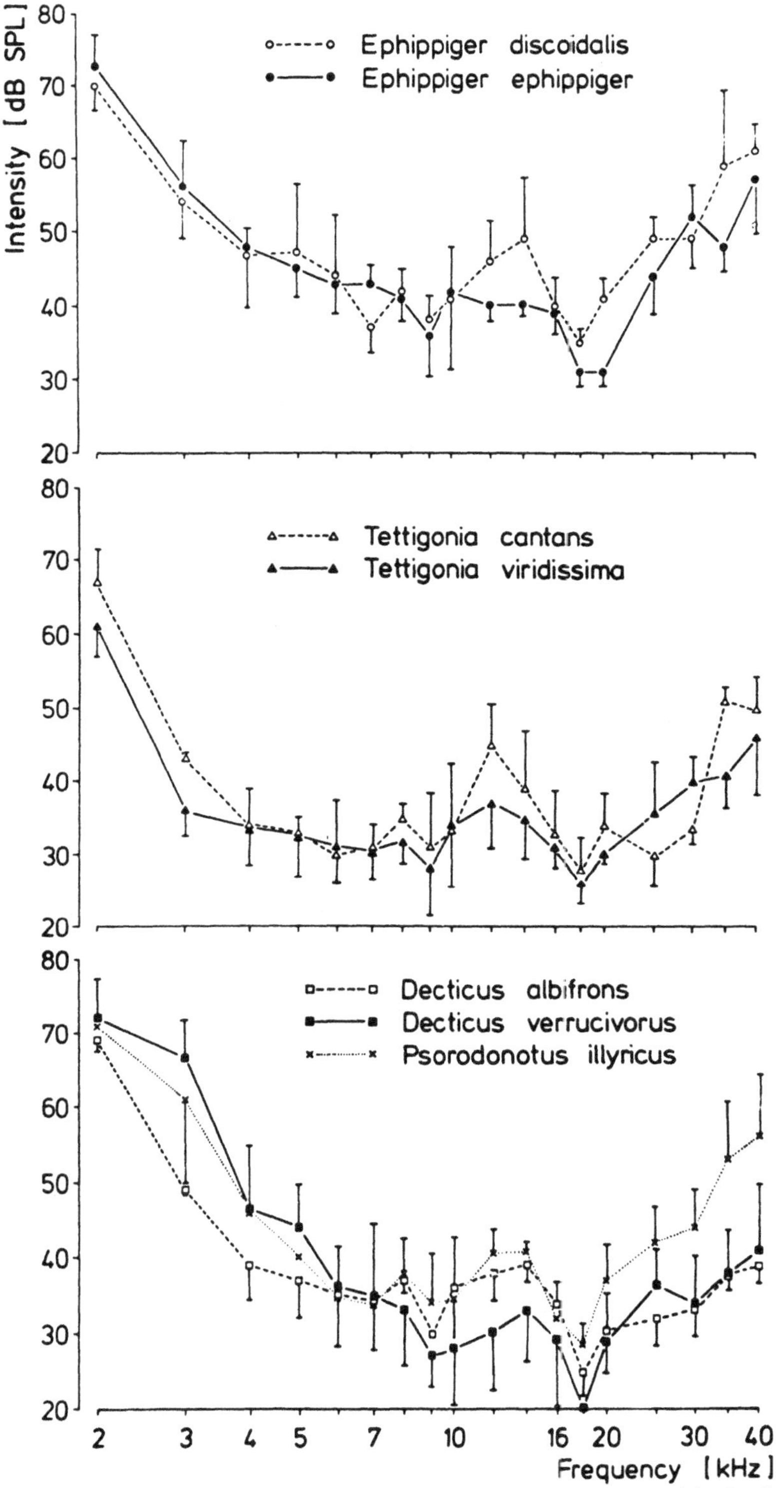

Fig. 8. "Hearing" threshold curves of the seven tettigoniid species, measured by hook electrode recordings from the tympanal nerve. Mean values of at least six experiments. (After Kalmring *et al.* 1987.)

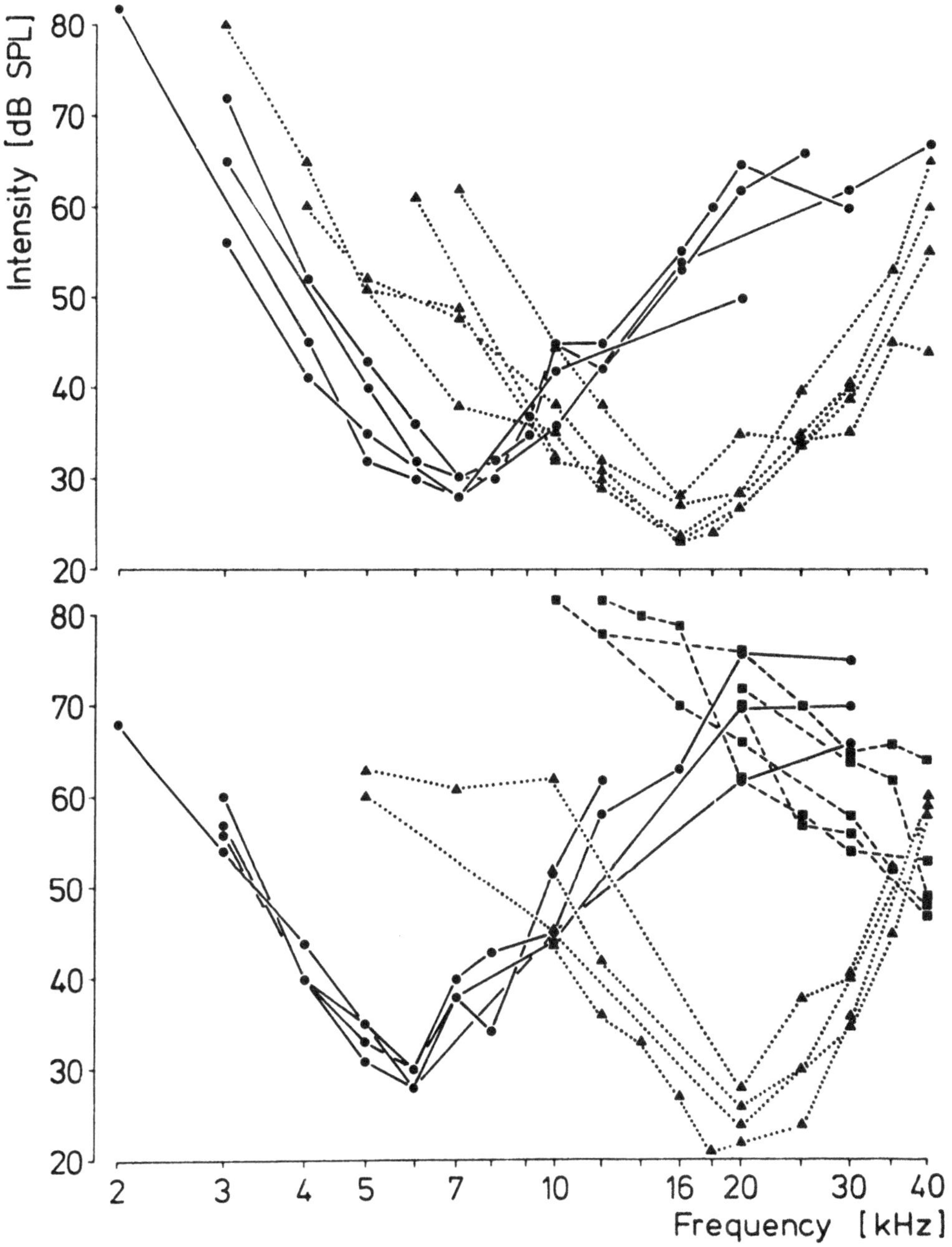

Fig. 9. Examples of threshold curves of five different functional types of acoustic receptor cells in tettigoniids. The 4 to 5 curves shown here for each type were obtained from cells in different species. (After Kalmring et al. 1987.)

nerve because of their fibre size.

Whereas the seven species show a great conformity with respect to the frequency range covered by their tympanal units, as well as their sensitivity, in some cases there are clear differences between the receptors of the different species as far as the response magnitude and the stimulus response latency in the suprathreshold range is concerned (Fig. 11). In their saturation range, the tympanal receptor cells of *D. albifrons* and the two tettigoniine species have the greatest average response magnitudes; about 10 spikes per stimulus or even greater for stimulus durations of 20 ms. The mean response magnitude of the acoustic receptors of

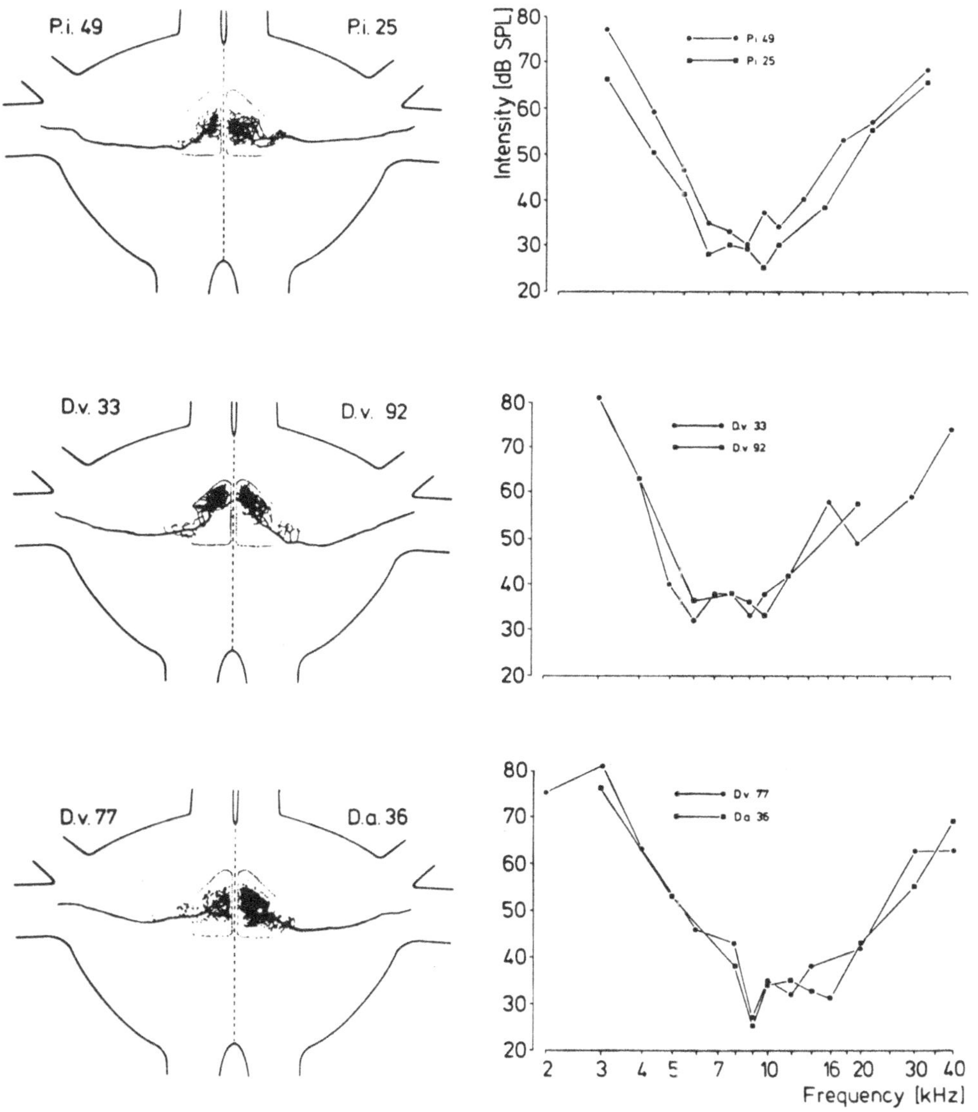

Fig. 10. Central projections of two functional types in *Psorodonotus illyricus* (P.i.), *Decticus verrucivorus* (D.v.), and *Decticus albifrons* (D.a.). Left: central projections of different receptor cells; right: their corresponding threshold curves. Note: differences in projection in P.i. 49 and P.i. 25 and in D.v. 77 and D.a. 36. The numbers indicate different experimental animals.

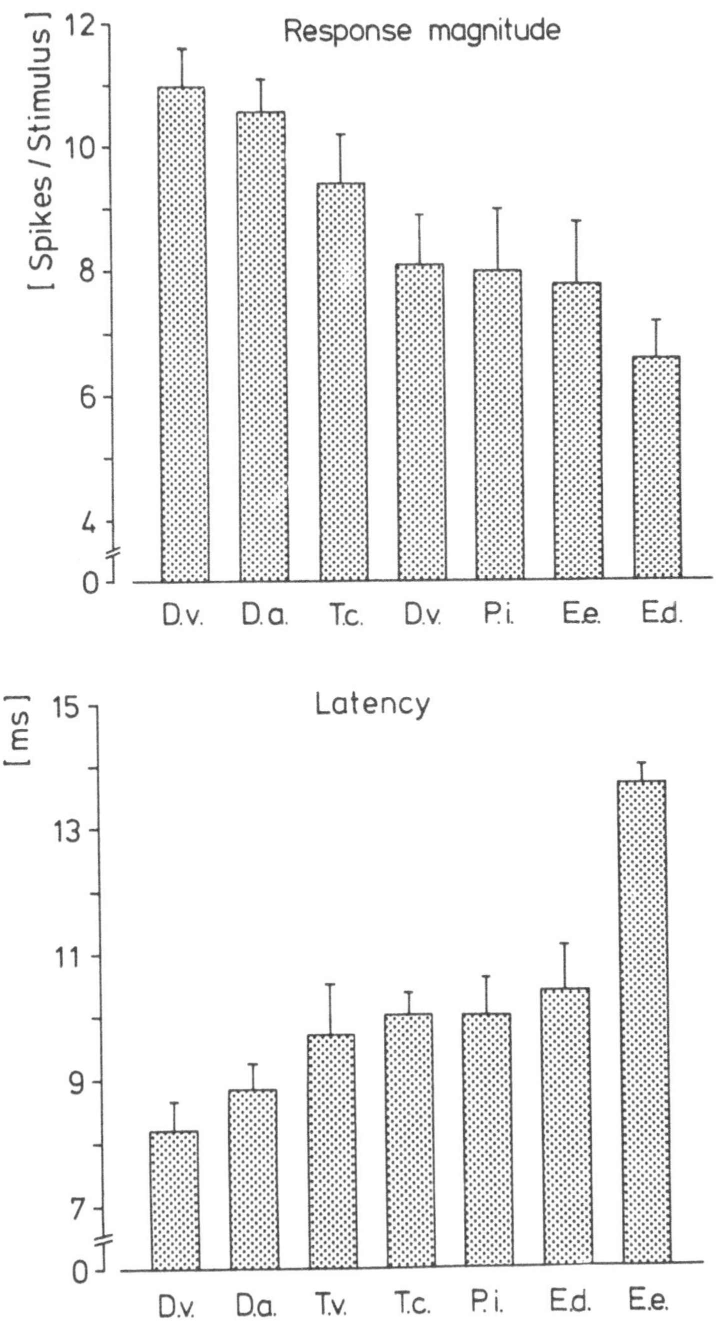

Fig. 11. Differences in the mean response magnitude and the mean latency between the acoustic receptor cells of the seven species: *Ephippiger ephippiger* (E.e.), *Ephippiger discoidalis* (E.d.), *Tettigonia cantans* (T.c.), *Tettigonia viridissima* (T.v.), *Decticus albifrons* (D.a.), *Decticus verrucivorus* (D.v.), and *Psorodonotus illyricus* (P.i.). The columns are mean values from all receptor cells recorded in one species. The data were determined at a suprathreshold range (81 dB SPL and characteristic frequency of the investigated cells). Stimulus duration 20 ms, rise and fall time of the stimulus 1 ms. (After Kalmring *et al.* 1987.)

P. illyricus, *D. verrucivorus* and *E. ephippiger* are at about 8 impulses per stimulus. In *E. discoidalis* there are only 7 impulses per stimulus. The stimulus response latencies at saturation show in some cases clear differences. For a stimulus intensity of 81 dB SPL the acoustic receptors of *D. verrucivorous* have a mean latency of about 8 to 9 ms, in *D. albifrons* 9 ms, the latencies of *T. cantans*, *T. viridissima*, *P. illyricus* and *E. discoidalis* are about 10 ms. The receptor units of *E. ephippiger* have by far the greatest latencies; measured at the recording site (in the acoustic neuropile of the prothoracic ganglion) they are about 13 to 14 ms.

Additionally, it is worth noting that in *E. ephippiger*, vibratory receptors appeared more numerous than those for airborne acoustic signals. Thus, we have the impression that the vibratory receptor system is, in comparison with the acoustic receptor system, better developed in this species than in the other six species (Kalmring *et al.* 1985). However, in none of the species did the receptor cells of the tympanal nerve respond exclusively to the time pattern of the conspecific song, or showed any significant habituation. For example the receptor cells of *P. illyricus* (a species with a very low syllable repetition rate) responded tonically and without any habituation to the songs of the two *Tettigonia* species that produce a higher syllable repetition rate. (Note: *T. viridissima* occurs syntopically with *P. illyricus*).

The synaptic transmission from the tympanal receptors to the central neurons has been investigated using the Current-Source-Density-Analysis (Breckow *et al.* 1982). The CSD technique provides information about "current sources" and "sinks" that result from synaptic activity in complex neuropiles. Single electrode recording in the region of the neuropile will only convey information about field potentials whose origin is uncertain. The special arrangement of the seven electrodes focussed at a local area (volume) of the neuropile provides not only activity as reflected through the summed action potential of the entire receptor array, but also current effects at the termination of these receptors, and from the postsynaptic potentials to the connecting inter-neurons (Breckow *et al.* 1982). We could find species-specific transmission processes in the acoustic neuropiles of the prothoracic ganglion in all species. That is the pure-tone stimulus with the frequency corresponding to the dominant frequency of the species-specific song evoked stronger current sources or sinks in the acoustic neuropile than stimuli of other frequencies. This is demonstrated by the averaged relative CSD-responses of several animals of each species (Fig. 12) to different pure tone stimulation frequencies. There is an enhanced synaptic activity in this neuropile when the animals are stimulated with the characteristic dominant frequency of their species-specific stridulatory song. That may be caused by an increased number of receptor cells tuned to that specific frequency (see above), or by these neurons having an increased number of arborisations within this area of the neuropile (Fig. 10). Nevertheless, in all species, information about heterospecific songs is transmitted synaptically by the tympanal receptor cells showing no apparent species "recognition". This result may be due to the broad banded frequency responses of the receptor neurons, which has been demonstrated in the acoustic threshold curves. Since the song spectra of the species examined in this study lie within their hearing range, each species is able to perceive and to process the songs on the others.

b) The subgenual organs are extremely sensitive vibratory receptors comprising 20 to 30 bipolar scolopidial cells, and with their satellite cells they hang like a sail in the haemolymph of the proximal parts of the tibia in all six legs (see Lakes & Schikorsky chapter 10). They are very sensitive to a wide frequency range of structure-borne sound signals. Their characteristic frequencies lie in two frequency bands 500-700 Hz or 1000-3000 Hz. In most species these receptor cells are able to respond to threshold accelerations as small as about 1 cm/s^2 (Fig. 13).

c) The intermediate organs are situated in the upper area of the *crista acustica* in the proximal parts of the tibia in all the six legs (see Lakes & Schikorsky chapter 10), and their function is as yet unknown. However, we believe that they are probably bifunctional vibration and airborne sound receptors, and in this way are similar to the subgenual receptors, but are sensitive to airborne sound of low frequencies between 1 and 10 kHz (Fig. 13) (Kalmring *et al.* 1978).

d) Campaniform sensilla are vibration receptors which respond in a phase- locked way to sinusoidal stimuli low frequencies up to about 150 Hz (Fig. 13). These receptors are transformed hair sensilla, whose cell innervate an elliptically shaped dome ("Ausdifferenzierung") of cuticle. These receptors are situated on each leg in groups (Grosch *et al.* 1985). In the species we have examined there are more than 100 of these receptors on each leg. The combined activity of auditory and vibratory receptors enable the tettigoniids to perceive structure-borne sound from 1 Hz to 5000 Hz, and airborne sound from 2 kHz to at least 70 kHz.

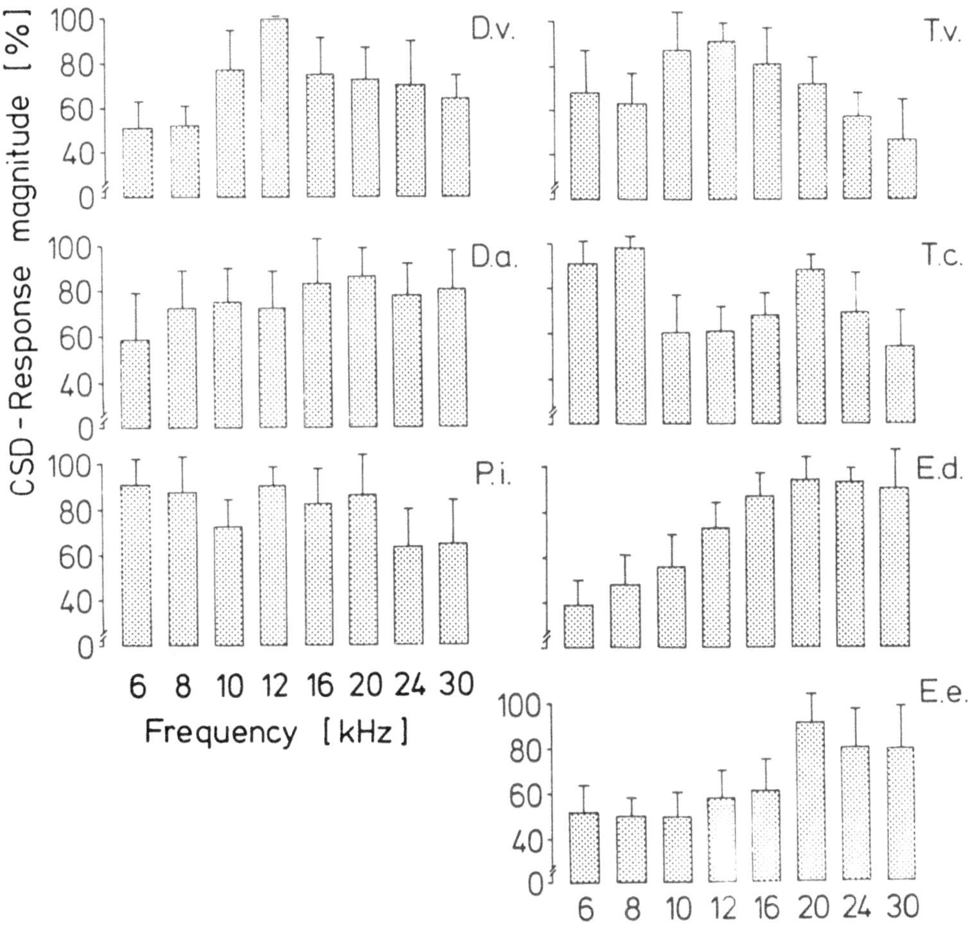

Fig. 12. The dependence of the relative CSD-response magnitude (evaluated by integration of CSD course) of the stimulus frequency for all the seven tettigoniid species. The abbreviations of the name of the species and stimulus parameters as in Fig. 11. (After Kalmring *et al.* 1987.)

Structure and Function of the Auditory and Vibratory Pathway in the Ventral Cord

The auditory and vibratory information from the different receptors of all six legs is convergently transmitted to the brain and to other higher nervous centres in the ventral cord. This is mediated by some 15 to 20 ascending fibres terminating in the supraoesophageal ganglion (Kalmring *et al.* 1985). These ventral cord interneurons, ascending to the brain, are influenced by both airborne sound and vibration. It could be demonstrated that they have no – or at least not exclusively – monosynaptic connections to the different receptors described above (Kalmring *et al.* 1979). Therefore segmental and/or intersegmental thoracic interneurons should be intercalated between the receptors and the ascending auditory-vibratory interneurons.

Thoracic first order interneurons are those with a direct connection to the primary afferent endings. Although the properties of many of these have been described, only in *T. viridissima* and *Mygalopsis marki* (Römer 1985, Römer & Bailey 1986) has a connection been demonstrated, and this to the omega-neuron. However, it can be argued that other first order neurons in the thoracic ganglia are likely to be present because of the demonstrated links between the vibratory receptors of all six legs and ventral nerve cord's neurons ascending to

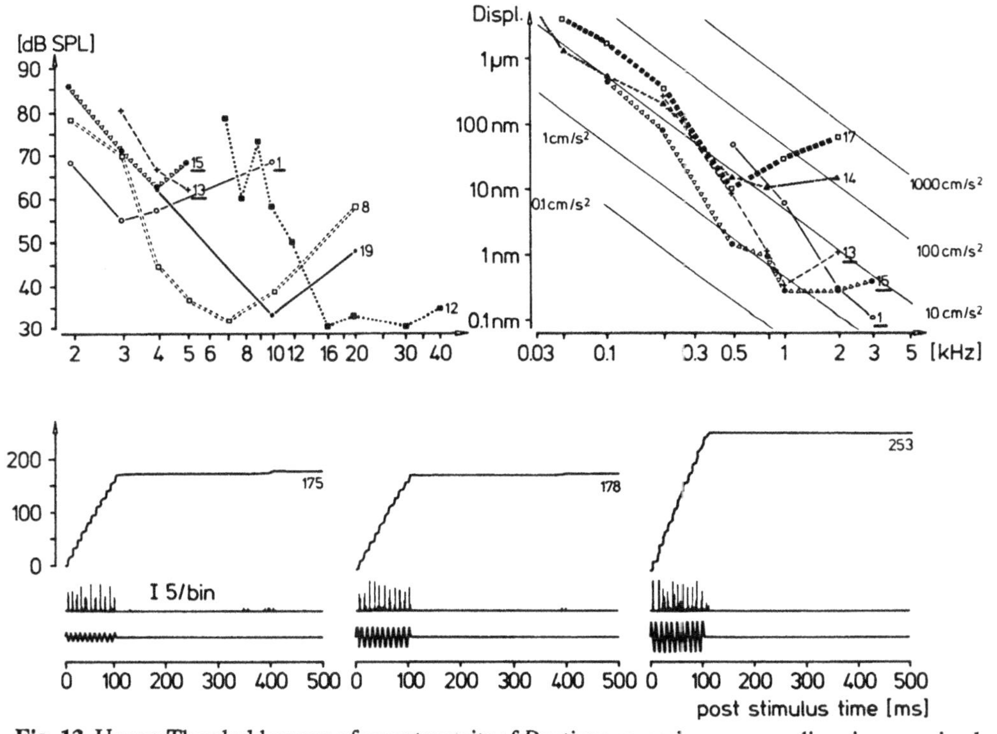

Fig. 13. Upper: Threshold curves of receptor units of *Decticus verrucivorus*; recordings in one animal from the same anatomical side; left: airborne sound stimuli: 6 different units (numbers 1, 8, 12, 13, 15, 19); right: vibration stimuli: 5 different units (numbers 1, 13, 14, 15, 170; note the bifunctional units 1, 13, and 15. Lower: Response characteristic of a campaniform sensillum in *Decticus verrucivorus*. Cumulative response curves (numbers of impulses are indicated at the end of the curves) and post-stimulus-time (PST)-histograms (15 presentations of the stimulus were used for the analysis in each case) are shown for a 100 Hz-stimulus of three different displacements (from left to right: 5.4 um, 17 um, 54 um). The pase-locked responsiveness of the campaniform sensilla is illustrated by the PST-histograms and the step-like slope of the curve of cumulative frequency distribution. The horizontal part of this curve indicates the lack of after discharges. (Changes after Kalmring *et al.* 1978.)

the supraesophageal ganglion. But not much is known about how the information is transmitted, perhaps from the receptors on mid and hindlegs to the ascending ventral cord neurons.

We provide a comparative study of the auditory-vibratory ventral-cord neurons ascending to the supraesophageal ganglion. In most cases these are "through"- neurons, with their dendritic structures situated within the thoracic ventral cord and their axons ascending to the supraesophageal ganglion as well as partly descending to the abdomen. In the different ganglia they possess multibranched collaterals that are probably information emitting structures. A typical example for the structure and the function of such a neuron is demonstrated in Kalmring *et al.* (1979).

We have classified the 15 to 20 auditory-vibratory ventral-cord neurons ascending to the supraoesophageal ganglion into three groups:

1) V-neurons: used for vibratory information (Kühne *et al.* 1980) reacting to supra-threshold vibratory stimuli. They are also influenced by the auditory receptors. (2) VS-neurons: reacting to vibratory as well as to acoustic stimuli with a supra-threshold response. The simultaneous presentation of both stimuli normally enhances their responses (Kalmring *et al.* 1983). (3) S-neurons: These show a supra-threshold response preferentially to acoustic stimuli. However, sub-threshold excitatory and/or inhibitory influences of vibratory receptors can be demonstrated in each of these units. Most of the auditory-vibratory ventral cord neurons ascending to the supraesophageal ganglion respond to stimuli between 1 Hz and to 70 kHz within the physiological range.

Within the different functional groups single neurons are assigned to quite unambiguous functional types. These functional types could be found in all seven species, and all react in a similar way in each species. The functional similarity in the combined auditory and vibratory pathways on the level of ventral cord is also remarkable even to the finer details of their response. Such a similarity is shown in the threshold curves as well as the supra-threshold response characteristics of each of the single neuron types. For example, in all species the VS4-neuron responds with an on-burst activity to auditory or vibratory stimuli of a duration of more than 30 ms. They even have the same frequency range, a functional correspondence of the auditory-vibratory pathway in the ventral-cord and also the situation is paralleled by the auditory-vibratory pathway of acridids, which has evolved in an entirely different way with quite a different morphology (Kalmring *et al.* 1985).

Since behavioural experiments show that bushcrickets are able to locate and to recognize the species-specific song (e.g. Busnel *et al.* 1956, Bailey & Robinson 1971), recognition processes must take place in the central nervous system. Kalmring *et al.* (1985) were able to show that in the central auditory-vibratory ventral-cord neurons of a range of bushcricket species, not only did they respond to the species-specific song, but equally well to heterospecific songs. In all species the same functional types of central ventral-cord neurons respond to the different songs with a very similar discharge pattern, and surprisingly heterospecific songs often elicit a more distinct response than the species-specific song! (Figs. 14 & 15). Such results indicate that the recognition of the species-specific song (1) does not take place at the level of the ventral cord but more likely in higher centre of the CNS (probably in the supraoesophageal ganglion), and/or (2) is due to the additive effect of all discharge patterns of the ascending ventral cord neurons. Neuronal filters that are sharply tuned to the parameters of the species-specific song do not exist at the single cell level. For this reason it is perhaps simplistic to attempt an explanation of such a complex process such as "recognition" using investigations of single cells alone. More complex neurophysiological methods have to be chosen that correspond to the complexity of the processes to investigate.

Such an approach could be for example through the application of Current-Source-Density-Analysis, which examines in real-time the activity of a few or even many neuronal elements.

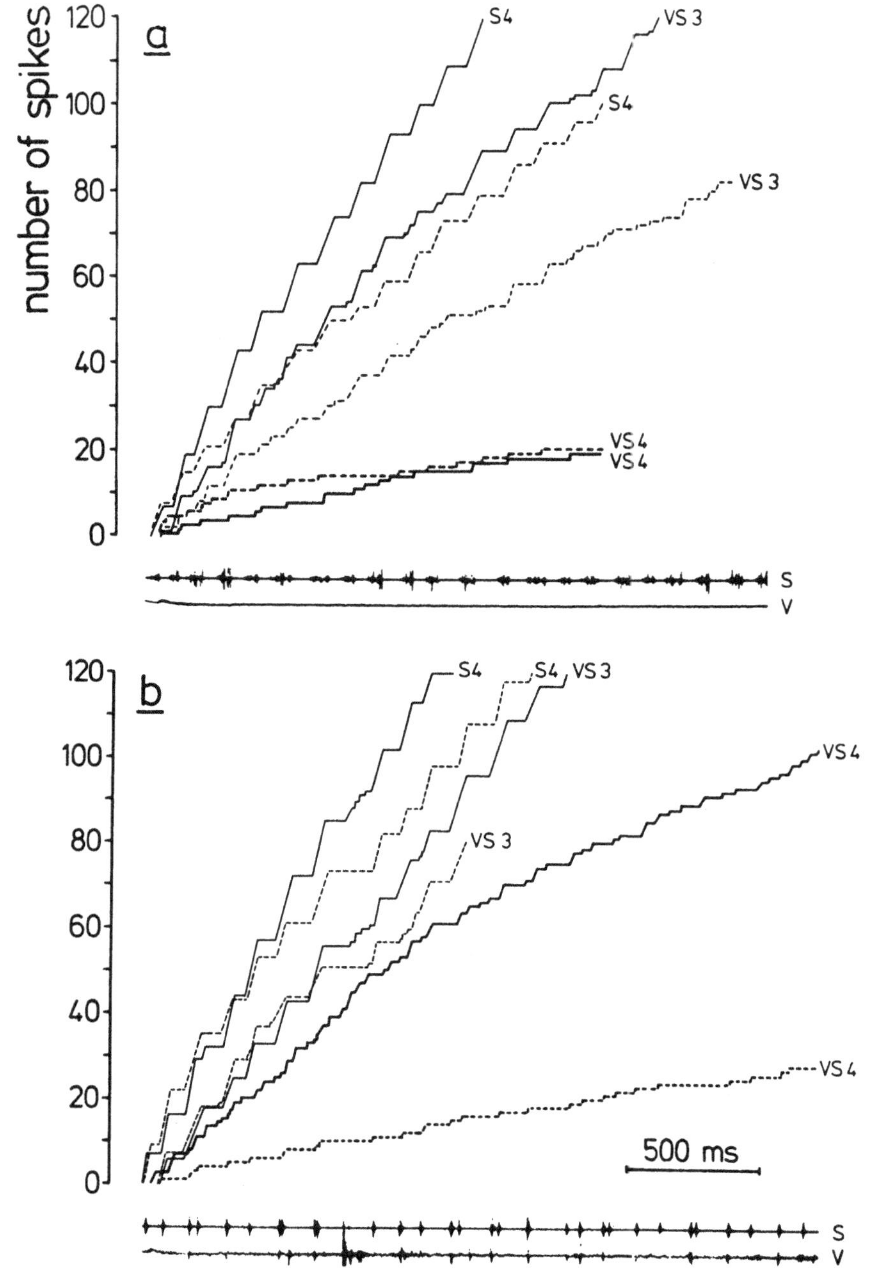

Fig. 14. Response characteristics (cumulative response curves) of three different tonically responding auditory-vibratory ventral nerve cord neurons (S4, VS3, and VS4) of *Decticus verrucivorus* when stimulated with the stridulatory songs of a) *Decticus verrucivorus* and b) *Decticus albifrons*. Solid curves: stimulation with the songs and the related vibratons simultaneously; dahsed curves: stimulation with the song alone. (Changed after Kalmring *et al.* 1985.)

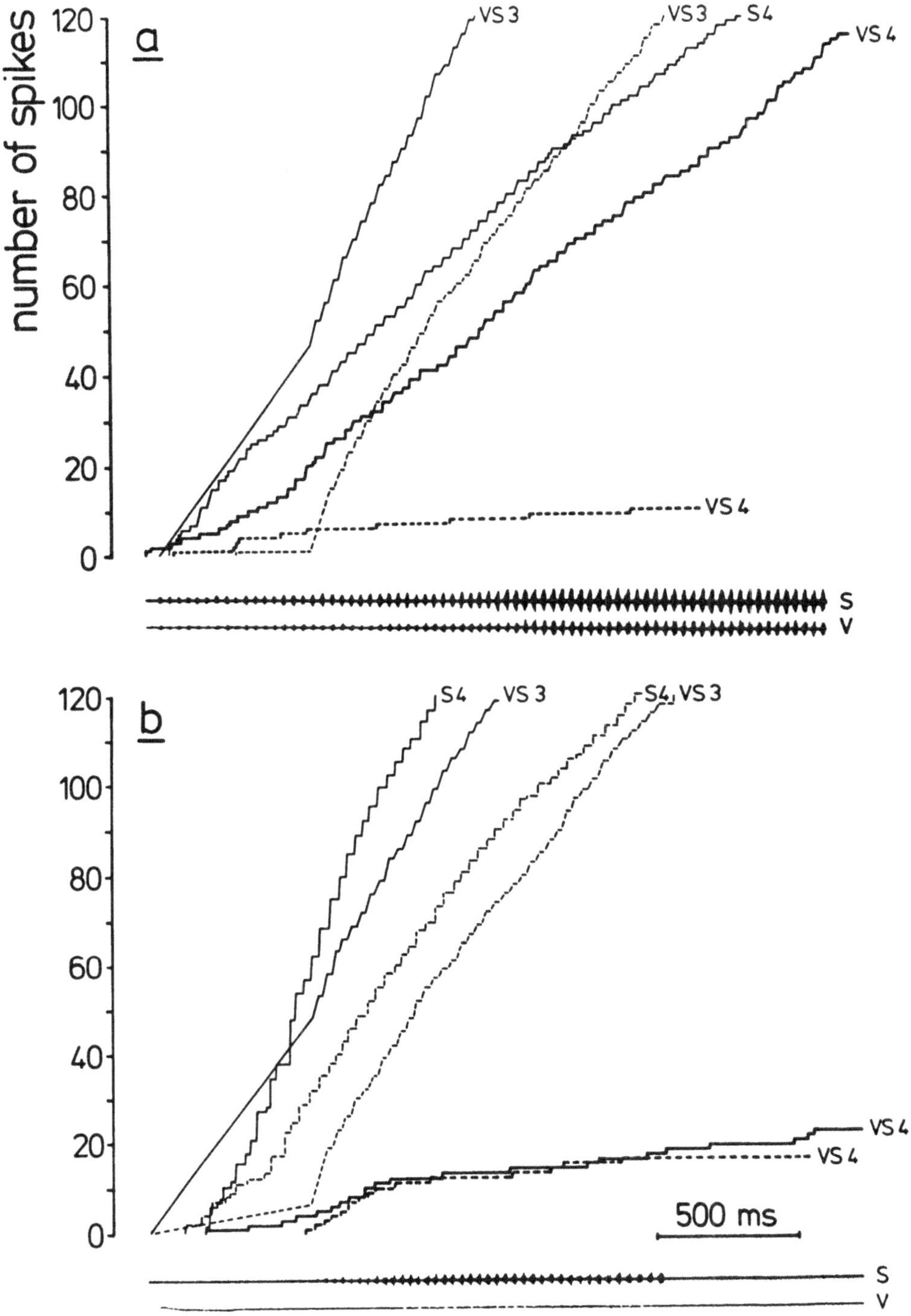

Fig. 15. Response characteristics (cumulative response curves) of three different tonically responding auditory-vibratory ventral nerve cord neurons (S4, VS3, and VS4) of *Decticus verrucivorus* when stimulated with the stridulatory songs of a) *Tettigonia cantans* and b) *Pholidoptera littoralis*. Solid curves: stimulation with the songs and the related vibrations simultaneously; dashed curves: stimulation with the song alone. (Changed after Kalmring *et al.* 1985.)

References

Bailey W.J. & D. Robinson. 1971. Song as a possible isolating mechanism in the genus *Homorocoryphus* (Tettigonioidea, Orthoptera). Anim. Behav. 19: 390-397.

Bailey W.J. & R.O. Stephen. 1984. Auditory acuity in the orientation behaviour of the bushcricket *Pachysagella australis* Walker (Orthoptera, Tettigoniidae, Saginae). Anim. Behav. 32: 816-829.

Breckow J., K. Kalmring & R. Eckhorn. 1982. Multichannel-recordings and real- time Current Source Density (CSD) analysis in the central-nervous system of insects. Biol. Cybern. 45: 115-121.

Brownell P.H. 1977. Compressional and surface waves in sand: Used by desert scorpions to locate prey. Science 197: 479-481.

Busness R.G. 1955. Sur certains rapports entre le moyen d'information acoustiques et le comportement acoustiques des Orthoptères. Coll. Acoust. Orthopt. Ann. Epiphyties I.N.R.A., pp. 281-306.

Busnel R.G., B. Dumortier & M.-C. Busnel. 1956. Recherches sur le comportement acoustiques des Ephippigeres (Orthoptères, Tettigoniidae). Bull. Biol. Fr. Belg. 90: 219-286.

Cokl A., C. Otto & K. Kalmring. 1985. The processing of directional vibratory signals in the ventral nerve cord of *Locusta migratoria*. J. Comp. Physiol. 156: 45-52.

Cremer L., M. Keckl & E.E. Ungar. 1973. Structure Borne Sound. Springer Verlag, Berlin, Heidelberg, New York.

Dumortier B. 1963. Ethological and physiological study of sound emissions in Arthropoda. In: Acoustic Behaviour of Animals. Ed. R.G. Busnel. Elsevier, Amsterdam, London, New York. pp. 583-654.

Grosch A., F. Callender, M. Petersen, A. Cokl & K. Kalmring. 1985. Vibration receptors of larvae and of images in locusts: Location on the legs central projections and physiology. In: Acoustic and Vibrational Communication in Insects. Eds. K. Kalmring & N. Elsner. Paul-Parey-Verlag, Berlin, Hamburg. pp 151-161.

Haupt R. 1973. Beitrag zum Problem der Lärmminderung durch Waldbestände. 1. Teil: Methodische Grundlagen. Arch. Naturschutz Landschaftsforsch. 13: 309-327.

Kalmring K., W. Kaiser, C. Otto & R. Kühne. 1985. Coprocessing of vibratory and auditory information in the CNS of tettigoniids and locusts. In: Acoustic and Vibrational Communication in Insects. Eds. K. Kalmring & N. Elsner. Paul-Parey- Verlag, Berlin, Hamburg. pp 193-202.

Kalmring K. & R. Kühne. 1980. The coding of airborne-sound and vibration signals in bimodal ventral-cord neurons of the grasshopper *Tettigonia cantans*. J. Comp. Physiol. 139: 267-275.

Kalmring K., R. Kühne & B. Lewis. 1983. The acoustic behaviour of the bushcricket *Tettigonia cantans*. III. Coprocessing of auditory and vibratory information in the central nervous system. Behav. Processes 8: 213-228.

Kalmring K., B. Lewis & A. Eichendorf. 1978. The physiological characteristics of the primary sensory neurons of the complex tibial organ of *Decticus verrucivorus* L. J. Comp. Physiol. 127: 109-121.

Kalmring K., H.G. Rehbein & R. Kühne. 1979. An auditory giant neuron in the ventral cord of *Decticus verrucivorus* (Tettigoniidae). J. Comp. Physiol. 132: 225-234.

Keuper A., K. Kalmring, A. Schatral, W. Latimer, & W. Kaiser. 1986. Behavioural adaptations of ground living bushcrickets to the properties of sound propagation in low grassland. Oecologia 70: 414-422.

Keuper A. & R. Kühne. 1983. The acoustic behaviour of the bushcricket *Tettigonia cantans*. II. Transmission of airborne-sound and vibration signals in the biotope. Behav. Processes 8: 125-145.

Keuper A., C. Otto, W. Latimer & A. Schatral. 1985. Airborne sound and vibration signals of bushcrickets and locusts; their importance for the behaviour in the biotope. In: Acoustic and Vibrational Communication in Insects. Eds. K. Kalmring & N. Elsner. Paul-Parey-Verlag, Berlin, Hamburg. pp 135-142.

Keuper A., S. Weidemann, K. Kalmring & D. Kaminski. 1988a. Sound production and sound emission in seven species of European tettigoniids. I. The different parameters of the song; their relation to the morphology of the bushcricket. Bioacoustics 1, 31-48.

Keuper A., S. Weidemann, K. Kalmring & D. Kaminski. 1988b. Sound production and sound emission in seven species of European tettigoniids. II. Wing morphology and the frequency content of the song. Bioacoustics 1, 171-186.

Kühne R., B. Lewis & K. Kalmring. 1980. The responses of ventral-cord neurons of *Decticus verrucivorus* (L) to sound and vibration stimuli. Behav. Processes 5: 55-74.

Latimer W. & A. Schatral. 1983. The acoustic behaviour of the bushcricket *Tettigonia cantans*. I. Behavioural responses to sound and vibration. Behav. Processes 8: 113-124.

Markl H. 1983. Vibrational communication. In: Neuroethology and Behavioural Physiology. Eds. F. Huber & H. Markl. Springer Verlag, Berlin, Heidelberg. pp 332-353.

Michelsen A. 1978. Sound reception in different environments. In: Sensory Ecology, Review and Perspectives. Ed. M.A. Ali. Plenum Press, New York. pp 345-373.

Michelsen A., F. Fink, M. Gogala & D. Traue. 1982. Plants as transmission channels for insect vibrational songs. Behav. Ecol. Sociobiol. 11: 269-281.

Michelsen A. & O.N. Larsen. 1983. Strategies for acoustic communication in complex environments. In: Neuroethology and Behavioural Physiology. Eds. F. Huber & H. Markl. Springer Verlag, Berlin, Heidelberg. pp 321-331.

Morris G.K., G.E. Kerr & J.H. Fullard. 1978. Phonotactic preferences of female katydids (Orthoptera: Tettigoniidae: *Conocephalus nigropleurum*). Can. J. Zool. 56: 1479-1487.

Oldfield B.P. 1983. Central projections of primary auditory fibres in Tettigoniidae (Orthoptera: Ensifera). J. Comp. Physiol. 151: 389-395.

Römer H. 1985. Anatomical representation of frequency and intensity in the auditory system of Orthoptera. In: Acoustic and Vibrational Communication in Insects. Eds. K. Kalmring & N. Elsner. Paul-Parey-Verlag, Hamburg, Berlin. pp 25-32.

Römer H. 1987. Representation of auditory distance within a central neuropil of the bushcricket *Mygalopsis marki*. J. Comp. Physiol. 161: 33-42.

Römer H. & W.J. Bailey. 1986. Insect hearing in the field. II. Male spacing behaviour and correlated acoustic cues in the bushcricket *Mygalopsis marki*. J. Comp. Physiol. 159: 627-638.

Schatral A., W. Latimer & B. Broughton. 1984. Spatial dispersion and agonistic contacts of male bushcrickets in the biotope. Z. Tierpsychol. 65: 201-214.

Schumacher R. 1973. Morphologische Untersuchungen der tibialen Tympanalorgane von neun einheimischen Laubheuschreckenarten (Orthoptera, Tettigoniidae). Z. Morphol. Tiere 75: 267-282.

Schwabe J. 1906. Beitrage zur Morphologie und Histologie der tympanalen Sinnesapparate der Orthopteren. Zoologica 50: 1-154.

Silver S., K. Kalmring & R. Kühne. 1980. The responses of central acoustic and vibratory interneurons in bushcrickets and locusts to ultrasonic stimulation. Physiol. Entomol. 5: 427-435.

Wiley R.H. & D.G. Richards. 1978. Physical constraints on acoustic communication in the atmosphere: implications for the evolution of animal vocalizations. Behav. Ecol. Sociobiol. 3: 69-94.

12. THE EAR OF THE BUSHCRICKET

Winston J. Bailey

Introduction

The calls of crickets and bushcrickets dominate the night noises of many terrestrial habitats around the world with either their musical tones, or rasping rattles and creaks. If our ears were open to higher frequencies, in some habitats these night noises would be almost deafening, mixing with the predatory cries of bats and the sharp cracks from disturbed foliage, or breaking twigs. The temporal structure of the calls of most tettigoniids is complex, and rarely do the calls follow the melodic pattern typical of most crickets. Also, unlike their grylline relatives, the songs are transmitted at high sonic (10-20 kHz), or even ultrasonic frequencies (Sales & Pye 1976).

We may consider the ear as having evolved for two reasons. One is for social communication between and within the sexes, and the other for a less specific awareness of the insect's environment. Included in this "acoustic environment" will be the noises made by predators (Belwood chapter 2) as well as other acoustic animals. Conflict of function may occur where the frequencies of calling conspecifics, and searching predators occupy similar frequency bands. In crickets, these two functions appear to have been separated, with lower frequencies used for social communication and the ultrasonic band for the detection and avoidance of bats (Nolan & Hoy 1986). However, most tettigoniids communicate with broad-band signals, where conflict between frequency channels will be inevitable. The constant interplay between predator detection, and mate location forms the underlying theme of this chapter. And although the review covers the controversy surrounding the mechanics of sound reception (Lewis 1974; Nocke 1975; Bailey & Stephen 1978; Hill & Oldfield 1981; Stephen & Bailey 1982; Bailey & Stephen 1984), it tries to maintain the relevance of structure and evolved function. I use the approach of comparative morphology. My thesis is that if selection has acted on hearing efficiency across a broad taxon in response to different behavioural needs, then because species have evolved under different social and environmental pressures, comparative morphology could provide clues as to the function of the ear itself. As evolution in any one taxon progresses at a different rate from a related taxon, then the observer should view a spectrum of events across taxa (Darwin 1874).

I have examined a wide range of morphological types, and from as many subfamilies as were available in both the field and museum. But, despite this extensive search, new material is constantly becoming available, often with associated biological data and particularly from the New World tropics (see Morris 1980; Morris & Beier 1982; Belwood chapter 2).

The survey of tettigoniid material has only been made possible by the close co-operation of the two principal museums where this study has taken place, the British Museum (Natural History), London and the Australian National Insect Collection, Canberra. Innumerable workers have offered material when requested, and I am grateful to these people as I am to Dr David Ragge and Dr David Rentz, curators of these collections. In fact it is appropriate that this work is finally being published in a commemorative volume. It was only through David Ragge's patience and encouragement, and the constant help of Ms Judith Marshall and Linda Pitkin over a number of visits to the British Museum, that this work was completed.

The techniques use for this study were simple. External structures of fresh material were drawn by *camera lucida*, and internal structures were examined by dissection. Clearing dried museum material in 10% KOH, allowed me to dissect the internal cuticular trachea. Where detail was required, structures were examined by SEM.

The Tettigoniid Ear

A description of the neuro-anatomy of the tettigoniid ear is provided by Lakes and Schikorski (Chapter 10), however, the following overview gives a framework of function for discussions on ear anatomy. In addition it stabilises certain morphological terms that have been used by various authors in describing the ear.

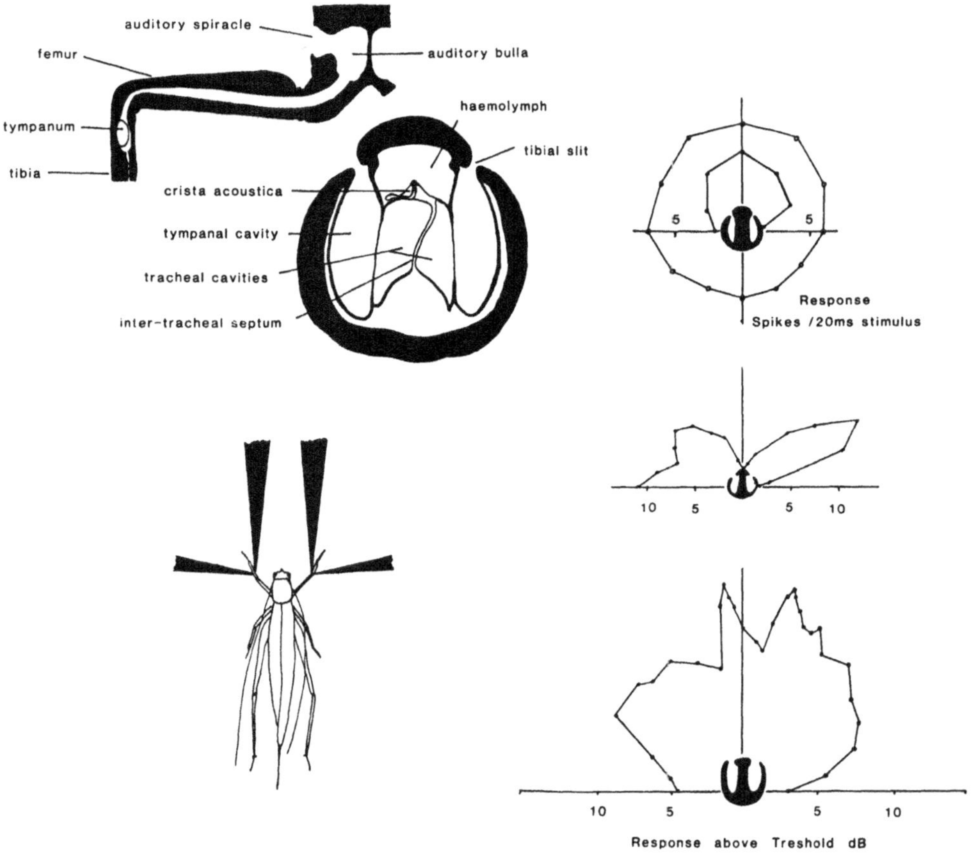

Fig. 1. Theory of slit-function in the tettigoniids – **A.** The ear and associated trachea (modified from Bennet-Clark 1985). **B.** Theoretically postulated areas of maximum sensitivity around the leg of *Tettigonia viridissima* (Redrawn from Autrum 1963). (B) Hearing threshold curve around the foreleg of *Decticus verrucivorus* (Redrawn from Rheinlaender 1975)) **C.** Measured suprathreshold curves at critical frequencies around the forelegs of three tettigoniid species, (i) *Mygalopsis marki* with the tracheal input controlled by the addition of a plastic pipe (Bailey & Stephen 1978); (ii) *Tympanophpora* sp. without tracheal control (Bailey & Stephen 1984); (iii) *Metaballus litus* without tracheal control (Bailey *et al.* 1988).

The "generalised" ear of the bushcricket consists of two tympanic membranes situated on the front and back of the proximal tibia of the forelegs. The tympana are backed by paired trachea, which meet along the midline to form a secondary central membrane (Fig. 1B). The *crista acustica*, or acoustic chordotonal organ, has its sense cells arranged in a linear series along the length of this membrane (Schwabe 1906; Schumacher 1979; Oldfield 1982). Distally the two trachea are reduced in cross section, and in some cases a small opening beneath the hearing organ leads through a much narrower trachea to the tibia and tarsus. Proximal to the hearing organ, the two trachea unite into a single trunk beneath the knee, and proceed along the femur to the thorax. In the thorax, the trachea opens into a cavity, which I will call the auditory bulla (the *vesicula acustica* of Graber (1876) and Zeuner (1936) and vesicle of Morris *et al.*1989). This in turn opens through the pleuron of the prothorax close to, but independent of the ventilatory spiracle. This opening is often a large hole posterior to the pronotal shield, and I call this the auditory spiracle (*femoral stigmata* of Zeuner (1936); *orifice acoustique*, Charpentier (1927); "Gehorstigma" hearing stigma, Ander 1939; acoustic stigma of Morris *et al.* 1989). The entire auditory trachea in the tettigoniids is dedicated to hearing.

In many genera the cuticle surrounding the external tympana is raised into folds that may form a cup around each tympanum. The opening they make from the tympanum to the outside may be shaped into a slit, or round hole.

The paired trachea beneath the knee form a central membrane, and it is on this structure that the sensillae are attached. In species where the cuticular folding outside the tympanic membranes is poorly developed, or completely lacking (naked tympana), the external tympana will react to pressure differences around the leg and beneath the membrane. Any movement of these tympana will produce a complex motion in the central membrane that will be transduced through shearing forces to the cells of the *crista acustica*. The action of this system is made more complex by the curtain of connective tissue suspending the *crista acustica* from the knee. (See Lakes & Schikorski chapter 10.)

Where both ends of the system are open through the auditory spiracle and the tympana, the ear will respond to a pressure gradient either side of the tympanic membranes. Factors affecting the magnitude of this gradient are far from simple, and relative pressure changes will be influenced by (i) the wavelength of the call, (ii) the distance of the knee from the auditory spiracle, (iii) its orientation relative to the sound source, and (iv) the length of the trachea. Also there will be more complex components related to, (v) the structure of the trachea's inner surface, and (vi) to resonances in the auditory bulla and connecting leg trachea. Even further complexity is produced by the acoustic properties of (vii) the often unequal cavities beneath the tympana, and finally (viii), where there are external subslit cavities, the resonances developed within these cavities.

Also any simple explanation of the way the ear works is not helped by the added complexity of the call itself! And whereas theoretical models of the cricket ear (Fletcher & Thwaites 1979; Hill & Boyan 1977) have been possible, largely because calculations can be developed around a fixed "species" wave length, such a model is not possible with the bushcricket ear. Even those models constructed for crickets, may be considered at best simplistic, and these have been surrounded by speculative discussion (Johnstone et *al.* 1970; Larsen & Michelsen 1978; Kliendienst 1980, 1981; Schmitz *et al.* 1983; Huber 1987; Larsen 1987; Larsen *et al.* 1989).

The system I have described monitors airborne sound. But there are, in association with the *crista acustica*, a series of acoustic-vibratory receptors that detect low frequency sounds and substrate vibration. The detection of substrate-borne vibration does not form part of the discussion in this chapter (see Kalmring *et al.* chapter 11).

Two Acoustic Inputs to the Ear – History and Controversy

Anatomists have described the external structures of the ensiferan hearing system in considerable detail (Graber 1876; Schwabe 1906; Charpentier 1924, 1927; Zeuner 1936; Ander 1939; Beier 1955; Ragge 1980). Ander (1939), provides the most complete work on this subject listing 22 dissected tettigoniid species. He discounts the earlier works by Berlese (1882) and Nietsch (1885,1905) as containing too many errors, and he even claims that much of Zeuner's work suffered in the same way. (It is strange how history seems to have repeated itself nearly 50 years later, with similar controversy over the tettigoniid hearing system, often with parallel vitriol!)

Some of these descriptions have been associated with the systematic organisation of the various subfamilies, and although instructive in their detail in that they reflect morphological variation, they provide little comment as to hearing function. However, early workers such as Graber (1876), did speculate on the function of different parts of the hearing system. For example he suggested that the auditory trachea could act as a sound guide to the internal aspect of the tympanal membranes, and Zeuner (1936) took this argument further by suggesting that at some future point in evolutionary time the external membranes could be lost altogether.

Tubes and resonators

It was Lewis (1974) who first demonstrated how the large tracheal tracts of the tettigoniid ear acted as sound paths to the tympanic membranes. He provided evidence to suggest that this was indeed the primary pathway for sound, at least for his experimental insect, *Ruspolia differens* (Serville) (= *Homorocoryphus vicinus* Walker) (Copiphorini). Concurrently, Nocke (1974,1975) published an account of experiments with the Australian phaneropterine *Acripeza reticulata* Guerin, where he discovered that the equivalent, though much expanded auditory bullae, acted as a resonator to the carrier frequency of the song. Although both acknowledged the importance of this input to the function of the ear, divergent theories of the precise function developed as new techniques were applied to the problem of acoustic function. Lewis, in suggesting that the trachea were the primary inputs to the tympanic membranes, remained cautious as to the influence the horn-shaped trachea could have on the ear's overall frequency characteristics (Lewis 1974). Nocke, on the other hand, invoked a more profound function to the auditory trachea in *A. reticulata*; the system acted as a pair of lightly damped tube-resonators. The characteristics of these resonators certainly appeared to influence the frequency tuning of the entire hearing system (Nocke 1975). Seymour *et al.* (1978), using *R. differens* and the tettigoniine *Tettigonia viridissima* L. were able to confirm Lewis' original views, but suggested that the system was rather broadly tuned and with a low acoustic efficiency.

Hill and Oldfield (1981) re-examined Nocke's ideas with experiments on three quite different tettigoniid species from two subfamilies. Again, by using alternative techniques to those adopted by Lewis, Nocke and Seymour *et al.*, they confirmed the importance of the trachea as a sound guide to the tympanic membranes. But they suggested the structure acted as a more efficient amplifier than had been proposed by Seymour *et al.* (1978). The lack of concordance with these workers was put down to differences in experimental technique: Seymour *et al.* (1978) used laser vibrometry, whereas Hill and Oldfield (1981) used direct measurements of sound pressure by microphone probes at the termination of the trachea. However, their conclusion was that the trachea acted as an amplifier to the higher frequency components within the song's sound spectrum.

The implications for such a high frequency emphasis are significant in the light of the behavioural experiments carried out on female phonotaxis by Latimer and Sippel (1987) and Bailey and Yeoh (1988). These workers found that the presence of high frequencies in the call influenced the female's preference for a male's call, and also the accuracy with which it located the signal (c.f. Latimer & Lewis 1986 in crickets).

Slits and subslit cavities

Emphasis on the auditory trachea as the primary sound route to the tympana countered the existing ideas of ensiferan hearing. These early ideas were based on a scenario developed by Autrum following his historic series of neurophysiological experiments with *Tettigonia viridissima* (1940,1963) (Fig. 1B). He suggested that sound interacted with the outer surface of the tympanic membranes, entering through the slits and subslit cavities. Thus, when the sound's direction was normal (perpendicular) to the slits, neural activity was highest: the membranes were most active. Thus, from general theory, ideas that bushcrickets detected the direction of the sound through these external slits held acceptance until the experiments of Lewis (1974).

However, in a rather different series of experiments (Bailey & Stephen 1978; Stephen & Bailey 1982; Bailey *et al.* 1988), it was possible to demonstrate that the external slits could function as sound guides. This was contrary to the finding of Rheinlaender (1975) (Fig 1C). However, by controlling both the phase of the sound reaching the internal and external parts of the membrane, and by using critical frequencies, polar plots around the leg were produced that had some similarity with Autrum's suggested model (Fig 1D).

Like most animals, bushcrickets have two "ears", one pair of tympana on each foreleg. Each ear is served by a single tympanic nerve running to the auditory neuropile of the prothorax, and emphasis on the directional properties of one input alone is only part of the binaural integration carried out in the central nervous system (see Rheinlaender & Römer chapter 13). It is undoubtedly this binaural input that provides auditory acuity to the orienting female (Rheinlaender & Römer 1980; Bailey & Stephen 1984: see also Wendler *et al.* 1980; Weber *et al.* 1981; Huber 1987 on crickets). However, in species where the cuticular folds around the membrane may act as more obvious sound guides to the tympana. The use of such a directional input may increase the differential between the ears thereby emphasising binaural sensitivity.

The Internal Sound Guide – The Auditory Trachea

Ventilation and cross tracheal connections

It is a characteristic of insect tympanal organs that they are backed by a trachea. In terrestrial insects at least, this allows the membrane to move with changes in air pressure infront of and behind the membrane (Michelsen & Nocke 1974; Seymour et al. 1978; Kleindienst *et al.* 1983; Bennet-Clark 1984). In most families of ensiferan Orthoptera, such as the mole crickets, crickets, rhaphidophorids, gryllacridids, stenopelmatids and haglids (Zeuner 1936; Hill & Boyan 1977; Field *et al.* 1980), the trachea open through spiracles at the base of the forelegs, and their opening is controlled in the usual insectan manner by valves. There appears to be no evidence that the opening and closing of these valves is linked to the acoustic behaviour of the searching female (Schmitz *et al.* 1983). Further, in crickets, the auditory trachea appear linked across the insect's mid-line. This provides for the possibility of an acoustic interaction between the ears (Hill & Boyan 1977; Kleindienst *et al.* 1983).

The connecting tube in crickets, has an internal septum (Zhantiev *et al.* 1975; Hill & Boyan

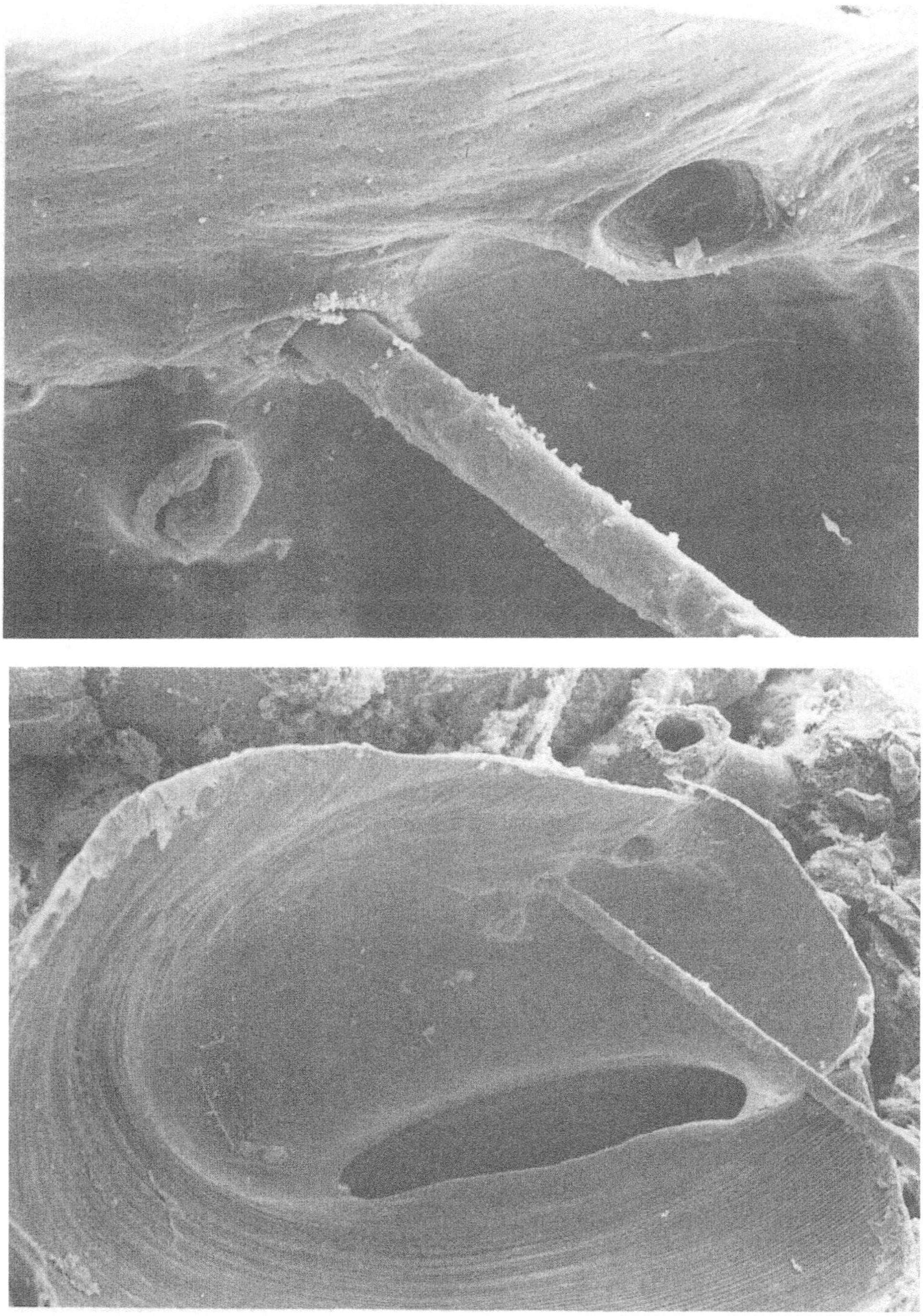

Fig. 2. The internal anatomy of the auditory bullae of *Pachysaga australis* showing the connecting trachea through which a bristle has been inserted demonstrating the open isthmus between the two bullae.

1977), and contrasts with that of the wetas (Stenopelmatididae) where the connecting trachea are open (Field *et al.* 1980). Ander (1939) observed that in certain cricket species, where the wings have lost their acoustic function, the diameter of the trachea is reduced. He therefore assumed that the trachea were important in sound detection. A simple test of the importance of this cross trachea was to sever the cross connection in chronic preparations. If of value to the orienting insect, acuity or left-right discrimination should be lessened – this was not so (Larsen *et al.* in 1989).

In the Stenopelmatidae, the first thoracic spiracle opens to three tracheal trunks, one leading to the head (cephalic trunk), a second travelling along the side of the thorax (lateral trunk), and the third, the median branch opens to a system divided into two parts incorporating the lateral-dorsal trunk and the anterior leg trachea. This anterior leg trachea connects to a transverse trachea to the opposite side of the body, and is equivalent to the auditory trachea of the other ensifera. The branch from the anterior trachea appears separate from this transverse commissure, linking through a complex pathway, to the lateral trunk trachea. The system has an obvious primary function of ventilation, and is described by Zeuner (1936) for *Cratomelus armatus* Blanchard and more fully by Ball & Field (1981) for the weta, *Hemideina crassidens* (Blanchard).

The haglids (Haglidae), are regarded as intermediate between the crickets (Gryllidae and Gryllotalpidae) and the bushcrickets. And in this group the auditory system appears to be linked to the ventilatory system with no separate auditory opening. Unlike the stenopelmatids and gryllids, the system appears to be more dedicated to hearing, in that there are fewer trunks leading to the rest of the thorax (Ander 1939). The trachea are continuous across the body with evidence that in some species the constriction at the mid-line is not as marked as in the gryllids. The trachea branch to minor tracheal trunks associated with the main respiratory system and the leg trachea opens behind naked asymmetrical tympana. The anterior tympanum is thickened, in a similar manner to that found in certain crickets (Young & Ball 1974).

Two variations of this basic structure have evolved in the Tettigoniidae. First, the auditory trachea are larger and in some cases appear to act as huge internal resonators (see Nocke 1974; 1975), or even hearing trumpets (Ander 1939; Lewis 1964; Seymour *et al.* 1978) (Fig 7; Table III). Second, the internal auditory sound guide has been separated from the prothoracic ventilatory spiracle and associated trachea (Siebold 1844).

The vestiges of the ventilatory system in tettigoniid families are present in the form of a cross connection between the two bullae. In most families this remains as a thread of cuticle, however, occasionally this connection is open, and quite complex in design. For example, in the sagine *Pachysaga australis* (Walker), the connecting elements open via a series of narrow channels from one bulla to the other, with the internal opening raised in a complex array of tubercules (Fig. 2). Sectioning the thorax, reveals that this thread is indeed open and forms a narrow isthmus. However, there is no evidence of an acoustic function.

The system does appear to be open in larval copiphorines (*Ruspolia differens*; WJB unpublished data), and is closed to a cuticular thread in later instars. Ander (1939) provides an extensive description of the larval development of the trachea and auditory spiracle in a number of genera.

In the somewhat anomalous stick-like and phytophagous tettigoniid, *Phasmodes ranatriformis* Westwood, where both sexes appear to be mute and the vestigial tympana are present only as indentations on the foretibia, the "auditory" tracheal elements are associated with the ventilatory system, in that they open through valved spiracles. The leg trachea divides into two, in the normal manner distal to the knee, but these two trachea then continue as a double tracheal element through the remaining tibia to the tarsus. In the region of the *crista acustica*, the paired trachea do not appear to meet at the midline beneath the sensilla. In representatives of similar stick-like tettigoniids belonging to the subfamily Zaprochilinae (Gen. nov. 22), the tracheal openings exhibit sexual dimorphism. In the female, the trachea opens through a quite usual auditory spiracle, whereas in the male the trachea opens through a three valved spiracle (Fig. 3). The auditory sensitivity of the male appears 5-10 dB lower that that of the female

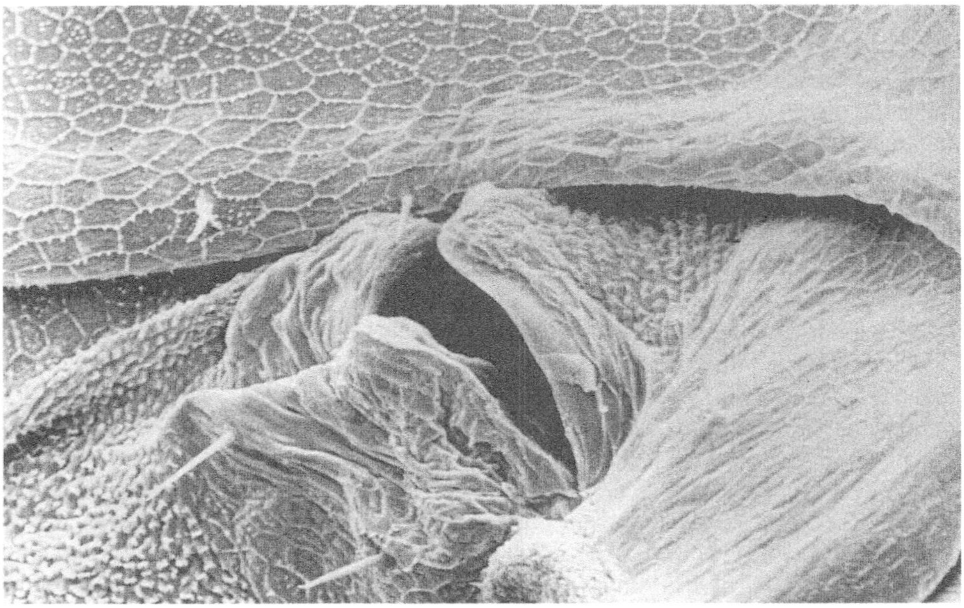

Fig. 3. Sexual dimorphism of the auditory spiracle in an undescribed genus of Zaprochilinae. **A** Male; **B** Female

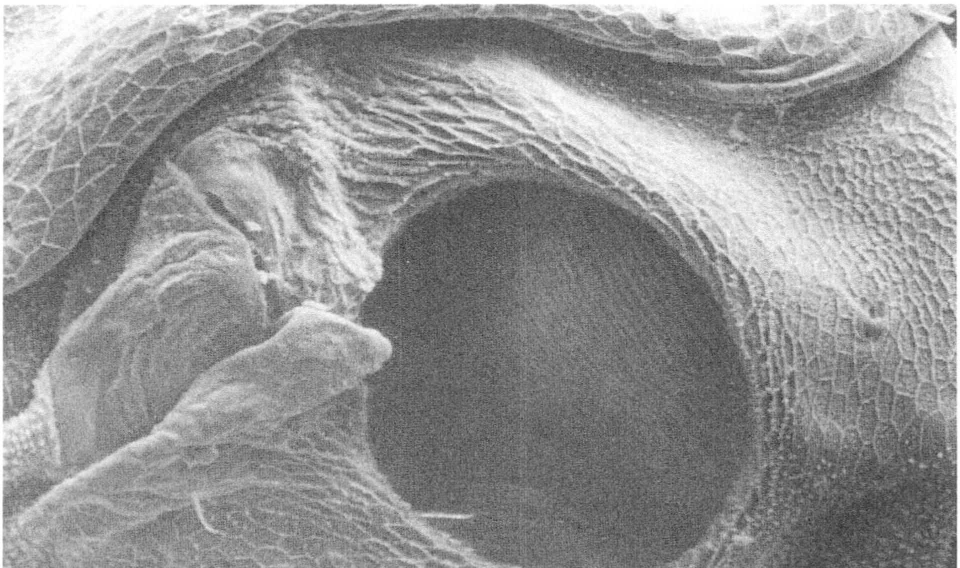

(Römer & Bailey unpublished data). Observations of male behaviour in the field are strongly suggestive that males do not respond to the sound made by conspecifics (Bailey & Simmons manuscript).

Tracheal closure – pronotal shields and hydraulics

Having established that the acoustic trachea play a major role as a sound guide to the internal surface of the tympanic membranes, an essential aspect of its function must be that it is open to changes in sound pressure. In a number of subfamilies, such as the Tettigoniinae, Listroscelidinae and Saginae the spiracle's opening is large and unimpeded by the posterior/ lateral extension of the pronotal shield (Fig 4), Table I. However, in other subfamilies such as the Pseudophyllinae, Phyllophorinae, many Phaneropterinae and some Conocephalinae, the lateral lobes extend over the auditory spiracle providing varying degrees of occlusion. Again, in some genera the lateral lobe fits snugly against the hind margin of the pleural sclerites, forming a near perfect seal (e.g. some Phaneropterinae). In a number of species within the genus *Orchelimum* (Conocephalinae: Conocephalini), the lateral lobes are even

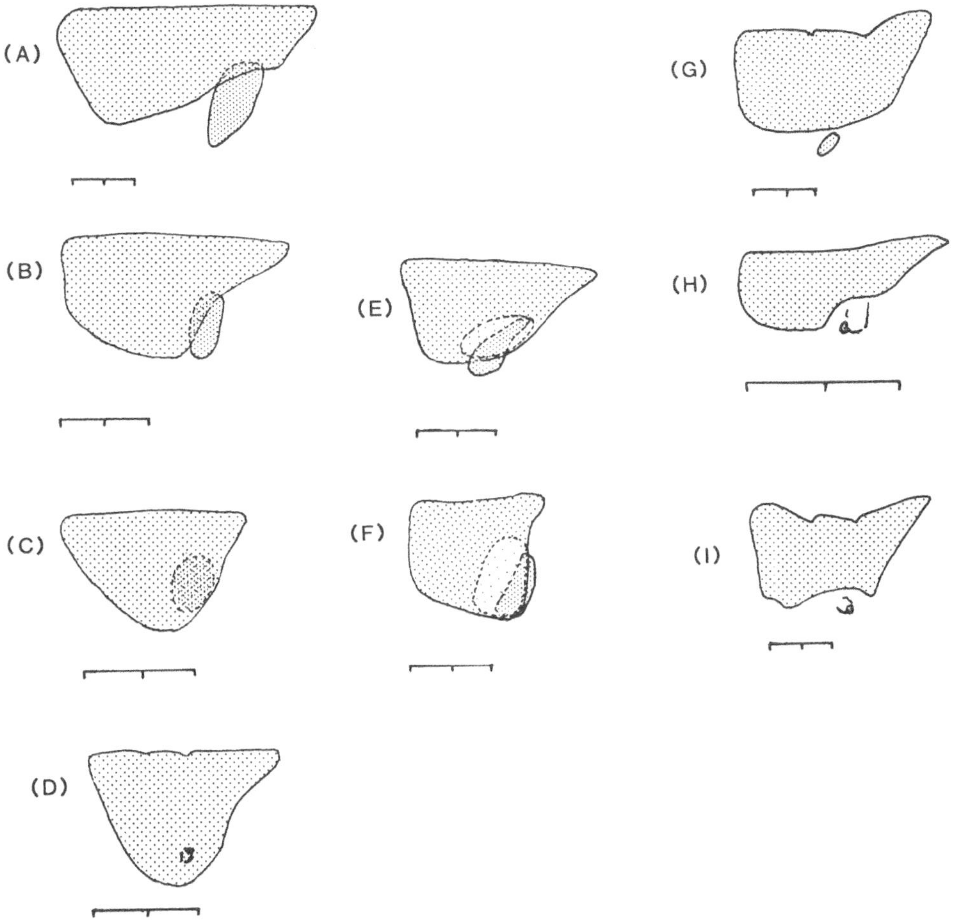

Fig. 4. The occlusion of the auditory spiracle by the prothoracic side keels in the Tettigoniidae. **A.** *Pachysaga australis*; **B.** *Tettigonia viridissima*; **C.** *Elephantodeta* sp.; **D.** *Mustius afzelli*; **E.** *Orchelimum carinatum*; **F.** *Azamia biplagiatta*; **G.** *Ephipiggerida zapatari*; **H.** *Phlugis mantispa*; **I.** *Nesoecia cooksoni*.

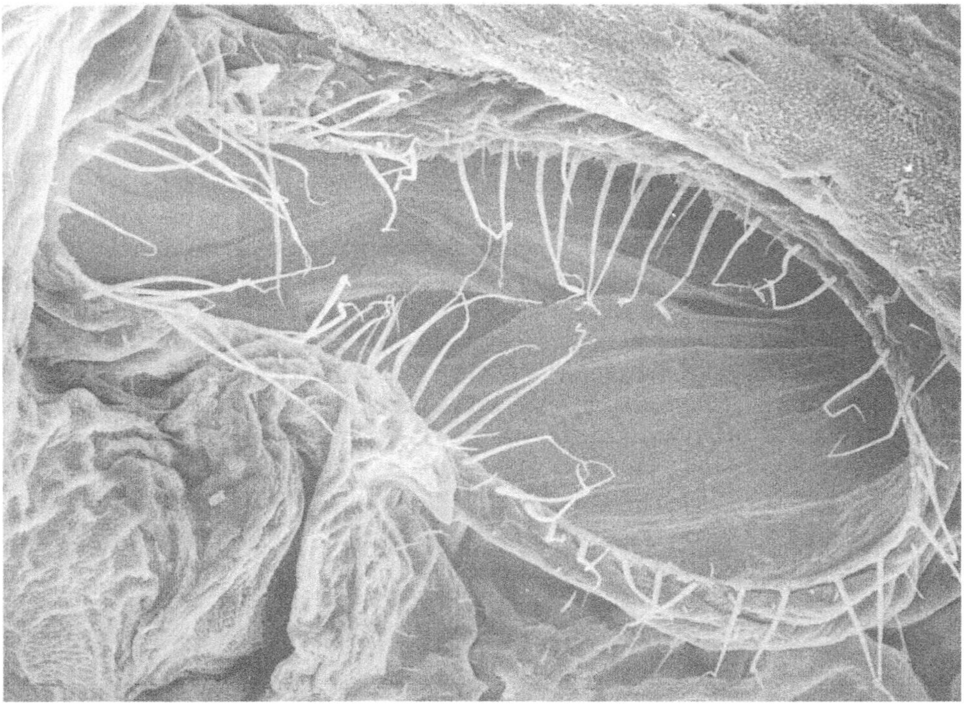

Habetia sp. (Conocephalinae Agraeciini)

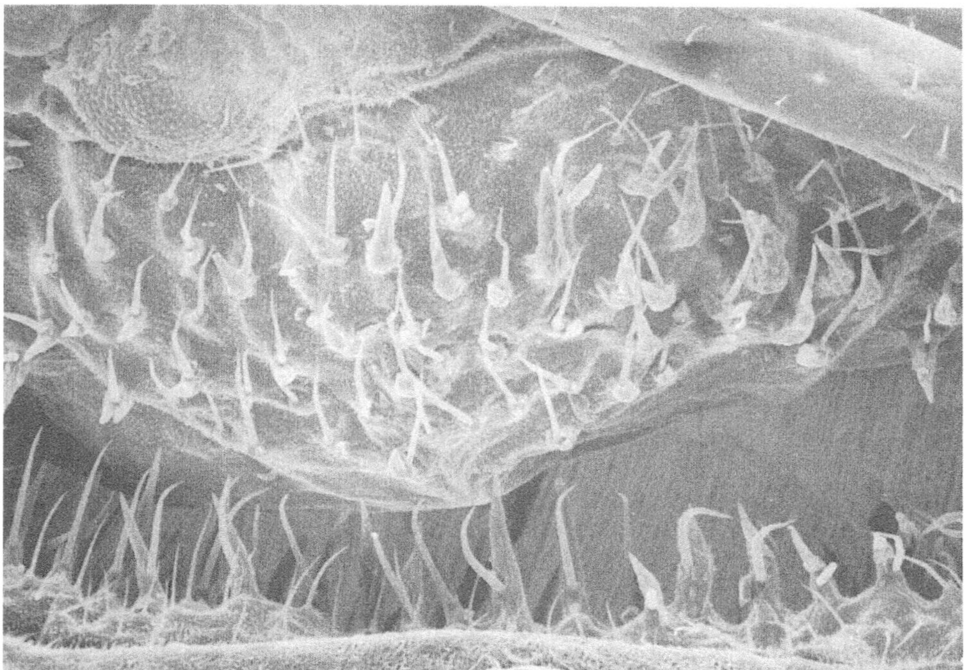

Anabrus simplex (Tettigoniinae)

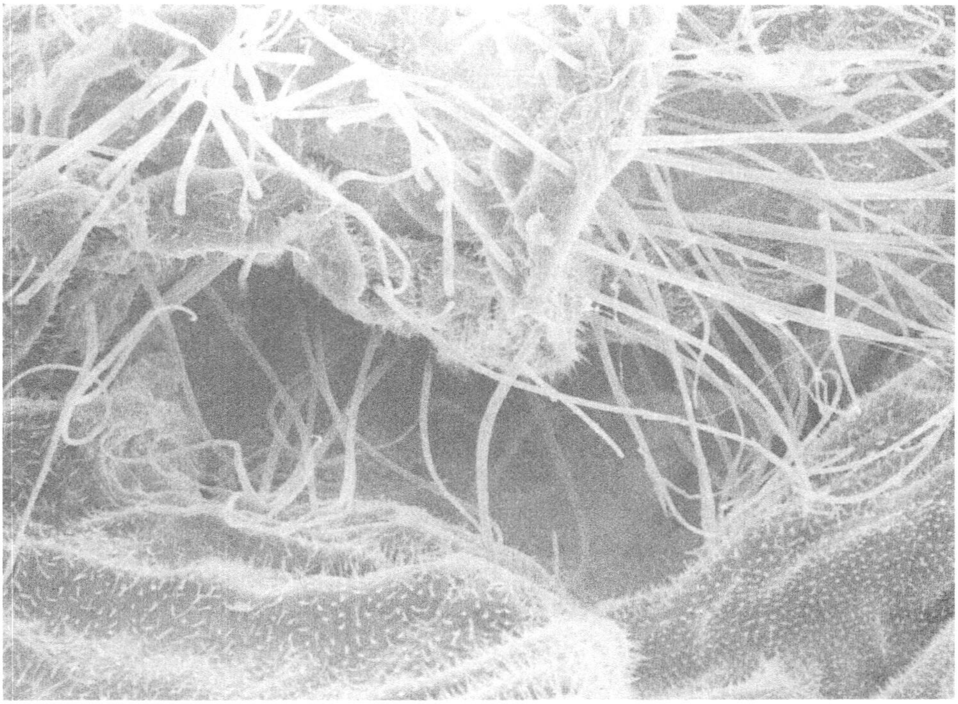

Oxyaspis sp. (Phyllophorinae)

Fig. 5 (opposite page, and above). SEM's of the auditory spiracle of three species of tettigoniids with cuticular spines protecting the opening of the auditory bullae **A**. *Habetia* sp. **B**. *Anabrus simplex* ; **C**. *Oxyaspis* sp.

formed into a thin "secondary tympanum". This appears to be a feature of a range of genera, and is to be found sporadically within other subfamilies (Table I). Without doubt, the form of this "secondary tympanum" over the auditory spiracle implies an acoustic function. In some species the covering is nearly transparent.

Where the lateral lobe fits closely around the auditory spiracle, the margin is often fringed with hairs or spines (Zeuner 1936; Ander 1939; Tinkham 1973) (Fig. 4). Similar spines occlude the opening of the spiracle (Fig. 5). This may offer some form of protection to the internal cavity. Bailey and McCrae (1978) describe an infestation of the tracheal mite belonging to the genus *Berlese* in the auditory trachea of the copihorine *Ruspolia flavovirens* (Karny), where the mite caused extensive damage to the internal wall of the trachea and tympanic membranes. The entire tube was obviously blocked with both mites and frass. The significance of such an infestation is paralleled by mite infestations of noctuid moths (Treat 1975). However, in examining the large range of material for this study I have only encountered one specimen with an infestation of what may be a tracheal mite, and this infestation was certainly not comparable to that found in *R. flavovirens*. Ander (1939) describes a number of specimens with mites, but he also mentions specimens in which the trachea contained sand grains!

A third, more dramatic closing mechanism has been described for the sagine *Pachysaga australis* (Fig. 6) (Stephen & Bailey 1982; Bennet-Clark 1984). Females were examined under a low power binocular microscope during phonotactic trials (n = 6), and in this active state, females were noted to shut down the wall of the auditory spiracle in the manner of a hydostatically controlled door. This could only be induced in those insects that were actively orienting to the male's call, and never with males (n = 10). Further examination of the morphology of the very flexible region of cuticle surrounding the auditory trachea in other live material certainly suggests that such a closing mechanism is at least possible by changes in the pressure of the haemolymph.

Hence, spiracular occlusion, either by the prothoracic lateral lobe, or complex spines, or indeed muscular contraction of the walls of the spiracle, may have a primary role of protecting the inner parts of the auditory system (defense). It is unclear as to whether selection has shifted this primary function of defence to secondary auditory functions that may affect the insect's sensitivity to either conspecifics or the noise of predators.

The auditory bullae – sexual dimorphism

Sexual dimorphism in the shape of auditory bullae is found in groups where both male and female acoustic signalling is known. The best examples described so far are in the Phaneropterinae with genera such as (*Caedicia*, *Polichne* (Hill & Oldfield 1981) and *Elephantodeta* (Fig 7). Thus, if females call with fundamental frequencies different from those of the male, and the auditory bullae do have some tuning characteristic linked to this frequency (Nocke 1975), this may be sufficient reason for such a dimorphism to evolve. Where known, the stridulating mechanism is different in females from that of the male and in most phaneropterine examples such as *Leptophyes punctatissima* and many of the Barbi-

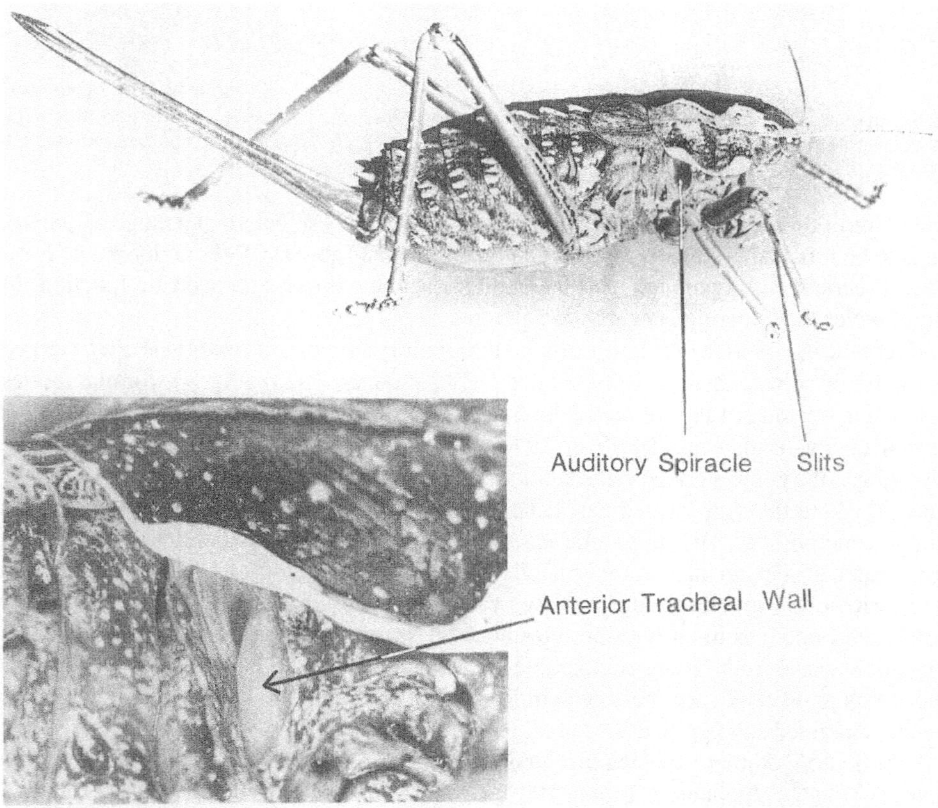

Fig. 6. The pneumatic slit closing system in female *Pachysaga australis*. **A.** The female showing the normal open spiracle. **B.** The auditory spiracle closed by the flexible anterior margin of the spiracle.

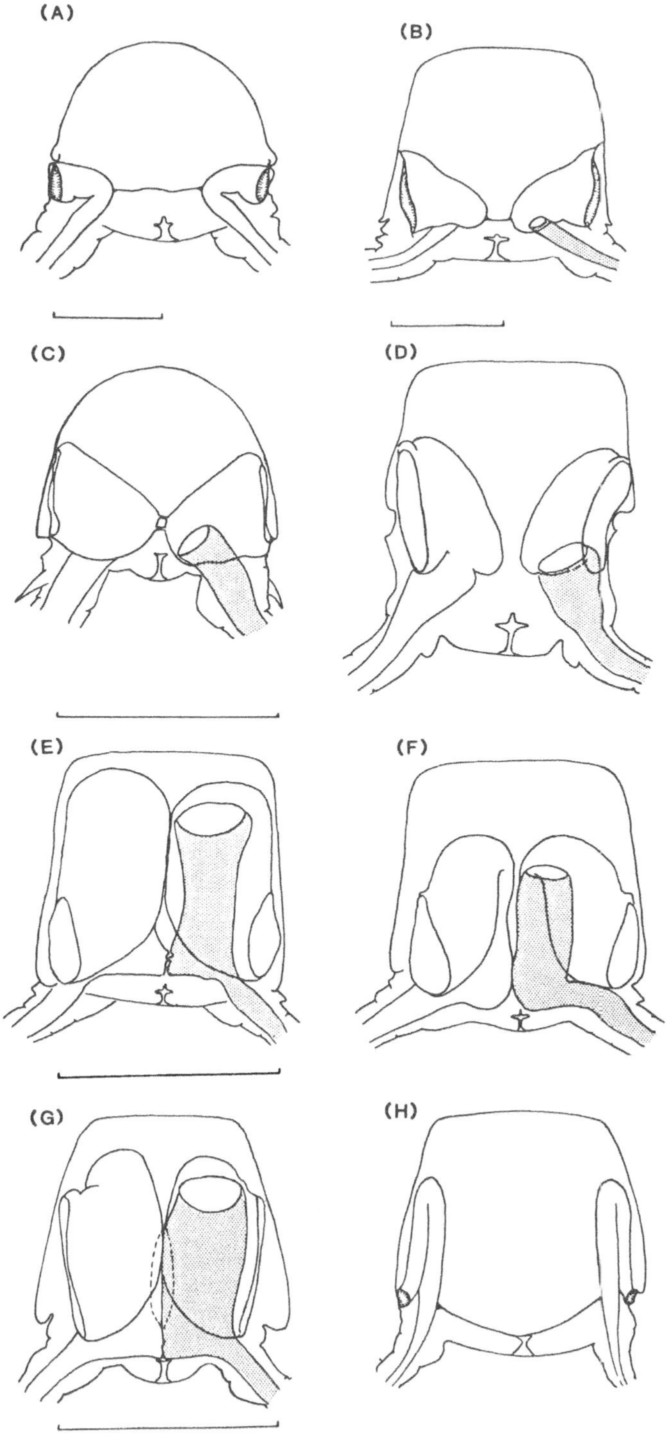

Fig. 7. The auditory bullae of tettigoniids diplaying the range of cavity structure and associated leg trachea. Sexual dimorphism is shown by *Elephantodeta* sp. in figures **E**. (Male) and **F**. (Female). **A**. *Bliastes* sp.(Pseudophyllinae); **B**. *Anabrus simplex* (Tettigoniinae); **C**. *Hexacentrus munda* (Listroscelidinae); **D**. *Metaballus litus* (Tettigoniinae); **E**. *Elephantodeta* sp. (Male); **F**. *Elephantodeta* (Female)(Phaneropterinae); **G**. *Oxycous lesnei* (Phaneropterinae); **H**. *Oxyaspis* sp. (Pseudophyllinae).

tistini, the female's call is much quieter than the males (Zimmerman *et al* 1989; Heller chapter 8), and such a difference may well lead to structural dimorphism.

Cavities and organ pipes

Table III arranges the morphology of the auditory bullae in 39 of the 132 species examined in this study into seven morphological types. They are assessed on the basis of increasing morphological complexity, from a simple direct tube connecting the auditory trachea to the leg trunk, to examples where the bullae are elaborately expanded, or formed into tubes. Such an arrangement implies a functional series and any implied function requires empirical data. Unfortunately the range of genera that have been subject to experiment are few, and indeed those insects that have been selected for study do not illustrate anything approaching the range of structure available in the family.

Simple horn-like tubes open from an expanded auditory spiracle to the leg trunk (e.g. *Bradyporus dilatus* (F) (Bradyporinae), *Bliastes* sp. (Pseudophyllinae), *Solomona* sp. (Conocephalinae; Agraeciini) (Zeuner 1936) and *Pseudorhynchus lessoni* (Conocephalinae; Copiphorini) (Table III; Fig 7), and such a system fulfills the requisites of a functional horn, amplifying sounds to the ear (Lewis 1974; Hill and Oldfield 1981). In many genera, however, the openings to the legs from the large bullae (perhaps over simplistically drawn by Zeuner (1936) as even and rounded connections) open to the leg trunk through holes with a distinct edge (Hill & Oldfield 1981). In other words, they are not always neatly tapered horns. In the leaf-like phytophagous taxa, the auditory bullae are variously expanded as spherical resonators, or elongated organ pipes, leading to the trachea of the leg. Such a tube-like structure may follow an extended path at the outer lateral limits of the pronotum (e.g. *Oxyaspis* (Fig. 7). Again, in genera where the bullae are expanded into complex cavities, it is difficult to ignore completely the notion developed by Nocke (1975) that the bullae have a distinct resonant function. Most genera that have well developed auditory trachea also have the opening of the ascending bullae towards the dorsal extremity. This increases the length of the tube within the confines of the thorax.

"Cross-talk"

Nocke (1975) discusses the question of "cross-talk" between the two halves of the acoustic system through the auditory bullae. He concludes that in *Acripeza reticulata*, there was no evidence of cross-talk at the frequencies he tested. Such negative evidence offers no explanation for systems where cross-talk is obviously possible, at least by implication from the proximity of the two bullae. Additional expansions of the bullae, such as described for *Zeuneria* sp. (Zeuner 1936) add credence to the idea that some systems may use crosstalk to enhance an as yet unknown aspect of their acoustic sensitivity. In this species the bullae are expanded into additional chambers that meet at the mid-line.

The External Sound Guides – Slits Cavities and Pinnae

For the most part, members of the Gryllidae, Gryllotalpidae and Haglidae have exposed tympana at one end of the auditory tract, and valve-controlled spiracular openings at the other. And, except in some eneopterine crickets, the structure of the cuticle surrounding the membrane shows little elaboration. However, within the Tettigoniidae, the variation of cuticular slits, and the range of auditory spiraclular openings and associated trachea has evolved to an extreme. And variation exists within genera, for even in the more primitive subfamilies such as the Hetrodinae, there is a range from slit bearing tympana, such as in *Eugaster gryoni* (Bolivar), to those with naked tympana. Also within the Listroscelidinae, the tympana are naked in such genera as *Phlugis mantispa* (Bolivar), whereas in the related

Table I. The Range of Structures Occluding the Auditory Spiracle of the Principal Subfamilies of Tettigoniididae

	Species	Notes
I. Prothoracic Side Keel		
COMPLETE		
Phyllophorinae	*Hyperhomala lanceolata* Brunner	
Phaneropterinae	*Azamia biplagiata* Bolivar	Side keel thinned and raised to a bulla.
Phaneropterinae	*Poecilimon haydari* Ramme	Secondary tympanum over the spiracle.
PARTIAL WITH SECONDARY TYMPANUM		
Conocephalinae	*Orchelimum carinatum* Walker	
	Orchelimum vulgare Harris	
	Orchelimum bradleyi Rehn & Hebard	
	Orchelimum agile (De Geer)	
Agraecini	*Labugama pearsoni*	
	Oxystethus lobatus Redtenbacher	
Phyllophorinae	*Cratioma occulatum* Karny	
Pseudophyllinae	*Mustius afzelii* (Stål)	
Phyllophorinae	*Phyllophora lanceolata* Brunner	Auditory opening small – equivalent size to the respiratory spiracle.
Phaneropterinae	*Arantia rectifolia* (Brunner)	
	Horatosphaga montivaga Sjöstedt	Raised to form a bulla but less distinct than in *H. meruensis*.
	H. meruensis Sjöstedt	
	Conchotopoda belcki Karsch	
	Peronura clavigera Karsch	
	Discomena ornata Brunner	Ventral margin raised to form a bulla.
Listroscelidinae	*Phlugis mantispa* Boliva	Pleurum of the prothorax extends as a fold over the auditory spiracle.
II. Partial Occlusion with Ventral Margin of the Side Keels Thinned to form a Tympanum		
Listroscelidinae	*Hexacentrus dorsatus* Redtenbacher	
Conocephalinae	*Karniella bullata* Rehn	
Phaneropterinae	*Azamia biplagiata* Bolivar	
Phaneropterinae	*Eurycorypha ornatipes* Karsch	
Ephippigerinae	*Ephippigerida cruciger* Fiebig	Occluded by pleural fold.
III. Closed or Tubiculous Spiracle of Comparable Dimension to the Respiratory Spiracle		
Agraeciini	*Habetia* sp.	
Pseudophyllinae	*Nesoecia cooksonia* (Butler)	
	Panoploscelis armata Scudder	
	Cymatomerella muta Beier	
	Tympanophyllum atroterminatum Brunner	
	Lacipoda immunda Brunner	
	Mustius afzelii (Ståhl)	Totally occluded
	Brochopeplus exultatus Walker	
	Cleandrus titan (White)	
	Oxyaspis congensis (Bolivar)	
	Zabalius aridus (Walker)	
IV. Spiracle Small – Of Comparable size to External Openings – Slits		
Pseudophyllinae	*Nastonotus reductus* (Brunner)	
Meconematinae	*Meconema thalassinum* (De Geer)	

Table II. Ear Structure in the Tettigoniidae

	Species	Notes
TYPE 1. Both Tympana open with no Cuticular Folding		
Phaneropterinae	*Caedicia* sp	
Hetrodinae	*Hemihetrodes* sp	
Phaneropterinae	*Acripeza reticulata* Guerin	
Phaneropterinae	*Elephantodeta* sp	
Meconematinae	*Amytta occidentalis* Karsch	The membranes are very close together behind a very bulbous area containing the chordotonal organ.
Meconematinae	*Meconema thalassinum* DeGeer	
Meconematinae	*Amytta* sp	
Listroscelidinae	*Xiphidiopsis elaphocerca* Karny	
	Phlugis mantispa Bolivar	
Acridoxeninae	*Acridoxena hewaniana* Smith	
Ephippigerinae	*Ephippigerida zapatari* (Bolivar)	
Ephippigerinae	*Ephippiger ephippiger* Fiebig	
Ephippigerinae	*Ephippigerida cunii* (Bolivar)	
Pycnogastrinae	*Pycnogaster cucullata* (Charpentier)	
Conocephalinae	*Karniella bullata* Rehn	
Conocephalinae	*Thyridorhoptrum baileyi* Pitkin	
Phaneropterinae	*Terpnistra zebrata* Serville	Slight development of lip.
Phaneropterinae	*Phaneroptera nana* Fieber	Slight lip on the anterior membrane.
Phaneropterinae	*Eulioptera disparidens* Ragge	A defined lip.
Phaneropterinae	*Pardalota hassi* Griffini	
Phaneropterinae	*Poecilimon haydari* Ramme	
TYPE 2. Both Tympana open but with Raised Cuticular Folding over the Anterior Membrane		
Agraeciini	*Habetia* sp	
Phaneropterinae	*Tropidophrys amydra* Karsch	
	Gelatopoia bicolor (Brunner)	Although well exposed membranes have internal lip.
	Oxycous lesnei Chopard	Marked sexual dimorphism.
	Oxycous apertus Ragge	No observed sexual dimorphism.
	Oxycous undulatus Ragge	(Also examined *O. magnus*, as in *O.undulatus*).
	Gravenreuthia saturata Karsch	
	Eurycorypha ornatipes Karsch	Internal fold not as marked as *Oxycous*.
	Morgenia melica Karsch	
	Plangiopsis schoutedeni Griffini	
	Poecilopsyra octoseriata (Haan)	
	Preussia lobatipes Karsch	
	Zeuneria melanopeza Karsch	
	Arantia rectifolia Brunner	
	Dapanera genuteres Karsch	

Table II. Ear Structure in the Tettigoniidae (cont.)

	Species	Notes
TYPE 3. Cuticular Ridges Forming under the Tympana but not Large Enough to Form Cups		
Phyllophorinae	*Phyllophora boschnai* de Jong	
Agraeciini	*Oxystellus lobatus* Redtenbacher	Intermediate between types 3 and 5
Phyllophorinae	*Phyllophora lanceolata* Brunner	
	Siliquofera grandis (Blanchard)	Distinct raised folds but are not large enough to form cups.
	Sasima aequalis Karny	Pronounced cups with the internal cup larger than the posterior cup.
Hetrodinae	*Enyaliopsis transvaalensis* Peringuey	Open tympana with distinct raised ridges.
	Cosmoderus maculatus (Kirby)	Ridges beneath the membranes are not so pronounced as in *H transvaalensis*.
Phaneropterinae	*Horatosphaga montivaga* Sjöstedt	Distinct cups.
	Horatosphaga meruensis Sjöstedt	
	Conchotopoda belcki Karsch	
	Diogena fausta (Burmeister)	
TYPE 4. Cuticular Folds Well Developed to Form Cup-like Cavities		
Pseudophyllinae	*Tympanoptera* sp	
Listroscelidinae	*Hexacentrus munda* (Walker)	
Copiphorini	*Copiphora cornuta* (De Geer)	Could be considered in Type 5.
Pseudophyllinae	*Zabalius apicalis* (Bolivar)	
	Zabalius aridus (Walker)	
	Oxyaspis congensis Brunner	
	Cratioma occulatum Karny	
	Cratiatum dilatatum Karny	
	Pseudophyllus titan White	
	Onomarchus leuconotus (Serville)	
	Brochopeplus exaltatus (Walker)	
	Phyllopectis crepitans (Brunner)	
	Mastophyllum scabricolle (Serville)	Although the are bulbous the openings are not large.
	Mustius afzelii Stål	
	Lacipoda immunda Brunner	
	Tympanophyllum atroterminatum (Brunner)	
	Cymatomerella muta Beier	
Agraeciini	*Scytoceroides ceylonensis*	
	Labugama pearsoni	
Listroscelidinae	*Phisis pallida* (Walker)	
	Decolya roseopicta (Uvarov)	
	Neobarrettia hakippa Cohn	Although cavities are present like *Cerberodon viridis* this is close to a lip-like structure.
	Hexacentrus dorsatus Redtenbacher	
Phaneropterinae	*Tylopsis lilifolia* (Burmeister)	Close to slits but well defined lips.

Table II. Ear Structure of the Tettigoniidae (cont.)

	Species	Notes
TYPE 5. Cuticular Folds Forming Narrow Slits – Cavities not Bulbous		
Tettigoniinae	*Anabrus simplex* Haldeman	
Phaneropterinae	*Azamia biplagiata* Bolivar	Although slit-like in structure the cuticular folds bulge to form cavities.
Listroscelidinae	*Hexacentrus munda* Walker	Slits are open and conform to the structure of the resonators in Type 4.
Bradyporinae	*Bradyporus dilatatus* (Stål)	Very small slit holes particularly for such a large animal.
Copiphorini	*Bucrates malivolans* (Scudder)	
Pseudophyllinae	*Panoploscelis armata* Brunner	
	Nesoecia cooksoni Butler	
	Nastonotus reductus Brunner	Intermediate between cups and slits.
Listroscelidinae	*Cerberodon viridis* Perty	Pronounced cavity close to cups.
Tettigoniinae	*Platycleis sabulosa* Azam	
	Metrioptera azami	
	Tettigonia cantans (Fuessly)	
	Tettigonia viridissima (L.)	
Hetrodinae	*Eugaster guyoni* Serville	Slits are almost totally occluded by thick cuticular lips over both tympana.
Phaneropterinae	*Azamia biplagiata* Bolivar	Well developed cavities.
	Drepanophyllum marmoratum Karsch	
	Peronura clavigera Karsch	

Table III. Organisation of the Auditory Trachea in the Tettigoniidae

I. Straight Tube to the Leg – Flared Horn Stage I		
Bradyporinae	*Bradyporus dilatatus* (F.)	
Hetrodinae	*Anepsceptus horridus* (Burmeister)	Tube with wide opening.
Copiphorini	*Pseudorhynchus lessoni* (Serville)	
Pseudophyllinae	*Bliastes* sp.	
II. Expanded Bullae – Flared Horn Stage II		
Conocephalini	*Conocephalus upoluensis* (Karny)	
	Orchelimum agile (De Geer)	
Copiphorini	*Mygalopsis marki* Bailey	
	Ruspolia differens Serville	
Tettigoniinae	*Decticus albifrons* (Fabricius)	
	Metaballus litus Rentz	
	Xederra sp	
Pseudophyllinae	*Tympanoptera* sp.	
	Bliastes sp.	

Table III. Organisation of the Auditory Trachea in the Tettigoniidae (cont.)

	Species	Notes
III Expanded Bullae – Bullae Much Larger than Auditory Spiracle – Resonator Stage I		
Hetrodinae	*Hemihetrodes* sp.	Resonator with small opening to leg.
Tettigoniinae	*Platycleis brevipennis* = *Metrioptera roeselii* (Hagenbach)	
	Anabrus simplex (Haldeman)	
Microtettigoniinae	*Microtettigonia* sp	Bullae meet at the midline and open through a reflexed opening to the leg.
Conocephalini	*Thyridorhoptrum baileyi* Pitkin	
Agraeciini	*Nicsara* sp.	
	Habetia sp.	
Meconematinae	*Amytta occidentalis* (Karsch)	
Saginae	*Hemisaga denticulata* (White)	Bullae close to midline.
	Pachysaga australis Walker	
Listroscelidinae	*Requena verticalis* Walker	
	Phisis pallida Walker	Constriction between bullae and leg trachea.
Phyllophorinae	*Phyllophora boschnai*	
	Azamia biplagiata	Bullae close to midline – note size of cavities formed by the side keels over the spiracular opening.
IV Expanded Bullae – Filling much of the Prothorax – Resonator Stage II		
Tympanophorinae	*Tympanophora diminuta* Riek	
Tettigoniinae	*Metaballus*	
Listroscelidinae	*Hexacentrus munda* (Walker)	Bullae close together at the midline.
Phaneropterinae	*Horatosphaga leggeii* (Kirby)	Round bullae in the dorsal prothorax.
V Expanded Bullae – Running to the top of the Prothorax – Resonator Stage III		
Phaneropterinae	*Torbia* sp	
	Polichne sp	Sexual dimorphism.
	Elephantodeta sp	Massive expanded bullae filling nearly the whole prothorax.
	Oxycous lesnei Chopard	
	Torbia	
	Caedicia simplex (Walker)	
	Acripeza reticulata Guerin	
VI Bullae Tube Like – Not Greatly Expanded – Resonator Stage IV		
Phyllophorinae	*Oxyaspis* sp.	

Decolya roseopicta (Uvarov) and *Phisis pallida* Walker the cuticular folds form large resonator-type cavities around the tympana. Morris *et al.* (1989) describes one Conocephaline species where the tympana are naked, an unusual condition for most genera within this subfamily. Further, there appears to be no relationship between the size or structure of the auditory bullae and the style of the tympanal cuticular covering.

There is, however, one general observation that is worthy of further consideration. Just as

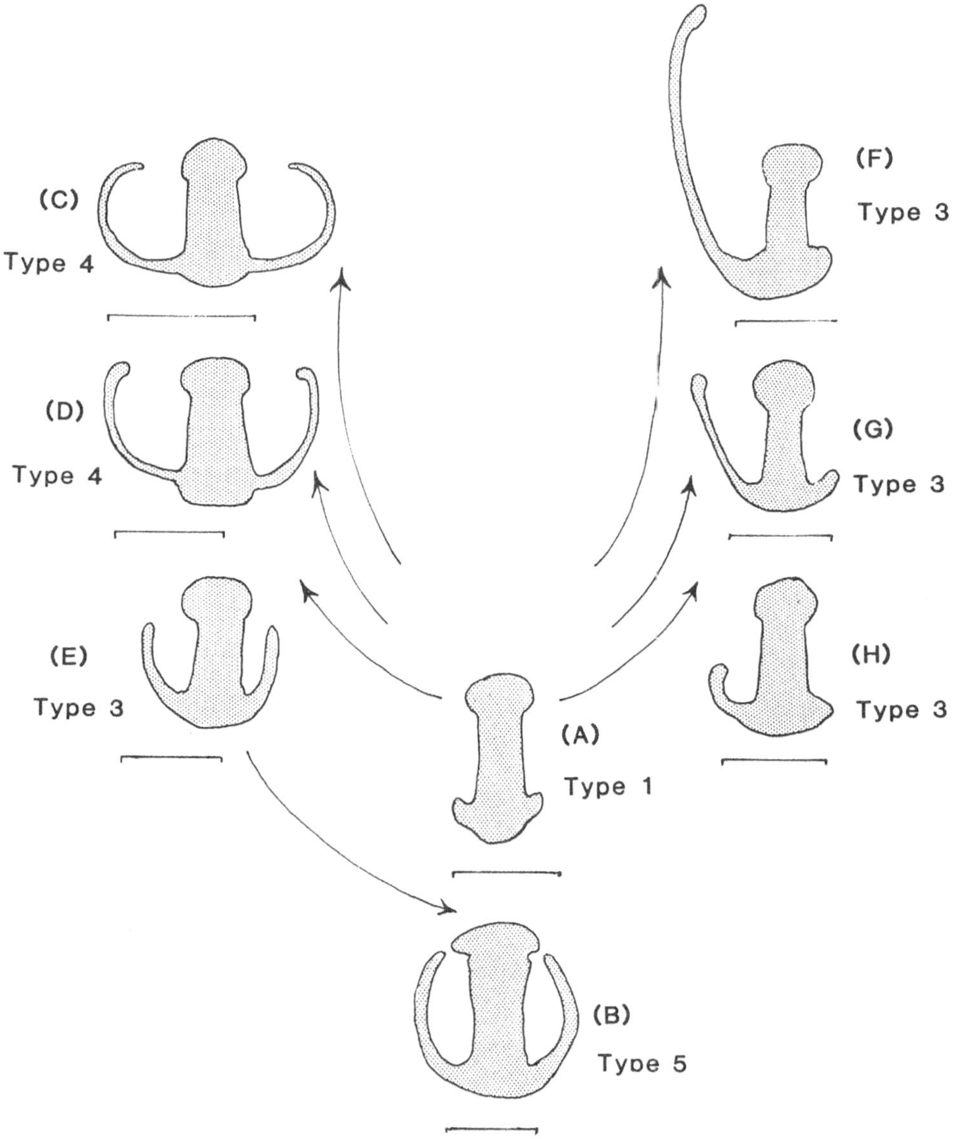

Fig. 8. The cuticular folds surrounding the tympana of the ears of tettigoniids arranged in three hypothetical lines and derived from the naked tympanum typical of many Phaneropterinae such as *Elephantodeta* sp.(A). Leading to the development of the anterior cuticular fold forming pinnae like structures in male *Oxycous lesnei* (F) through *Lacipoda munda* (D) and *Horatosphaga meruensis* (E), to "Helmholtz - type" resonators in *Phisis pallida* (C) through *Gravenreuthia saturata* (H) and *Zeuneria melanopeza* (G). The more typical tettigoniine ear is illustrated by *Decticus verrucivorus*.

Rentz (1975) suggested that tettigoniids subject to high predation rates in rainforests appear to have extraordinarily long antennae compared to more temperate species, those species with well developed cuticular folds forming expanded subslit cavities or pinnae are mainly found in the tropical forest systems, and seldom in temperate zones and those not living in rainforests. This may be due to the fact that the most elaborate tettigoniids, the Pseudophyllinae, are predominantly forest dwelling insects. However, it may be instructive to note that in this entire sub-family the auditory trachea are virtually closed.

Comparison between the more "typical" cricket tympana, and the naked tympana of the tettigoniid groups suggests that in the latter, the membranes appear to be nearly parallel, whereas in the Gryllidae and Haglidae that have been examined, the membranes are set at an obtuse angle to each other. Again in crickets and hagliids one membrane is thickened, with quite obviously different acoustic properties, and in the tettigoniids both membranes appear to have the same thickness and are of nearly equal in size. In *Amytta* sp. and *Xiphidiopsis elaphocerca* Karny the membranes are completely parallel and close to the mid-line of the tibia. Even the membranes within the enclosed cavities of the slits, or resonators tend to be parallel (e.g. *Decticus verrucivorus* (Graber 1876)).

Tympanic coverings – slits, resonators and pinnae

The covering of the tympanum by cuticular folding follows three broad patterns in the Tettigoniidae (Fig 8). (i) The anterior membrane is covered or partially covered by a fold forming a pinna-like structure (Type 2) (Table II), (ii) both membranes are covered forming cups, which in some cases appear as hollow resonators (type 3 & 4), and (iii) the cuticular folds form an almost enclosed cavity around the membrane opening through a narrow slit (type 5).

The formation of an anterior pinna varies between species from the cuticle being ridged into a small fold, ventral to the membrane, to the cuticle forming a full cup. In at least two tropical pseudophylline species this cup forms an effective pinna (e.g. *Oxycous lesnei* Chopard (Fig 8 & 9)(Chopard 1969) from Africa and *Mimetica* sp. from Costa Rica). In *O. lesnei*, the female has by comparison a reduced pinna-like structure, whereas in related species such as *O. apertus*, the male's pinna, although slightly more developed than the female, is of comparable size. In *O. undulatus* and *O. magnus* there is no evidence of sexual dimporphism. It is noticeable that in this subfamily females call in response to the male, inducing male phonotaxis and selection may be on the male to increase auditory acuity.

It would be reasonable to predict that the development of a pinna, shielding sound from the inner membrane, would increase binaural separation. In species with naked, or near naked tympana, the lip under the anterior membrane develops into a distinct cup before any similar development over the posterior membrane.

The formation of cuticular folds surrounding the tympana would appear to have been subject to quite independent selection within some species of crickets. For example two eneopterine genera *Calyptotrypus* and *Madasumma*, both have elaborate cuticular folds around the posterior membrane (Ander 1939; Chopard 1969; Heller 1984). This situation is strangely paralleled in certain Neotropical moths of the family Notodontidae. The ears of these moths are served by a single peripheral neuron, and two large species *Antaea lichyi* and *Hapigia curvilinea*, have cup-like projections acting as pinnae around the ear. Fullard (1988) describes an experiment where removal of these cuticular pinnae reduced the directionality of the system. He suggests that these cups act as "high frequency enhancers" and by removing the cups he demonstrated that the system is less sensitive to frequencies above 30 kHz.

In the tettigoniids, the cuticular folding over both membranes ranges in form from a

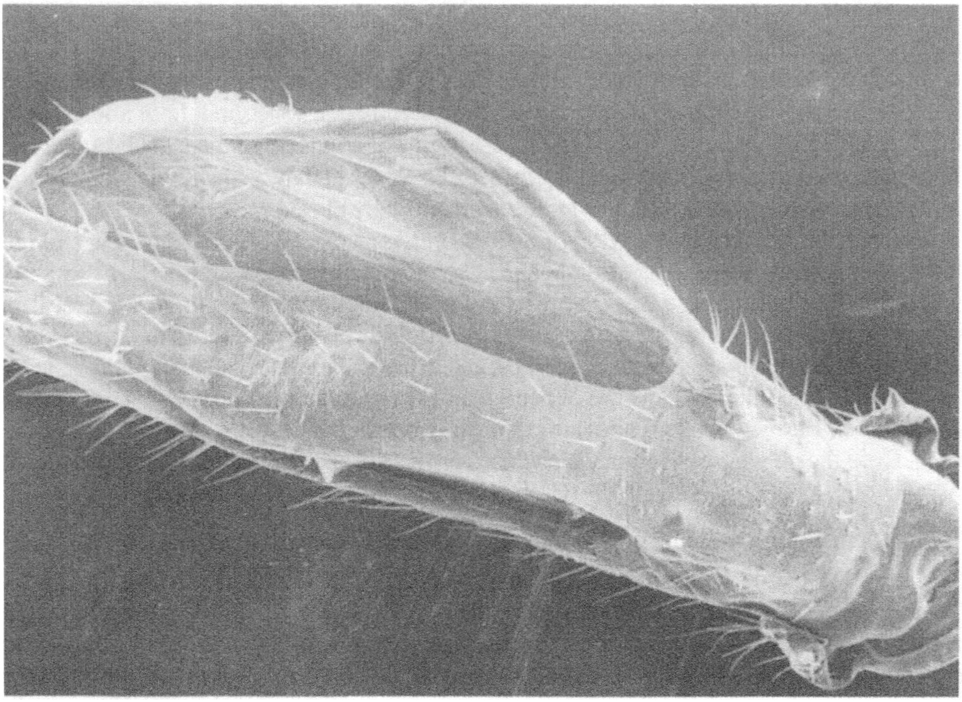

(A)

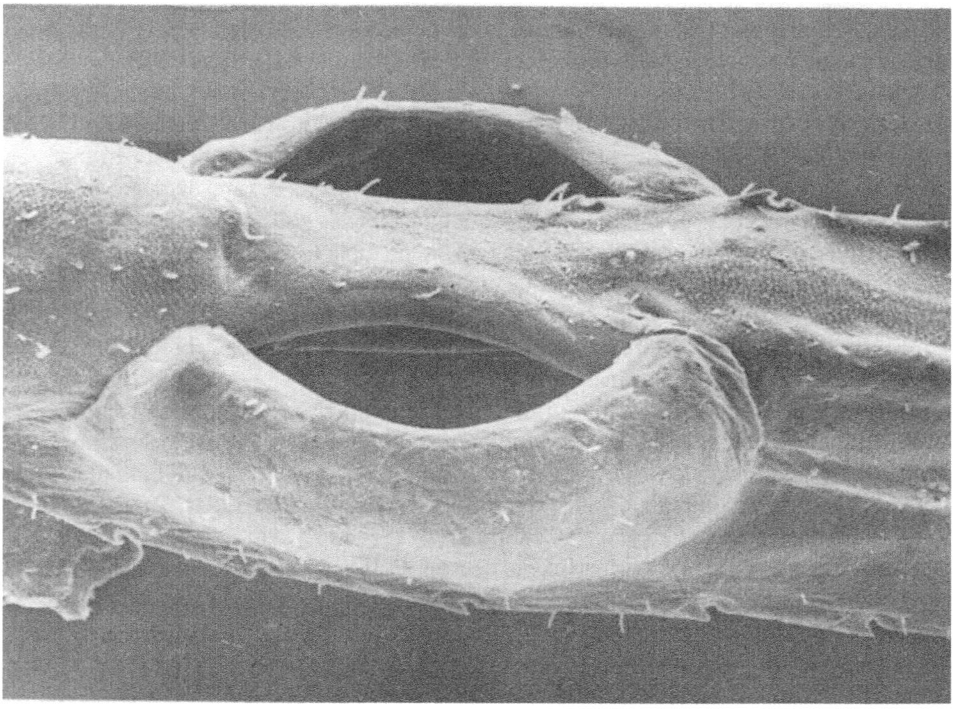

(B)

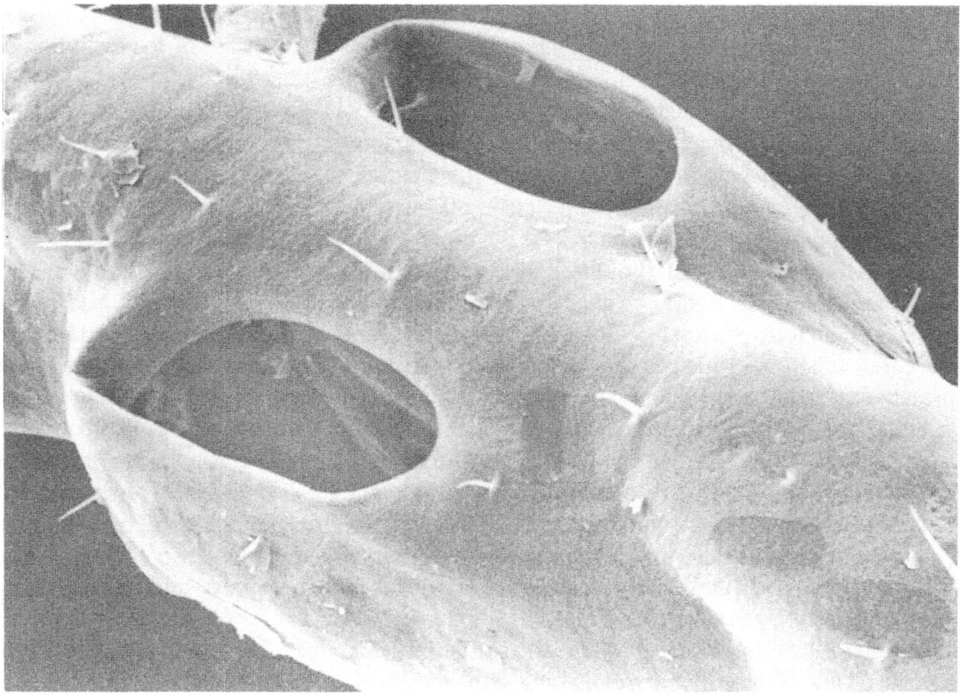

(C)

Fig. 9 (this page and opposite). SEM's of the ear of three species of tettigoniid. **A.** *Oxycous lesnei*; **B.** *Mustius* sp.; **C.** *Phisis pallida*.

capacious resonator, with cup-like openings to folds forming smaller chambers with slit-like openings. There are a range of genera from various subfamilies that range between these extremes (Fig 8). As with the development of the pinna, there is frequent asymmetry between posterior and anterior tympana. The anterior cup is always the larger of the two. Cup formation reaches a peak in genera such as *Phisis* and *Decolya* (Listroscelidinae), where they form true "Helmholtz-like" resonators. In *Lacipoda immunda* Brunner (Pseudopyhylinae) they remain as quite open forwardly projecting cups (Fig 9). In those species with less well developed sub-slit cavities, the slit opening appears to subtend a wider angle from the leg's mid-line. Although the slit cavities of the tettiginiine *Metaballus litus* Rentz are not nearly so well developed as some of the more tropical species, Bailey *et al.* (1988) found that the angle of sound incidence into the slit cavities was narrower than the more obtusely angled *Mygalopsis marki* Bailey (Bailey & Stephen 1978) (Fig 1). In other words, there is sufficient evidence to suggest that under certain conditions, sound may enter the sub-slit cavities, and where these cavities open along the midline of the tibia, it is likely that they both may function in concert (see above).

Hence, I predict that in species with forwardly directed Helmholtz-like cavities, the entire system will be highly tuned to certain frequencies within the song of either the male, or in some cases the female. Also that this highly tuned system will influence the insect's acoustic acuity; the precision with which it can locate its partner.

The question remains as to the function of the input through the auditory spiracle. If predation is a powerful selective influence, as suggested in tropical forest systems (Belwood & Morris 1987), then the separation of the predator avoidance "channel", from that used for conspecific recognition, may be sufficient to lead to the development of a unique extension of resonators surrounding the tympanic membranes. (For a discussion of the evolution of calling strategies in forest systems see Morris 1980; Morris & Beier 1982; Morris *et al.* 1989).

An alternative hypothesis has been suggested earlier in this chapter. That the sub-slit cavities may merely increase the sharpness of the tuning provided by the tracheal system, thereby effectively matching the impedance either side of the membrane(s). If this were the case then there would be little need for elaborate openings, or indeed for these openings to be symmetrical along the median axis of the tibia.

The development of nearly closed slits appears to be a feature of the Bradyporinae, Tettigoniinae (*sensu* Rentz, chapter 1) and Conocephalinae. If selection operated in the manner predicted by Zeuner (1936), in that the sound entering the trachea would so dominate the entire function of the hearing system as to reduce the contribution of the slits to mere pressure escapement devices, one may predict that the slits would close to a small hole. The bradyporine, *Bradyporus dilatatus* (F) does indeed have such a minute cuticular opening in front of the membranes, but then the tracheal opening is similarly small. In *Ephippiger ephippiger* (Fiebig) the spiracular openings are again minute, and the tracheal opening has a diameter of less the 1mm, which is in fact smaller than the maximum dimensions of the slits. (It is worth noting that the measured auditory sensitivity of *E. ephippiger* to airborne sound is lower than that of other European tettigoniids with more open auditory spiracles and wider slits (see Kalmring *et al.* chapter 11).)

In many phaneropterines, and some listroscelidines (e.g. *Hexacentrus munda*), the slits are narrow (Table II, type 5) but open to quite expanded cavities beneath. Such an insect would be a useful example to test the initial hypothesis of matched impedance due to the radiation of sound from the tympanic membranes into the subslit cavities; the small slit openings could act as escapement ports.

Good From Bad Sounds

Multi-modal senses

Two of the main functions of the auditory system of insects are to detect predators and to detect the signals of calling conspecific mates (Doherty & Hoy 1985); in other words they are under selection in both social and non- social contexts (c.f. West-Eberhard 1984; Belwood & Morris 1987; Belwood chapter 2). A third, less clearly defined role is to respond to, or at least be aware of "habitat" noises including the calls of other animals (Römer *et al.* 1989). The separation of these non-conspecific sounds from the detection of the calls of suitable mates may be through the utilisation of different receptor systems, or by the differentiation of more than one component of the noise within one receptor system. Hence, predators that produce low frequency disturbance sounds may be detected by the subgenual organs as they disturb the substrate (Kalmring *et al.* Chapter 11). Or if the predator produces additional sounds in the very low frequency range, the signal may be detected through an entirely different system such as the anal cerci (Camhi & Tom 1978 – Blattodea; Kämper 1984 – Gryllidae). In contrast the receptors of the tympanal organ are tuned to sounds well above these frequencies, and here the separation of "good" from "bad" information may be sorted by more complex processes within the auditory neuropile itself (Nolen & Hoy 1986). Thus flying female crickets can move towards calling males but away from flying bats.

Increasing selective pressure on auditory sensitivity (a lowering of the auditory threshold) may have had as its primary target the detection of predatory noises. However, parallel to this could be the evolution of sexual signals within a social context. For example, in the Acrididae, although the tympanal hearing organ is remarkably consistent in its gross structure throughout the entire taxon (Otte 1977), the mechanism for sound production involves a variety of structures and organs, and appears to have evolved independently a number of times (Riede 1987). The recognition of signals within a social context in grasshoppers has an undoubted independent origin from that of predator detection. Hence, visual signalling with limbs in an essentially diurnal taxon, with its role in species specific pattern recognition, is enhanced in many groups by an auditory input (Otte 1977); the auditory receptors have a divided and quite antithetical function, attraction rather than flight.

Thus in the Ensifera, two vibrational inputs with quite different functional histories, act synergistically under certain prescribed conditions. Several authors have shown that when the temporal modulation of an airborne sound is also transmitted through the substrate, the entire auditory system has an increased sensitivity to that pattern (Bell 1980, 1981; Kalmring & Kühne 1980; Kalmring *et al.* 1983; Latimer & Schatral 1983; Keuper & Kühne 1983). Keuper and Kühne (1983) argue that acoustic signals, if of sufficient energy, will evoke standing waves within the substrate, and if provided in the species specific pattern, will give additional information to the female as to the proximity of the male. In other words, what must be considered a mechanism used primarily for the detection of predators, now adds information to the social signal.

The second social function is the detection of the dominance status of a close male. In this context substrate vibration is often used. For example the bounce behaviour of crickets (Bailey & Stoddart 1982), and tremulation in many tettigoniids (Busnel *et al.* 1955; Morris 1980; Kalmring *et al.* chapter 11).

The auditory trachea of the tettigoniids, at least in those few species examined, appear to respond to a broad band of high to ultra-high sound. If there is any parallel with cricket hearing, one may predict that such ultrasound, which is well beyond the fundamental frequency of the song, may indicate an approaching predator. This could be from a bat, a foraging insectivore such as a shrew, or more likely as transients produced by vegetation disturbance. Such auditory input is rather non-specialised, but provides sufficient binaural information to detect from which side any predator may be approaching. If one adds to this the social signals of calling mates, (males in most species but occasionally also females) then one should expect the receptor system to become rather more specialised in order to differentiate "good" from "bad" sound.

A route for such specialisation would be to narrow the frequency band of transmission frequencies with a concomitant reduction in the band-width of the receptor system. This is certainly possible in the gryllids, where the social signal is clearly separated from the detection of ultrasound (Moissef *et al.* 1978; Moissef & Hoy 1983). A too narrow frequency band in the call may be energetically advantageous, but has quite serious problems in terms of temporal pattern transmission in complex habitats, and a too narrowly tuned receptor system means that the hearing organ is entirely dedicated to one function. Perhaps an extreme example of this is the predator detection function of the moth receptor with its very limited number of receptor cells (Roeder 1966).

Crypsis and predator detection

An insect using crypsis as its main defence may have a relaxed selective pressure on an auditory input that was once initially used for the detection of diurnal predators. In addition, such crypsis, at least in terms of shape, may even be effective against some nocturnal predators. This pressure would be further relaxed if the social signal itself evolved an anti-predatory function, such as complex alternation within aggregations, or even the production of loud chorusing signals as in cicadas (Simmons *et al.* 1971).

Within the tettigoniids, this development is best seen in the leaf-like bushcrickets, the Phyllophorinae, Pseudophyllinae and Phaneropterinae. Few workers have studied the calling strategies of the first two subfamilies (see Morris *et al.* 1989), it is now well established that many of the phaneropterines not only produce complex call patterns, but also have both male alternation and male/female duetting (see both Robinson and Heller in this volume). In contrast, active, largely carnivorous members of subfamilies such as the Saginae, Listroscelidinae and many Tettigoniinae, tend to be less cryptic and are often diurnal in their calling patterns. Here one would predict that the detection of predators (including cannibalism) has a higher selective value compared to cryptic phytophagous groups. Species exposed in this way would evolve a highly sensitive predator defense system.

The development of a dedicated tracheal system, with its high-pass filtering characteristic may have had an initial role of predator detection, leaving the input through the outside of the tympanic membranes as a detector and locater of conspecific signals. Thus, the physical expansion of the auditory spiracle, and the horn-like nature of the auditory trachea, may have raised the sensitivity of the entire system to the point where in some forms it appears to entirely dominate the input over the external tympanic membranes (c.f. Lewis 1974; Hill & Oldfield 1981). Here, as in the acridid system, side to side discrimination, which is enhanced by the shadowing effects of the body, provides cues as to the direction the *predator* was approaching and thereby the best way to escape.

Two modifications of this system have occurred in the more cryptic phytophagous species. First, there is a reduction of the spiracular input as illustrated by many, though not all Phaneropterinae, and all Pseudophyllinae and Phyllophorinae. Second, the development of enlarged and highly specialised auditory bullae with distinct resonances (Nocke 1975). Hence, I suggest that with the reduction in selection on a predator detector system in cryptic phytophagous groups, the initial role of this highly sensitive input has been taken over by social signalling. The consistent occurrence of sexual dimorphism of the auditory bullae within these subfamilies further supports the hypothesis that there has been a shift in the functional roles of this input, and the development of a highly specialised species (sex) detecting system.

What is the role of secondary tympana over the auditory spiracle formed by the lateral lobe of the pronotum, evolving quite independently in a number of subfamilies? Here there is no suggestion of a cross correlation between its occurrence in any one species, and crypsis or food preference, at least when viewed from the museum desk. Neither is there any obvious correlation between its presence and the shape of the cuticular fold around the tympana, or indeed to the shape of the trachea beneath. Mechanically, such a tympanic covering could act as a sounding board for vibration across a wide spectrum, and pressure changes beneath the thinned cuticular membrane would be amplified via the auditory trachea to the acoustic sensilla. An obvious role would be the detection of broad-band disturbance sounds produced either by conspecifics or predators.

Acknowledgments

This project has occupied my "spare" time for more than a decade, and hence it is impossible to acknowledge those many people who have contributed through discussion. Some of these ideas were initiated during an interesting period of research with Dick Stephen on hearing function in the tettigoniids. His insights on the way the system worked undoubtedly influenced many of my conclusions. I have given credit to those museums that have loaned and sacrificed specimens in the introductory section of the chapter. Klaus- Gerhard Heller drew my attention to the slit bearing eneopterine crickets. Julie Price and Paul Yeoh assisted with much of the anatomical studies, dissecting, preparing and sectioning fresh and loaned material, also to Tom Stewart for the SEM's. I thank Andrea Schatral, Ruth Bailey, Tom Walker and Klaus-Gerhard Heller, Eldon Ball, Ken Hill and David Rentz for reviewing the manuscript at different stages of writing. Financial support has been in part from the Australian Grants Committee.

References

Ander K. 1939. Vergleichend-anatomische und phylogenetische Studien über die Ensifera (Saltatoria). Opusc. Entomol. Suppl. 306 pp

Autrum H. 1940. Über Lautäusserungen und Schallwahrnehmung bei Arthopoden. II. Das Richtungshören von Locusta und Versuch einer Hörtheorie für Tympanalorgane von Locustidentyp. Z. Vergl. Physiol. 28: 326-352.

Autrum H. 1963. Anatomy and physiology of sound receptors in invertebrates.In: Acoustic Behaviour of Animals. Ed. R.-G. Busnel. Elsevier, Amsterdam. pp 412 – 433

Bailey W.J. (1990) Mate finding: selection on sensory cues. In: Reproductive Behaviour of Insects: Individuals and Populations. Eds. W.J. Bailey & J. Ridsdill-Smith. Chapman & Hall London (in press)

Bailey W.J. & A.W.R. McCrae. 1978. The general biology and phenology of swarming in the East African tettigoniid *Ruspolia differens* (Serville) (Orthoptera). J. Nat. Hist. 12: 259-288.

Bailey W.J. & R.O. Stephen. 1978. Directional and auditory slit function: A theory of hearing in bushcrickets. Science 201: 633-634.

Bailey W.J. & R.O. Stephen. 1984. Auditory acuity in the orientation behaviour of the bushcricket *Pachysagella australis* Walker (Orthoptera, Tettigoniidae, Saginae). Anim. Behav. 32: 816-829.

Bailey W.J., R.O. Stephen & P.B. Yeoh. 1988. Signal transmission in noisy environments: Auditory masking in the tympanic nerve of the bushcricket *Metaballus litus* (Orthoptera: Tettigoniidae). J. Acoust. Soc. Am. 83: 1828-1832.

Bailey W.J. & J.A. Stoddart. 1982. A method for the construction of dominance hierarchies in the field cricket *Teleogryllus oceanicus* (Le Guillou). Anim. Behav. 30: 216-230.

Bailey W.J. & P.B. Yeoh. 1988. Female phonotaxis and frequency discrimination in the bushcricket *Requena verticalis* (Tettigoniidae: Listroscelidinae). Physiol. Entomol. 13: 363-372.

Ball E.E. & L.H. Field. 1981. Structure of the auditory system of the weta *Hemideina crassida* (Blanchard 1851) (Ortoptera: Ensifera: Gryllacridoidea: Stenopelmatidae). 1. Morphology and histology. Cell Tissue Res. 217: 321-343.

Beier M. 1955. Embioidea und Orthopteroidea. In: Klassen und Ordnungen des Tierreichs. Arthropoda. Ed. H.G. Bronns. III Insecta. Leipzig Akademische Verlagsgesellschaft. pp 1-304.

Bell P.D. 1980. Multimodal communication by the black-horned tree cricket, *Oecanthus nigricornis* (Walker) (Orthoptera: Gryllidae). Can. J. Zool. 58: 1861-1867.

Bell P.D. 1981. Transmission of vibrations along plant stems: implications for insect communication. J. New York Ent. Soc. 88: 210-216

Belwood J.J. 1990. Anti-predator defences and ecology of Neotropical forest katydids, especially the Pseudophyllinae. In: The Tettigoniidae: Behaviour, Systematics, Evolution. Eds. W.J. Bailey & D.C.F. Rentz. Crawford House Press, Bathurst.

Belwood J.J. & G.K. Morris. 1987. Bat predation and its influence on calling behavior in Neotropical katydids. Science 238: 64-67.

Bennet-Clark H.C. 1984. Insect Hearing: Acoustics and transduction. In: Insect Communication. Ed. T. Lewis. Academic Press, London. pp 49-82.

Berlese A. 1882. Osservazione sulla anatomia descrittiva del *Gryllus campestris* L. Atti. Soc. Veneto-Trentina Sci. Nat. Anno 1881 Padova.

Busnel R.-G., F. Pasquinelly & B. Dumortier. 1955. La trémulation du corps et la transmission aux supports des vibrations en résultant comme moyen d'information à courte portée des Ephippigères mâle et femelle. Bull. Soc. Zool. Fr. 80: 18-22.

Camhi J. & W. Tom. 1978. The escape behaviour of the cockroach *Periplaneta americana*. I. Turning response to wind puffs. J. Comp. Physiol. 128: 193-201.

Charpentier F. 1924. Sur le double stigmate de quelques Orthoptère. Bull. Soc. Entomol. Belg. T. VI Bruxelles.

Charpentier F. 1927. Sur les trachees de la base des pattes et des âiles de la sauterelle vert. Ann. Soc. Scientif. de Bruxelles. Ser. B T. 47 Louvain.

Chopard L. 1969. Orthoptera Vol 2. Grylloidea. In: The Fauna of India and adjacent countries. R.B. Seymour Sewell (Ed). Zoological Survey of India, Calcutta. Baptist Mission Press. Calcutta xviii – 421 pp.

Darwin C. 1874. The descent of man, and sexual selection in relation to sex. Murray, London.

Doherty J. & R.R. Hoy. 1985. Communication in Insects: III. The auditory behaviour of crickets: Some views of genetic coupling, song recognition, and predator detection. Quart. Rev. Biol. 60: 457-472.

Field L.H., K.G. Hill & E.E. Ball. 1980. Physiological and biophysical properties of the auditory system of the New Zealand weta, *Hemideina crassidens* (Blanchard 1851) (Ensifera: Stenopelmatidae). J. Comp. Physiol. 141: 31-37.

Fletcher N.H. & S. Thwaites. 1979. Acoustical analysis of the auditory system of the cricket *Teleogryllus commodus* (Walker). J. Acoust. Soc. Am. 66: 350-357.

Fullard J.H. 1984. External auditory structures in two species of neotropical notodontid moths. J. Comp. Physiol. 155: 625-632.

Graber V. 1876. Die tympanalen Sinnesapparate der Orthopteren. Denkschr. k. Akad. Wiss. Math. Naturwiss. Cl. Bd. 36. Wien.

Heller K.-G. 1984. Zur Bioakustik und Phylogenie der Gattung Poecilimon (Orthoptera, Tettigoniidae, Phaneropterinae). Zool. Jb. Syst. 111: 69-117.

Hill K.G. & G.S. Boyan. 1977. Sensitivity to frequency and direction of sound in the auditory system of crickets (Gryllidae). J. Comp. Physiol. 121: 79-97.

Hill K.G. & B.P. Oldfield. 1981. Auditory function in Tettigoniidae (Orthoptera: Ensifera). J. Comp. Physiol. 142: 169-180.

Huber F. 1987. Plasticity in the auditory system of crickets: phonotaxis with one ear and neuronal reorganisation within the auditory pathway. J. Comp. Physiol. 161: 583-604.

Johnstone B.M., J.C. Saunders & J.R. Johnstone. 1970. Tympanic membrane response in the cricket. Nature 227: 625-626.

Kalmring K. & R. Kühne. 1980. The coding of airborne-sound and vibration signals in bimodal ventral-nerve cord neurons of the grasshopper *Tettigonia cantans*. J. Comp. Physiol. 139: 267-275.

Kalmring K., R. Kühne & D.B. Lewis. 1983. The acoustic behaviour of the bushcricket *Tettigonia cantans*. III. Coprocessing of auditory and vibratory information in the CNS. Behav. Processes 8: 213-228.

Kämper G. 1984. Abdominal ascending interneurons in crickets: reponse to sound at the 30 Hz calling song frequency. J. Comp. Physiol. 155: 507-520.

Keuper A. & R. Kühne. 1983. The acoustic behaviour of the bushcricket *Tettigonia cantans*.

II. Transmission of airborne-sound and vibration signals in the biotope. Behav. Proc. 8: 125-145.

Kleindienst H., D.W. Wohlers & O. Larsen. 1983. Tympanic membrane motion is necessary for hearing in crickets. J. Comp. Physiol. 151: 397-400.

Lakes R. & T. Schikorsky. 1990. The neuroanatomy of the tettigoniids. In: The Tettigoniidae: Behaviour, Systematics, Evolution. Eds. W.J. Bailey & D.C.F. Rentz. Crawford House Press, Bathurst.

Larsen O.N. 1987. The cricket's anterior tympanum revisited. Naturwissenschaften 74:

Larsen O.N. & A. Michelsen. 1978. Biophysics of the ensiferan ear. III. The cricket ear as a four-input system. J. Comp. Physiol. 123: 217-227.

Larsen O.N., H.-U. Kleindienst & A. Michelsen. (manuscript). Biophysical aspects of sound reception. In: Behavioural strategies and underlying physiological mechanisms in crickets. Eds. F. Huber, T.E. Moore & W. Loher.

Latimer W. & D.B. Lewis. 1986. Song harmonic content as a parameter determining acoustic orientation behaviour in the cricket *Teleogryllus oceanicus* (Le Guillou). J. Comp. Physiol. 158 : 583-591.

Latimer W. & A. Schatral. 1983. The acoustical behaviour of the bushcrickets *Tettigonia cantans*. I. Behavioural responses to sound and vibration. Behav. Process. 8: 113-124.

Latimer W. & M. Sippel. 1987. Acoustic cues for female choice and male competition in *Tettigonia cantans*. Anim. Behav. 35: 887-910.

Lewis D.B. 1974. The physiology of the tettigoniid ear. I – III. J. Exp. Biol. 60: 853-859.

Michelsen A. & O.N. Larsen. 1978. Biophysics of the ensiferan ear. I. Tympanal vibrations in bushcrickets (Tettigoniidae) studied with laser vibrometry. J. Comp. Physiol. 123: 193-203.

Michelsen A. & H. Nocke. 1974. Biophysical aspects of sound communication in insects. Adv. Insect Physiol. 10: 247-296.

Moiseff A. & R.R. Hoy. 1983. Sensitivity to ultrasound in an identified auditory interneuron in the cricket: a possible neural link in phonotactic behavior. J. Comp. Physiol. 152: 155-167.

Moiseff A., G.S. Pollack & R.R. Hoy. 1978. Steering responses of flying crickets to sound and ultrasound: Mate attraction and predator avoidance. Proc. Nat. Acad. Sci. USA 75: 4052-4056.

Morris G.K. 1980. Calling display and mating behaviour of *Copiphora rhinoceros* Pictet (Orthoptera: Tettigoniidae). Anim. Behav. 28: 42-51.

Morris G.K. & M. Beier. 1982. Song structure and description of some Costa Rican katydids (Orthoptera: Tettigoniidae). Trans. Am. Entomol. Soc. 108: 287-314.

Morris G.K., D.E. Klimas & D.A. Nickle. 1989. Acoustic signals and systematics of false leaf katydids from Equador (Orthoptera, Tettigoniidae, Pseudophyllinae). (Manuscript)

Nietsch V. 1894. Über das Tracheensystem von *Locusta viridissima*. Verh. kk. zool-botan. Ges. Wein. Bd. 44. Versaml.

Nietsch V. 1905. Das Tracheensystem der Insekten. Mitteil. Naturwiss. Ver. f. Steiermark Jahrg. 1904

Nocke H. 1974. The tympanal trachea an an integral part of the ear in *Acripeza reticulata* Guerin (Orthoptera: Tettigoniidae). Z. Naturforsch. 29: 652-654.

Nocke H. 1975. Physical and physiological properties of the tettigoniid ear. J. Comp. Physiol. 100: 25-57.

Nolen G.T. & R.R. Hoy. 1986. Phonotaxis in flying crickets. II. Physiological mechanisms of two-tone suppression of the higher frequency avoidance steering behavior by the calling song. J. Comp. Physiol. 159 : 441-456.

Oldfield B.P. 1982. Tonotopic organisation of auditory receptors in Tettigoniidae (Orthoptera : Ensifera). J. Comp. Physiol. 147: 461-469.

Otte D. & W. Cade. 1976. On the role of olfaction in sexual and interspecies recognition in crickets (*Acheta* and *Gryllus*). Anim. Behav. 24: 1-6.

Otte D. 1977. Communication in Orthoptera. In: How Animals Communicate. Ed. T. Sebeok. Indiana, Bloomington. pp 334-361.

Ragge D.R. 1980. A review of the African Phaneropterinae with open tympana (Orthoptera: Tettigoniidae). Bull. Brit. Mus. (Nat. Hist.) Entomol. 40: 67-192.

Rentz D.C. 1975. Two new katydids of the genus *Melanonotus* from Costa Rica with comments on their life history strategies (Tettigonidae: Pseudophyllinae). Entomol. News 86: 129-140.

Rheinlaender J. 1975. Transmission of acoustic information at three neuronal levels in the auditory system of *Decticus verrucivorus* (Tettigoniidae, Orthoptera) J. Comp. Physiol. 97: 1-53.

Rheinlaender J. & H. Römer. 1990. The neuroethology of sound reception in bushcrickets. In: The Tettigoniidae: Behaviour, Systematics, Evolution. Eds. W.J. Bailey & D.C.F. Rentz. Crawford House Press, Bathurst.

Rheinlaender J. & H. Römer. 1980. Bilateral coding of sound direction in the CNS of the bushcricket *Tettigonia viridissima* L. (Orthoptera, Tettigoniidae). J. Comp. Physiol. 140: 101-111.

Riede K. 1987. A comparative study of mating behaviour in some Neotropical grasshoppers (Acridoidea). Ethology 76: 265-296.

Robinson D.A. 1990. Communication between the sexes in bushcrickets. In: The Tettigoniidae: Behaviour, Systematics, Evolution. Eds. W.J. Bailey & D.C.F.Rentz. Crawford House Press, Bathurst.

Roeder K.D. 1966. Acoustic sensitivity of the noctuid tympanic organ and its range for the cries of bats. J. Insect. Physiol. 12: 843-859.

Römer H.,Bailey W.J. & I.R. Dadour 1989. Insect hearing in the field. III. Masking by noise. J. Comp. Physiol. 164: 609-620

Sales G. & D. Pye. 1974. Ultrasonic Communication in Animals. Chapman and Hall, London.

Schmitz B., H. Scharstein & G. Wendler. 1983. Phonotaxis in *Gryllus campestris* L. (Orthoptera, Gryllidae) J. Comp. Physiol. 152: 257-264.

Schumacher R. 1979. Zur funktionellen Morphologie des auditiven Systems der Laubheuschrecken (Orthoptera: Tettigonoidea). Entomol. Ges. 5: 321-356.

Schwabe J. 1906. Beitrag zur Morphologie und Histologie der tympanalen Sinnesapparate der Orthoptera. Zoologica (Stuttg.) 50: 1-154.

Seymour C., D.B. Lewis, O. Larsen & A. Michelsen. 1978. Biophysics of the ensiferan ear. II. The steady state gain of the hearing trumpet in bushcrickets. J. Comp. Physiol. 123: 205-216.

Siebold C.T. von. 1844. Uber das Stimm – und Gehörogan der Orthoptera. Arc. f. Naturgesch. X. Jahrg. Berlin

Simmons J.A., E.G. Wever & J.M. Pylka. 1970. Periodic cicada: Sound production and hearing. Science 171: 212-212.

Stephen R.O. & W.J. Bailey. 1982. Bioacoustics of the ear of the bushcricket *Hemisaga* (Saginae). J. Acoust. Soc. Am. 72: 13-25.

Tinkham E.R. 1973. Nearctic desert Decticidae (Orthoptera). III. The true tympanum in certain genera with key. Great Basin Nat. 33: 197-200.

Treat A.E. 1975. Mites of moths and butterflies. Cornell University Press, Ithaca, New York. 362 pp.

Weber T., J. Thorson & F. Huber. 1981. Auditory behaviour of the cricket. I. Dynamics of compensatory walking and discrimination paradigms on the Kramer treadmill. J. Comp. Physiol. 141: 215-232.

Wendler D.W., M. Dambach, B. Schmitz & H. Scharstein. 1980. Analysis of the acoustic orientation behaviour in crickets (*Gryllus campestris* DeGeer). Naturwissenschaften 67: 99-101.

West-Eberhard M.J. 1984. Sexual selection, competitive communication and species-specific signals in insects. In: Insect Communication. Ed. T. Lewis. Academic Press, London. pp 283-324.

Young D. & E. Ball. 1974. Structure and development of the auditory system in the prothoracic leg of the cricket *Teleogryllus commodus* (Walker). I. Adult structure. Zellforsh. mikrosk. Anat. 147: 293-312.

Zeuner F.E. 1936. The prothoracic tracheal apparatus of *Saltatoria* (Orthoptera). Proc. R. Entomol. Soc. Lond. 11: 11-21.

Zhantiev R.D., I.N. Kalimkina & V.S. Chukanov. 1975. Characteristics of the directional sensitivity of the tympanal organs in the cricket *Gryllus bimaculatus* Deg. (Orthoptera: Gryllidae). Entomol. Obozr. 54: 249-257.

Zimmermann U., Rheinlaender J. & D. Robinson (1989). Cues for male phonotaxis in the duetting bushcricket *Leptophyes punctissima*. J. Comp. Physiol. 164: 621-628

13. ACOUSTIC CUES FOR SOUND LOCALISATION AND SPACING IN ORTHOPTERAN INSECTS

Jurgen Rheinlaender and Heiner Römer

Introduction

A bushcricket, resting in a meadow may use both visual and acoustic channels for communication. However, given the dense nature of the habitat visual communication over any distance may remain quite limiting. By comparison, sound waves propagate through air, bypassing obstacles such as leaves, bushes or even small topographical features in the habitat. Furthermore, unlike chemical signals acoustic signals fade out rapidly in air, allowing the transmission of temporal information. Hence, it is of little surprise to find acoustic signals involved in a variety of biological situations such as mate recognition, territorial defense, capture of prey and predator avoidance. There are, however, some disadvantages in relying on such a signal. Firstly, communication does not take place in total silence. When several species are calling at the same time animals must coevolve filter mechanisms for sound production and perception in order to transmit and detect their own conspecific signal. We call the sum of all the emitted individual signals biological noise. This noise is superimposed on a background of abiotic noise produced by wind, foliage movement or other extraneous factors. Secondly, the signal degrades as it travels through the environment so that the intensity, the time structure and the frequency content may differ drastically from the original broadcast signal.

Until recently little thought has been given to the effects of sound degradation on the listening insect, or to the ways in which the sender obviates these difficulties. (For a review on sound propagation in the natural habitat see Michelsen (1978), and on insect signals Michelsen and Larsen (1983). The majority of studies on insect acoustic communication have been performed in the undistorted conditions of the laboratory. This chapter will introduce the reader to some important aspects of sound communication in the field, and considers the consequences for processing acoustic signals by the nervous system. We want to show how both the hearing organs as well as the central nervous system cope with requirements for detecting and localising a conspecific's call.

Frequency Selectivity and Spacing

When investigating calling insects in nature, one basic observation is that singing individuals tend not to be randomly distributed within a population, but instead they maintain a more or less regular distance from their neighbours. To maintain this distance both singing and hearing is absolutely necessary (e.g.Bailey & Thiele 1983 and see chapter 6). When measuring the distances between singing male *Mygalopsis marki* Bailey (Copiphorini) in a population close to Perth, Western Australia we found a distribution of inter-male distances with a range between 1 and 27m shown in Fig. 1, where the mean nearest neighbour distance was 11.5 m (Römer & Bailey 1986).

How do males maintain this distance? On the basis of similar observations on other animals that communicate by sound, some scientists have speculated that this distance between singing neighbours might correspond to the maximum possible hearing distance. Although at least theoretically unlikely, males could distance themselves from their immediate neighbours until their neighbours' calls are just barely audible (Brenowitz *et al.*, 1984; Wilczynski and Brenowitz, 1988). If the auditory threshold was used for spacing, only part of the population would loose contact with their neighbours and consequently the regular distribution of calling insects would be lost. However, a more convincing test of the hypothesis is by field experiment using the insect's own hearing system as a way of assessing its perception of its calling neighbours. We have therefore developed a portable recording device (Rheinlaender and Römer, 1986) which enabled us to monitor the nervous activity of single auditory cells of insects within natural populations. The anatomy of this cell, the Omega neuron, is shown in Fig. 2. The special advantage of using this cell is that its threshold is almost equivalent to the threshold of the entire ear, as we have shown for the bushcricket

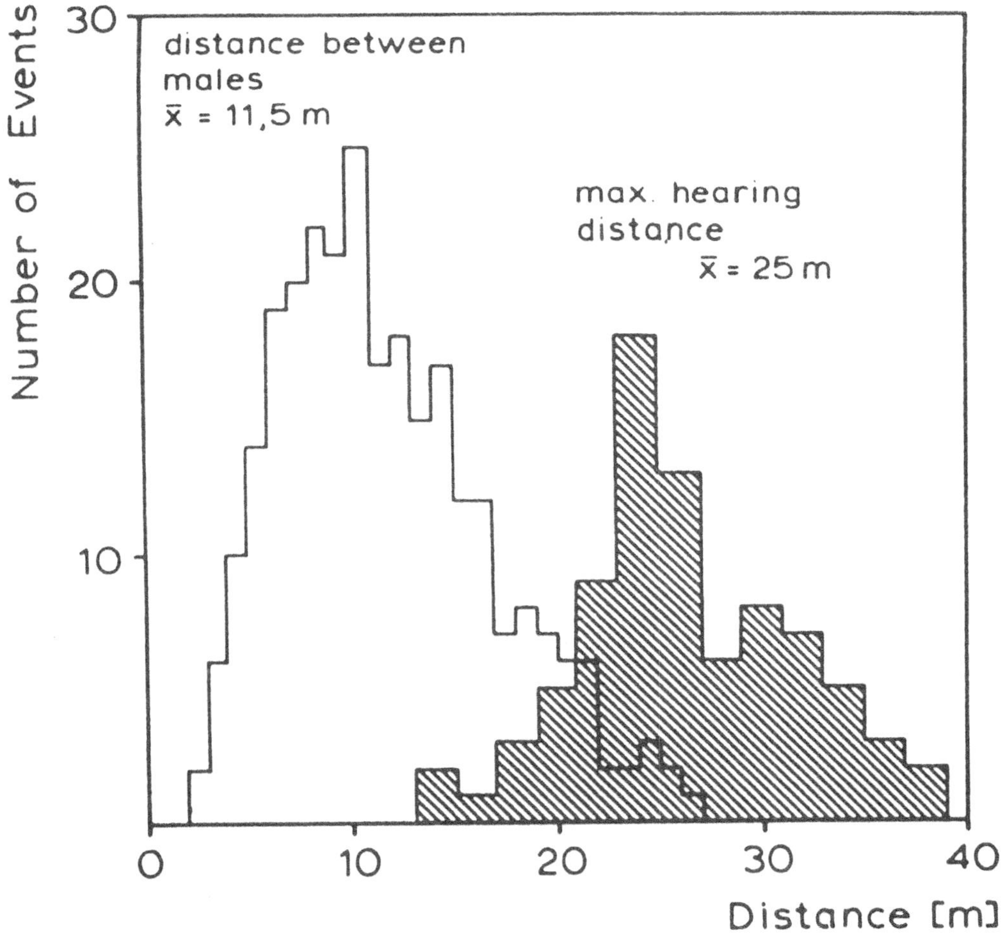

Fig. 1. Distribution of distances between singing males in a population of *Mygalopsis marki*. For comparison, the distribution of maximum hearing distances is given in the same habitat, measured neurophysiologically with the portable recording unit (after Römer and Bailey, 1986).

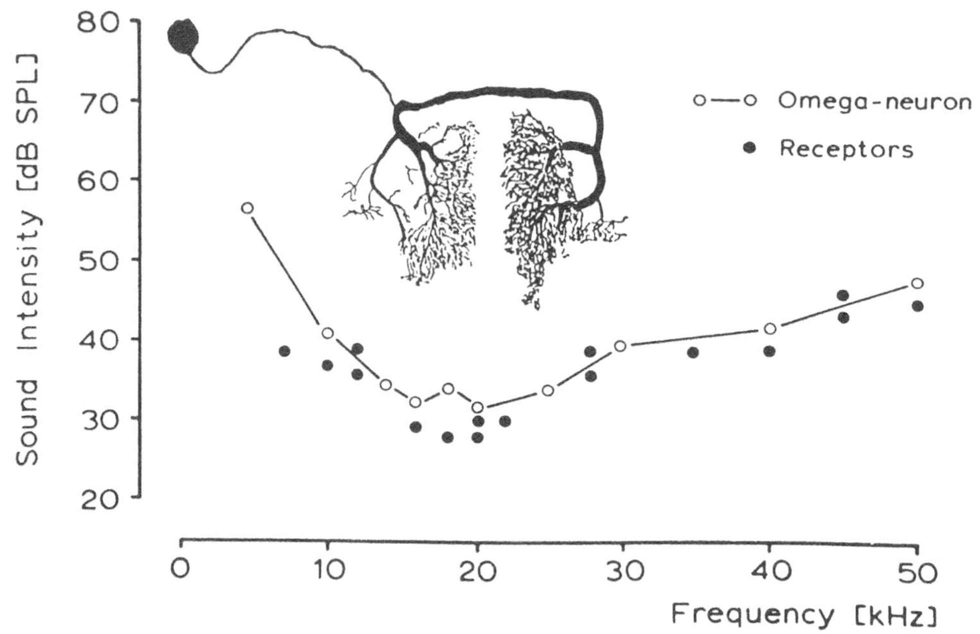

Fig. 2. Anatomy and threshold curve of the omega-neuron in the prothoracic ganglion of the bushcricket *Tettigonia viridissima*. A homologous neuron also exists in all bushcricket species so far examined; it is connected monosynaptically to most auditory receptors (Römer *et al.*, 1988). For comparison, the sensitivity of single auditory receptors at their characteristic frequency is given as black dots.

Tettigonia viridissima (L.) in Fig 2. By positioning the preparation at increasing distances from the singer, we were able to determine the maximum hearing distances of each male in the same population where measurements of the distances to the nearest neighbours had been made. The results in Fig. 1 clearly demonstrate that the maximum hearing distance is much greater than that inter-male distance (compare the means of 26 m with 11.5 m).

Gerhardt *et al.* (1988), following their behavioural experiments on frogs, indicated that acoustic spacing on supra-threshold signals may well be a general mechanism used by individuals within a chorus, and our studies confirm this. However, the sensory mechanism for maintaining this distance is still unknown, and a first step is to understand the properties of the hearing organ and each individual receptor. It is the receptor system that perceives and processes the sound before passing on the encoded information to higher centres of the nervous system.

For a long time insect sound communication seemed to be based only on amplitude modulation alone, i.e. the temporal pattern of a stereotyped sound signal. Then in 1960, Horridge was able to show by recording the activity of central nerve cells of the locust, that the hearing organ was capable of discriminating the pitch of a sound. His findings were later extended by experiments directed at the peripheral hearing organ of the same animal (Michelsen, 1971; Römer, 1976), and similar results have since been found for both bushcrickets (Zhantiev, 1971; Rheinlaender, 1975; Kalmring *et al.*, 1978; Oldfield, 1982) and crickets (Esch *et al.*, 1980; Oldfield *et al.*, 1986).

The hearing organs of bushcrickets are located in the tibia of the forelegs, just below the knee. Schwabe (1906) described the basic anatomy of the ear, consisting of an array of about 30 individual sensory cells, organized linearly in a structure called the *crista acustica* (see also Lakes & Schikorsky chapter 10) where the size of the receptor cells within the array decreases towards the distal part of the leg. Even at this early stage Schwabe suggested that this anatomical configuration might correlate with the different tuning characteristics of the receptors. Recently, this hypothesis was confirmed by Oldfield (1982), by recording the activity of identified cells within the organ (Fig. 3A) where he demonstrated that the characteristic frequency (CF) of receptors increases with their proximal to their distal location in the *crista acustica*. In this respect the bushcricket ear is analogous to the vertebrate ear in as much as the position of a receptor correlates with its characteristic frequency.

The nerve fibres of the auditory receptors run through the leg nerve into the prothoracic ganglion of the ventral nerve cord where they bifurcate and form presynaptic endings in a specialized area of the ganglion, called the auditory neuropil. Oldfield (1983) and Römer (1983) demonstrated independently of each other that the tonotopic arrangement in the ear is maintained at the level of the CNS in two species of bushcrickets. One example of a single receptor is shown in Fig. 3B, where the reconstruction is based on an intracellular staining of the receptor with the fluorescent dye Lucifer Yellow. The endings of the fibre are limited to a rather small area on the ipsilateral side of the ganglion only. The specific target area of each individual receptor, however, can be demonstrated most clearly when analyzing sagittal

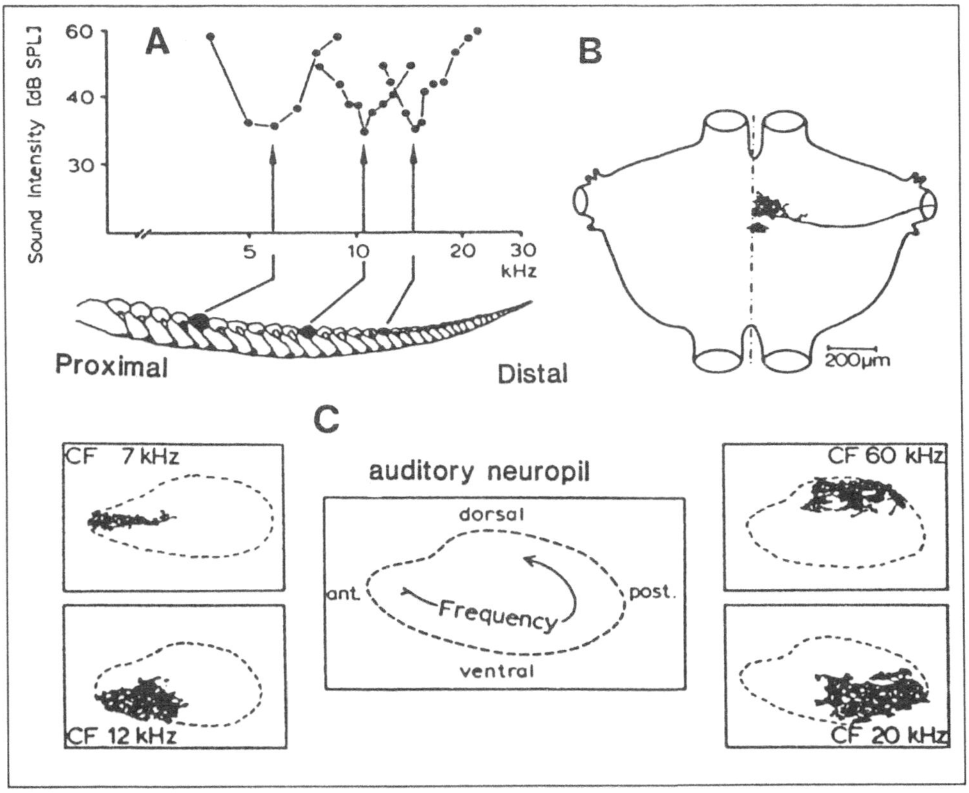

Fig. 3. A. Threshold curves of three auditory receptors in the tympanic organ of *Mygalopsis marki*. Their specific location within the hearing organ (*crista acustica*) is indicated by arrows (after Oldfield, 1982). **B.** Projection area of a single auditory receptor in the prothoracic ganglion. **C.** Target area of 4 individual receptors in the auditory neuropil of the prothoracic ganglion. as viewed from lateral, about 30 *fm* from the ganglionic midline (see arrowhead in B). The borderline of the neuropil is given as a broken line. Note the specific target area of each receptor according to its characteristic frequency (CF). The tonotopic arrangements is indicated by the arrow running counter-clockwise in the schematic diagram.

sections through the auditory neuropil close to the ganglionic midline (indicated by the arrowhead in Fig. 3B). Four examples of such sections are shown in Fig. 3C. All receptors differ in their characteristic frequency, the one with a CF of 4 kHz is represented by a small band of endings in the very anterior and intermediate part of the neuropil, and the 20 kHz receptor has the largest projection area, lying in the postero-ventral region and overlapping to some extent with the target area of the 10 kHz receptor. The fibre tuned to 60 kHz has branches exclusively in the dorsal and more anterior part of the neuropil. Thus, we can conclude that the frequency information in the bushcricket ear is represented in an ordered manner in the central neuropil. Each area is designated with a particular frequency that increases anti-clockwise in both the antero-posterior and the dorso-ventral axis, as shown schematically in the diagram of Fig. 3C (and Fig. 8 Lakes & Schikorsky chapter 10). The direction of the arrow in this figure indicates increasing characteristic frequencies.

The basic question is however, in which way does the frequency selectivity we have described and the tonotopic arrangement within the nervous system contribute to the insect's ability to maintain a consistent distance from its neighbours? Firstly, we should consider the fact that both the overall amplitude and the shape of the spectrum of the signal changes as the distance between sender and receiver increases. Generally higher frequencies are attenuated more with distance than lower frequencies (Keuper and Kühne, 1983; Römer and Bailey, 1986; Keuper *et al.*, 1986). This phenomenon is shown in Fig. 4 where a signal close to the sender (2 m) contains the full frequency spectrum of the song, whereas at 25 m all higher components except one around 10 kHz are missing. The extent of this filtering depends of course on several factors, such as the density of the vegetation, the height of the sender and/or the receiver, and the climatic conditions. Thus, there are two obvious features of sound that are affected by distance, the overall loss in amplitude and in addition the change in the spectral composition. Both may be used by the receiver to estimate the distance to the sender.

In Fig. 4 the tuning curves of 3 different tympanic receptors are superimposed on the song spectra at selected distances from the sender. The characteristic frequencies of these receptors are 10, 20 and 40kHz respectively. Remember that the ear contains about 30 receptors, each with a different CF between about 2 and 60 kHz (see Fig. 3A). From the example given in Fig. 4 it is obvious that a signal perceived at 25 m can only excite the receptor with a CF of 10 kHz. At an intermediate distance of 15 m, the spectrum is broader and the overall amplitude has increased so that now both the 10 kHz and 20 kHz receptors respond, but the 20 kHz receptor less than the 10 kHz. Finally, at a close distance of 2 m, the signal energy is high enough to stimulate all receptors. The superposition of the song spectra, and the tuning curves of auditory receptors in the example of Fig. 4, demonstrate that in order to get information about the afferent activity at different communication distances, it is quite incorrect to consider each unique sound parameter (frequency, intensity or temporal pattern) separately. Instead it is the mix of parameters that is characteristic for a certain sender/receiver relationship. Therefore, we used computer techniques to generate complex sound signals, which later served as stimuli in a series of neurophysiological experiments in the laboratory. These stimuli are digital replicates of analogue sound recordings made in nature at different distances from a singing male and played to the nerve preparation in an anechoic room. Hence we were able to present to the preparation controlled stimuli representing sound from known distances. The results of these experiments are summarized in Fig. 5A. The diagram of the response magnitude of 7 receptors (with different CF's) at various distances illustrates that at the maximum distance only one receptor (CF 10 kHz) responds to the signal at all. The closer the "distance" to the signaller, the stronger is the response in this receptor, and also in the others. The most sensitive receptors saturate at medium distances, however,

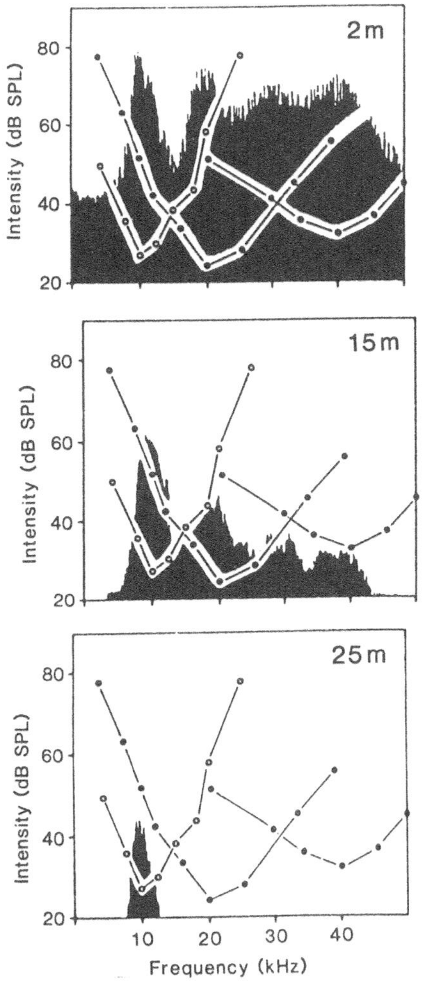

Fig. 4. Frequency spectrum of the song of the bushcricket, *Tettigonia viridissima* measured at distances of 2, 15 and 25 m in a bushland habitat. Note the decrease of high frequency components with increasing distance to the signaller. For comparison, the threshold curves of 3 auditory receptors (CF 10, 20 and 40 kHz) are superimposed on the spectra.

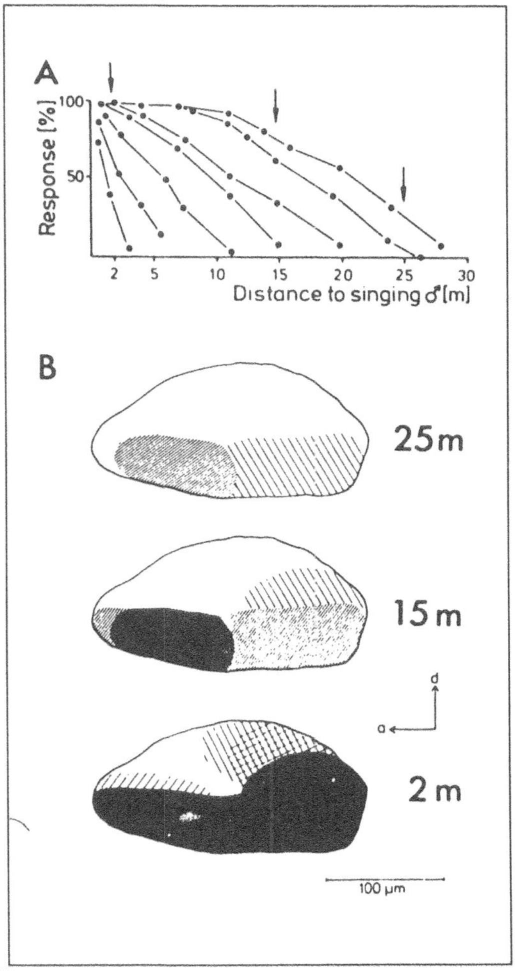

Fig. 5. A. Distance characteristics of seven auditory receptors of *Mygalopsis marki* in response to sound representing different distances from a signaller. B. Profile of receptor activity in the auditory neuropil at 3 different distances from a signaller as indicated (same lateral view as shown in Fig. 1C). Note that the activity profile is dependent on communication distance. For further explanations see text (Römer 1987).

as less sensitive receptors start to respond, the dynamic range of responses is maintained up to 1-2 m. Thus, the cooperation of all the receptors guarantees that afferent information, in a graded form, provides a cue for distance over the whole communication range (Römer, 1987).

This graded level of receptor activity, changing with distance, is transmitted to the auditory neuropil of the prothoracic ganglion (compare with Fig.3B). Due to the tonotopic organisation of the neuropil (Fig 3C), a change of distance between sender and receiver also affects the spatial distribution of the spike activity of receptors within the neuropil. This is shown schematically in Fig. 5B. At a distance of 25 m only the antero-ventral area of the neuropil is activated, as it is at this location that the 10 kHz receptor bifurcates. With decreasing distance, the amount of activity in this area not only increases but also spreads into other areas of the neuropil. Thus, the receptor information concerning auditory distance is converted at the central nervous level into a spatial distribution of sensory activity. It is obvious that this kind of ordering should facilitate the read-out of such information by single higher-order nerve cells (Römer, 1987).

Since the discovery that insect auditory receptors are tuned, scientists have been primarily concerned with aspects of the frequency analysis by the ear. The picture we present here demonstrates how this frequency selectivity can be used to discriminate the relative distance to a signaller.

Sound Localisation in the Laboratory and Outdoors

Behavioural studies in the laboratory

Although there are many incidental reports about the acoustic orientation behaviour of orthopteran insects, only a few laboratory studies describe this behaviour quantitatively. One of the reasons may be that insect sound signals in general have been found to be species-specific and, consequently, much effort has been employed to the problem of signal detection and pattern recognition rather than to the aspects of sound localisation behaviour (for review see Huber, 1983). Furthermore, some researchers may have been discouraged when trying to design phonotactic experiments for an insect, as it is hard to fulfill the conditions of the classical orientation paradigm: Firstly, it requires that the results derive from single orientation trials performed independently from each other. Secondly, the orientation must occur under so-called open-loop conditions, i.e. that the sound stimulus is off when the phonotactic movement begins.

In orthopteran insects two basic strategies are used in order to bring male and female together. The first is when the male is stationary and produces a calling song over a considerable time period, sometimes for many hours during the day and night, and it is the female that localises and approaches the singing male. In the ideal case each song element elicits a turning movement in the female – she moves forward a few steps until the next sound element and then makes another turn. Thus the female approaches the male through a successive series of course adjustment. Although a number of quantitative measures can be taken from such a phonotactic path, the results are basically descriptive and do not answer the important questions of whether the insect is able to discriminate between different angles of sound incidence. Angle discrimination has been described in humans and barn owls by Blauert (1983) and Knudsen (1981) respectively.

In some species of grasshoppers and bushcrickets mate finding is by both sexes engaging in a duet (see review by Robinson chapter 7). A stridulating male elicits an acoustic reply from the stationary female, to which the male then responds by phonotaxis. This duetting is always initiated by the male and ensures that communication between the sexes is maintained throughout the whole phonotactic approach. The orientation of the male appears to be highly stereotyped and is therefore a suitable candidate for behavioural studies of sound localisation. The female can easily be replaced by a sound dummy emitted from a speaker.

The basic experimental paradigm is illustrated in Fig. 6. A motivated male is sitting on a small platform, so that the synthetic reply of a female can be broadcast via a speaker at any angle of sound incidence in the horizontal plane. By means of a video system, the stimulus angle α, the turn angle γ and the temporal relationship between signal and response can be analyzed. In the grasshopper *Chorthippus biguttulus*, for example, the male turns after a delay of about 500 ms with respect to the onset of the female reply. This behaviour persists even when the song (normally lasting for about 1 s) is shortened to 400 ms (see inset in Fig. 6), so that the experiments occur under open-loop conditions. Therefore, given this open-loop paradigm we can study the basic problem of angular discrimination in a localising insect. A further advantage of the method is that the stimulus angle α can be varied independently from test to test.

The accumulated data of the turning responses of 9 males of *C. biguttulus* are shown in Fig. 7, where stimulus angle α is plotted versus turn angle γ. Each dot represents a single turn. If a one-to-one relationship between stimulus angle and turn angle existed, all dots would follow an "ideal" regression line as shown below in Fig. 8. Obviously, this is not the case, the males do not align their longitudinal body axis to the target axis, but turn with an angle of about 80 degrees to the stimulated side, quite independently of whether the stimulus angle is e.g. 60 or 120 degrees. In front of the animal there is a narrow but distinct sector of ambiguity in the order of ±10 degrees. Here, almost equal numbers of turns to the correct and incorrect side occur, which means that the insects do not lateralise correctly.

The results documented in Fig. 8 are representative for the orientation behaviour of those crickets and bushcrickets in which the female moves towards the stationary male (Murphey & Zaretsky, 1972; Bailey & Thompson, 1977; Oldfield, 1980; Bailey & Stephen, 1984). This example shows accumulated data of six female *Tettigonia cantans* (Fuessly) orienting to the male song which lasts for several seconds. There is a clear linear relationship between stimulus angle and turn angle γ ($r = 0.89$). The conclusion, however, that the females can discriminate between different angles of incident sound is not compulsory: (i) females can still operate under closed-loop conditions in this paradigm, (ii) it is completely unknown whether the phonotactic path is influenced by the integration over several orienting processes (remember that both conditions hold for the experiment shown in Fig. 7).

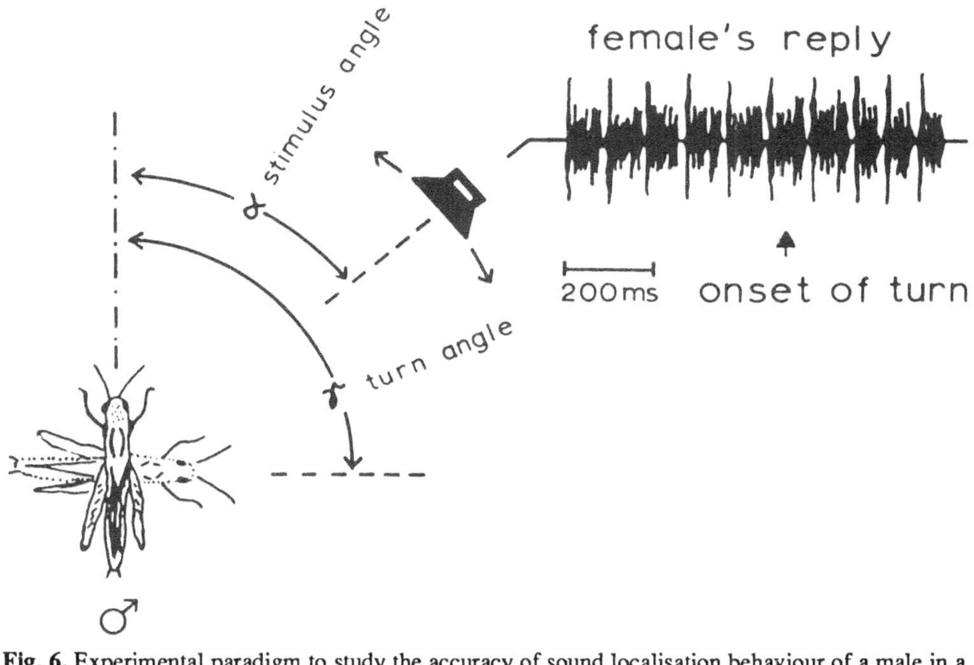

Fig. 6. Experimental paradigm to study the accuracy of sound localisation behaviour of a male in a duetting grasshopper species. The female is replaced by a speaker broadcasting a mimic of her acoustic reply to the male's song. The behaviour is monitored by a video system for the analysis of stimulus angle and turn angle.

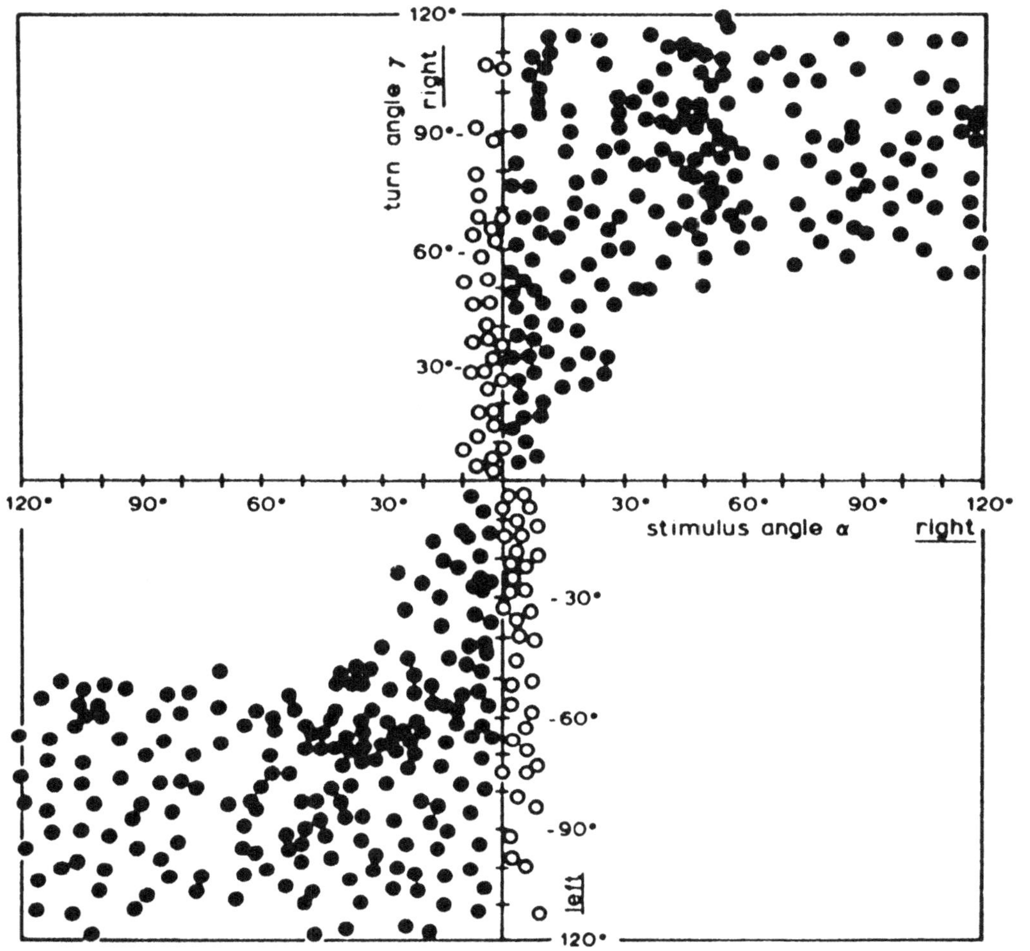

Fig. 7. Relationship between stimulus angle and turn angle of 9 males of *Chorthippus biguttulus*. Each dot represents a single turn. Note the discrepancy between angular dependent orientation behaviour and the documented results, indicating right-left discrimination only (after Rheinlaender and D. von Helversen, unpublished).

Independently of the species studied, the phonotactic path of all insects differs from a straight line, rather it resembles a zigzag course (Fig. 9A), in which the animal finally reaches the sound source due to successive course adjustments. The accumulated data of 10 phonotactic approaches by *C. biguttulus* are shown in Fig. 9B. They illustrate that the insects makes a minimal deviation from the target, taking a more or less direct path to the speaker. This observation can also be confirmed by measuring the distance covered by the insect; it is only about 30% greater than a straight line between starting point and speaker. A recent behavioural analysis has shown that the minimum interaural intensity difference for eliciting a turning reaction is in the order of 1-2 dB (von Helversen & Rheinlaender, 1988).

The orientation pattern is similar both in crickets and bushcrickets and can be regarded as fundamental for the acoustic orientation of orthopteran insects (c.f. zigzagging behaviour to pheromone trails by moths. All the studies that have been made so far reveal a pronounced zigzag behaviour while orienting to sound, which may also be interpreted as a means to overcome the frontal zone of ambiguity, which in *C. biguttulus* is in the order of ± 10x (Fig.

7, see also Latimer & Lewis, 1986). Furthermore, we should keep in mind that these studies were performed under laboratory conditions, with a completely flat experimental arena whereas in the natural environment the path to a signaller may be strewn with obstacles, so that an orienting insect is forced to deviate from a straight course. The phonotactic pattern we have described contrasts to other orientational procedures such as the escape response of a cockroach or prey-catching of the owl, in which precise and immediate localisation of the target (predator or prey) is absolutely necessary for survival (Camhi, 1981; Knudsen and Konishi, 1979).

Neurophysiological correlates of directional hearing

Orientation behaviour as represented at the neuronal level differs from that of pattern recognition in several aspects. Pattern recognition of species-specific characters requires a rather complicated neuronal network which seems to be located in the brain (Huber, 1983). In contrast, the correlation between directional behaviour and the way single nerve cells respond to the direction of the sound can be found at lower stations of the auditory pathway.

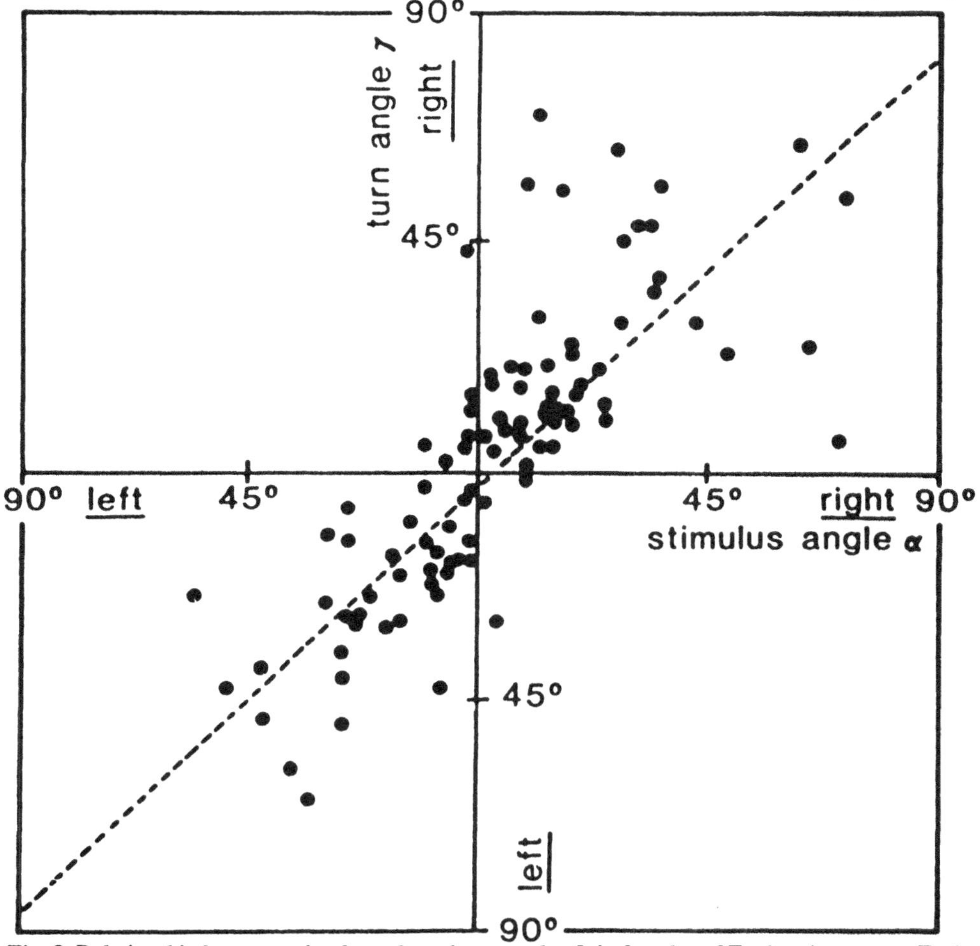

Fig. 8. Relationship between stimuls angle and turn angle of six females of *Tettigonia cantans*. Each dot represents a single turn. The correlation coefficient of the regression line (broken line) is 0.89. Note that the data indicate angular dependent orientation behaviour, in contrast to those shown in Fig. 7 (after Hardt and Rheinlaender, unpublished).

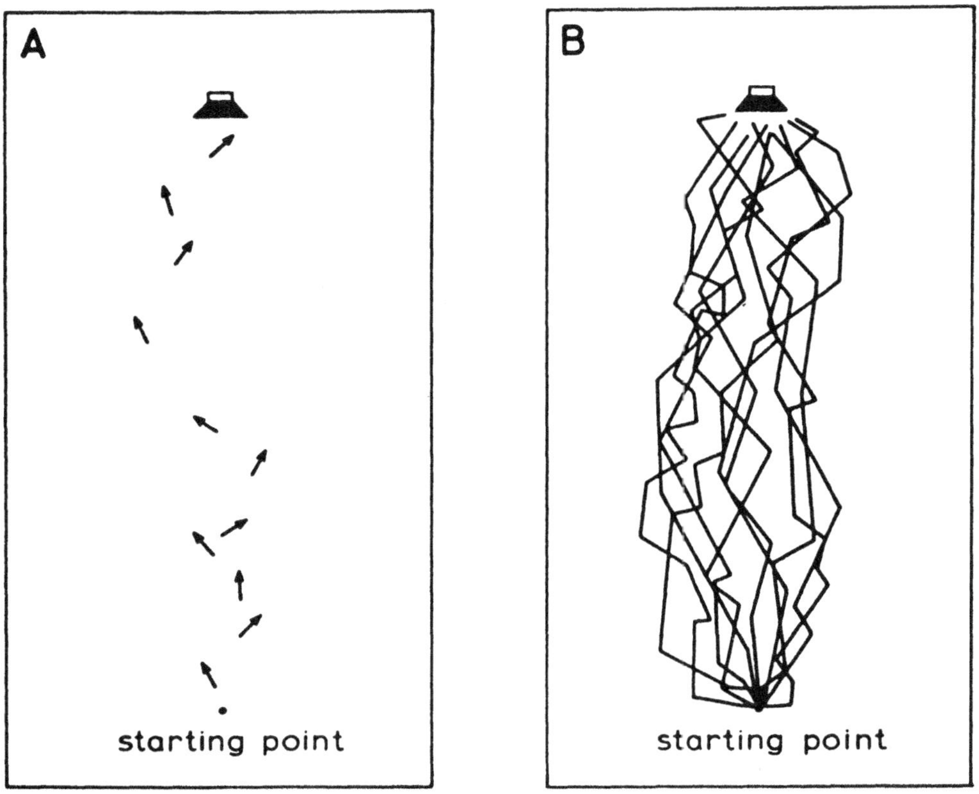

Fig. 9. Phonotactic approach pattern of a male of *Chorthippus biguttulus* to the sound source over a distance of 1 m (arrows indicate longitudinal body axis of the grasshopper just before each turn). B Accumulated date of 10 phonotactic approaches (after Rheinlaender and Dörrscheidt, unpublished).

In this article, we restrict ourselves to the description of neuronal responses at the ventral nerve cord, as here the basic mechanisms of directional hearing in insects can be found (for review of peripheral processes see Lewis, 1983). We describe first the results obtained in the undisturbed laboratory situation before considering the problem associated with directional hearing in the field.

One important advantage of the insect nervous system is the reliability with which recordings from identified interneurons can be performed. The detailed knowledge of their anatomy (for example the course of axons or main dendrites) provides, in addition, the possibility of recording simultaneously the responses from homologous neurons on the right and left side. The comparison of the paired responses gives a direct insight into the performance of the CNS in response to a directional stimulus.

Fig. 10 documents an experiment in which the responses of a pair of directionally sensitive interneurons is recorded in one animal as the sound source is moved in front of an insect from one side to the other (the anatomy of these interneurons in the prothoracic ganglion of the bushcricket is given in the insets). The responses of each neuron reach a maximum for sound from the ipsilateral side, and a minimum response occurs for sound from the opposite side with a strong response decrement in the frontal zone. It is remarkable that the dynamic range of the responses is within a sector of ±30°. The pair of neurons exhibits symmetrical response characteristics, and the curves of the averaged responses intersect with frontal sound

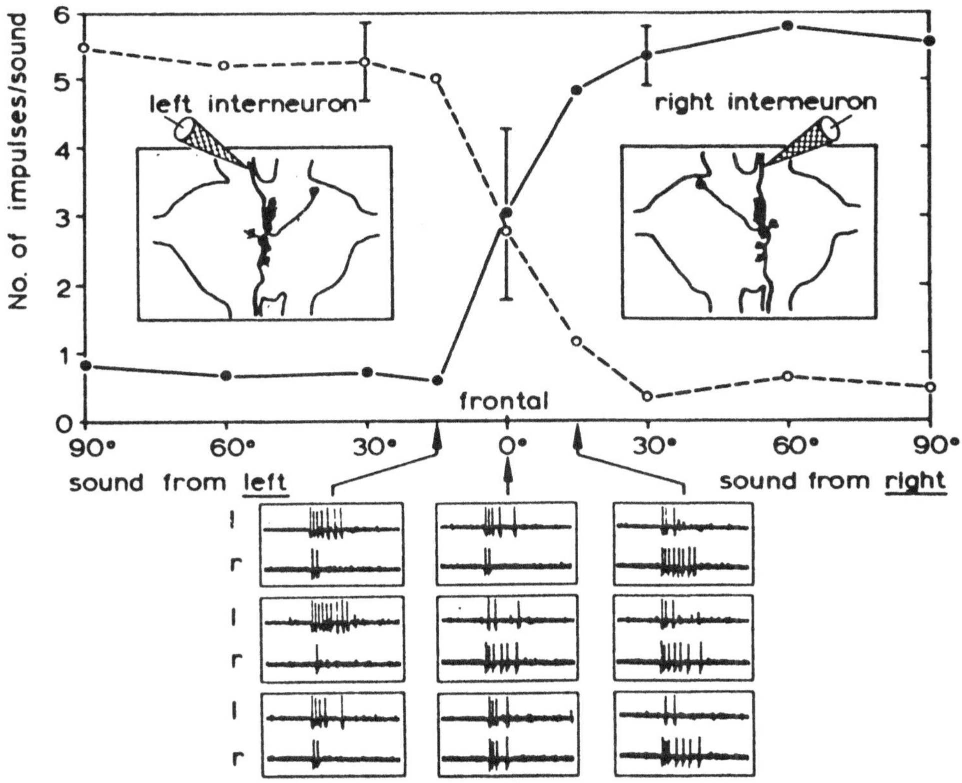

Fig. 10. Averaged directional responses of a pair of auditory interneurons at different angles of sound incidence. The insets show the anatomy of the two neurons in the prothoracic ganglion. Note the steep slope of the two curves close to frontal stimulation and the flat, angular-independent characteristics at more lateral speaker positions. In contrast to the averaged data, the two neurons rarely discharge symmetrically with frontal stimulation (see spike recording examples at 0°) (after Rheinlaender and Römer, 1980).

presentation (speaker position 0°). This is exactly what we would expect from a system that has evolved to discriminate right from left stimuli.

The averaged data in Fig. 10 tend to suggest a perfectly symmetrical response to frontal stimulation. However, when we compare single pairs of responses separately, this is rarely the case as shown by the three examples in this figure (Rheinlaender & Römer, 1980). More often, the two neurons discharge asymmetrically with either the one or the other leading in response magnitude. This "unstable" response changes as soon as the sound source is moved by only 15° laterally, with the ipsilateral neuron now firing with high reliability much more strongly than its counterpart. These results correlate with the orientation behaviour of the insect, in that with frontal stimulation the animals turn with equal probability to either side (compare with Fig. 7), whereas with stimulation outside this frontal zone, the insect moves consistently to the more stimulated side.

The phenomenon of lateral inhibition is the basis for this pronounced right-left discrimination found in these neurons. This inhibition appears when we record the synaptic activity of directionally sensitive interneurons with an intracellular electrode (Römer *et al.*, 1981). Representative examples of two such neurons in the CNS of the locust are given in Fig. 11. Both neurons originate in the metathoracic ganglion of the ventral nerve cord, which is the first level of synaptic processing in the locust's auditory pathway. They are both excited when the sound stimulus is on the ipsilateral side and inhibited by sound from contralateral. The inhibitory potential (see downward deflection of the membrane potential in Fig. 11) results in the pronounced decrease of the spike response, as was already observed in the extracellular recording of Fig. 10. However, the temporal relationship between excitation and inhibition differs in the two neurons, and can be seen most clearly with frontal stimulation. In the example given in Fig. 11A, the excitation precedes the inhibition, so that only spike number but not the response latency is affected by sound direction. In contrast, the example of Fig.

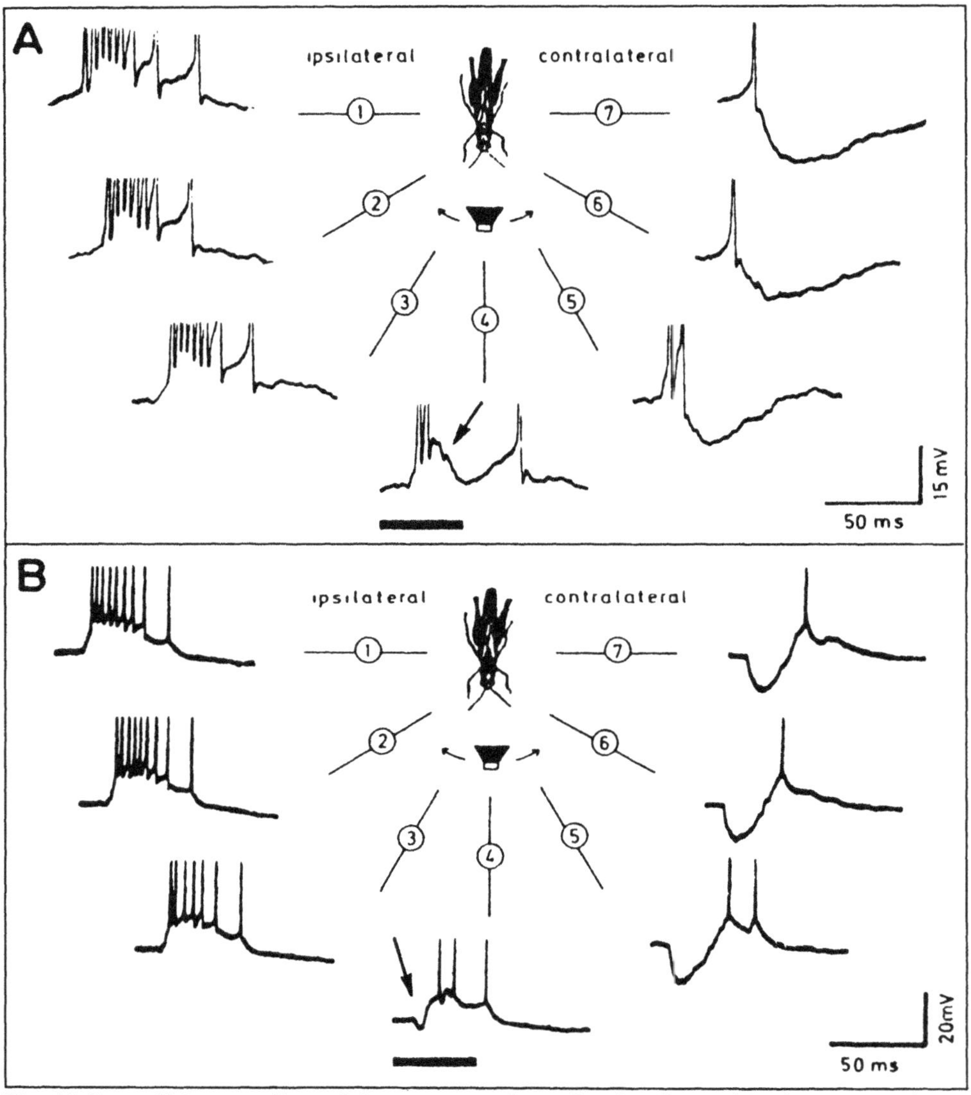

Fig. 11. Intracellular recordings of the responses of two directionally sensitive interneurons in the metathoracic ganglion of the locust with the sound source rotated in 30° steps in front of the animal. Both neurons are excited ipsilaterally, and inhibited by contralateral sound. In A the inhibition (negative deflection of membrane potential) follows the exitation, whereas in B this temporal relationship is reversed (see first appearance of inhibition with frontal speaker position in both examples). The black bar represents the sound stimulus (after Römer *et al.*, 1981).

11B shows the inhibitory potential preceding the excitatory potential at corresponding speaker positions, arrowed in the figure. Hence, both spike count and the latency of response are dependent on sound direction. These examples represent two basic modes of synaptic interaction between excitation and inhibition. They are found in insects as well as in vertebrates and can be regarded as fundamental features of the directional processing in the CNS.

Directional hearing outdoors

Of course, localisation of a signaller in the natural environment is more complex than that in the laboratory. In order to understand the performance of auditory nervous processing under natural conditions two experimental strategies may be followed. One is to imitate the complex acoustic conditions of a habitat in the laboratory, for example by installing plants into the experimental arena or by adding biotic and abiotic noise in addition to the signal. However, such a procedure is very hard to perform in that the experimental paradigm is but a rough mimic of the natural situation, and the effort for copying the relevant environmental conditions is enormous. We have therefore chosen a second possibility. By using a portable recording unit for a pair of nerve cells, we were able to study the processing of directional information in the field (Rheinlaender & Römer 1986). In the following we present one example showing the insect's capacity to encode directional information in its habitat.

Fig. 12 shows the directional characteristics of a pair of auditory interneurons recorded in dense bushland at a distance of 10m from the sound source. (It is the same pair of neurons we have used in the laboratory experiments, compare with Fig. 10). The vegetation was about 1.5m high and we simulated three receiving positions, at ground level, at 0.75m and near the top of the plants at 1.5m. At each of these positions we turned the preparation in the horizontal plane, from 90° right to 90° left in order to examine the responses of the two neurons at different angles of sound incidence. With the preparation on ground, the two neurones were unable to distinguish the direction of the sound: there is no systematic change of the neuronal responses regarding different angles of sound incidence. However, each sound pulse elicits a response and we can therefore conclude that although the insect perceives the signal it is unable to localise it. This situation changes when the preparation is raised by 0.75 m above ground, where despite the recording position being within the vegetation, the right-left difference in spike number shows a clear preference for the neuron at the stimulated side. With a height of 1.5 m above the ground the crossing point of the curves (see arrow) now coincides with frontal stimulation. The result demonstrates that with increasing height of the receiver in the vegetation the precision of right-left discrimination increases. This finding correlates with field observations, in which phonotactically orienting bushcrickets moved to their target at the upper levels of the vegetation. This will however, expose the orienting insect to greater levels of predation.

Although these results exhibit some important features of insect hearing in the field, their general applicability is limited in that vegetation, topographic characteristics and local climatic conditions differ between habitats. The physical condition will have important consequences to the quality of sound propagation (for review see Michelsen, 1978). Using a similar portable recording device as described here, Römer and Bailey (1986) reported the maximum hearing distance of an Australian bushcricket differing between insects in the same habitat by a factor of 3 (see Fig. 1, this chapter).

Conclusions

Most studies on both the acoustic behaviour of insects and the nervous processing within the auditory pathway have been carried out in the laboratory. Invariably the stimuli used in such experiments have been artificial, representing sounds which the insects never experience in nature and with background noise levels deliberately reduced. It is from these kinds of studies that we can learn how an insect is able to perform under an ideal set of circumstances. For example, what are the minimum detectable differences in intensity or time the auditory

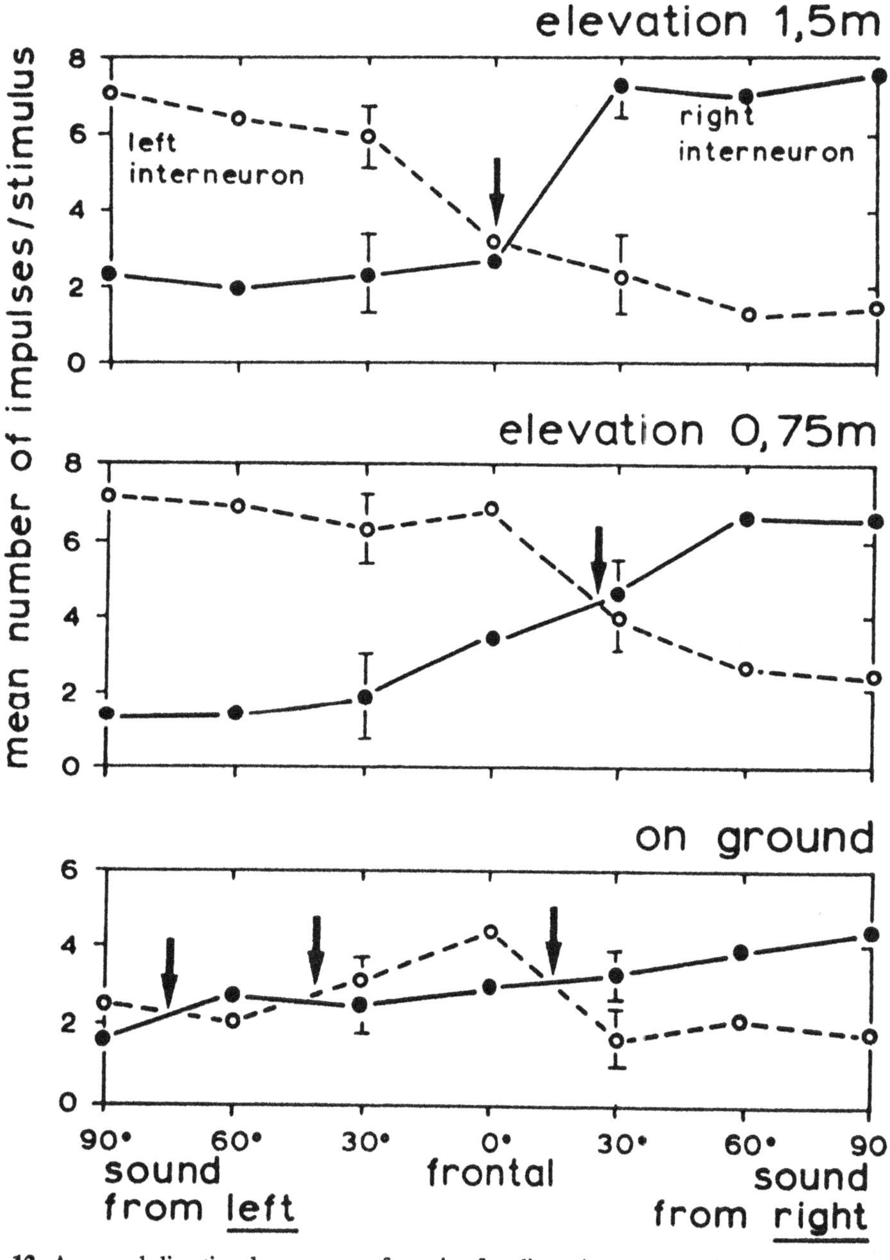

Fig. 12. Averaged directional responses of a pair of auditory interneurons (anatomy see Fig. 10) recorded in a bushland habitat at different elevations of the preparation above the ground (distance between sound source and the preparation 10 m). Note the loss of directional sensitivity on the ground (after Rheinlaender and Römer, 1986).

pathway can discriminate. Similar experiments conducted under natural conditions might show a far greater variation in discriminating ability when compared to the laboratory situation. Furthermore, we should keep in mind that it was under the disturbed natural conditions that both the physiological and anatomical properties of nerve cells evolved. Therefore, we must perform both field and laboratory experiments concurrently in order to achieve a clearer understanding of such a classical communication system.

REFERENCES

Bailey W.J. & D.R. Thiele. 1983. Male spacing behaviour in the Tettigoniidae: An experimental approach. In: Orthopteran Mating Systems: Sexual Competition in a Diverse Group of Insects. Eds. D.T. Gwynne & G.K. Morris. Westview Press, Boulder, Colorado. pp 163-184.

Bailey W.J. & R.O. Stephen. 1984. Auditory acuity in the orientation behaviour of the bushcricket *Pachysagella australis* Walker (Orthoptera, Tettigoniidae, Saginae). Anim. Behav. 32: 816-829.

Bailey W.J. & P. Thomson. 1977. Orientation in the cricket *Teleogryllus oceanicus*. J. Exp. Biol. 67: 61-75.

Blauert J. 1983. Spatial Hearing. MIT Press, Cambridge, Mass.

Brenowitz E.A., W. Wilczinski & H. Zakon. 1984. Acoustic communication in spring peepers. Environmental and behavioural aspects. J. Comp. Physiol. 155: 585-592.

Camhi J.M. 1981. Die Fluchtreaktion der Schabe. Spek. Wiss. 244: 111-120.

Esch H., Huber F., & D.W. Wohlers. 1980. Primary auditory neurons in crickets: physiology and central projections. J. Comp. Physiol. 137: 27-35.

Hardt M. 1988. Zur Phonotaxis von Laubheuschrecken: Eine vergleichende verhaltensphysiologische und neurophysiologisch/neuroanatomische Untersuchung. PhD Thesis, Bochum.

Helversen von D. & J. Rheinlaender. 1988. Interaural intensity and time discrimination in an unrestrained grasshopper: a tentative behavioural approach. J. Comp. Physiol. 162: 333-340.

Horridge G.A. 1960. Pitch discrimination in Orthoptera (Insecta) demonstrated by responses of central auditory neurones. Nature 185: 623-624.

Huber F. 1983. Neuronal correlates of orthopteran and cicada phonotaxis. In: Neuroethology and behavioural physiology. Eds. F. Huber & H. Markl. Springer Verlag, Berlin, Heidelberg, New York, Tokyo. pp 108-135.

Kalmring K., B. Lewis & A. Eichendorf. 1978. The physiological characteristics of the primary neurons of the complex tibial organ of *Decticus verrucivorus* L. (Orthoptera, Tettigoniidae). J. Comp. Physiol. 127: 109-127.

Keuper A. & R. Kühne. 1983. The acoustic behaviour of the bushcricket *Tettigonia cantans*. II. Transmission of air-borne sound and vibration signals in the biotope. Behav. Processes 8: 125-145.

Keuper A., K. Kalmring, A. Schatral, W. Latimer & W. Kaiser. 1986. Behavioural adaptations of ground living bushcrickets to the properties of sound propagation in low grassland. Oecologia 70: 414-422.

Knudsen E.I. 1981. The hearing of the barn owl. Sci. Am. 245: 82-91.

Knudsen E. & M. Konishi. 1979. Mechanisms of sound localisation in the barn owl (*Tyto alba*). J. Comp. Physiol. 113: 13-21.

Latimer W. & D.B. Lewis. 1986. Song harmonic content as a parameter determining acoustic orientation behaviour in the cricket *Teleogryllus oceanicus* (Le Guillou). J. Comp.

Physiol. 158: 583-591.
Lewis D.B. Directional cues for auditory localisation. In: Bioacoustics. Ed. D.B. Lewis. Academic Press, London, New York, Paris.
Michelsen A. 1971. The physiology of the locust ear. I. Frequency sensitivity of single cells in the isolated ear. J. Comp. Physiol. 71: 49-62.
Michelsen A. 1978. Sound reception in different environments. In: Sensory Ecology. Ed. M.A. Ali. Plenum, New York, London. pp 345-373.
Michelsen A. & O.N. Larsen. 1983. Strategies for acoustic communication in complex environments. In: Neuroethology and Behavioural Physiology. Eds. F. Huber & H. Markl. Springer Verlag, Berlin, Heidelberg, New York, Tokyo pp 321-331.
Murphey R.M. & D. Zaretzky. 1972. Orientation to calling song by female crickets, *Scapsipedus marginatus* (Gryllidae). J. Exp. Biol. 56: 335-352.
Oldfield B. 1980. Accuracy of orientation in female crickets, *Teleogryllus oceanicus* (Gryllidae): dependence on song spectrum. J. Comp. Physiol. 141: 93-100.
Oldfield B. 1982. Tonotopic organization of auditory receptors in tettigoniidae (Orthoptera Ensifera). J. Comp. Physiol. 147: 461-469.
Oldfield B. 1983. Central projections of primary auditory fibers in Tettigoniidae (Orthoptera: Ensifera). J. Comp. Physiol. 151: 389-395.
Oldfield B.P., H.-U. Kleindienst & F. Huber. 1986. Physiology and tonotopic organization of auditory receptors in the cricket *Gryllus bimaculatus* DeGeer. J. Comp. Physiol. 159: 457-464.
Rheinlaender J. 1975. Transmission of acoustic information at three neuronal levels in the auditory system of *Decticus verrucivorus* (Tettigoniidae: Orthoptera). J. Comp. Physiol. 97: 1-55.
Rheinlaender J. & H. Römer. 1980. Bilateral coding of sound direction in the CNS of the bushcricket *Tettigonia viridissima* L. (Orthoptera, Tettigoniidae). J. Comp. Physiol. 140: 101-111.
Rheinlaender J. & H. Römer. 1986. Insect hearing in the field. I. The use of identified nerve cells as "biological microphones." J. Comp. Physiol. 158: 647- 651.
Römer H. 1976. Die Informationsverarbeitung tympanaler Rezeptorelemente von *Locusta migratoria* (Acrididae, Orthoptera). J. Comp. Physiol. 109: 101-122.
Römer H. 1983. Tonotopic organization of the auditory neuropil in the bushcricket, *Tettigonia viridissima*. Nature 306: 60-62.
Römer H. 1987. Representation of auditory distance within a central neuropil of the bushcricket *Mygalopsis marki*. J. Comp. Physiol. 161: 33-42.
Römer H., J. Rheinlaender & R. Dronse. 1981. Intracellular studies on auditory processing in the metathoracic ganglion of the locust. J. Comp. Physiol. 144: 305-312.
Römer H. & W.J. Bailey. 1986. Insect hearing in the field. II. Male spacing behaviour and correlated acoustic cues in the bushcricket *Mygalopsis marki*. J. Comp. Physiol. 159: 627-638.
Römer H., V. Marquardt & M. Hardt. 1988. The organization of a sensory neuropile in the auditory pathway of two groups of Orthoptera. J. Comp. Neurol. 275: 201-215.
Schwabe J. 1906. Beiträge zur Morphologie und Histologie der tympanalen Sinnesapparate der Orthopteren. Zoologica 50: 1-154.
Wilczynski W. & E.A. Brenowitz. 1988. Acoustic cues mediate inter-male spacing in a neotropical frog. Anim. Behav. 36: 1054-1063.
Zhantiev R.D. 1971. Frequency characteristics of tympanal organs in grasshoppers (Orthoptera, Tettigoniidae). Zool. Zh. 50: 507-514.

14. SPECIATION IN THE GENUS *MYGALOPSIS* IN SOUTHWESTERN AUSTRALIA:
A New Look at Distinguishing Modes of Speciation

William Barendse

Introduction

Biological diversity is generated fundamentally by the processes of evolution within lineages. This evolution of a lineage consists of two processes, the first is speciation or cladogenesis which is the splitting of one species into several daughter species, and the second is anagenesis which is the genetic change within a species over time (Rensch, 1956; Mayr, 1963). Although these two processes are related, the degree of anagenetic change, which does occur in a lineage, is not generally related to the amount of speciation. This division is in some ways artificial since speciation is not thought of as an instantaneous event and is a change in the genetic structure of the population over time, but it is heuristic because it separates two intrinsically different processes in the production of biological diversity.

This relation between these two processes is the basis of several problems in the study of speciation. If speciation is considered to be the acquisition of genetic – including behavioural – barriers to gene flow between two populations then the study of anagenesis does little to enhance our knowledge of the processes involved. Since the event is likely to occur over a period of time longer than the lifespan of any observer, and the pairs of incipient species that are going to complete speciation are unlikely to be correctly chosen by any observer, the chance of direct observations being made on the changes over time in the relatively rare process of speciation is largely non-existent. This leaves only the inference of modes of speciation from the population genetic structure, geographic distributions, and natural histories of the species under consideration.

The diverse evidence used in the inference of the mode of speciation are generally orchestrated within a lineage using the comparative method (e.g. Mayr, 1947; Hairston and Pope, 1948; Mayr, 1963; Endler, 1977; White, 1978). Levels of divergence for each data type are ranked and then the most plausible mode of speciation is chosen within the limits set by the appropriate population genetic theories. However, there are several flaws in the comparative method. First, no single case of speciation can be individually studied since the method relies upon the ensemble of evidence with the data points being the differences between species. Second, if the limits set by the population genetic theories are changed, or different theories are used, more and qualitatively different modes of speciation become plausible for the same data (Bush, 1975; Endler, 1977; Templeton, 1980a). The comparative method, although a great success in providing a framework within which speciation can be discussed, fails to satisfy when questions are posed about the mode of speciation in any particular instance.

The alternative way of viewing the problem is to admit that there are many modes of speciation which make predictions and have consequences for the genetic structures, geographic distributions, and natural histories of the species under consideration. Furthermore, a cluster of these predictions can lead to the distinguishing of modes of speciation in the appropriate species (Bush, 1975; Templeton, 1980b). As an example, imagine two species that are distributed allopatrically to each other, that is, no parts of the geographic ranges of these two species overlap. These species are less likely to have arisen by a mode of sympatric speciation, that is, with one of the daughter species forming due to the exploitation of a niche that occurs within the range of the parental species (Maynard Smith, 1966; Bush, 1969). In this example, all the processes of speciation occur while there is the potential for interbreeding between the two forms that occupy the different niches. Sympatric speciation is unlikely because of the many more steps that have to be proposed including the range movement of the two new species from a condition of overlapping ranges to that of completely separated geographic ranges. In combination with other concurring predictions, an origin through sympatric speciation could be ruled out for this example. The major flaw of this method is that in many instances no evidence would be able to discriminate between several modes of speciation.

A good study of speciation, then, is one which so marshals population genetics, geographic distributions, geological change, climatology and natural history that all but one mode of speciation can be rejected as being the cause of speciation. Naturally, different evidence is required to separate allopatric speciation from parapatric speciation than is required to separate allopatric speciation from sympatric speciation. In addition, groups should be screened beforehand so as to minimize the chance that the mode of speciation will remain unresolved.

One promising group for such an analysis is the genus *Mygalopsis*, which consists of five species, *M. thielei*, *M. sandowi*, *M. pauperculus*, *M. marki* sp. A and sp. B (Bailey, 1979; Sandow, 1980; Thiele, 1981; Dadour & Johnson, 1983; Barendse, 1984, 1987). The species are copiphorine tettigoniids which occur in the southwestern part of Western Australia. They are characterized by low potential vagility due to brachyptery and non-saltatorial hindlimbs, crypsis, and by occurring as geographical replacements of one another, generally in the coastal districts. This is sufficient to suggest that sympatric speciation is unlikely since it would require more steps of divergence to provide geographical replacements than would other modes of speciation. The presence of (1) a hybrid zone, (2) one of the few true parapatric zones – a zone where the ranges of the two species meet but do not overlap even though the individuals may be able to move easily across the zone (Endler, 1977; White, 1978) – in insects, and (3) very little range overlap occurring between the other species, strengthens this conclusion. In addition, there is no evidence of niche separation and little genetic differentiation between seasonal morphs within *M. marki* sp. A (Dadour & Johnson, 1983; Barendse, 1987), the most well studied of these species. This leaves some combination of allopatric and parapatric speciation. Here allopatric speciation is considered as speciation that occurs in total geographic isolation between the pairs of incipient species, usually with a geographic barrier between them (Mayr, 1963). Parapatric speciation is considered as speciation that occurs while the ranges of the two species are in contact but do not overlap so that speciation occurs in the face of gene exchange between the incipient species (Endler, 1977; White, 1978). Speciation then occurs through the development of genetic barriers to gene flow, barriers caused by selection favouring two different morphs in two different parts of the species range and selection against the intermediates. Over time there may be some combination of these two modes of speciation, due to fluctuations in the ranges of the incipient species.

Only a very limited repertoire of evidence can distinguish between allopatric and parapatric speciation (Bush, 1975; Endler, 1977; Templeton, 1980b). Allopatric speciation requires only that the two species are free to diverge because there is an absence of gene exchange between the incipient species, this absence being provided by geographic isolation (Mayr, 1963). Speciation is recognized after incipient species diverge sufficiently so that either gene exchange between populations is not possible or when allopatric isolates are so divergent in their genotypes that it is unlikely that gene exchange would occur between them (Mayr, 1963). The divergence may occur in many different ways.

Alternatively, parapatric speciation in general requires that groups of populations are pushed by selection pressure in opposite adaptive directions (e.g. Ehrlich & Raven 1969), that this leads to differences in fitness so that natural selection occurs against the intermediates. The zone of intermediates then becomes a hybrid zone and later a contact zone between two species (Endler, 1977). This is only possible if there is a low level of gene flow between populations, as a high level of gene flow would swamp the effects of selection and the species could become in effect a panmictic unit. There are thus two ways in which allopatric and parapatric speciation could be distinguished, that is, through the analysis of gene flow and population genetic structure and through an analysis of the origins of any hybrid zones. The analysis of selection pressures would be irrelevant since allopatric speciation could occur through divergence produced by selection while the populations were in isolation. Further, the analysis of hybrid zones is weak evidence since the formation of a hybrid zone is unlikely to be directly observed.

In this context, the natural history of each species of the genus *Mygalopsis* shows that aggregations are not continuous, but are rather discrete populations. Since the individuals are brachypterous these aggregations can be seen as stepping stones along which gene flow could occur. Therefore, there is great potential for the species to be subdivided into races if the gene flow is of the appropriate magnitude (Wright, 1940, 1943; Kimura, 1955; Kimura & Weiss, 1964; Endler, 1977).

However, the differentiation between hypotheses requires more precise considerations since there is a range of optimum values within which parapatric speciation occurs while certain population structures may make it less likely or impossible. The data necessary for such a study are (1) the historical sequence of speciation so that appropriate pairs of species are compared, (2) the level of gene flow within species and their population structure, (3) an analysis of the hybrid zone which may provide supporting evidence, and (4) a survey of the geology and geography of the appropriate regions. The substance for the subsequent discussion is essentially that in Barendse (1987). Only the crucial details are reported here.

The Historical Sequence of Speciation

Knowing the historical sequence of speciation is important for two very separate reasons. Firstly, if the sequence is known then closeness of origin can be compared to proximity of the geographic range. If the most closely related species are the geographically most close then the simplest explanation is that the speciation has been due to the fragmentation of an ancestral species range (e.g. Hillis, 1985). Such a pattern is indispensable for parapatric speciation, while allopatric speciation can be associated either with fragmentation or with other patterns of the formation of species. Secondly, the sequence also determines which species are most closely related and directs attention away from spurious comparisons between species, as well as from spurious mechanisms of or possible geological factors in speciation (Brundin, 1972; Ball, 1976; Platnick & Nelson, 1978; Cracraft, 1982).

The historical sequence of speciation was determined using allozyme data which were coded so that presence or absence of each allele for each species was noted. The characters that were shared between two or more species were found and then used to nest the species in inclusive groups to minimize the amount of homoplasy (Wiley, 1981). Morphological characters were not used in conjunction with this analysis firstly because of the great similarity in morphology among the species and secondly because a phylogenetic analysis shows great stability for a range of types of data (Mickevich & Johnson, 1976; Mickevich, 1978; Sokal, 1983a,b). Two shortest trees were derived and these were shown to be significantly shorter than all other possible trees for these taxa (Barendse, 1987). These tests of significance are based upon the jack-knife and bootstrap (Efron, 1979; Efron & Gong, 1983; Felsenstein, 1985) methods of resampling the data so that a distribution of tree lengths could be obtained (see Barendse, 1987). The two trees differ in only one detail, that is in the order in which *M. sandowi* and *M. thielei* enter the tree (Figure 1). Both show that the geographically most proximate species are the most closely related without exceptions. The biogeographical interpretation for such a pattern would be that an ancestral range fragmented, and the difference between the two trees may indicate that the ancestors of some of the species – *M. pauperculus*, *M. sandowi*, *M. thielei* – arose through a three way split.

These events can be placed in temporal perspective by the estimation of divergence times from genetic divergence between species. Nei's (1978) unbiased genetic distances (D) were calculated, averaged within species, and then related to divergence time by two formulas. The

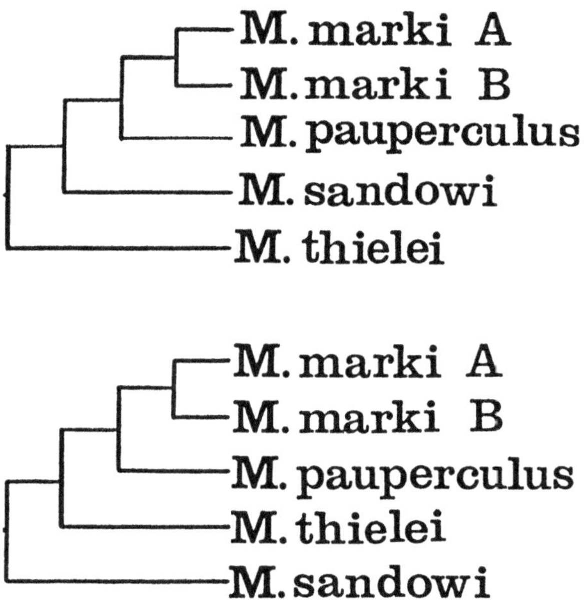

Fig. 1. The two shortest cladograms for the genus *Mygalopsis*. They are both 84 steps long and are significantly shorter than all other possible cladograms.

first gives a very slow rate of divergence and is determined from correlations of biogeographical events, fossils, and minimum possible dates for speciation with genetic divergence in vertebrates (e.g. Case *et al.*, 1975; Sarich, 1977; Maxson & Maxson, 1979) while the second shows a fast rate of divergence and is determined from similar correlations of genetic divergence in Hawaiian *Drosophila*. The upper limit was calculated from 1D = 18 m.yrs b.p. (Avise & Aquadro, 1982) and the lower limit from 1D = 2 m.yrs. b.p. (Carson, 1976). The divergence times and genetic distances show two major pulses of speciation. The first occurred sometime between 10 and 2 m.yrs b.p., and produced the ancestors of *M. sandowi*, *M. pauperculus*, and *M. thielei*, with the ancestors to the two *M. marki* species produced between 8 and 1 m.yrs b.p. The second pulse occurred much later, producing two species which are morphologically indistinguishable, *M. marki* sp. A and sp. B sometime between 1 m.yrs b.p. and 100 000 yrs b.p.

The Analysis of Gene Flow

The analysis of gene flow to distinguish between parapatric and allopatric speciation is no simple matter. Although allopatric speciation would be associated with high levels of gene flow and parapatric speciation with lower levels, the occurrence of parapatric speciation does not simply follow from lower gene flow. The population structure must be such that the changes due to directional selection will accrue in groups of populations. Two factors have a direct bearing upon this effect. Populations with small effective population sizes will dissipate these changes because of random sampling of gametes from one generation to another (Wright, 1969; Rohlf & Schnell, 1971; Endler, 1977; Wright, 1978; Cohan, 1984; Wade, 1984). The movement along an adaptive landscape may be facilitated, but this movement will not occur for groups of populations if the level of gene flow between populations is too low. Many populations would then find separate adaptive peaks, races would be unlikely to form and parapatric speciation would be correspondingly unlikely.

The measurement of gene flow is difficult because the level of interpopulation migration or the proportion of migrants in a population cannot be easily ascertained. This difficulty arises because tagging and following animals does not indicate the distance that the genes travel in one generation. Moreover, effective population size is an important consideration and it is even more difficult to measure because the effective population size is only loosely represented by the actual census and depends upon the variance in reproductive success and the ratio of successfully reproducing males to females in the population (Wright, 1940; Crow & Kimura, 1970). It is usually lower than the census.

Fortunately, gene flow is relatively easy to measure as Nm, the product of effective population size (N) and the proportion of migrants (m). Nm has been subject to several theoretical analyses and can be inferred from the distribution of alleles and allele frequencies in the different populations of a species. When Nm is large, that is, both N and m are large, so that Nm is of the order of 10 to 50, then parapatric speciation is unlikely because the species is virtually one panmictic unit. When Nm is very low (less than 0.5), then each population, irrespective of the population size or the proportion of migrants, diverges independently (Wright, 1940). If N is small, that is less than 200, then it is likely that part of this divergence will be due to the sampling of gametes from one generation to another (Wright, 1948, 1965; Rohlf & Schnell, 1971). This effect is the stronger the smaller the size of the population. Parapatric speciation is unlikely, since groups of populations are unlikely to diverge in the same adaptive direction. Therefore, between the values of Nm of 0.5 and 10 and in populations with sizes much greater than 100, parapatric speciation is possible, although allopatric speciation cannot be rejected for these values of Nm. Allopatric speciation cannot be rejected by a study of gene flow.

The level of gene flow was determined from allele frequencies in the two most recently formed species, *M. marki* sp. A and sp. B, using Slatkin's methods (1980, 1981, 1985; see Barendse, 1987). Essentially this requires that for all polymorphic loci examined in the species, the conditional average frequencies of its alleles are calculated – the frequency of the allele averaged only over those populations in which it occurs – and that these frequencies are graphed as a function of the proportion of populations in which they occur. This graph expresses the level of gene flow and the value of the gene flow is obtained by comparing the shape of this curve with the expected curves for different levels of gene flow. For these two species, the observed level is compared with the expected curves for different levels of gene flow in Fig. 2.

The level of gene flow found between these populations of *M. marki* sp. A and sp. B is unlikely to be greater than 0.25 which is equivalent to 1 migrant every 4 generations. At such a level of gene flow, independent differentiation of populations is encouraged. Further, the aggregations of calling males usually have less than 100 males calling per night and sometimes fewer than 10. These densities are stable over time and individuals hold their station for weeks during the breeding season (Thiele, 1981). This indicates that the effective size of the population is likely to be low. The actual value cannot be ascertained since these

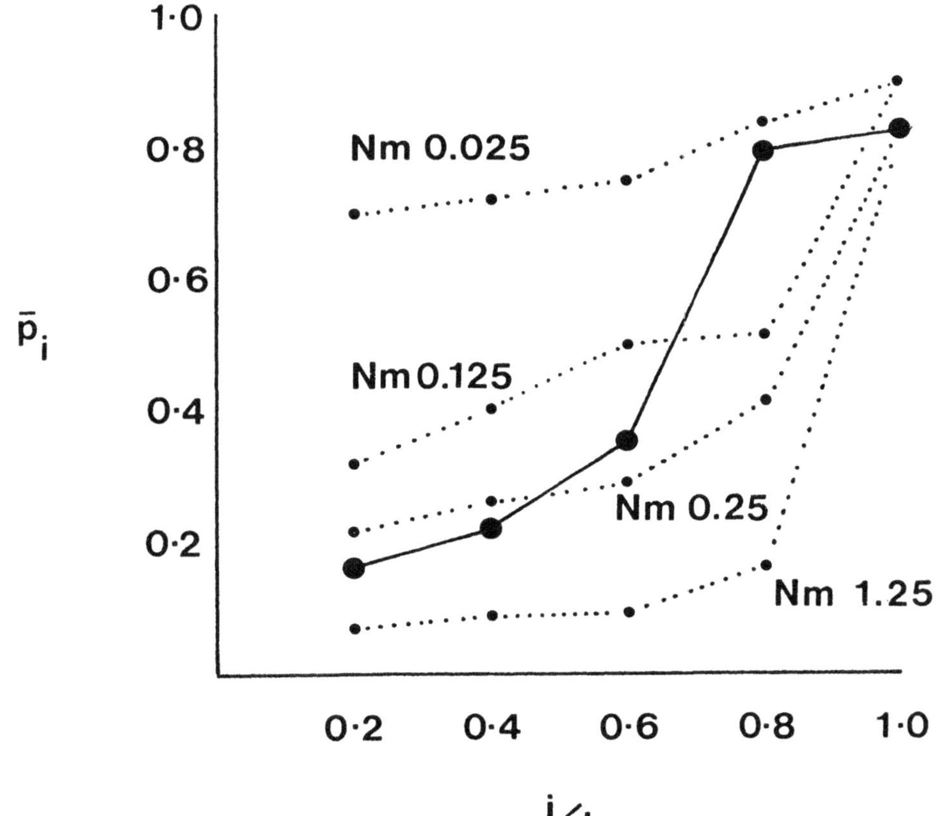

Fig. 2. The conditional average frequency of an electromorpg (p) as a function of the proportion of populations (i/d) in which it occurs. The expected curves for different values of Nm are included (Slatkin 1981).

are cryptic species and the reproductive performance of each individual could not be calculated. However, with so few animals calling it is unlikely that the effective population size is greater than 100 and may be much lower. At such low population sizes, the stochastic sampling of gametes from one generation to another can vitiate much directional selection, even large selection coefficients may promote divergence between local populations (Cohan, 1984), and it is unlikely that races could form.

This low level of gene flow can be verified by other means. A low level of gene flow encourages chaotic differentiation of populations so that the similarity between populations are unlikely to be related to the geographic distance between them. Alternatively, if the gene flow is large enough and selection is effective, then there will be a relationship between genetic similarity of populations and their geographic proximity (e.g. Schaal, 1974; Nozawa *et al.*, 1975; Selander & Kaufman, 1975; Johnson & Brown, 1980). After the allele frequencies for the populations of *M. marki* sp. A and sp. B were transformed into Nei's (1978) genetic identities and the distances between these samples were measured and correlated to each other, no relationship was found between genetic identity and geographic distance although there are significant differences between some of the genetic identities. Thus, the divergence of populations appears chaotic (Barendse, 1987).

Other tests of low levels of gene flow, small population size, and chaotic divergence between populations can be performed at a microgeographic level.

The Levels of Microgeographic Differentiation

The factor which plays the most important part in determining the scope for differentiation at a microgeographic level is the size of the neighbourhood. The neighbourhood is the area within which the parents for the next generation will be drawn, although the matings need not be at random (Wright, 1946, 1951). If the neighbourhood includes only a few individuals then the degree of stochastic sampling of genes will be high. A large degree of differentiation may then be found within an aggregation that is large enough to contain several neighbourhoods. In *Mygalopsis*, where the differentiation between populations appears chaotic, the differentiation within a local aggregation may show a relationship with distance. Moreover, with small effective population size and low gene flow, significant differentiation could be expected on a microgeographic scale. An estimation of the size of the neighbourhood and the differentiation in gene frequencies within a population could confirm the importance of stochastic differentiation in these species.

The size of the neighbourhood cannot be easily or accurately measured by direct means. Several obstacles exist, noticeably the inability to measure the movement of genes from one generation to another in a natural population, a distance different to the movement of individuals during their lifetimes. Further, although difficulties exist in measuring the distance that an animal traverses before mating, these are increased in a cryptic species where only the calling males can be regularly and confidently observed. Lastly, the most important animals are the very few that travel the farthest, and these are the most difficult to observe.

However, the neighbourhood also corresponds to the area within which there is no significant increase in the variance in gene frequencies as a function of distance. This variance in gene frequencies is measures as Fst (Wright, 1978), the standardised variance, corrected for the sampling variance due to sample size. To estimate the variance in allele frequencies and measure the change in variance as distance increases, a large high-density aggregation was subsampled. Ten samples were taken so that 'holes' were produced in the aggregation of calling males when all the individuals where removed from that part of the aggregation. These 'holes' remained over the two days of sampling, suggesting that the numbers of non-calling males was low. Thiele (1981) found that such holes could remain for extended periods. The distances between samples where measured while the numbers of individuals per square meter were counted in the most and least dense parts of the aggregate. The allele frequencies showed significant differences between sample sites and the corrected variance of the allele frequencies was also significant although there was no clinal relationship between allele frequencies and position in the aggregation (Barendse, 1987). Average values for Fst were calculated by averaging values of Fst between samples when the samples were within the following ranges: (1) 0 to 20 m., (2) 21 to 40 m., (3) 41 to 60 m., (4) 61 to 80 m., and (5) 81 to 100 m. When the variance is plotted as a function of distance within several classes of distance (Figure 3), the variance in allele frequencies increases gradually as the distance between the samples increases. At a distance equal to a separation of 40 to 60 meters between samples, the variance increases significantly over that for samples separated by 20 to 40 meters, suggesting that the radius of the neighbourhood is approximately 20 meters. From the range of densities found at the site, the number of calling males in a neighbourhood was between 20 and 68. The effective population sizes would be correspondingly minute.

With such small neighbourhood sizes two effects can be expected. The first is a great deal of stochastic differentiation within an aggregation and the second is some inbreeding. Evidence for inbreeding is hard to obtain as it is observed as a decrease in the number of heterozygotes in natural populations but there are many causes for a decrease in the observed number of heterozygotes. These are (1) the sampling of several neighbourhoods that have different gene frequencies – the Wahlund's effect, (2) spatial mixing of populations with different gene frequencies, (3) selection against heterozygotes, (4) temporal mixing of populations with different gene frequencies, (5) the sampling of cryptic species, (6) the

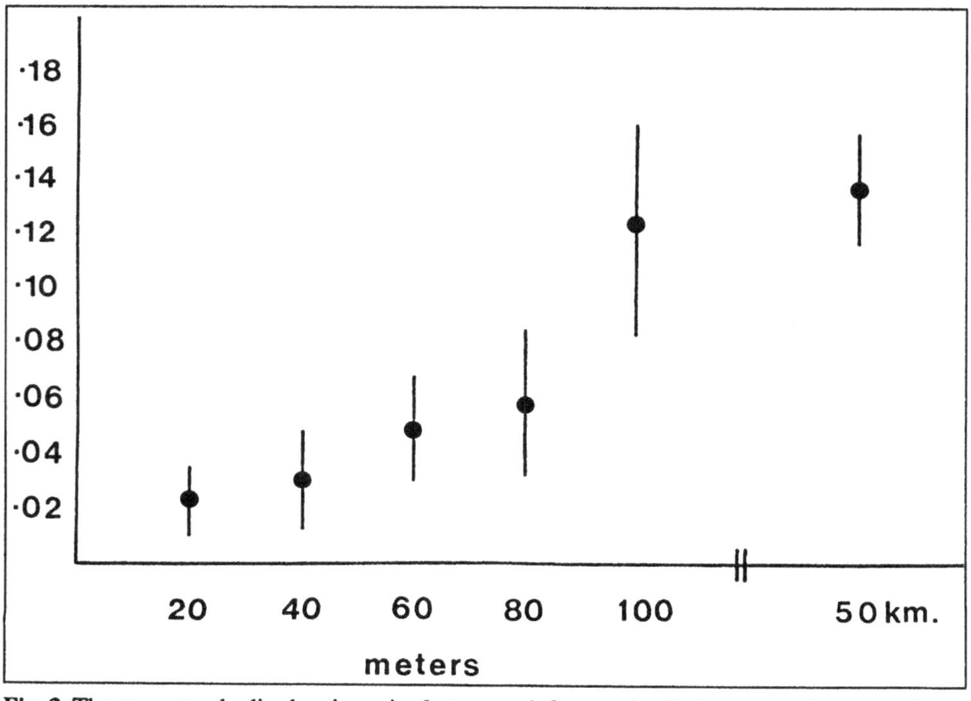

Fig. 3. The mean standardised variance in electromorph frequencies Fst between pairs of samples as a function of distance. The confidence interval of each mean is shown.

presence of null alleles, and (7) assortative mating. Except for the sampling of cryptic species, these factors affect single loci, so the presence of pervasive deficits in the number of heterozygotes simultaneously at several loci suggests inbreeding if cryptic species can be ruled out.

To calculate the deficits in the number of heterozygotes at several loci simultaneously, populations were examined in which more than five loci were polymorphic. The proportion of loci which showed deficits in the number of heterozygotes were calculated and plotted as a stem and leaf (Tukey) diagram after the proportion of heterozygotes was adjusted for bias from small sample sizes (Levene, 1949; Kirby, 1975). A sign test (Snedecor & Cochran, 1969) was applied to the data to determine if a significant majority of these populations showed a majority of loci with a deficit in the number of heterozygotes (Barendse, 1987). Such a majority is present but it is found only when all the populations are considered together. The populations from the hybrid zone do not show this phenomenon when considered by themselves, suggesting that the presence of cryptic species is not likely to be a cause of this phenomenon. Inbreeding is the only likely explanation which cannot be rejected.

The Hybrid Zone between *M. marki* sp. a and sp. a

The analysis of hybrid zones is unlikely to provide evidence for or against any mode of speciation if the hybrid zone is at equilibrium since the equilibrium characteristics of hybrid zones do not depend upon the mode of speciation (Endler, 1977). However, if the hybrid zone has not yet reached equilibrium, or it is out of equilibrium it may provide insights, along with other data, into the mode of speciation. Generally, if the hybrid zone occurs through allopatric means, it begins narrower than at equilibrium because it would begin as a contact zone. Alternatively, if the hybrid zone occurs through parapatric means, it begins broader than at equilibrium because it would begin as a zone in which there were many individuals of lower fitness. However, a zone that was broader than at equilibrium may indicate other, disturbing factors. Therefore, the time since divergence of the two species must be approximately known, and if it is greater than the time needed to go to an equilibrium width, then a hybrid zone broader than expected would indicate disturbing influences on the hybrid zone. In short, the actual width of the hybrid zone must be compared to an equilibrium width calculated from gene flow parameters and the divergence time for the two species must be compared to the time that would be necessary for the hybrid zone to reach an equilibrium width.

The hybrid zone was measured in these two morphologically indistinguishable species by sampling the allozymes of populations. One marker exists, alternative electromorphs at the phosphoglucomutase locus, and it is the only known diagnostic character.

The equilibrium width of the zone was estimated from the relationship (1) $W = l/(-h)$ (May *et al.*, 1975) and (2) $W = (2M/s)$ (Barton, 1979) in which W is the equilibrium width, l is the gene flow distance, $-h = s$ is the selection against hybrids or heterozygotes, and M is the variance in progeny position per generation. The value of s against F2 and backcross hybrids is 1.00 and there is a minimum viability selection against F1 hybrids of 0.34 (Barendse, 1987). The gene flow distance, l, is calculated from the equation $l = x(m)$ (Endler, 1977) where x is the average distance between demes and m is the proportion of each population that migrates every generation. The value of the variance in progeny position, M, can be calculated in the stepping stone model from the equation $M = xm2$ where x is the average distance between demes. Here x was estimated from the average distance between samples in the hybrid zone – an overestimate. The range of effective population sizes is between 10 and 100, the value of Nm is 0.25, so the proportion of migrants per generation is between 0.025 and 0.0025. The average distance between samples in the zone is 20.96 km (s.d. = 14.2, n = 14). Therefore, the range of equilibrium widths for the hybrid zone between *M. marki* sp.A and sp. B for the model of May *et al.* (1975) is 1.4 to 4.6 km and for the model of Barton (1979) is 0.016 to 0.16 km. A smaller population size gives a wider zone.

The actual width of the zone is between 35 and 55 km; the smaller value is the distance between the two hybrid zone populations which are farthest apart while the larger value is the distance between the two populations, one from each species, that are not hybrid populations and are the most proximate. Such a width for the hybrid zone is at least an order of magnitude larger than would be expected. Moreover, such a large width for a hybrid zone is contrary to the intuitive expectations for these species.

The time that would be necessary for a hybrid zone to reach an equilibrium width in these species if allopatric speciation occurred is inversely proportional to the gene flow distance (Endler, 1977) and would require less than one generation (Barendse, 1987). It is obvious that these species have had more than enough time for the hybrid zone between them to reach an equilibrium in width as they have been studied for more than 10 years and have an estimated divergence time of at least 100 000 years. Thus, the hybrid zone is likely to be broader than expected because of factors other than the mode of speciation.

The factor that is most likely to have caused this is the population structure of these species. There is a great deal of divergence between local aggregations, and a hybrid zone is more aptly described as a zone of hybrid populations. The zone of hybrid populations could broaden if stochastic forces are important and if selection is not strong across the zone for alternative morphs or species (Barton, 1979; Barton & Hewitt, 1985). These species are morphologically indistinguishable, and there is no ecotone which could gives rise to strong selection across the hybrid zone. Therefore, the spread of the hybrid zone is likely to occur and probably occurs when a gravid female from a hybrid population migrates to a non-hybrid population. An explanation based on the disturbance of the habitat is unlikely to be correct since the parapatric zone between *M. pauperculus* and *M. marki* sp. B occurs in a disturbed habitat and it has remained less than the width of a street.

Discussion

The two most recently formed species in the genus *Mygalopsis*, *M. marki* sp. A and sp. B, are most likely the result of allopatric speciation. The evidence indicates that they have small effective population sizes – between 10 and 100, chaotic divergence between local populations while there is significant divergence within local populations, and low levels of gene exchange of the order of 1 migrant in 4 generations between populations. These are characteristics of species that are unlikely to have groups of populations that diverge together, which suggests that parapatric speciation is unlikely for these two species. It is most likely that speciation would occur in a single population in isolation when the species range was divided and one of the parts went through a population bottleneck. The hybrid zone formed where the two new species came into contact.

The great width of the zone of hybrid populations in species with low levels of gene flow suggests that the zone has become wider since its foundation as there has been far more than sufficient time for the hybrid zone to reach equilibrium. If so, parapatric speciation is unlikely because a hybrid zone would be expected to become narrower over time. This is not strong evidence since a hybrid zone may expand if the selective constraints on it are lifted. However, it supports the evidence from the population structure of the species and so argues against parapatric speciation.

Irrespective of the genetic mechanism by which this speciation occurs, that is, through a genetic transilience, or as a consequence of adaptive change as a response to the environment, isolation of the new species appears to have been necessary for speciation. Since there are no barriers to gene flow on the Swan Coastal Plain where these species occur, the only explanation which is likely is for some temporary barrier to have existed in the past.

The Swan Coastal Plain is the product of dune building during the Pleistocene and it has been inundated by the sea when there were high sea levels produced by eustatic changes in the sea level during the interglacial periods (McArthur & Bettenay, 1974; Playford & Leech, 1977). Several times this coastal plain was divided into two unequal parts by the sea: the southern part was usually much smaller than the northern part and those populations went through a population bottleneck. Speciation may have occurred with this isolation. The last time that the sea level was high enough to split the plain completely was 100 000 yrs b.p., a time which corresponds well with the minimum time for the divergence of these two species.

A temporary barrier may also occur in a region without topographic barriers through climatic change so that populations become isolated. However, the climatic changes which have occurred in Western Australia over the past million years have been dominated by the periodic ice ages of the Pleistocene, which have led to changes in the local temperatures and the amount, frequency and location of rainfall, changes which would have increased the number and amount of semi-arid habitats in Western Australia (Hopper, 1979;Bint, 1981). Since the Swan Coastal Plain has a very gentle, monotonic change in rainfall and temperature with latitude, with differences in vegetation only in the more northerly part of the range of *M. marki* sp. A, a change in the amount of rainfall and heat is likely to shift the vegetation zones so that the plain became more arid or humid. Such a change is unlikely to split the range of an ancestral species even if it does cause changes in the range of that species.

Since the characteristics of brachyptery, subdivided populations, non-saltatorial hind limbs, and crypsis are characteristics of the other species of this genus, the levels of gene flow are likely to be similar. The aggregations that occur are of similar size, although those of *M. pauperculus* appear more dense while those of *M. sandowi* and *M. thielei* appear less dense (Barendse, 1987). Further, these species show similar patterns of chaotic differentiation in which the genetic similarity of populations is not related to the geographic proximity between them (Barendse, 1987).

Given similar population structures among these species it is likely that speciation followed similar lines. Since there are no permanent barriers present between these species, speciation is likely to have occurred under similar circumstances. These species appear to have diverged from each other in the late Pliocene to early Pleistocene, prior to the eustatic fluctuations in sea-level caused by the ice ages. Then the level of the sea was 250 m higher than at present (Fairbridge, 1961), and if these are translated into a height above sea level then the Swan Coastal Plain disappears, a large island is formed south of the main continental mass, and the Stirling Mountain Range encroaches on the sea. Such an interpretation is valid since Western Australia is one of the tectonically most stable areas on earth (Fairbridge, 1961). This results in the coastal districts of southwestern Australia being divided into three isolates through temporary barriers, which corresponds with the three species or isolates in the initial fragmentation of the genus. Furthermore, no changes in sea level subsequent to this would have split the ranges of these isolates while there is no evidence that further speciation has occurred in these regions. Interestingly, the speciation event since then corresponds to the production of new areas of land and to the subdivision of this new land by the fluctuation in sea levels; the formation of the Swan Coastal Plain, its colonisation, and subsequent inundation leading to the formation of first the ancestors of *M. marki* and then the subdivision of this species into *M. marki* sp. A and sp. B.

These interpretations explain several anomalies in the distribution of these species. First, it postulates that the ancestral populations of *M. sandowi* occurred along the coast like the other species, but with the production of a new coastal plain and a novel species, this species became the only non-coastal species in the genus. Second, it postulates that the isolates of *M. sandowi* that occurs in the Stirling Ranges, several hundred kilometers from other populations of *M. sandowi*, are relicts from the time when the ancestral populations occupied the southern coastal districts. With the climatic changes and the drop in sea level, they were either excluded from the southern coast by ancestors to *M. pauperculus* or they were inviable in the newly emerged, colder southern regions. The extent of differentiation between these isolated populations and the main body of the species is not known although they possess the standard morphology and male mating call structure of the species (Sandow, 1980).

Although the incursions of the sea, and the population structure of these species, appear to require both isolation and the presence of population bottlenecks for speciation to occur, the speciation in this genus may not have been tied to these events. Other, unknown temporary barriers may have existed in the past in south western Australia, but the incursions of the sea provide the simplest and most obvious source of temporary barriers.

The orchestration of diverse evidence in this example of speciation, especially the use of population genetic parameters, has distinguished between two modes of speciation in the formation of a pair of sister species. This has occurred without inferences from patterns of divergence of several related groups, the backbone of studies of speciation. As has been pointed out earlier, the comparative method would not have been able to discriminate between allopatric and parapatric speciation. Furthermore, it is evident that such a method would have favoured the interpretation of parapatric speciation in this genus when the geographic distributions of these species are considered in relation to the climatic zones, the historical sequence of speciation, the locomotory adaptations of these species, and the appearance of speciation through the fragmentation of an ancestral species in a topographically featureless region divided into several subregions of distinct types of vegetation (Prider, 1966; Churchward & McArthur, 1980; Heddle *et al.*, 1980). However, such an hypothesis is untenable in the face of the evidence from the population structures of these species and the measurements of gene flow. Such evidence also removes many plausible hypotheses of the route that speciation followed, and provides more specific hypotheses about speciation in a particular group.

Within the broader considerations of the fauna of southwestern Australia, the hypotheses that speciation could occur in this region only over large geographic scales, and that fragmentation of ancestral species could not occur to produce the modern fauna, have to be replaced by other hypotheses (e.g. Barendse, 1984; Maxson & Roberts, 1984). Furthermore, temporary barriers due to incursions of the sea appear to have had an important impact on the number of species in one genus. Similarly, these incursions may have provided effective barriers for other species and large-scale fragmentation may have contributed significantly to the diversity of the fauna. An extension of this work would be to determine whether other groups were affected by these barriers and to determine whether the historical sequence of splitting of areas has lead to similar historical sequences in the speciation in susceptible genera.

References

Avise J.C. & C.F. Aquadro. 1982. A comparative summary of genetic distances in the vertebrates: patterns and correlations. Evol. Biol. 15: 151-185.

Bailey W.J. 1979. A review of the Australian Copiphorini (Tettigoniidae: Orthoptera). Aust.

J. Zool. 27: 1015-1049.
Ball I.R. 1976. Nature and formulation of biogeographic hypotheses. Syst. Zool. 24: 407-430.
Barendse W. 1984. Speciation in the genus *Crinia* (Anura: Myobatrachidae) in southern Australia: a phylogenetic analysis of allozyme data supporting endemic speciation in southwestern Australia. Evolution 38: 1238-1250.
Barendse W.J. 1987. Speciation in the genus *Mygalopsis* (Orthoptera: Tettigoniidae). PhD Thesis, University of Western Australia, Nedlands.
Barton N.H. 1979. The dynamics of hybrid zones. Heredity 43: 341-359.
Barton N.H. & G.M. Hewitt. 1985. Analysis of hybrid zones. Ann. Rev. Ecol. Syst. 16: 113-148.
Bint A.N. 1981. An early Pliocene assemblage from Lake Tay, south-western Australia, and its phytogeographic implications. Aust. J. Bot. 29: 277-291.
Brundin L. 1972. Phylogenetics and biogeography. Syst. Zool. 21: 69-79.
Bush G.L. 1969. Sympatric host race formation and speciation in frugivorous flies of the genus *Rhagoletis*. Evolution 23: 237-251.
Bush G.L. 1975. Modes of animal speciation. Ann. Rev. Ecol. Syst. 6: 339-364.
Carson H.L. 1976. Inference of the time of origin of some *Drosophila* species. Nature 259: 395-396.
Case S.M., P.G. Haneline & M.F. Smith. 1975. Protein variation in several species of *Hyla*. Syst. Zool. 24: 281-295.
Churchward H.M. & W.M. McArthur. 1980. Landforms and soils of the Darling system, Western Australia. In: Atlas of Natural Resources, Darling system, Western Australia. Dept. of Conservation and Environment, Western Australia.
Cohan F.M. 1984. Can uniform selection retard random genetic divergence between isolated conspecific populations? Evolution 38: 495-504.
Cracraft J. 1982. Geographic differentiation, cladistics, and vicariance biogeography: reconstructing the tempo and mode of evolution. Am. Zool. 22: 411-424.
Crow J.F. & M. Kimura. 1970. An Introduction to Population Genetic Theory. Harper and Row, New York.
Dadour I.R. & M.S. Johnson. 1983. Genetic differentiation, hybridization and reproductive isolation in *Mygalopsis marki* Bailey (Orthoptera: Tettigoniidae). Aust. J. Zool. 31: 353-360.
Efron B. 1979. Bootstrap methods: another look at the jackknife. Ann. Statist. 7: 1-26.
Efron B. & G. Gong. 1983. A leisurely look at the bootstrap, the jackknife, and cross-validation. Am. Statist. 37: 36-48.
Ehrlich P. & P.H. Raven. 1969. Differentiation of populations. Science 165: 1228-1232.
Endler J.A. 1977. Geographic Variation, Speciation, and Clines. Princeton University Press, Princeton.
Fairbridge R.W. 1961. Eustatic changes in sea level. Physics and Chemistry of the Earth 4: 99-185.
Felsenstein J. 1985. Confidence limits on phylogenies: an approach using the bootstrap. Evolution 39: 783-791.
Hairston N.G. & C.H. Pope. 1948. Geographic variation and speciation in Appalachian salamanders (*Plethodon jordoni* Group). Evolution 2: 266-278.
Heddle E.M., D.W. Loneragan & J.J. Havel. 1980. Vegetation complexes of the Darling system, Western Australia. In: Atlas of Natural resources, Darling System, Western Australia. Dept. of Conservation and Environment, Western Australia.

Hillis D.M. 1985. Evolutionary genetics of the Andean lizard genus *Pholidobolus* (Sauria: Gymnophthalimidae): phylogeny, biogeography, and a comparison of tree construction techniques. Syst. Zool. 34: 109-126.

Hopper S.D. 1979. Biogeographical aspects of speciation in the southwest Australian flora. Ann. Rev. Ecol. Syst. 10: 399-422.

Johnson M.S. & J.L. Brown. 1980. Genetic variation among trait groups and apparent absence of close inbreeding in grey-crowned babblers. Behav. Ecol. Sociobiol. 7: 93-98.

Kimura M. 1955. Stochastic processes and distribution of gene frequencies under natural selection. Cold Spring Harbour Symp. Quant. Biol. 20: 33-53.

Kimura M. & G.H. Weiss. 1964. The stepping stone model of population structure and the decrease of genetic correlation with distance. Genetics 49: 561-576.

Kirby G.C. 1975. Heterozygote frequencies in small populations. Theor. Popul. Biol. 8: 31-48.

Levene H. 1949. On a matching problem arising in genetics. Ann. Math. Statist. 20: 91-94.

McArthur W.M. & E. Bettenay. 1974. The development and distribution of the soils of the Swan coastal plain, Western Australia. C.S.I.R.O. Soil Publication no. 6.

Maxson L.R. & J.D. Roberts. 1984. Albumin and Australian frogs: molecular data a challenge to speciation model. Science 225: 957-958.

Maxson L.R. & R.D. Maxson. 1979. Comparative albumin and biochemical evolution in plethodontid salamanders. Evolution 33: 1057-1062.

May R.M., J.A. Endler & R.E. McMurtrie. 1975. Gene frequency clines in the presence of selection opposed by gene flow. Am. Nat. 109: 659-676.

Maynard Smith J. 1966. Sympatric speciation. Am. Nat. 100: 637-650.

Mayr E. 1947. Ecological factors in speciation. Evolution 1: 263-288.

Mayr E. 1963. Animal Species and Evolution. Harvard University Press, Cambridge, Massachusetts. p 797.

Mickevich M.F. 1978. Taxonomic congruence. Syst. Zool. 27: 143-158.

Mickevich M.F. & M.S. Johnson. 1976. Congruence between morphological and allozyme data in evolutionary inference and character evolution. Syst. Zool. 25: 260-270.

Miller R.G. 1974. The jackknife – a review. Biometrika 61: 1-15.

Nei M. 1978. Estimation of average heterozygosity and genetic distance from a small sample of individuals. Genetics 89: 583-590.

Nelson G. & N.I. Platnick. 1980. A vicariance approach to historical biogeography. Bioscience 30: 339-343.

Nozawa K., T. Shotake & Y. Okura. 1975. Blood protein polymorphisms and population structure of the Japanese macaque, *Macaca fuscata fuscata*. In: Isozymes, Vol. 4. Ed. C.L. Markert. Academic Press, New York. pp 225-241.

Platnick N.I. & G. Nelson. 1978. A method of analysis for historical biogeography. Syst. Zool. 27: 1-16.

Playford P.E. & R.E.J. Leech. 1977. Geology and hydrology of Rottnest Island. Geol. Surv. Western Australia Report 6.

Prider R.T. 1966. The lateritized surface of Western Australia. Aust. J. Sci. 28: 443-451.

Rensch B. 1956. Evolution above the Species level. Methuen and Company, London. p 419.

Rohlf F.J. & G.D. Schnell. 1971. An investigation of the isolation-by-distance model. Am. Nat. 105: 295-325.

Sandow J.D. 1980. The morphology, distribution and acoustic repertoire of the genus *Mygalopsis* (Orthoptera: Tettigoniidae). MSc Thesis, University of Western Australia, Nedlands.

Sarich V.M. 1977. Rates, sample sizes, and the neutrality hypothesis for electrophoresis in evolutionary studies. Nature 265: 24-28.

Schaal B.A. 1974. Isolation by distance in *Liatris cylindracea*. Nature 252: 703.

Selander R.K. & D.W. Kaufman. 1975. Genetic population structure and breeding systems. In: Isozymes, Vol. 4. Ed. C.L. Markert. Academic Press, New York. pp 27-48.

Slatkin M. 1980. The distribution of mutant alleles in a subdivided population. Genetics 95: 503-524.

Slatkin M. 1981. Estimating levels of gene flow in natural populations. Genetics 99: 323-335.

Slatkin M. 1985. Gene flow in natural populations. Ann. Rev. Ecol. Syst. 16: 393-430.

Snedecor G.W. & W.G. Cochran. 1969. Statistical Methods. 6th ed. Iowa State University Press, Ames, Iowa. p 593.

Sokal R.R. 1983a. A phylogenetic analysis of the caminalcules. II. Estimating the true cladogram. Syst. Zool. 32: 185-201.

Sokal R.R. 1983b. A phylogenetic analysis of the caminalcules. IV. Congruence and character stability. Syst. Zool. 32: 259-275.

Templeton A.R. 1980a. The theory of speciation via the founder principle. Genetics 94: 1011-1038.

Templeton A.R. 1980b. Modes of speciation and inferences based on genetic distances. Evolution 34: 719-729.

Thiele D.R. 1981. Mechanisms of acoustic spacing within populations of the tettigoniid *Mygalopsis marki*. PhD Thesis, University of Western Australia, Nedlands.

Wade M.J. 1984. Change in group selected traits that occur when group selection is relaxed. Evolution 38: 1047-1058.

White M.J.D. 1978. Modes of Speciation. Freeman, San Fransisco.

Wiley E.O. 1981. Phylogenetics: the Theory and Practice of Phylogenetic Systematics. John Wiley and Sons, New York. p 439.

Wright S. 1931. Evolution in mendelian populations. Genetics 16: 97-159.

Wright S. 1940. Breeding structure of populations in relation to speciation. Am. Nat. 74: 232-248.

Wright S. 1943. Isolation by distance. Genetics 28: 114-138.

Wright S. 1946. Isolation by distance under diverse systems of mating. Genetics 31: 39-59.

Wright S. 1948. On the roles of directed and random changes in gene frequency in the genetics of populations. Evolution 2: 279-294.

Wright S. 1951. The genetical structure of populations. Ann. Eugenics 15: 323-354.

Wright S. 1965. The interpretation of population structure by F – statistics with special regard to systems of mating. Evolution 19: 395-420.

Wright S. 1969. Evolution and the Genetics of Populations, Vol. II. The theory of gene frequencies. The University of Chicago Press.

Wright S. 1978. Evolution and the Genetics of Populations. Vol. IV. Variability within and among natural populations. The University of Chicago Press.

15. THE PREDATORY SAGINAE

A.P. Kaltenbach

Introduction

Within the Tettigoniidae the subfamily Saginae appears to be one of the most ancient of recent groups, perhaps evolving more recently than the related Tettigoniinae. The Saginae are medium-sized to very large Saltatoria with a body shape ranging from slender to very bulky. The head also varies between genera from orthognathous to hypognathous. The antennae are usually filiform and thickened proximally and do not extend much beyond the tip of abdomen. The fastigium of the vertex is not elongated as in some families and the eyes are large and prominent. All species have powerful jaws. The fore and middle legs are long and robust in the more bulky species, and the ventral surfaces are equipped with strong spines. By comparison the hind femora is less inflated. Only the genus *Clonia* Stål contains a number of fully winged species, and, in most sagines, the wings are abbreviated or even vestigial. The incurved male cerci are hooked apically, but without a distinct internal tooth, and the concealed genitalia of the known species are not sclerotized. The dorsal and ventral margins of the ovipositor are serrate. The Saginae are defined by a combination of characters. A single character, which is highly variable, for example, the absence of the dorsal apical spur of hind tibia, cannot be regarded as a definite criterion.

Fifteen genera were described originally within the Saginae mainly based on the absence of apical spurs on the dorsal surface of the hind tibia. But the presence or absence of the spurs appears to be an unreliable character. The results of a comparative analysis of 22 morphological features of the Tettigoniidae indicate that the Australian genera and the African *Pseudosaga* Brancsik should be excluded from the Saginae.

The genera which have been or should be eliminated from the Saginae are listed below. Some of them are known as synonyms of genera previously described. Other genera differ in so many morphological characters from *Saga* that they have to be transferred to other subfamilies of the Tettigoniidae. However, where to place the genera *Terpandrulus* Brunner, *Pachysaga* Brunner and *Psacadonotus* Redtenbacher remains doubtful. They appear to be wrongly placed within the Saginae. The reasons why *Megalotheca* (South Africa and Madagascar) should be not transferred from the Conocephalinae to the Saginae are explained in Kaltenbach (1986).

Genera here excluded from Saginae

1. *Microsaga* Saussure 1888: junior synonym of *Banza* Walker 1870, Conocephalinae (Syn. Kirby 1906).

2. *Xiphosaga* Saussure 1888: junior synonym of *Hemisaga* Saussure 1888 (syn.: Saussure 1892); cf. *Hemisaga.*

3. *Hemisaga* Saussure 1888: to Listroscelidinae (this paper).

4. *Miosaga* Saussure 1892: junior synonym of *Psacadonotus* Redtenbacher 1891, (syn. Rentz, pers. comm. 1982).

5. *Terpandrus* Stal 1874: to Listroscelidinae (transf. Rentz, pers.comm. 1982).

6. *Terpandroides* Ander 1957: to Listroscelidinae (transf. Rentz 1979).

7. *Terpandrulus* Brunner von Wattenwyl 1893. (Kaltenbach 1986). 8. *Pachysaga* Brunner von Wattenwyl 1893: (Kaltenbach 1986) cf. *Pachysagella*.

9. *Pachysagella* Hebard 1922: junior synonym of *Pachysaga* Brunner von Wattenwyl 1893 (syn. Kaltenbach 1986); to Tettigoniinae(Uvarov 1924), to Saginae (Rentz 1985), excluded from Saginae (Kaltenbach 1986).

10. *Pseudosaga* Brancsik 1897: to Mecopodinae (Kaltenbach 1986).

11. *Psacadonotus* Redtenbacher 1891, originally included in the Conocephalinae but which should be excluded from Saginae for same reasons given in Kaltenbach, 1986 for *Pachysaga*.

SPATIAL RELATIONSHIPS

Zoogeography – Horizontal Distribution

The present distributional range of the Saginae is restricted to parts of Ethiopian Africa and to the West-Palaearctic zoogeographic region.

For several reasons (see Kaltenbach 1971) I regard southern central Africa as a primary center of distribution of the apparently primitive genus *Clonia*. The species of the subgenus *Clonia* are widely distributed in southern, central and east Africa. Many species of this subgenus are fully winged, but the short winged *C.* (*C.*) *kenyana* Uvarov reaches the northernmost distribution of all African Saginae at Marsabit. *Clonia* (*Leptoclonia*) *minuta* (Haan) and *C.* (*Hemiclonia*) *melanoptera* (*L.*) are endemics to the southwestern Cape. *Clonia* (*Leptoclonia*) *vittata* (Thunberg) and *C.* (*L.*) *vansoni* Kaltenbach are widespread in the Karoo-Kalaharian subregion, living mainly south of the Tropic of Capricorn. The former species extends into Ovamboland (Namibia). It is mostly macropterous but brachyptery becomes more common to the south in the southern Cape Province. The subgenera *Xanthoclonia* Kaltenbach and *Hemiclonia* (except *H. melanoptera* as specified above) inhabit the eastern part of South Africa surrounded by the large arch of the Limpopo River.

Only three specimens of the genus *Cloniella* Kaltenbach are known. These few records are not sufficient for defining the true range of this genus.

Peringueyella Saussure is well represented in the central and southern parts of southern Africa but it apparently does not extend into the Cape. Some old specimens of *P. jocosa* Saussure in the museums of Geneva and Vienna were labelled afterwards with "Cap b. sp.". This printed inscription was incorrect and a handwritten label, attached to a specimen of the original series, indicates "Delagoa-Bay" in Mozambique as the locality where the specimens really were collected (cf. Kaltenbach 1971).

The species of *Saga* Charpentier inhabit suitable areas in southwest Asia (except Arabia), southern Siberia, extending into the region east of Lake Zaysan, E. Kazakhstan (84°57' E), the Caucasus region, Trans-Caspian region and central and southern Europe including the Iberian Peninsula. The northern-most record of the genus *Saga* (*S. pedo* Pallas) is located near Sverinogolovskoye, Kurgan district, USSR (54°30'N).

The centre of distribution of *Saga* appears to be the east coast of the Mediterranean Sea. The ancestors of *Saga* doubtless were of African origin.

Vertical Distribution

The biotopes of the species of Saginae are situated mostly in areas from sea level to altitudes of about 1500 m. Some species of *Clonia* and *Saga*, however, occur between 1500 and 2000 m. The African altitudinal record belongs to *Clonia* (*Hemiclonia*) *lalandei* Saussure from the Golden Gate National Park, Orange Free State 1980 m (Rentz and Brodey). *Saga ephippigera* Fischer de Waldheim was found at 2400 m, below the Kara Dağ, E. Turkey (Bilgutay). The following species were recorded from high areas between 1500 and 1700 m (Kaltenbach 1967, 1971): *Clonia* (*C.*) *multispina* Uvarov (1570 m, Mitwaba, Shaba/Zaire (Ross and Leech)); *C.* (*C.*) *zernyi* Kaltenbach (1500-1700 m, Ugano, Ruvuma/Tanzania (Zerny)); *C.* (*Hemiclonia*) *assimilis* Kaltenbach and *C.* (*H.*) *charpentieri* Kaltenbach (1600 m nr Kokstad, Natal/South Africa (Rentz and Brodey)); *Saga natoliae* Serville (1700 m, Sultan Dağ Mts., W. Turkey (Willemse)); *S. rhodiensis* Salfi (1600 m, Tahtali Dağ, Antalya, W. Turkey (Karamanoğlu)).

Climatic Factors

The Saginae seem to react indifferently to sunlight but sound production seems to be dependent on the onset of dusk. They show neither positive nor negative responses to light in the field, but they are attracted to artificial lights at night.

All species of Saginae are thermophilic insects, and temperature appears to be a limiting factor influencing their horizontal and vertical distribution. There is also a profound effect of temperature on rate of embryonic development and on the duration of larval stages. Eggs of the obligate parthenogenetic *Saga pedo* commences development without a diapause at temperatures above 25°C (see Hartley chapter 4). The eggs contained embryos after 40 to 43 days at 30°C and after 85 days at 20°C. Beneath these temperatures diapause sets in and may be broken by a period of cooling. At 12°C eggs hatched only after a diapause. Hartley and Warne (1972) provide the following hatching times following a defined cooling period. Thus first instars emerged from eggs after (i) 84 days after 70 and 83 days of cooling, (ii) after 113 days after 46 and 48 days and (iii) after 131 days after 30 to 43 days, (all at 16°C).

Saga natoliae attains maturity in southern Macedonia (Greece) at an average temperature of 25°C in June, about 70 days after hatching (Kaltenbach 1970). For postembryonic development under laboratory conditions (28°-36°C day, 16°-18°C night) this species requires 39 to 43 days (♂♂) or 42 to 45 days (♀♀) (Sänger 1980).

The influence of temperature on the time of maturation is most significant (cf. Kaltenbach 1970). An index of development rates can be gauged from the Effective Temperature (ET) (see Hartley this volume) which can be obtained from the equation

$$ET = \frac{8T+14AR}{AR+8}$$

(T = mean annual temperature, AR = annual range difference between the warmest and the coldest months).

Thus, the first adults of *S. ephippigera* do not appear in Tbilisi/Trans-Caucasia (ET 13.5) before July, while in Tiberias/Israel (ET 16.6) they emerge during the first days of June.

The occurrence of melanic forms of *Saga* varies in any one population, and its occurrence also appears linked to the index ET. This differs between species. The degree of melanism depends, of course, on genotypic determination, which differs among the various species. Below an environmental ET of 15°C, up to 46% of the specimens of *S. ephippigera* show melanic patterns. Between 15° and 17°C ET the rate of melanic forms decreases to 21% maximum. It is suggested that melanism permits a greater utilization of the solar radiation by the increased capacity of absorption. The localization of the melanins is specific in each of the species of *Saga* (Kaltenbach 1970).

Most Saginae seem to be mesohygric insects. There is no truly xerophilous species in this subfamily, but some species can live in drier biotopes. *Clonia* (*Leptoclonia*) *vittata* Thunberg lives in the semiarid conditions of Cape Karoo. *Saga pedo* prefers the slopes of xerothermic hills in Central Europe. Although true hygrophilous Saginae are unknown some species of *Clonia* and *Saga* occupy the shrub-vegetated slopes of river gorges.

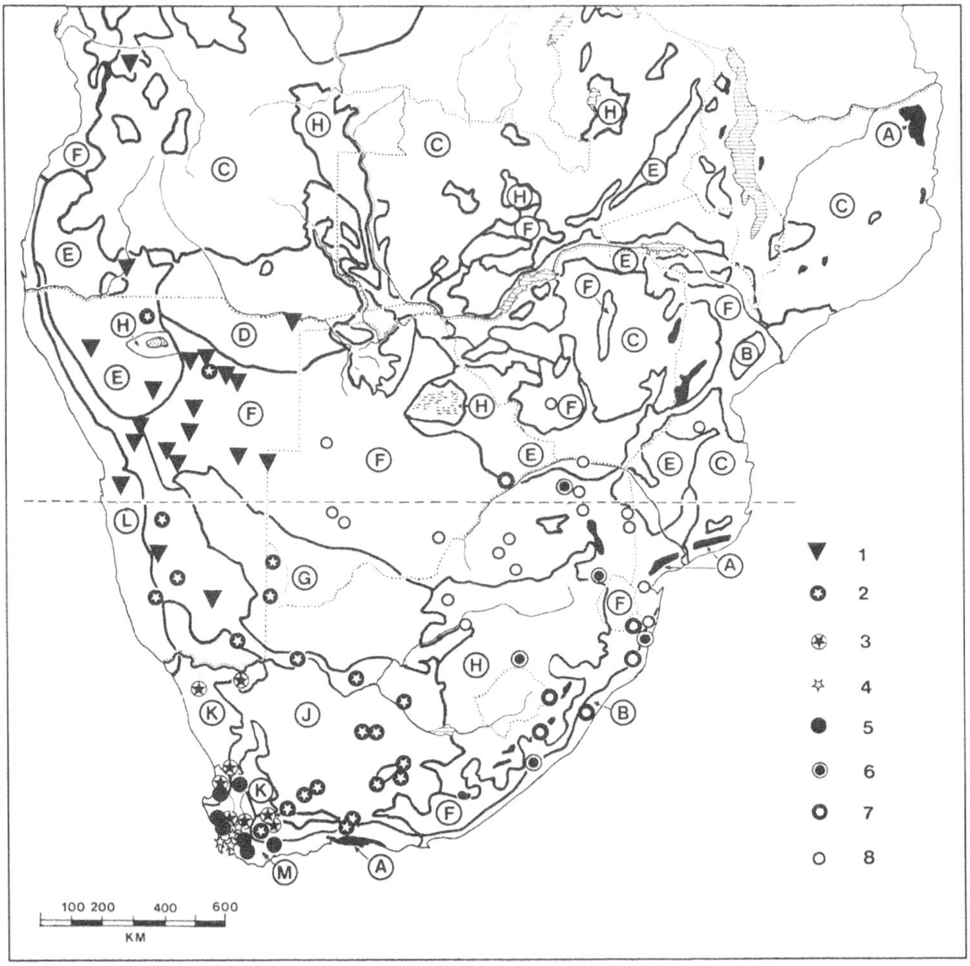

Fig. 1. Plant associations and known distribution of some species of Saginae in southern Africa: **A**, forest; **B**, forest-savanna mosaic; **C**, Miombo vegetation; **D**, *Baikiaea*-tree savanna; **E**, Mopane vegetation; **F**, Bushveld; **G**, mixed tree and shrub savanna; **H**, Grassveld; **J**, dwarf shrub savanna and Karoo; **K**, succulent dwarf shrub vegetation; **L**, desert vegetation of the Namib; **M**, *Macchia* (Fynbos vegetation). - 1, *Clonia* (*C.*) *caudata* Uvarov; 2, *Clonia* (*Leptoclonia*) *vittata* (Thunberg); 3, *Clonia* (*L.*) *vansoni* Kaltenbach; 4, *Clonia* (*L.*) *minuta* (Haan); 5, *Clonia* (*Hemiclonia*) *melanoptera* (Linné); 6, *Clonia* (H.) *lalandei* Saussure; 7, *Clonia* (H.) *assimilis* Kaltenbach; 8, *Peringueyella jocosa* Saussure.

Plant Associations

The plant communities of different vegetation types in east, central and southern Africa harbour various species of Saginae. None of these species inhabits the true desert or the rain forest. A number of Ethiopian plant associations and their associated Saginae are listed below (letters in brackets refer to the plant associations in the map, Fig. 1):

1. Forest (A): No species
2. Open areas in the Forest-Savanna Mosaic (B): *Clonia* (*Clonia*) *zernyi*, *C.* (*C.*)

wahlbergi, *C.* (*Hemiclonia*) *lalandei*, *C.* (*H.*) *assimilis*, *Peringueyella jocosa*, *P. zulu*.

3. Open areas in the Miombo Vegetation (C): *Clonia* (*C.*) *multispina*, *C.* (*C.*) *zernyi*, *C.* (*C.*) *wahlbergi*, *C.* (*C.*) *dewittei*, *C.* (*C.*) *burri*.

4. Open areas in the Baikiaea-Tree Savanna (D): *Clonia* (*C.*) *caudata*.

5. Open areas in the Mopane Vegetation (E): *Clonia* (*C.*) *wahlbergi*, *C.* (*C.*) *intermedia*, *C.* (*C.*) *caudata*, *C.* (*Leptoclonia*) *vittata*.

6. Bushveld (F): *Clonia* (*C.*) *wahlbergi*, *C.* (*C.*) *intermedia*, *C.* (*C.*) *kalahariensis*, *C.* (*C.*) *caudata*, *C.* (*Leptoclonia*) *vittata*, *C.* (*Hemiclonia*) *lalandei*, *C.* (*H.*) *assimilis*, *Cloniella praedatoria*, *Peringueyella jocosa*, *P. rentzi*.

7. Mixed Tree and Shrub Savanna (southern Kalahari) (G): *Clonia* (*C.*) *kalahariensis*, *C.* (*Leptoclonia*) *vittata*.

8. Grassveld (H): *Clonia* (*C.*) *saussurei*, *C.* (*C.*) *kalahariensis*, *C.* (*C.*) *uvarovi*, *C.* (*Hemiclonia*) *lalandei*, *C.* (*H.*) *assimilis*, *C.* (*H.*) *charpentieri*, *Peringueyella jocosa*.

9. Dwarf Shrub Savanna and Karoo (J): *Clonia* (*C.*) *caudata*, *C.* (*Leptoclonia*) *vittata*, *C.* (*L.*) *vansoni*.

10. Succulent Dwarf Shrub Vegetation (K): *Clonia* (*Leptoclonia*) *vansoni*, *C.* (*Hemiclonia*) *melanoptera*.

11. Desert Vegetation of the Namib (L): *Clonia* (*C.*) *caudata* 1♀ recorded from Gobabeb near the Kuiseb River), *C.* (*Leptoclonia*) *vittata* (one record from the *Stipagrostis*-grass plains near Aus).

12. Macchia (Fynbos Vegetation) (M): *Clonia* (*L.*) *vansoni*, *C.* (*Leptoclonia*) *minuta*, *C.* (*Hemiclonia*) *melanoptera*.

The distribution of some species of Saginae in various southern Africa plant association is marked in Figure 1. The stenotopic *Clonia* (*Leptoclonia*) *vittata* inhabits almost exclusively the Cape Karoid dwarf shrub and the open shrub vegetation in Namibia. *Clonia* (*L.*) *minuta* (Haan) is restricted to the fynbos macchia in southwestern Cape. *Clonia* (*L.*) *vansoni* Kaltenbach and *C.* (*Hemiclonia*) *melanoptera* (Linnaeus) are dwellers of the fynbos and the adjacent succulent dwarf shrub vegetation.

The subgenus *Clonia* has species that are more or less eurytopic. They live in open areas of the tree savannas as well as in the bushveld and grassland. The latter veld types are also the predominant monotop of *Clonia* (*Hemiclonia*) *lalandei* and *C.* (*H.*) *assimilis* and of the species of *Peringueyella*.

Saga species inhabit both dry and wet meadows, pastures, shrubby hillsides and gorges as well as fallow grain fields and vineyards in southern Europe and western Asia. In Central Europe *Saga pedo* shows a remarkable disjunct distribution, being confined to xerothermic hills which frequently are formed with a limestone subsoil.

Habitat

The species of the subgenera *Clonia* and *Leptoclonia*, the genera *Cloniella*, and *Peringueyella* and most species of *Saga* live in dense, tall grass or low shrubs. *Clonia* (*C.*) *wahlbergi* Stål, erroneously reported as *vittata*, also occurs on young maize plants and in weeds on the outskirts of the mealie fields (Distant 1892; Akerman 1932). *Hemiclonia* and some *Saga* species take refuge frequently in spiny scrubs. In Greek Macedonia, *S. natoliae* occurs in spiny hollyoak bushes (*Quercus coccifera*). *Saga pedo* is found in dense herbaceous vegetation, usually well hidden among long, green grass stems. On the xerothermic hills in Central Europe some of the major components of the vegetation in *Saga* habitats are: *Festuca* spp., *Sesleria* spp., *Poa* spp., *Centaurea* spp., *Eryngium campestre*, *Juniperus communis*, *Crataegus monogyna* and *Quercus pubescens*.

Relationships in Time

Annual Periodicity: Phenology

Our knowledge of the life cycles of the African Saginae is poor and, regrettably, nothing is known of the biology of Ethiopian species. In east and central Africa most individuals of *Clonia* (*C.*) *multispina* Uvarov reach maturity at the beginning of the rainy season in September. Adults were recorded between September and January. The main adult period of the closely related *C.* (*C.*) *wahlbergi* Stål lasts from November to March, corresponding to the adult period of *C.* (*Leptoclonia*) *vittata* and *Peringueyella jocosa* Saussure. Depending on the local climatic conditions, adult African Saginae rarely occur during the dry season. Nymphal stages, however, can be found during most times of the year.

Both field observations and laboratory studies (Kaltenbach 1970; Sanger 1980) have provided much information on several species of *Saga*. In Macedonia oviposition of *S. natoliae* Serville occurs from July to the end of September. The eggs overwinter and nymphs

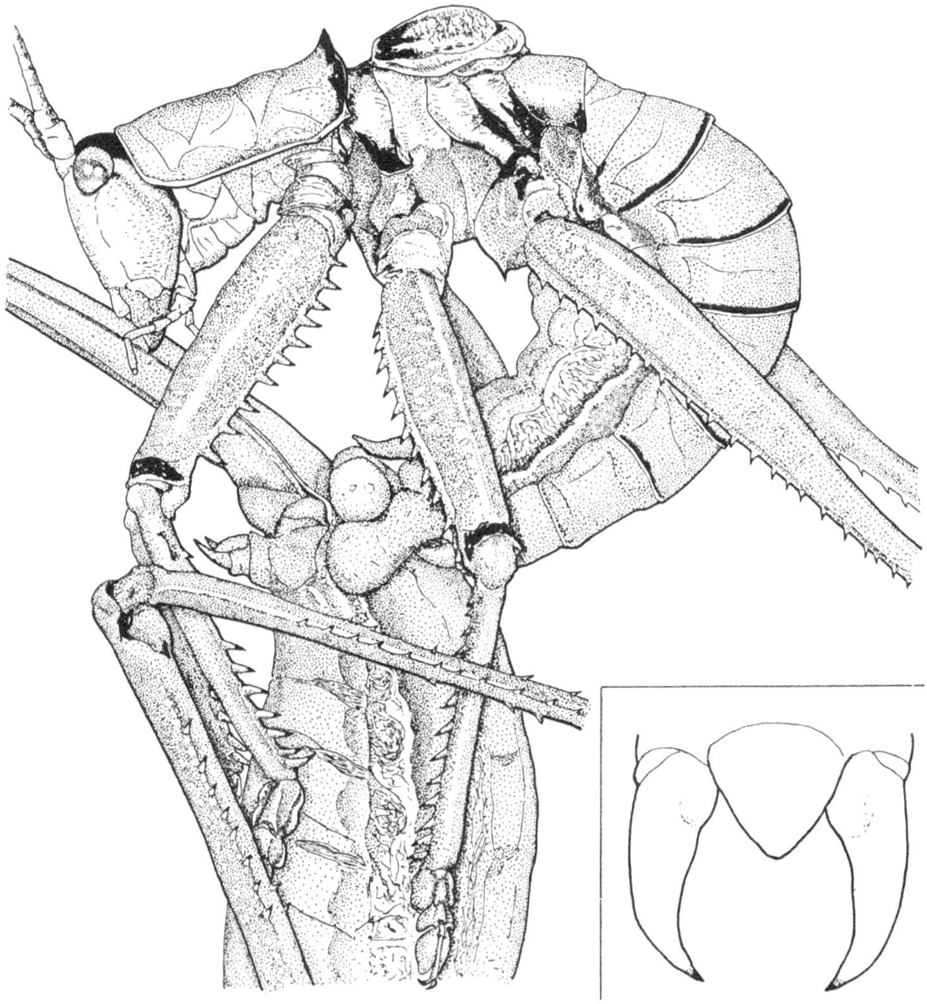

Fig. 2. Copulation in *Saga natoliae* Serville. The male is clasping the ovipositor of the female with his mandibles. – Inset: Cerci of a male of natoliae without internal padding.

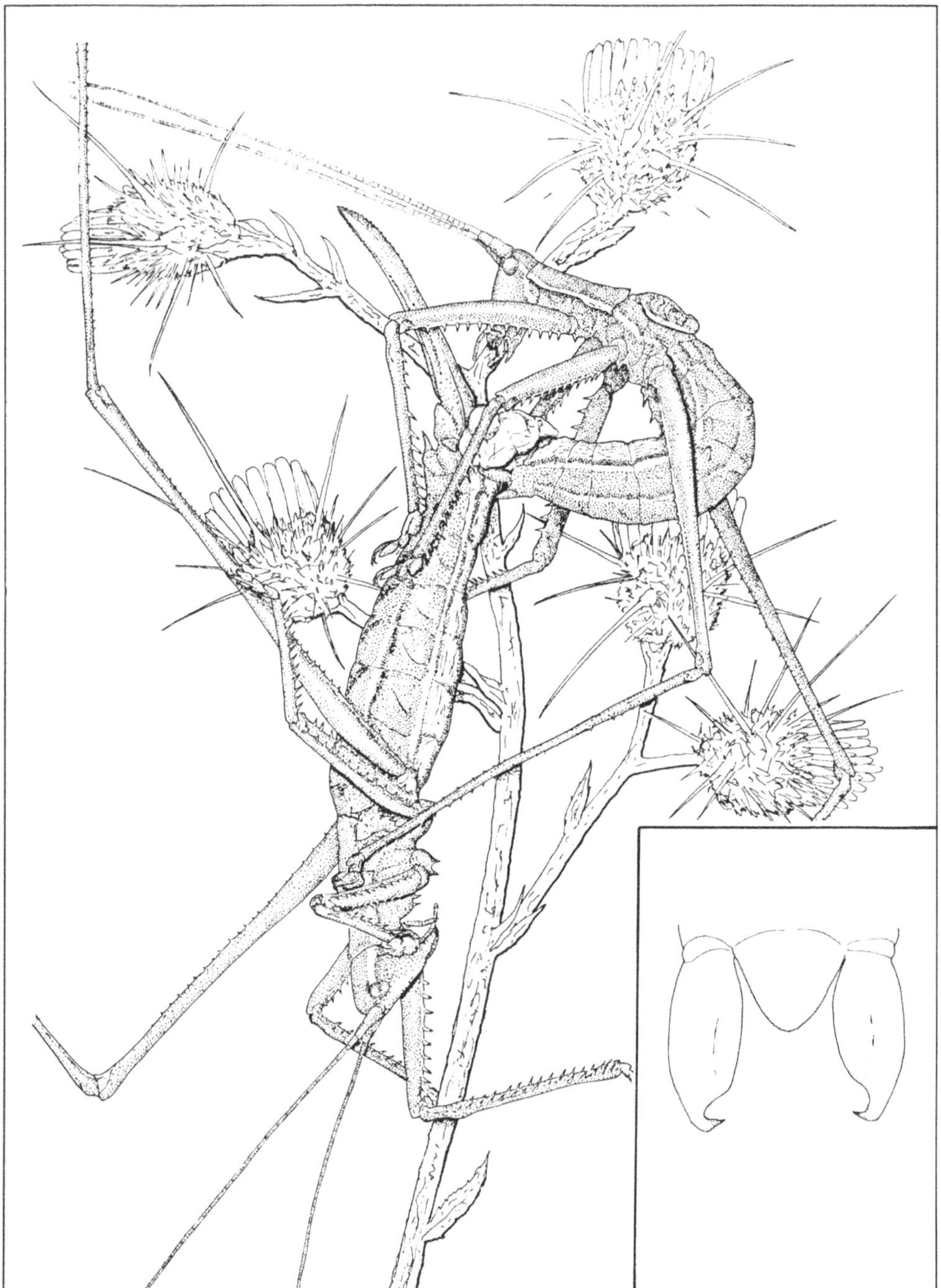

Fig. 3. Copulation in *Saga campbelli* Uvarov. The male does not clasp the ovipositor of the female with his mandibles. – Inset: Cerci of a male of *campbelli* with internal padding.

emerge during the first half of April. Adults usually do not appear before the last 10 days of June but females of *natoliae* can be found as late as October. These emergence times coincide to some extent with the life cycle of *S. pedo* (Pallas) but in Central Europe oviposition of the latter species does not occur before August and hatching is not known to occur before May with the first adults appearing in July. Females of *S. ephippigera* Fischer de Waldheim lay their eggs early in June, the nymphs emerging the following March and reaching maturity in May. This is also true for populations in Iraq. *S. cappadoica* Werner (Central Anatolia) has nearly the same phenology as *S. natoliae* but there are no records of specimens collected as late as October (Kaltenbach 1970).

Laboratory studies of *S. natoliae*, *S. rhodiensis* Salfi and *S. hellenica* Kaltenbach reveal a two and/or more year life cycle, due to the variable duration of the initial embryonic diapause. This initial diapause occurred independently of day length (Ingrisch 1986).

Daily Periodicity: Activity Rhythm

The most obvious activities of the adult are locomotion, stridulation, mating and oviposition. Sagines are predacious and mainly nocturnal insects (Akerman 1932; Kaltenbach 1970) and during the daytime some species of *Saga* sit motionless in dense grasses or bushes. After sunset they search for prey. Males of *S. natoliae* begin stridulating very slowly in late afternoon (Macedonia: July at 5 p.m.). The song is frequently interrupted by long periods of silence. Stridulation becomes more intensive at dusk where the pauses between calls are shorter, and towards midnight the song stops altogether. Males of *S. natoliae*, kept in darkened cages stridulated independently of the hour of the day. On the other hand *S. ornata* has been observed to be acoustically active in early morning (D.R. Ragge, pers. comm.).

The individual chirp lasts between 2.0 to 4.2 seconds with the pauses from 10 to 30 seconds or more in the afternoon and from 7 to 8 seconds after sunset. While stridulating the fore wings execute very quick movements, in *S. natoliae* these wings strokes range from 16 to 17 per second (Kaltenbach 1970).

Mating and oviposition seem to begin in late afternoon and last into the night.

After the hottest part of the season, the sagines extend their activities more into daytime periods, but as the season progresses activity of the remaining adults declines.

Biotic Relationships

Reproductive Behaviour

In the Saginae courtship is only a short prelude to mating and involves male stridulation, and once the pair have made contact, mutual antennation. *Saga* males increase the vigor of their stridulation at the sight of a female. This kind of song is very similar to the alarm sound produced when males are distressed. It is different again from the song produced during copulation as described below. However, we have no direct evidence that the male's calling song serves to attract the female (Kaltenbach 1970) and may have a territorial or singular function in male-male aggression. Three types of songs (cf Kaltenbach 1970): (i) Calling song, (ii) Alarm song, (iii) Soft song during mating.

Sexual behaviour is known only in a few species of *Saga*. Females of *S. natoliae* and *S. ephippigera* are usually not sexually responsive for 24 hours after the final moult. When the male is aware of the presence of the female he will reach the female by a sudden leap. Eventually, the male attains a position with his head at the base of the ovipositor and the tip of his body towards her head. He then bends his abdomen inwards until he is able to clasp the end of the female's abdomen with his cerci. The grip of his claspers seems to stimulate the opening of the female's subgenital plate. During this pre-copulatory period the male often produces a soft song, which consists of short bursts of sounds followed by long intervals.

In a successful copulation the spermatophore requires only 4.5 to 5 minutes to be produced. The male evaginates the membranous phallus and deposits the spermatophore upon the female gonopore. After 3-24 hours after copulation the female removes and consumes the spermatophylax.

Two distinct types of copulation have been observed in *Saga* species. In one type the male clasps the proximal part of the ovipositor with his mandibles during the entire act (Fig. 2). This form was observed regularly only in *S. natoliae* and may be a behavioural response to the lack of internal padding on the male cerci to stabilize him during copulation. Copulation in this species lasts from 8-15 minutes; once separated the male remains near the female for a further 60-90 minutes. In the second copulatory type males do not clasp the base of the ovipositor with their mandibles and the internal surface of the male cercus is well inflated or 'padded'. This form of copulation has been observed in *S. rhodiensis*, *S. ephippigera* and *S. campbelli* (Fig. 3). Spermatophore formation is short and requires 2.5-6 minutes, with the male separating from the female immediately following spermatophore transfer (Boldyrev 1915; Burr, Campbell & Uvarov 1923; Kaltenbach 1970; Sänger 1987).

The Saginae are bisexual with one well known exception. In *Saga pedo* reproduction is by obligatory parthenogenesis, which was explained by Matthey (1941, 1946, 1948) as a tetraploid thelyotoky. The polyploidy of the germ cells of *S. pedo* is evident from the duplication of the number of chromosomes (4n=68) in relation to the chromosome numbers of some bisexual species of *Saga* males (*S. ephippigera*: 2n=33; *S. cappadoica*: 2n=31 and *S. ornata*: 2n=31). Males of *S. pedo* are unknown. All published records concerning males of this species are due to misidentifications and refer to female nymphs or to males of various other species. Facultative parthenogenesis is not known in Saginae but may possibly occur in some of the bisexual species (Matthey 1946, 1950; Kaltenbach 1967, 1986).

Egg-laying, Development and Growth

Bisexual females of *Clonia* and *Saga* lay eggs singly in the ground about a fortnight after their first copulation. The parthenogenetic *S. pedo* begins egg-laying three to four weeks after its final ecdysis. Even in females of the bisexual species, oviposition is not restricted to mated females. Embryonic development of these eggs has not been observed.

Prior to oviposition the female tests the soil by palpating with the antennae and probing with the tip of the ovipositor. The large sword-shaped ovipositor, serrated distally, is used as a digging instrument. Five to eight eggs, arranged in a double row, are deposited in the same site by inserting and withdrawing the ovipositor and finally using it to cover up the site. Oviposition occurs during the entire adult life of the female. *Saga natoliae* produces 50 to 100 eggs. The number of eggs laid by *S. pedo* ranges from 25 to 80 (Burr, *et al.* 1923).

Saga natoliae is the only species for which the complete life cycle is known (Sänger 1980). There are six nymphal stages in this species (vermiform larva not included), as determined in the laboratory at optimal climatic and trophic conditions. The number of moults and the number of stages does not differ between males and females. From the first to the fifth instar each stage lasted from five to eight days. The penultimate instar lasts 10 to 12 days before reaching maturity. The average body weight was approximately doubled from one moult to the next.

Rearing of *S. pedo* from egg to adult has not been successful with a complete succession of all nymphal stages. Based on earlier failed attempts to breed this species we assume eight immature instars (Bérenguier 1905; Zacher 1928).

The ultimate moult is a most critical phase in postembryonic life because these large, normally ferocious insects are basically defenceless at this time. The final moult usually occurs at night or in the early morning hours when the threat of predation is reduced and relative humidity is higher.

Locomotor activity, mating and food requirements reduce during the second half of adult life. Towards the end of life sagines exhibit clear symptoms of senile decay. The paralysed, withered tarsomeres may be shed or nibbled off. Dying is a gradual cessation of all physiological activities as it is usual in insects (Kaltenbach 1970). Adults of *Saga natoliae* were kept alive in cages for three to almost four months but the total postembryonic life span in nature appears to last from four to six months.

Abundance

Despite their relatively low population density, the Saginae are not as rare as earlier entomologists assumed. Their adaptive coloration and predominantly nocturnal activity contributed to their absence from early collections. An assessment of the real abundance of sagine tettigoniids in their habitat is probably not possible without removing the concealing vegetation and this would mean a complete devastation of the habitat. Their apparent lack of abundance to the casual observer is not always a true reflection of reality (Kaltenbach 1970).

Cannibalism

Like other predatory insects Saginae sometimes seize and consume individuals of their own kind. Records concerning this behaviour are quoted in Kaltenbach (1970). Cannibalism may occur in nature but, due to generally low population density may never be significant.

Heterotypic Relations

Trophic Relations

The Saginae are exclusively carnivorous insects and feed on a wide spectrum of various arthropods, consisting mainly of insects. They attack every living animal they can subdue up to their own size. Animals less than one tenth of the length of the predator are seldom accepted as prey. Inedible prey may be captured, tested with the palps and then rejected. Sagines usually do not feed on dead animals and they never accept plant food but do palpate and drink the fluid of fresh fruits (without consuming the fruit pulp). *Saga* species show a clear preference for grasshoppers, locusts and tettigoniids. Nadig (1987) reported a population of *Saga rhodiensis* Salfi showing preference for cicadas as food. *Clonia wahlbergi* (as *C. vittata*) was recorded as feeding on caterpillars of the mealie stalk-borer, *Busseola fusca*, in Natal (Akerman 1932). *Cloniella praedatoria* (as *Hemisaga praedatoria*) was found lurking among the tops of tall flowering grasses and catching the butterfly *Danaus chrysippus* (Distant 1892). The water requirements of Saginae are low if sufficient food is available (Kaltenbach 1970; Sänger 1980).

The Saginae are not "sit-and-wait" predators like many Mantodea. During their activity period they move about in search of prey tracking their prey through substrate-borne vibrations produced by walking or jumping (see Kalmring *et al.* chapter 11). After a short leap *Clonia* and *Saga* species overcome the victim by clasping it with their spined fore legs. In addition, they often use their mid legs for holding their catch. Nevertheless, the prey may succeed in escaping because sagines appear to be very clumsy captors. A captured victim is killed in a similar manner each time by biting into its throat (Burr, *et al.* 1923, Kaltenbach 1970).

Saga campbelli and *S. pedo* usually eat their prey at the point of capture. *Saga natoliae* frequently climbs up a twig holding its victim in its mandibles. Feeding occurs in a peculiar, unexpected hanging position, clinging only by the hind legs (Fig. 4).

Sagines consume their prey almost completely. They generally reject the heavily sclerotized parts such as mandibles and the ovipositor, legs and wings and even the gut and its contents (Burr, *et al.* 1923, Kaltenbach 1970, Sänger 1980).

Commensals

An unidentified fly of the family Milichiidae was observed escorting a female of *Saga pedo* and sometimes sitting upon her (Schremmer 1959). Some species of Milichiidae (e.g. *Desmometopa* spp.) have been observed on spiders, mantids, assassin bugs and robber flies. They lick the haemolymph from wounds of prey.

Predators and Protective Factors

Natural predators of Saginae are spiders, centipedes, carnivorous insects, frogs and toads, lizards, insect-eating birds, insectivores, various rodents and smaller carnivores. It seems likely that some African primates also may feed on them. The large adults are sometimes rejected when attacked and seized. There is a record of a larger bird which caught and swallowed a female *Peringueyella* sp. and immediately vomited the spiny prey.

Ectoparasitic mites are rarely found on *Saga*. Endoparasitic larvae of the flesh-fly *Sarcophaga carnaria* occasionally infest *S. pedo*. Hymenopterous egg parasites are unknown. Mermithoid worms have also not been recorded from sagines.

The green, tawny, brown or (more rarely) greyish coloration of the *Saga* species demonstrates a remarkable resemblance to the surrounding plants of the habitat. The cryptic effect is not produced by a mere colour resemblance but depends also upon the principle of obliterative shading (countershading of back and abdomen). Many specimens bear a

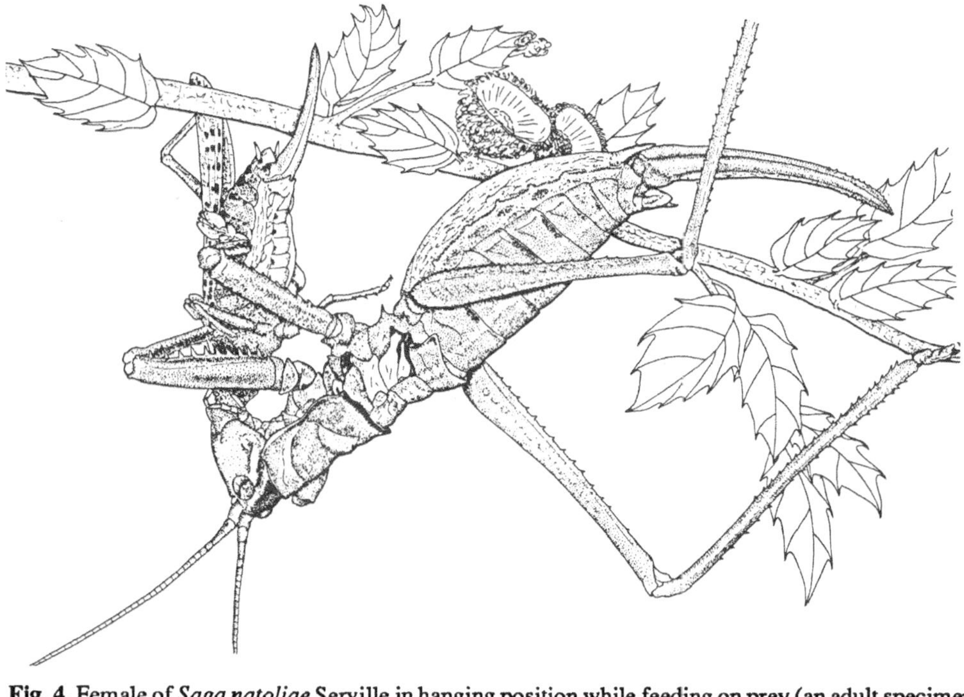

Fig. 4. Female of *Saga natoliae* Serville in hanging position while feeding on prey (an adult specimen of *Decticus verrucivorus* Linné).

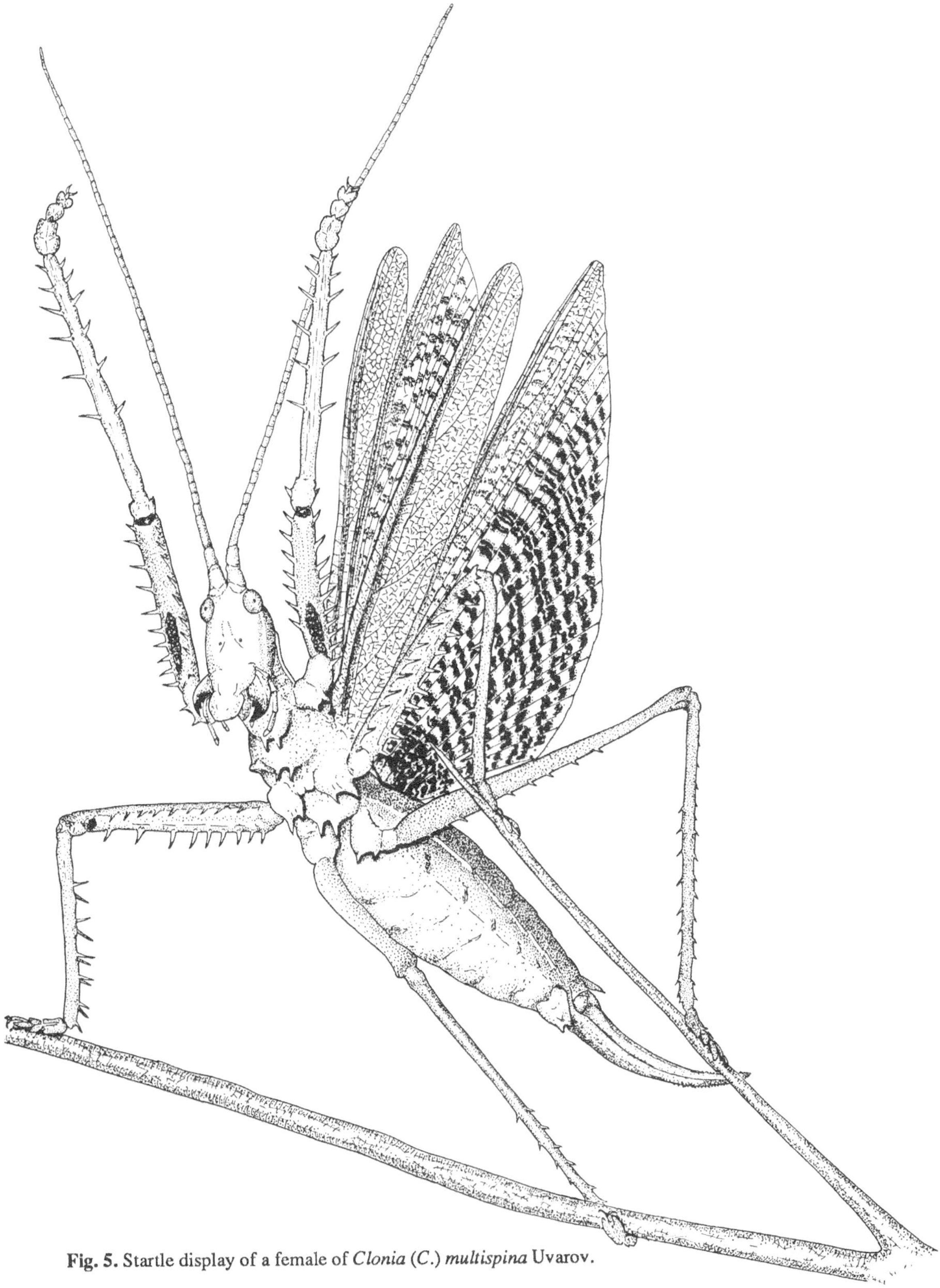

Fig. 5. Startle display of a female of *Clonia* (*C.*) *multispina* Uvarov.

conspicuous black and white pattern which has a disruptive function in their natural surroundings. Adaptive morphological colour change has been recorded several times in *Saga*. It has even been observed in young adults. The power of colour adjustment diminishes during the last weeks of adult life (Kaltenbach 1970).

To our eyes the cryptic coloration of *Saga* spp. is almost perfect. Like other animals with protective coloration, *Saga* individuals do not remain continually on a homochromous substrate. The shade and pattern (or lack of a pattern) can have a protective value even against animals lacking colour vision. This is true with regard to many predators that hunt at night when sagines are active.

If danger threatens, sagines at first remain motionless, relying on their natural camouflage. When approached or attacked by an enemy they try to escape by running or with a sudden leap into dense scrub. Agitated males often produce a sharp chirp before they launch themselves into flight. When cornered, *Clonia* and *Saga* adopt a peculiar startle attitude, similar to the defensive display of preying mantids, some Listroscelidinae and large spiders. Facing the aggressor, they elevate their body by rearing up on mid legs. The large spined fore legs are stretched out, exposing to view the black markings on the inner surface of femora. The mandibles are frequently held agape. While displaying, an excited female of *Clonia multispina* once vibrated the tongue-like hypopharynx, exposed by reflexing the labrum and spreading the mandibles. Fully winged species of *Clonia*, such as *C. multispina*, unfold and often raise their wings, presenting the black and white mottled hind wings (Fig. 5). A defensive behaviour was observed in individuals of *Saga pedo* where the insect fell to the ground and turned upside down presenting to the aggressor spread mandibles and stretched spiny fore and mid legs. This frightening stance inhibits smaller aggressors from attacking. Inexperienced larger predators are often momentarily bewildered. This short delay may permit the escape of the sagine.

A violently attacked *Saga* can enter a cataleptic state characterized by complete immobility. This motionless state can be produced by a hard pressure on the thoracic sternum and on the abdomen. It lasts for about 10 to 20 minutes and is followed by a very slow recovery, initiated by trembling and swaying of all legs. There is no evidence of a protective effect resulting from this kind of behaviour which occurs when the potential predator has already grasped the sagine. Death feigning prior to the attack may deceive animals which attack exclusively upon moving prey. However, such behaviour was not observed in the Saginae.

INTERSPECIFIC RELATIONS: CLASSIFICATION

Key to Known Genera and Subgenera of Saginae

1. Vertex presenting a conical appearance (Fig. 6). Prosternum unarmed or, at most, armed with a pair of short pegs. Wings very abbreviated or rudimentary. Slender species .... 4 *Peringueyella* Saussure

1'. Vertex not elongate or only slightly so (Figs 2-5, 10). Prosternum bispinose 2

2(1'). Wings well developed or abbreviated. Fore and mid tibiae with 6 spines on each margin of ventral surface .... 1. *Clonia* Stål 3

2'. Wings very reduced or rudimentary. Fore and mid tibiae with at least 8 spines on each margin of ventral surface 6

3(2) Slender species. Anal area of hind wings with black and white or black and yellow bands, concentrically arranged. Hind femur almost always more than 4.2 times the length of the pronotum 4

3' Mostly larger and more robust species. Anal area of hind wings entirely dark brown to brownish black (rarely with shiny patterns). Hind femur usually less than 4.2 times

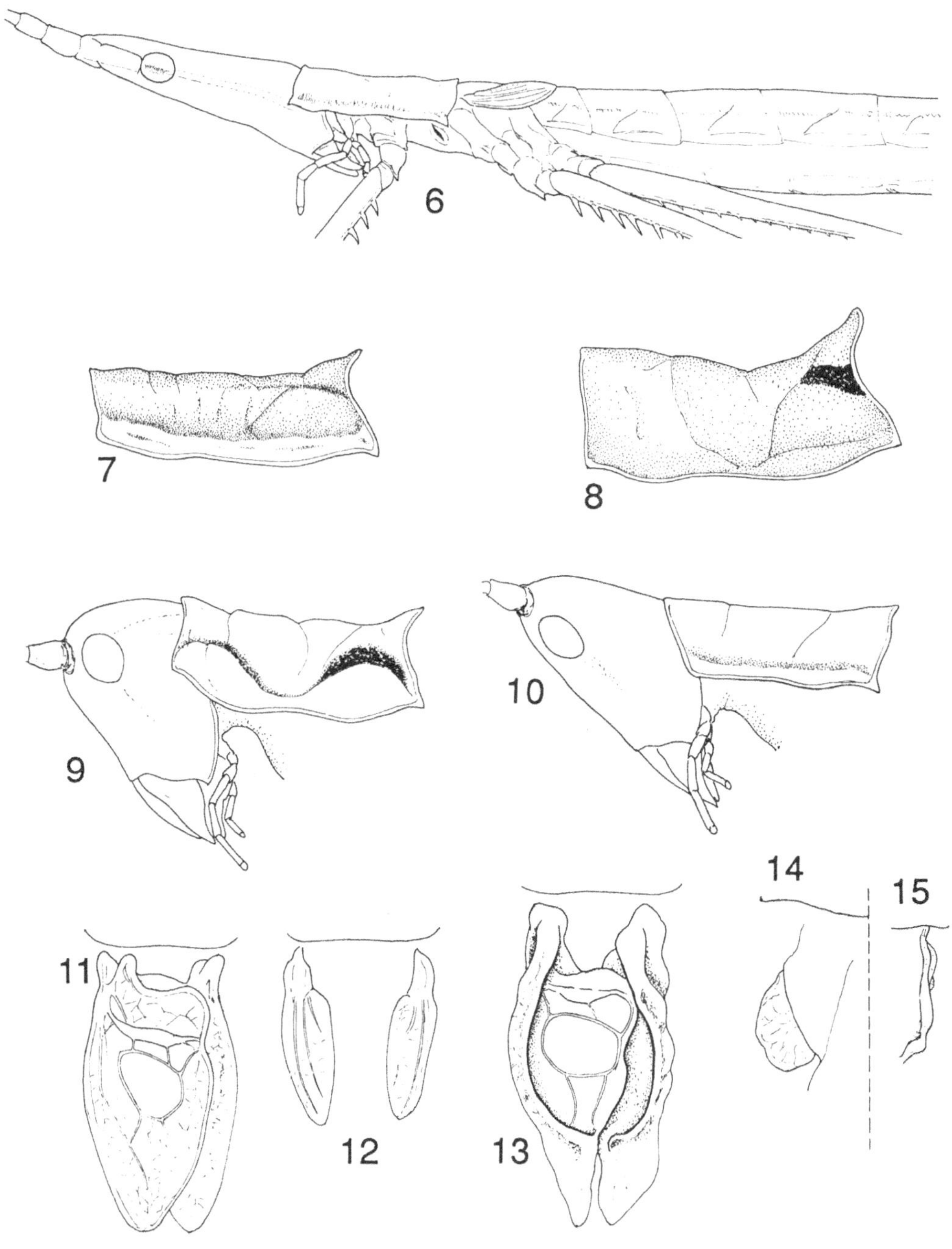

Figs. 6-15. Diagnostic structures in Saginae: 6, female of *Peringueyella j. jocosa* Saussure, showing conical prolongation of vertex and reduced wings; 7, male of *Clonia* (*C.*) *multispina* Uvarov, lateral view pronotum; 8, male of *Clonia* (*C.*) *wahlbergi* Stal, same view; 9, female of *Clonia* (*Leptoclonia*) *vittata* (Thunberg), lateral view head and pronotum; 10, female of *Cloniella praedatoria* (Distant), same view; 11, male of same species, dorsal view tegmina; 12, female of same species, reduced tegmina; 13, male of *Saga cappadoica* Werner, dorsal view tegmina; 14, female of *Saga natoliae* Serville, dorsal view left tegmen; 15, female of *Saga c. campbelli* Uvarov, dorsal view mesonotum (right flank) with lateral carina instead of tegmen.

the length of the pronotum..1d. Subgen. *Hemiclonia* Kirby

4(3) Hind wings with black and yellow patterns. Wings abbreviated ....
....1c. Subgen. *Xanthoclonia* Kaltenbach

4' Hind wings with black and whitish patterns (Fig. 5). Wings well developed or abbreviated 5

5(4') Sulcus of the lateral lob e of pronotum formed as a nearly straight line or at most slightly curved in the metazona (Figs 7,8)..............................1a. Subgen. *Clonia* Stål

5' Sulcus of the lateral lobe of pronotum clearly bisinuate (Fig. 9)..............................
....1b. Subgen. *Leptoclonia* Kaltenbach

6(2') Discoidal area of male tegmina elevated. Both tegmina combined forming a surrounding enclosure (bell-mouth) (Fig. 13). The tiny rudiments of female tegmina broader than long or wing rudiments absent (Figs 14,15).......................3. Gen. *Saga* Charpentier

6' Discoidal area of male tegmina not elevated (Fig. 11). Female tegmina vestigial, lobiform, longer than broad (Fig. 12)............................2. Gen. *Cloniella* Kaltenbach

Checklist of the Hitherto Known Species and Subspecies of Saginae

1. Genus *Clonia* Stål

Clonia Stål, 1855: 352. Type species: *Clonia wahlbergi* Stål by monotypy.

1a. Subgenus *Clonia* Stål s. str.

Clonia (*C.*) *multispina* Uvarov, (Figs 5,7): Southern central Africa to W.Tanzania.
Clonia multispina Uvarov, 1942: 59, 62, part.

Clonia (*C.*) *zernyi* Kaltenbach: E. and S. Tanzania; Malawi.
Clonia (*C.*) *zernyi* Kaltenbach, 1971: 421, 432.
Clonia multispina Uvarov, 1942: 59, part.

Clonia (*C.*) *wahlbergi* Stål (Fig. 8): Eastern part of southern Africa.
Clonia wahlbergi Stål, 1855: 352.

ssp. *maculosa* Walker: W.Zambia; Zimbabwe; W.Transvaal, Cape; Botswana; Namibia.
Saga maculosa Walker, 1869: 294.
Clonia (*C.*) *wahlbergi variabilis* Kaltenbach, 1971: 435.

Clonia (*C.*) *saussurei* Kaltenbach: South Africa.
Clonia (*C.*) *saussurei* Kaltenbach, 1971: 421, 423, 436.
Clonia (*C.*) *intermedia* Kaltenbach: Southern central Africa; Angola; Namibia.
Clonia (*C.*) *intermedia* Kaltenbach, 1971: 421, 423, 436.

Clonia (*C.*) *kalahariensis* Kaltenbach: South Africa; Botswana.
Clonia (*C.*) *kalahariensis* Kaltenbach, 1971: 421, 423, 437.

Clonia (*C.*) *caudata* Uvarov: Angola; Namibia.
Clonia caudata Uvarov, 1942: 58, 61, 62.

Clonia (C.) uvarovi Kaltenbach: Transvaal; Cape ?
Clonia (C.) uvarovi Kaltenbach, 1971: 421, 423, 438.

Clonia (C.) dewittei Kaltenbach: Southern and eastern Zaire.
Clonia (C.) dewittei Kaltenbach, 1971: 421, 425, 439.

Clonia (C.) burri Uvarov: E. Angola.
Clonia burri Uvarov, 1942: 59, 62.

Clonia (C.) jagoi Kaltenbach: SW Tanzania
Clonia (C.) jagoi Kaltenbach, 1971: 423, 441.

Clonia (C.) kenyana Uvarov: Kenya.
Clonia kenyana Uvarov, 1942: 60, 62.

Clonia (C.) angolana Kaltenbach: SW Angola.
Clonia (C.) angolana Kaltenbach, 1971: 421, 443.

1b. Subgenus *Leptoclonia* Kaltenbach

Leptoclonia Kaltenbach, 1971: 415. Type species: *Locusta vittata* Thunberg by original designation.

Clonia (L.) vittata (Thunberg): Namibia; Cape, Orange Free State. (Fig. 9)
Locusta vittata Thunberg, 1789: 102.
Clonia turneri Uvarov, 1942: 60, 62.

Clonia (L.) vansoni Kaltenbach: Western Cape.
Clonia (L.) vansoni Kaltenbach, 1971: 425, 446.

Clonia (L.) minuta (Haan): Southwestern Cape.
Locusta (Saga) minuta Haan, 1842: 209.

1c. Subgenus *Xanthoclonia* Kaltenbach

Xanthoclonia Kaltenbach, 1971: 417. Type species: *Clonia tessellata* Saussure by monotypy.

Clonia (X.) *tessellata* Saussure: South Africa.
Clonia tessellata Saussure, 1888: 140, 143

1d. Subgenus *Hemiclonia* Kirby

Hemiclonia Kirby, 1906: 226. Type species: *Gryllus melanopterus* Linnaeus by original designation.

Clonia (H.) melanoptera (Linnaeus): Southwestern Cape.
Gryllus (Tettigonia) melanopterus Linnaeus, 1758: 430.
Gryllus Tettigoniae azurea Stoll, 1813: 27.
Conocephalus spinigerus Thunberg, 1815: 277.
Saga elegans Blanchard, 1840: 29.

Clonia (H.) lalandei Saussure: Eastern South Africa.
Clonia lalandaei Saussure, 1888: 140, 142.

Clonia (H.) assimilis Kaltenbach: Botswana; Natal.
Clonia (H.) assimilis Kaltenbach, 1971: 426, 452.
Clonia lalandei: Saussure, 1892: 15. (nec *C. lalandaei* Saussure, 1888)

Clonia (H.) charpentieri Kaltenbach: Natal; Cape ?
Clonia (H.) charpentieri Kaltenbach, 1971: 426, 452, ♂ (nec ♀)

Clonia (H.) ignota Kaltenbach: Cape ?
Clonia (H.) ignota Kaltenbach, 1981: 189.
Clonia (H.) charpentieri Kaltenbach, 1971: 426, 453, ♀ (nec ♂)

2. Genus *Cloniella* Kaltenbach

Cloniella Kaltenbach, 1971: 418. Type species: *Hemisaga praedatoria* Distant by monotypy.

Cloniella zambesica Kaltenbach: Zambia.
Cloniella zambesica Kaltenbach, 1971: 427, 453.

Cloniella praedatoria (Distant): Transvaal (Figs 10-12)
Hemisaga praedatoria Distant, 1892: 65, 258.
Peringueyella praeditoria (err. impr.) auct.

3. Genus *Saga* Charpentier

Saga Charpentier, 1825: 95. Type species: *Locusta serrata* Fabricius (= *Gryllus pedo* Pallas) by monotypy.
Tettigopsis Fischer de Waldheim, 1830: 1-16. Type species: *Gryllus pedo* Pallas (Kirby 1906).

Saga natoliae Serville: Mediterranean east coast, south and west Turkey, southern Balkans except Greece south of Macedonia (Figs 2, 4, 13, 14).
Saga natoliae Serville, 1839: 541.
Gryllus Tettigoniae Onos Stoll, 1813: 24 (*nomen oblitum*).
Saga gigantea Herrich-Schäffer, 1840: 15, 27 (nec *Gryllus giganteus* Villers 1789).
Saga synophrys Charpentier, 1841: 319.
Saga brunneri Saussure, 1888: 131, 133.

Saga rhodiensis Salfi: SW.Turkey; Rhodes.
Saga rhodiensis Salfi, 1929: 214.

Saga ephippigera Fischer de Waldheim: Mediterranean east coast; East Turkey; Caucasus; NW Iran.
Saga ephippigera Fischer de Waldheim, 1846: 185.
Saga monstrosa Krauss, 1878: 509, 510.

Saga ephippigera var. de Syrie Saussure, 1888: 135.
? *Saga uvarovi* Ramme, 1951: 190.

ssp. *syriaca* Lucas: N. Syria; Iraq.
Saga syriaca Lucas, 1864: 5

Saga hellenica Kaltenbach: Greece; Macedonia; Albania.
Saga hellenica Kaltenbach, 1967: 28, 61.
Saga serrata auct. part. (nec *Locusta serrata* Fabricius, 1793).
Saga vittata auct. part. (nec *Tettigopsis vittata* Fischer de Waldheim, 1830).
Saga italica: Ramme, 1951: 180 (nec *Saga italica* Costa, 1836-1857).

Saga cappadoica Werner: Central and S. Turkey.
Saga cappadoica Werner, 1903: 528.
Saga serrata auct. (nec *Locusta serrata* Fabricius, 1793).

Saga rammei Kaltenbach: Macedonia; S. Bulgaria.
Saga rammei Kaltenbach, 1965: 484 (1967; 34, 70).
Saga ornata (?): Uvarov, 1923: 151 (nec *Saga ornata* Burmeister,1839).
Saga campbelli: Ramme, 1951: 182 (nec *Saga campbelli* Uvarov, 1921).

Saga puella Werner: Turkey.
Saga puella Werner, 1901: 298.

Saga campbelli Uvarov: S. Macedonia.
Saga campbelli Uvarov, 1921: 158.

ssp. *gracilis* Kis: N. Aegean Islands; Thrace; Dobrogea.
Saga italica gracilis Kis, 1962: 255.
Saga campbelli gracilis Kaltenbach, 1967: 35, 77.

Saga beieri Kaltenbach: Syria; E. Turkey.
Saga beieri Kaltenbach, 1967: 39, 78.
Saga lucasi auct. (nec *Saga lucasii* Saussure 1888)

Saga longicaudata Krauss: SE. Turkey
Saga longicaudata Krauss, 1878: 508, 510.

Saga ledereri Saussure: Lebanon; SE. Turkey ?
Saga ledereri Saussure, 1888: 131, 134.
Saga kraussi Kirby, 1906: 224 (nom. nov. pro *S. ornata*: Krauss, 1878, nec *ornata* Burmeister 1839).

Saga ornata Burmeister: Israel.
Saga ornata Burmeister, 1839: 717.
Saga lucasii Saussure, 1888: 135.
Saga savignyi Saussure, 1888: 132.
Saga gracilipes Uvarov, 1924: 179 (In: Buxton and Uvarov, 1924).

Saga pedo (Pallas): W. Siberia; Caucasus; southern Russia; Balkans (except central and southern Greece); central and S. Europe extending into the Iberian Peninsula. Introduced into USA (Michigan), but no new record since 1972.
Gryllus (Tettigonia) pedo Pallas, 1771: 467, Zf.
Gryllus Tett. giganteus De Villers, 1789: 451.
Locusta serrata Fabricius, 1793: 43.
Tettigopsis nudipes Fischer de Waldheim, 1830: 15.
Tettigopsis vittata Fischer de Waldheim, 1830: 16.
Saga undulata Fischer de Waldheim, 1846: 188.
Saga italica Costa, 1836-1857: 7.
Saga quadrisignata Philippi, 1863: 235.

4. Genus *Peringueyella* Saussure

Peringueyella Saussure, 1888: 129, 150. Type species: *Peringueyella jocosa* Saussure (Kirby 1906 designated).

Peringueyella macrocephala (Schaum): Zimbabwe; Mozambique; Botswana ?
Saga macrocephala Schaum, 1853: 778.

Peringueyella jocosa Saussure: SE. Africa.
Peringueyella jocosa Saussure, 1888: 152.

ssp. *multispina* Kaltenbach: Transvaal; Orange Free State; Botswana.
Peringueyella jocosa multispina Kaltenbach, 1971: 458.

Peringueyella rentzi Kaltenbach: Transvaal.
Peringueyella rentzi Kaltenbach, 1981: 190.

Peringueyella zulu Kaltenbach: Zululand.
Peringueyella zulu Kaltenbach, 1971: 427, 458.

Incertae sedis

Saga quadrisignata Philippi: Locality ? (Certainly not Chile!)
Saga quadrisignata Philippi, 1863: 235 (= *Gryllus pedo* Pallas ?).

Saga savignyi Saussure: Locality ?
Saga savignyi Saussure, 1888: 132 (= *Saga ornata* Burmeister ?).

Saga uvarovi Ramme: Israel (Gallilee).
Saga uvarovi Ramme, 1951: 190 (= *Saga ephippigera* Fischer de Waldheim?).

Emptera indica (Herbst): India (probably incorrect)
Locusta indica Herbst, 1786: 193.
Saga indica: Burmeister, 1839: 714
Emptera indica: Saussure, 1888: 148.

References

Akerman C. 1932. On the Carnivorous Habits of the Long-Horned Grasshopper, *Clonia vittata* Thunberg. Ann. Natal Mus. 7: 43-144.

Ander K. 1957. Zwei neue Saginen (Orth., Tettigoniidae). Entomol. Tidskr. 78: 1-5.

Bérenguier P. 1905. La Magicienne dentelée "*Saga serrata*". Bull. Soc. Sci. Nat. 33: 145-152.

Blanchard E. 1840. Hist. Nat. 3: 1-673.

Bodenheimer F.S. 1935. Ökologisch-zoogeographische Untersuchungen über die Orthopterenfauna Palastinas. Arch. Naturgesch. 4(B)-1: 88-142, 2: 145-216.

Boldyrev B.Th. 1915. Contributions à l'étude de la structure des spermatophores et des particularites de la copulation chez Locustodea et Gryllodea. Horae Soc. Entomol. Ross. 41: 1-245.

Brancsik C. 1897. Series Orthopterorum novorum. Jh. Naturwiss. Ver. Trencséner Com. 19-20: 52-87.

Brunner von Watteneyl C. 1893. Révision du Système des Orthopteres. Ann. Mus. Civ. Stor. Nat. 33: 3-230.

Burr M., B.P. Campbell & B.P. Uvarov. 1923. A contribution to our knowledge of the Orthoptera of Macedonia. Trans. Entomol. Soc. Lond. 1923: 110-169.

Buxton P.A. & B.P. Uvarov. 1924. A contribution to our knowledge of Orthoptera of Palestine. Bull. Soc. Entomol. Égypte 6: 167-214.

Cantrall I.J. 1972. *Saga pedo* (Pallas) (Tettigoniidae: Saginae), an Old World Katydid, new to Michigan. Great Lakes Entomol. 5: 103-106.

Caudell A.N. 1916. Orthoptera Fam. Locustidae Subfam. Saginae. Genera Insectorum. Ed. P. Wytsman. Bruxelles. 167: 1-10.

Charpentier T. 1825. Horae Entomologicae, adjectis tabulis novem coloratis. Wratislaviae. pp 1-261.

Charpentier T. 1841. Einige Bemerkungen die Orthopteren betreffend insbesonders in Bezug auf Burmeisters und Servilles Schriften uber diese Insektenabteilung. Ed. Germar. Leipzig. Z. Entomol. 3: 283-321.

Chopard L. 1937. Les Sagiens. Encyclopedie Francaise 5(2).

Chopard L. 1946. Quelques notes sur les meurs de la *Saga* (Orth. Tettigoniidae). Bull. Soc. Entomol. Fr. 51: 126-128.

Costa O. 1836-1857. Fauna del Regno di Napoli. Ortotteri/Locust. Napoli. pp 1-24.

Distant W.L. 1892. A Naturalist in the Transvaal. London. pp 1-277.

Fabricius J.C. 1793. Entomologia systematica. Hafniae. 1-519.

Fischer de Waldheim G. 1830. Notice sur le *Tettigopsis*, nouveau genre d'Orthopteres de la Russie. Moscow. pp 9-16.

Fischer de Waldheim G. 1846. Entomographie de la Russie. IV. Orthoptères. Moscow. 1-413.

De Haan W. 1842. Bijdragen tot de Kennis der Orthoptera. Verh. nederland-sche Overz. Bezitt. Insecta pp 45-238.

Hartley J.C. & A.C. Warne. 1972. The developmental biology of the egg stage of Western European Tettigoniidae (Orthoptera). J. Zool. 168: 267-298.

Hebard M. 1922. Studies in Malayan, Melanesian and Australian Tettigoniidae (Orthoptera). Proc. Acad. Nat. Sci. Phila. 74: 121-299.

Herbst J.F. 1786. Fortsetzung des Verzeichnisses meiner Insekten-Sammlung. Zweite

Klasse. Fuessly, Arch. Ins.-Gesch., Zurich. 7/8: 183-196.
Herrich-Schäffer G. 1840. Nomencl. Entomol. Regensburg. 2 (Verz. Europ. Ins./Orth.): 1-31.
Ingrisch S. 1986. The plurennial life cycles of the European Tettigoniidae (Insecta: Orthoptera). 1. The effect of temperature on embryonic development and hatching. Oecologia 70: 606-616.
Jaus I. 1934. Ein Beitrag zur Biologie von Saga *serrata* F. Konowia. Wien 13: 171-177.
Kaltenbach A. 1964. Zur Systematik und Verbreitung der Raubheuschrecken (Tettigoniidae-Saginae), insbesondere der europäischen Arten der Gattung Sag*a* Charpentier. Z. Arbeitsgem. Österr. Entomol. 16: 68-82.
Kaltenbach A. 1965. Dictyoptera und Orthopteroidea von Nordost-Griechenland und der Insel Thasos. Ann. Naturhist. Mus. Wien 68: 465-484.
Kaltenbach A. 1967. Unterlagen für eine Monographie der Saginae. I. Superrevision der Gattung Sag*a* Charpentier (Saltatoria: Tettigoniidae). Beitr. Entomol. 17: 3-107.
Kaltenbach A. 1969. Subfamilia Saginae. In: The Orthoptera of Europe. I. Ed. K. Harz. The Hague. pp 478-490.
Kaltenbach A. 1970. Unterlagen für eine Monographie der Saginae. II. Beiträge zur Autökologie der Gattung *Saga* Charpentier (Saltatoria: Tettigoniidae). Zool. Beitr. (n.F.) 16: 155-245.
Kaltenbach A. 1971. Unterlagen für eine Monographie der Saginae III. Die Saginae der äthiopischen Region (Saltatoria: Tettigoniidae). Beitr. Entomol. 21: 403-476.
Kaltenbach A. 1972. An illustrated key for identifying African Saginae (Saltatoria: Tettigoniidae). Ann. Natal Mus. 21: 281-295.
Kaltenbach A. 1981. Neue und wenig bekannte Saginae aus Südafrika (Salta-toria-Tettigoniidae). Sitzungsber. Akad. Wiss. Wien Math.-Naturwiss. Kl. (I) 190: 187-192.
Kaltenbach A. 1986. Saginae (Saltatoria-Tettigoniidae). In: Das Tierreich (The Animal Kingdom). Berlin, New York. 103: 1-92.
Kirby W.F. 1906. A Synonymic Catalogue of Orthoptera. London. 2(1): 1-562.
Kis B. 1962. *Saga italica gracilis* eine neue Unterart aus Rumänien (Orthoptera, Tettigoniidae). Ann. Hist.-Nat. Mus. Nat. Hung. 54: 255-258.
Krauss H. 1878. Die Orthopterenfauna Istriens. Sitzungsber. Akad. Wiss. Wien Math.-Naturwiss. Cl. (I) 78: 451-544.
Lang A. 1930. Über das Vorkommen von *Saga serrata* F. (Orthoptera) in der Umgebung von Deutsch-Altenburg, N.-Osterreich. Z. Ver. Naturbeobachter u. Sammler Wien 5/1: 1-4.
Linnaeus C. 1758. Systema Naturae. 10th ed. Holmiae. pp 1-824.
Lucas H. 1864. Bull. Entomol. (séance du 27 Janvier 1864). Ann. Soc. Entomol. Fr. (4) 4, 5.
Matthey R. 1941. Etude biologique et cytologique de *Saga pedo* Pallas (Orthoptères-Tettigoniidae). Rev. Suisse Zool. 48: 91-142.
Matthey R. 1946. Démonstration du caractère géographique de la parthénogènèse de *Saga pedo* Pallas et de sa polyploidie, par comparaison avec les éspéces bisexuées *S. ephippigera* Fisch. et *S. gracilipes* Uvar. Experientia (Basel) 2/7: 1-3.
Matthey R. 1948. Données nouvelles sur les chromosomes des Tettigonides et la parthénogénèse de *Saga pedo* Pallas. Rev. Suisse Zool. 55: 45-56.
Matthey R. 1950. Les chromosomes de *Saga cappadocica* Werner (Orthoptera – Tettigoniidae). 9. Jahresber. schweiz Ges. Vererbungsforsch.-Arch. Julius Klaus- Stiftung f. Vererbungsforsch. etc. Zürich 25: 44-46.
Nadig A. 1987. Zur Lebensweise von *Saga rhodiensis* Salfi. Articulata Würzburg II/10: 359-

361.
Pallas P.S. 1771. Reise durch verschiedene Provinzen des Russischen Reiches. Petersburg. 1: 1-504.
Philippi R.A. 1863. Verzeichnis der im Museum von Santiago befindlichen Chilenischen Orthopteren. Z. Ges. Naturwiss. Berlin 21: 217-245.
Ramme W. 1951. Zur Systematik, Faunistik und Biologie der Orthopteren von Südost-Europa und Vorderasien. Mitt. Zool. Mus. Berl. 27: 3-431.
Rentz D.C. 1979. Comments on the Classification of the Orthopteran Family Tettigoniidae, with a Key to Subfamilies and Description of Two New Subfamilies. Aust. J. Zool. 27: 991-1013.
Rentz D.C. 1985. A Monograph of the Tettigoniidae of Australia. Vol. I. The Tettigoniinae, CSIRO, Australia. pp. 1-384.
Sänger K. 1980. Zur Larvalentwicklung der Laubheuschrecken-Art Saga natoliae (Orthoptera: Tettigonioidea). Entomol. Gen. 6: 65-69.
Sänger K. 1987. The Role of Temporary Immobility in the Mating Behaviour of *Saga natoliae* Serville and *Saga ephippigera* Fischer de Waldheim (Ensifera: Tettigonioidea). Zool. Anz. 219: 66-69.
Salfi M. 1929. Ortotteri. In: Ricerche Faunistiche nelle Isole Italiane dell'Egeo. Arch. Zool. Ital. 13/1-2: 211-219.
Saussure H. 1888. Synopsis de la Tribu des Sagiens. Ann. Soc. Entomol. Fr. 8: 127-160.
Saussure H. 1892. Note Supplémentaire a la Synopsis de la Tribu des Sagiens. Ann. Soc. Entomol. Fr. 61: 5-16.
Schaum H. 1853. Ubersicht der von ihm ["Peters"] in Mossambique beobach-teten Orthopteren etc. Monatsber. dt. Akad. Wiss. Berlin 1853: 775-780.
Schremmer F. 1959. Ein gespenstiges Insekt. Universum für Natur und Technik Wien 14: 215-217.
Serville J.G. 1839. Histoire naturelle des Insectes. Orthoptères. Paris. pp 1-777.
Stål C. 1855. Entomólogiska Notiser. Hemipt., Orthopt., Forfic. fån Cafferlandet. Ofversigt r. VetenskAkad. Förh. Stockholm 12: 348-353.
Stoll C. 1813. Natuurlijke en naar het Leven nauwkeurige gekleurde Afbeeldingen en Beschrijvingen der Spoken, Wandelnde Bladen, Zabelspring- hanen etc., Zabelspr.: pp.1-28; Trek-Springh.: pp.1-42 (Amsterdam).
Thunberg C.P. 1789. Dissertatio entomologica novas insectorum species sistens. Uppsala. 5: 85-106.
Thunberg C.P. 1815. Hemipterorum maxillosorum genera illustrata. Mem. Acad. St. Petersb. 5: 211-301.
Uvarov B.P. 1921. Descriptions of four new Orthoptera from Macedonia. Entomol. Rec. J. Var. 33: 145-164.
Uvarov B.P. 1924. Notes on the Orthoptera in the British Museum. 3. Some less known or new genera and species of the subfamilies Tettigoniinae and Decticinae. Trans. Ent. Soc. Lond. 1923: 492-536.
Uvarov B.P. 1942. Revision of the Genus Cl*onia* Stål 1855 (Orthoptera, Tettigoniidae). Proc. R. Entomol. Soc. Lond. B 11/4: 57-62.
Villers C. 1789. Caroli Linnaei Entomologia. Lugduni. pp 1-766.
Walker F. 1869. Catalogue of the specimens of Dermaptera, Saltatoria in the collection of the British Museum.British Museum of Natural History, London. 2: 1-423 (290-294).
Werner F. 1901. Die Dermatopteren-und Orthopterenfauna Kleinasiens. Sitzungsber. Akad.

Wiss. Wien Math.-Naturwiss. Cl. (I) 110: 259-306.
Werner F. 1903. Neue Locustiden aus Westasien. Zool. Anz. 26: 528-531.
Zacher F. 1928. Züchtung von Orthopteren. In: Handbuch der biologischen Arbeitsmethoden, Sect. 9. Ed. E. Abderhalden. Berlin, Wien. 1/II, 1: 89-190.

16. KARYOTYPES AND MEIOSIS OF THE AUSTRALIAN TETTIGONIINAE

Norihiro Ueshima
D.C.F. Rentz

Introduction

The chromosome systems of the Tettigoniinae were reviewed by Hewitt (1979) for studies published to 1976. In addition, Mesa and Ferreira (1977), Ueshima and Rentz (1979), Camacho *et al.* (1981), Warchalowska-Sliwa (1984) and Ueshima (1986) have published on the group. However, only one Australian species, *Lanciana albidicornis* Walker, has ever been reported in the literature, by Ferreira (1969). We now have data on the karyotypes of the majority of the known species of Australian Tettigoniinae.

All of the material used in this study is referenced in Vol. 1 of Rentz (1985). It is expected that future contributions in that series will have karyotype data included with the taxonomy.

Materials and Methods

The specimens were collected by DCFR and each specimen was given an individual cytological preparation number (here an "R" number) which correlates the actual insect and its testes. The "R" numbers are listed by Rentz (1985) as "cytol. preps." and therefore most details of the collecting localities, dates etc. can be found under the appropriate species in the monograph. If not, a field trip stop number is recorded and the user can refer to the "Specimens examined" section of Rentz (1985) for precise locality data. Locality data for specimens collected subsequent to the publication of the above work are given in full. Testes were fixed in Carnoy's fluid, refrigerated, and later sent to NU for study.

Cytological observations were made on squash preparations stained with alcohol-carmine. Modern techniques of karyotype analysis, such as G-banding and C-banding methods, have not been used since the fixed materials had been kept for several years.

The magnification of figures is indicated by a scale which is 10µm on each.

Observations

Ectopistidectes daptes Rentz (Fig. 1)

Material examined: R78-193 (holotype), R78-177

2n♂ =23

The karyotype comprises two pairs of metacentric autosomes, nine pairs of telocentric autosomes, and a large telocentric X chromosome (Figs 1a, 1b).

At meiosis, metacentric bivalents usually have a chiasma in each arm, the one in the long arm being interstitial and that in the shorter arm terminal. Telocentric bivalents usually form a single terminal or distal chiasma. At first metaphase, there are 11 autosomal bivalents (2 metacentrics and nine telocentrics) and the telocentric X univalent (Fig. 1c). At first anaphase, the X moves to one pole without dividing, as is usual in Orthoptera (Fig. 1d). There are two types of second metaphase: 11A+X (Fig. 1e) and 11A (Fig. 1f).

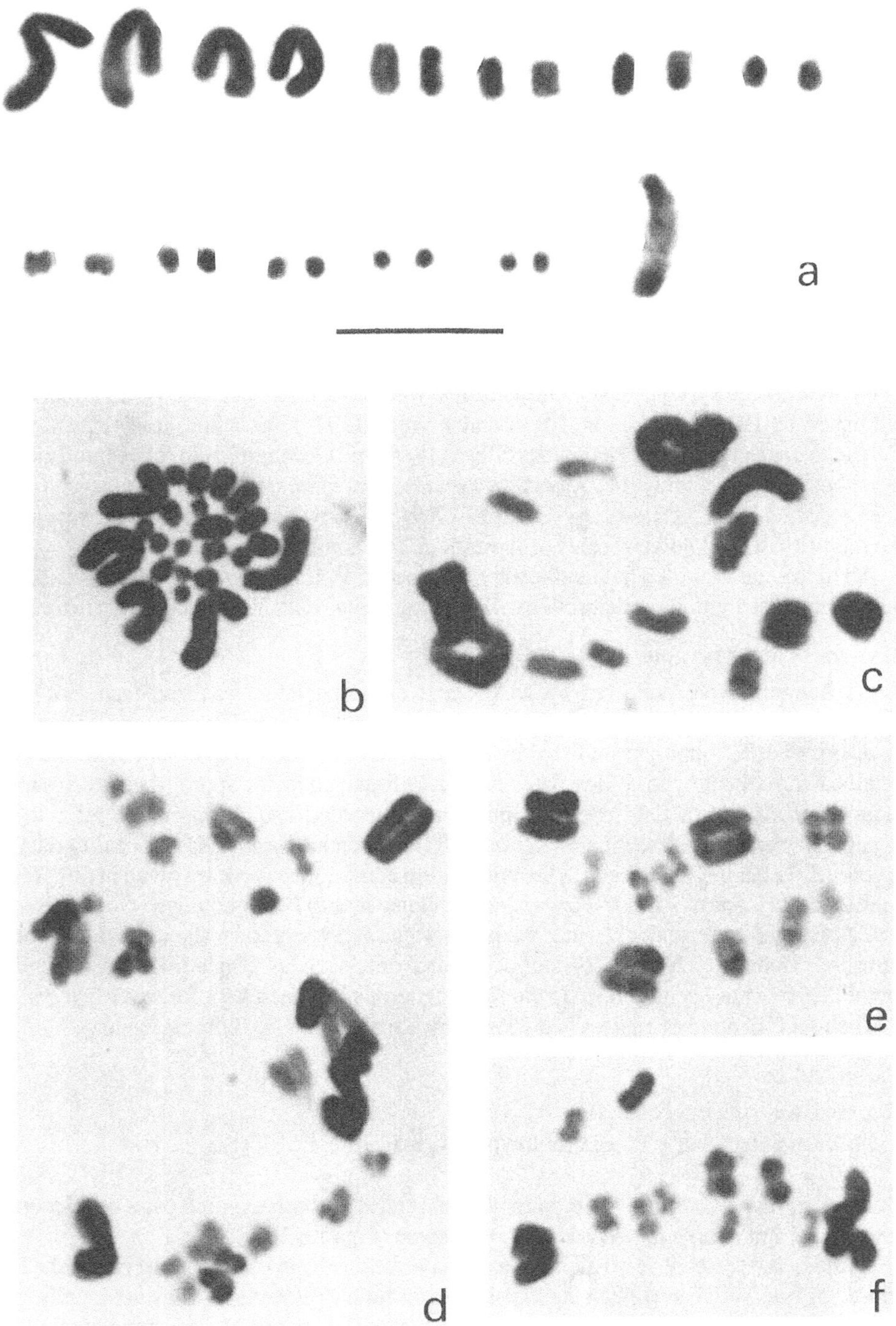

Fig. 1. *Ectopistidectes daptes* Rentz. A: Male karyotype, b. Spermatogonial metaphase, c: First metaphase, D: First anaphase, E: Second Metaphase with X, and f: Second etaphase without X.

Ectopistidectes viridis Rentz (Fig. 2)

Material examined: R81-17 (holotype), R81-12, -16, -18, -39

2n♂ = 23

The 2n♂ = 23 (Fig. 2a,b). The chromosome arrangement is similar to that of *E. daptes*. At first metaphase, there are 11 autosomal bivalents and the X univalent (Fig. 2c). At first division, the X moves to one pole without dividing (Fig. 2d). As a result of first division, two types of second metaphase are produced: 11A+X (Fig. 2e) and 11A (Fig. 2f).

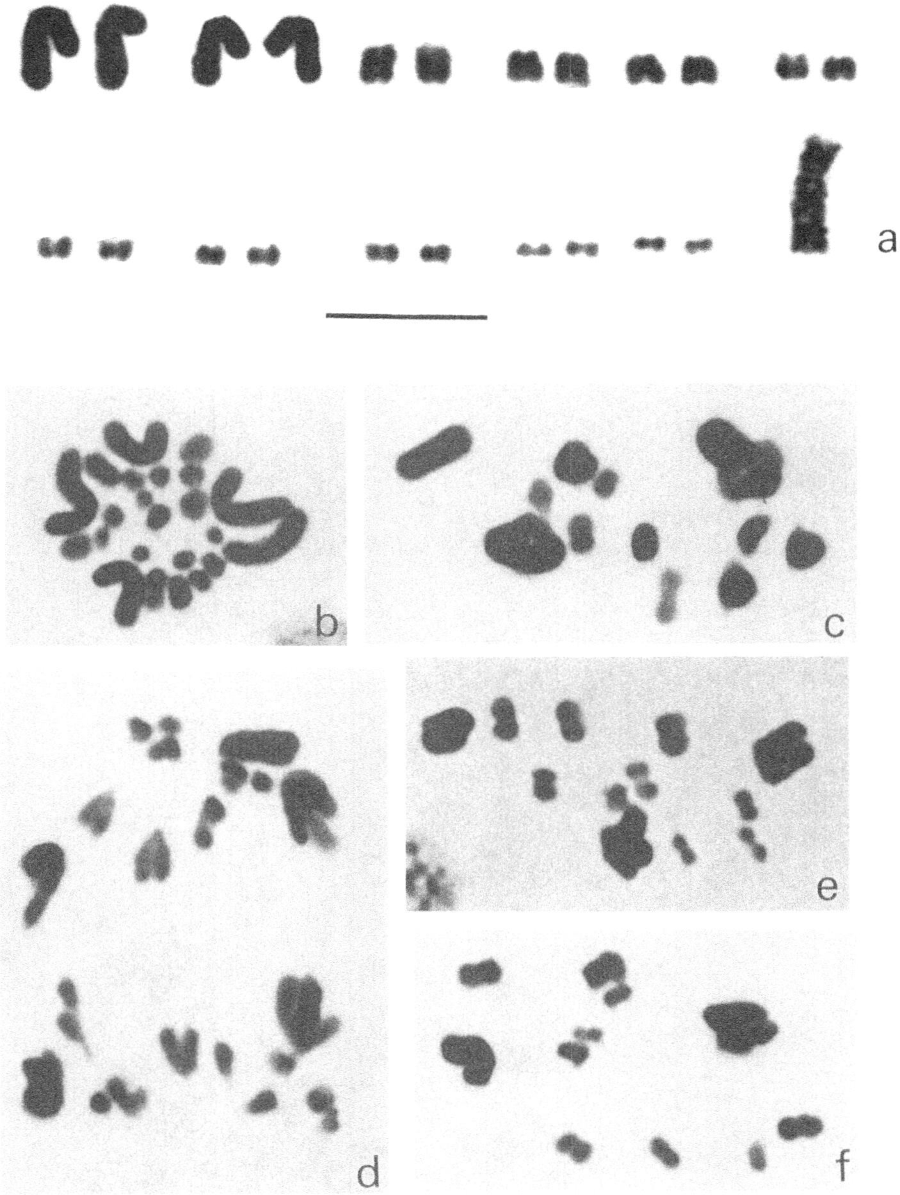

Fig. 2. *Ectopistidectes viridis* Rentz. A: Male karyotype, b: Spermatogonial metaphase, c: First metaphase, d: First metaphase, d: First anaphase, e: Second metaphase with X, and f: Second metaphase without X.

Chlorodectes baldersoni Rentz (Fig. 3)

Material examined: R80-8 (holotype)

2n♂ =25

The karyotype comprises one pair of metacentric autosomes, 12 pairs of telocentric autosomes and a metacentric X chromosome (Figs 3a, 3b). One limb of the metacentric autosome is about twice the length of the other. The arms of the metacentric X are almost equal in length.

During diplotene, there are 11 autosomal bivalents and a heteropycnotic X chromosome (Fig. 3c). At meiosis, a metacentric bivalent usually has a chiasma in each arm, the one in the long arm being interstitial, and the one in the short arm terminal. Telocentric bivalents usually form a single terminal or distal chiasma; however, they form a proximal chiasma as well. At first metaphase, there are 12 autosomal bivalents and the metacentric X univalent (Figs 3d, 3e). At first anaphase, the X moves to one pole without dividing (Fig. 3f).

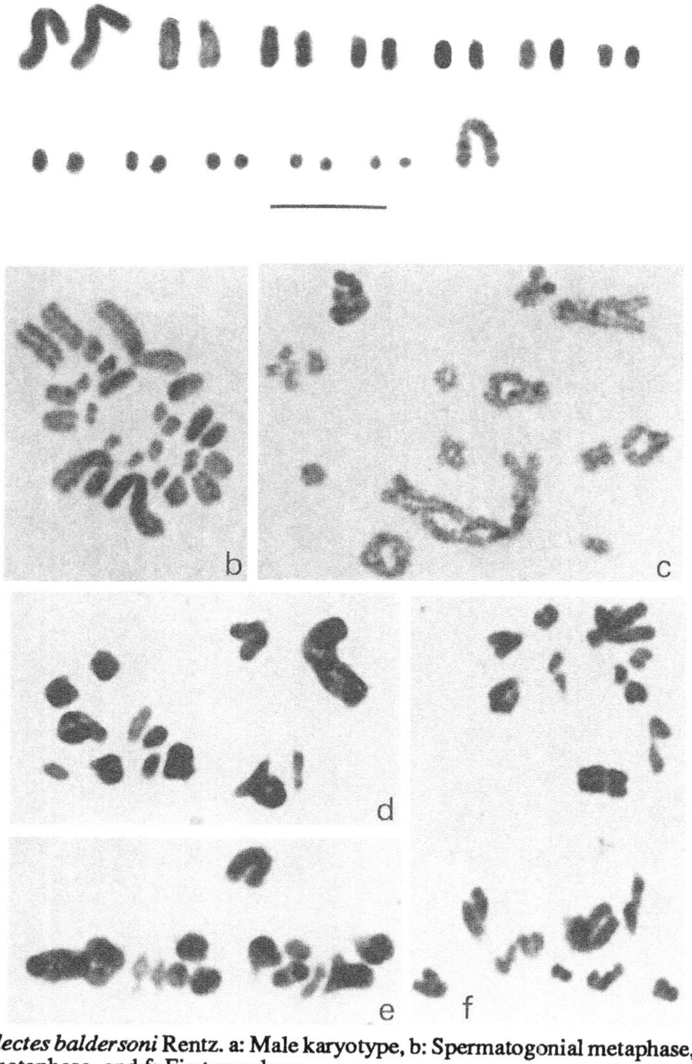

Fig. 3. *Chlorodectes baldersoni* Rentz. a: Male karyotype, b: Spermatogonial metaphase, c: Diplotene, d and e: First metaphase, and f: First anaphase.

Chlorodectes ligaenus Rentz (Fig. 4)

Material examined: R82-184 (holotype), R82-200, -207, R83-170

2n♂ = 23

The karyotype comprises one pair of metacentric and ten pairs of telocentric autosomes and a metacentric X (Figs 4a, 4b). One arm of the metacentric autosome is about twice the length of the other. The arms of the X are unequal in length.

During diplotene, there are 11 autosomal bivalents and a heteropycnotic X chromosome (Fig. 4c). At first metaphase, there are 11 autosomal bivalents and the X (Figs 4d, 4e).

Chlorodectes loquax Rentz (Fig. 5)

Material examined: R77-37 (holotype), R77-28, -31, -32, -33, -34

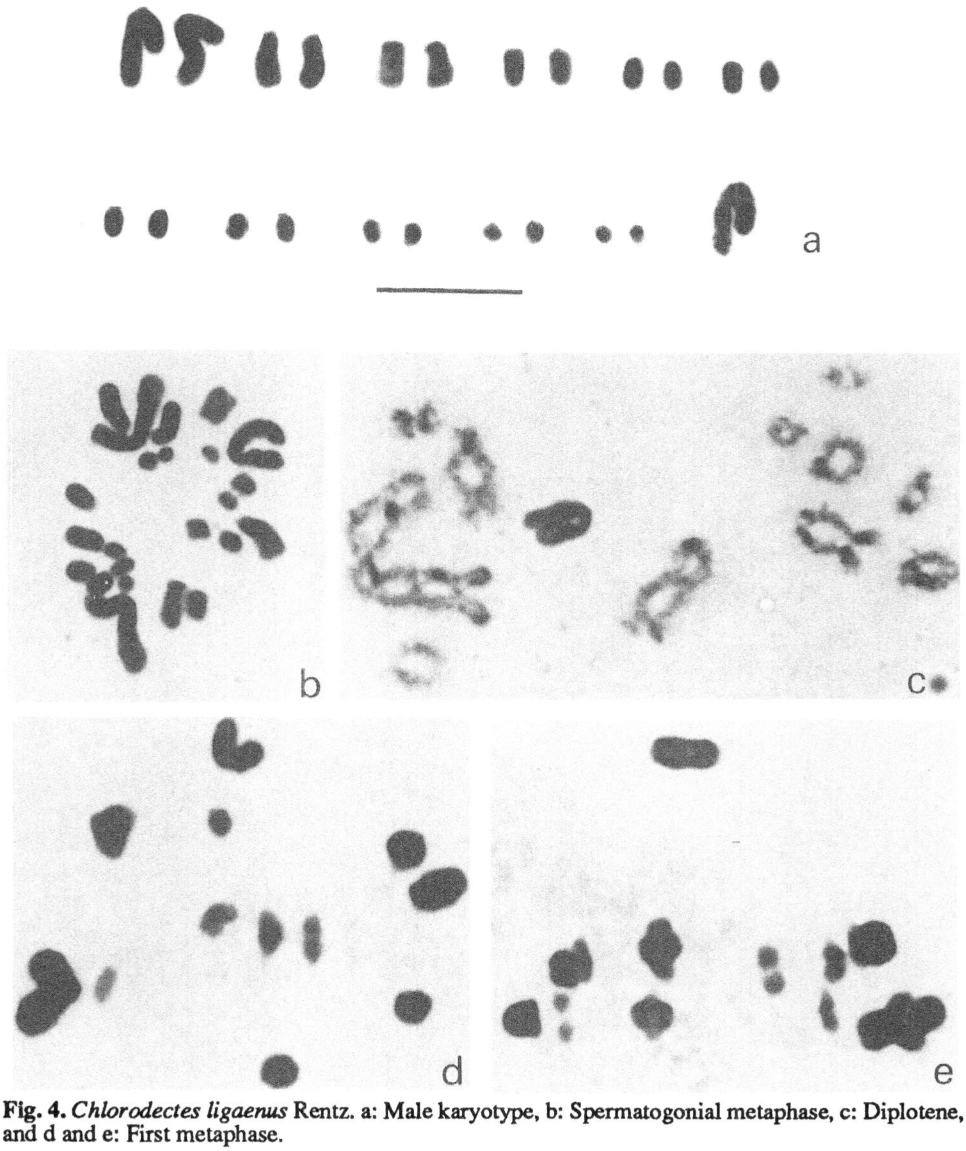

Fig. 4. *Chlorodectes ligaenus* Rentz. a: Male karyotype, b: Spermatogonial metaphase, c: Diplotene, and d and e: First metaphase.

2n♂ = 27

The spermatogonial metaphase of this species bears 27 chromosomes, consisting of one pair of metacentric autosomes, 12 pairs of telocentric autosomes, and a metacentric X chromosome (Figs 5a, 5b). One limb of the metacentric autosome is about three times the length of the other, and one limb of the X is also about three times the length of the other. The 12 pairs of telocentric autosomes show a gradation in size.

During diplotene, there are 13 autosomal bivalents and a heteropycnotic X chromosome (Fig. 5c). At first metaphase, there are 13 autosomal bivalents and metacentric X (Figs 5d, 5e).

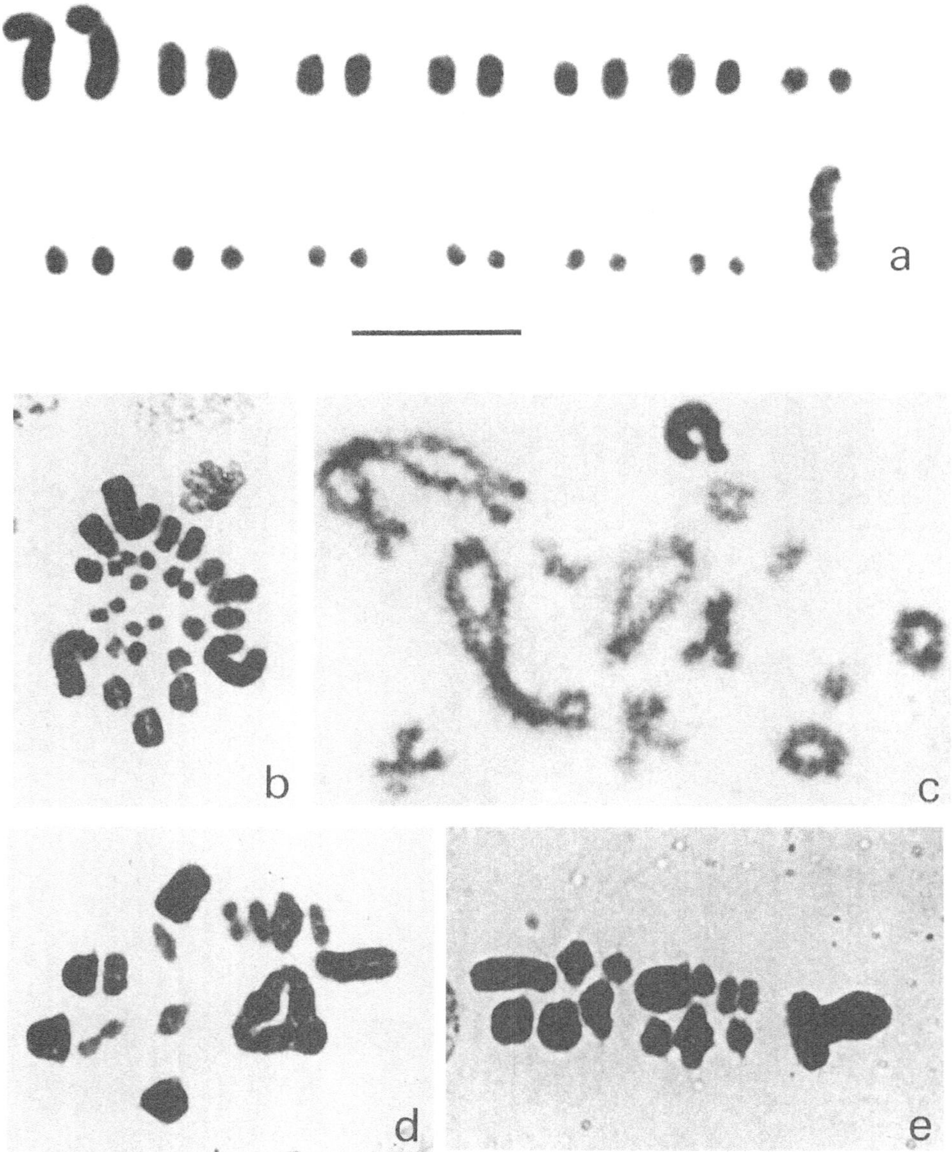

Fig. 5. *Chlorodectes loquax* Rentz. a: Male karyotype, b: spermatogonial metaphase, c: Diplotene, and d and e: First metaphase.

Chlorodectes montanus Rentz (Fig. 6)

Material examined: R86-11

2n♂ = 25

No spermatogonial metaphase of this species was obtained. However, based on meiosis, the chromosome complement can be assumed as 25 (24A+X) in the male. The karyotype of the species would be one pair of metacentric and 11 pairs of telecentric autosomes and the metacentric X. At first metaphase, there are 12 autosomal bivalents and the X (Fig. 6a). As a result of first division, two types of second spermatocytes are formed: 12A+X (Fig. 6b) and 12A (Fig. 6c).

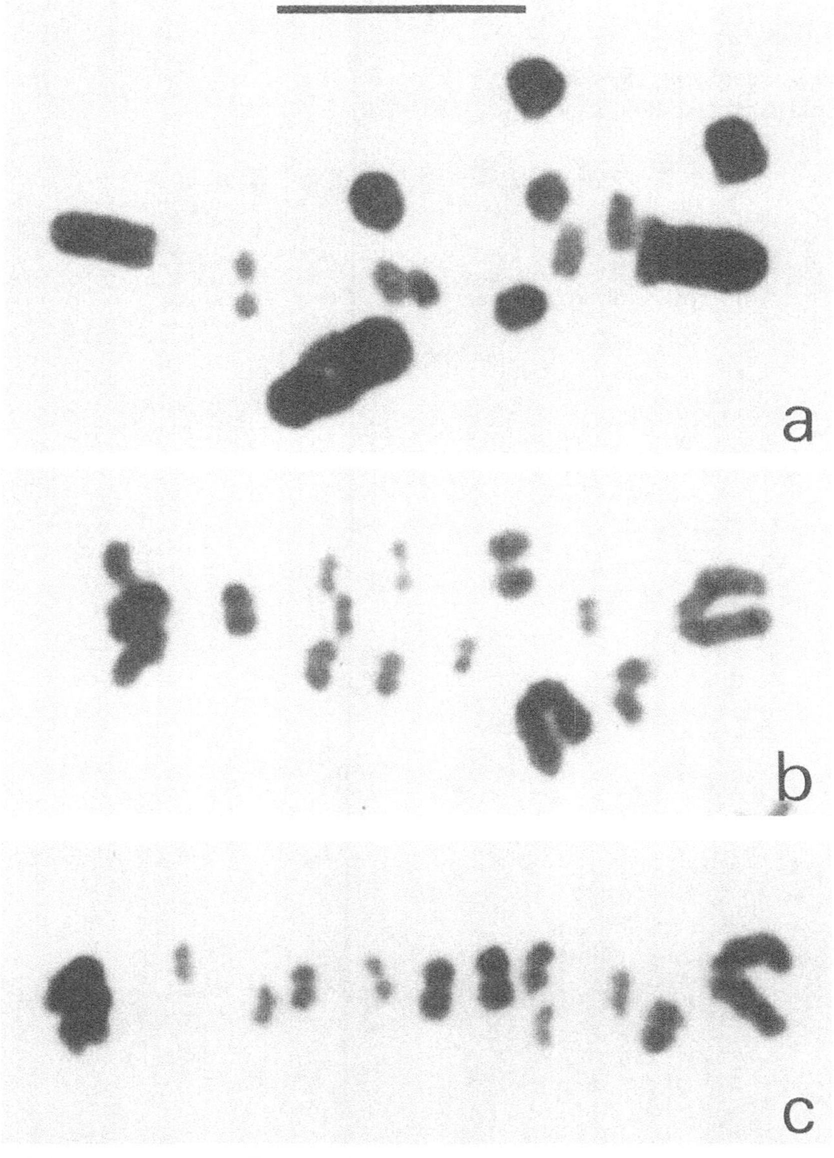

Fig. 6. *Chlorodectes montanus* Rentz. a: First metaphase, b: Second metaphase with X and c: Second metaphase without X.

Metaballus Herman

2n♂ = 27 in all species studied.

The karyotypes of all known species consist of 13 pairs of telocentric autosomes and a large telocentric X (Figs 7a, 7b, 8a, 8b, 9a, 10a, 10b, 11a, 12a, 12b, 13a, 13b, 14a, 15a, 15b, 16a).

At meiosis, large telocentric bivalents each usually form a single proximal chiasma; however, they may also form a distal, terminal chiasma as well (Figs 7c, 8c, 10c, 11b, 13c, 14b, 16b). At first metaphase, there are 13 autosomal bivalents (Figs 7d, 8d, 9b, 10d, 11c, 12c, 14c, 15c, 16c). At first anaphase, the X moves to one pole without dividing (Figs 7e, 10e). As a result of first division, two types of second spermatocytes are formed, 13A+X (Fig. 7f) and 13A (Fig. 7g).

Metaballus sagaeformis Herman (Fig. 7)

Material examined: R77-58, -60, -62, -119, -120

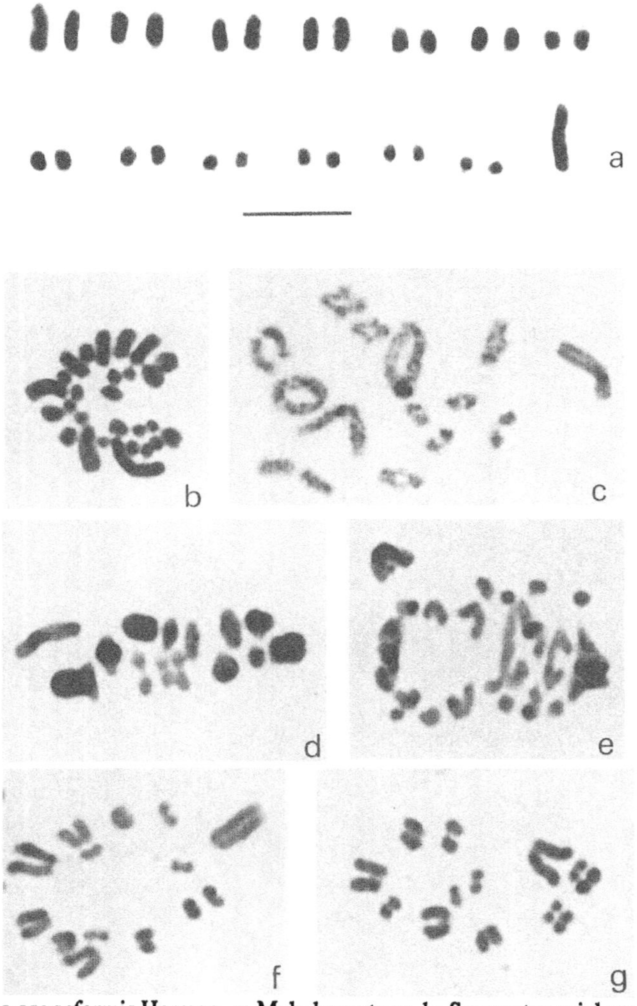

Fig. 7. *Metaballus sagaeformis* Herman. a: Male karyotype, b: Spermatogonial metaphase, c: Diplotene, d: First metaphase, e: First anaphase, f: Second metaphase with X, and g: Second metaphase without X.

Metaballus brevipennis Rentz (Fig. 8)

Material examined: R77-65 (holotype), R77-66, -69, -72, -73, R81-290, -294, -298

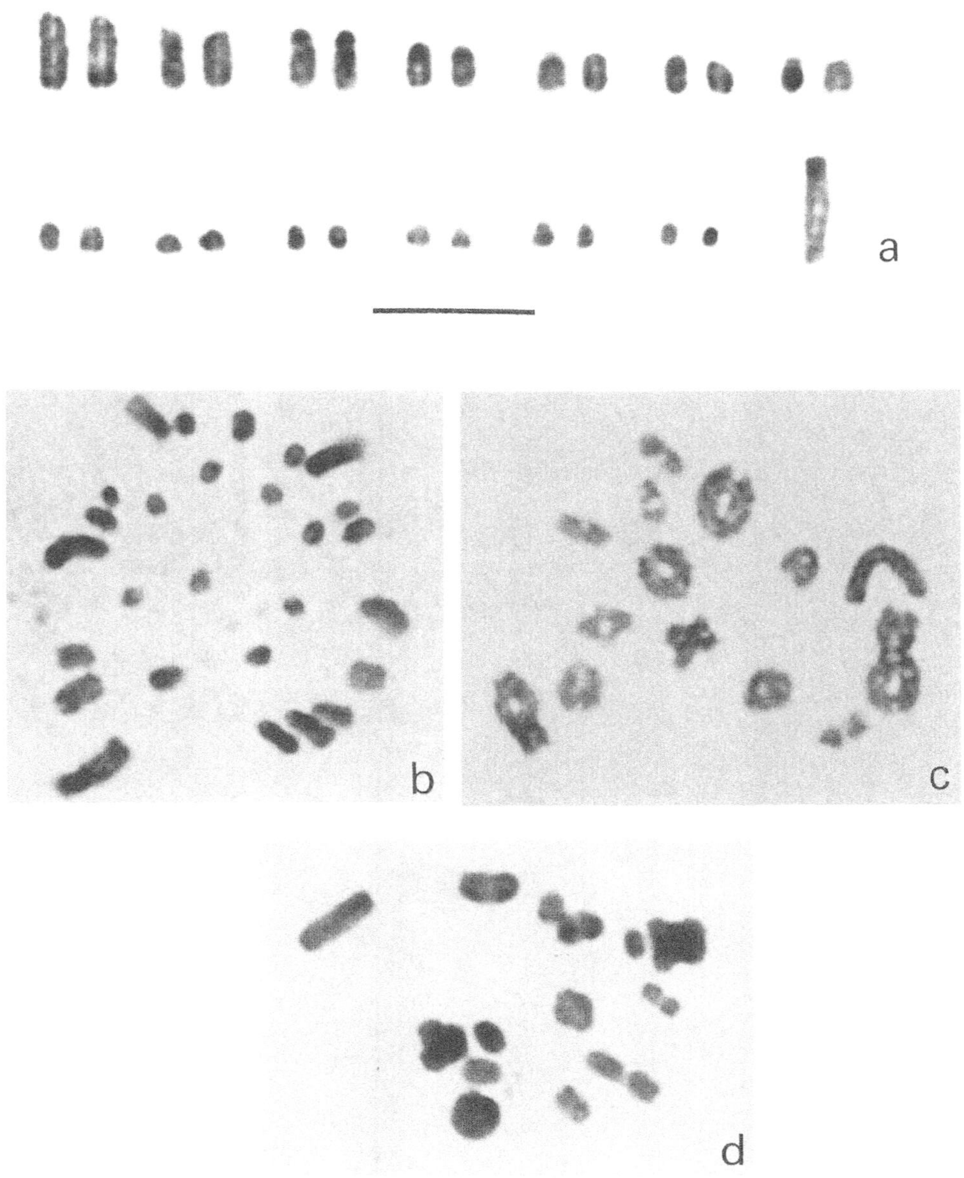

Fig. 8. *Metaballus brevippennis* Rentz. a: Male karyotype, b: Spermatogonial metaphase, c: Diplotene, and d: First metaphase.

Metaballus alatus Rentz (Fig. 9)
Material examined: R77-114 (holotype), R77-115, -116, -117, -118, R82-195

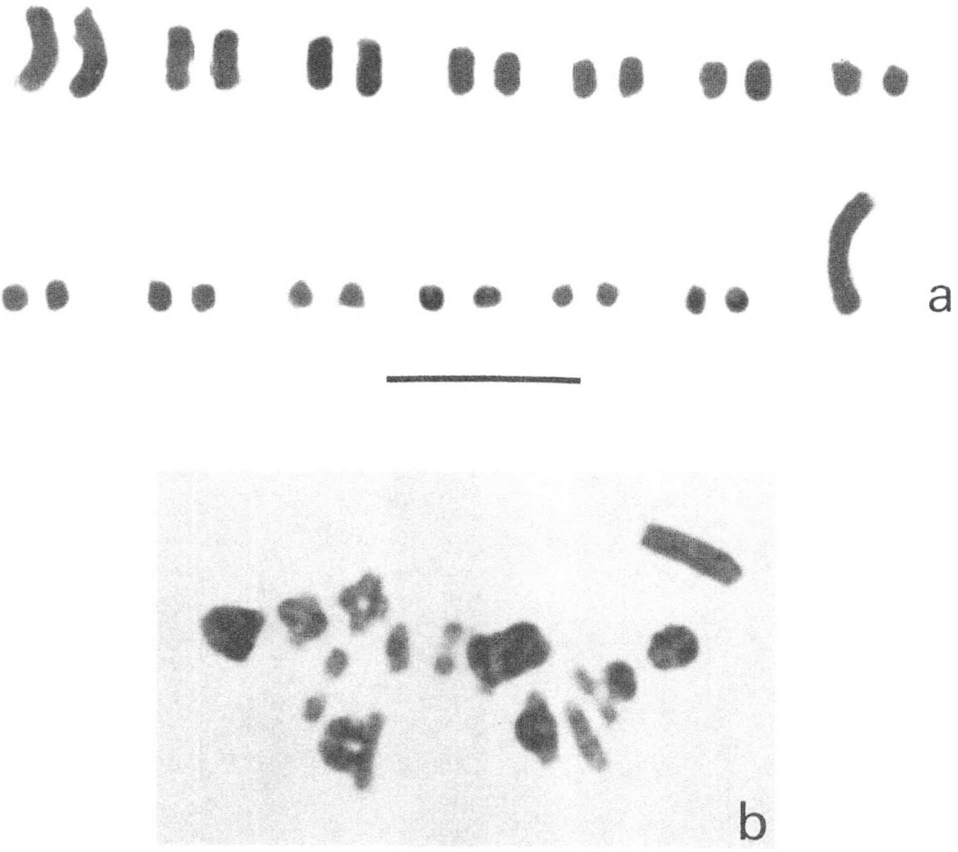

Fig. 9. *Metaballus alatus Rentz*. a: Male karyotype, and b: First metaphase.

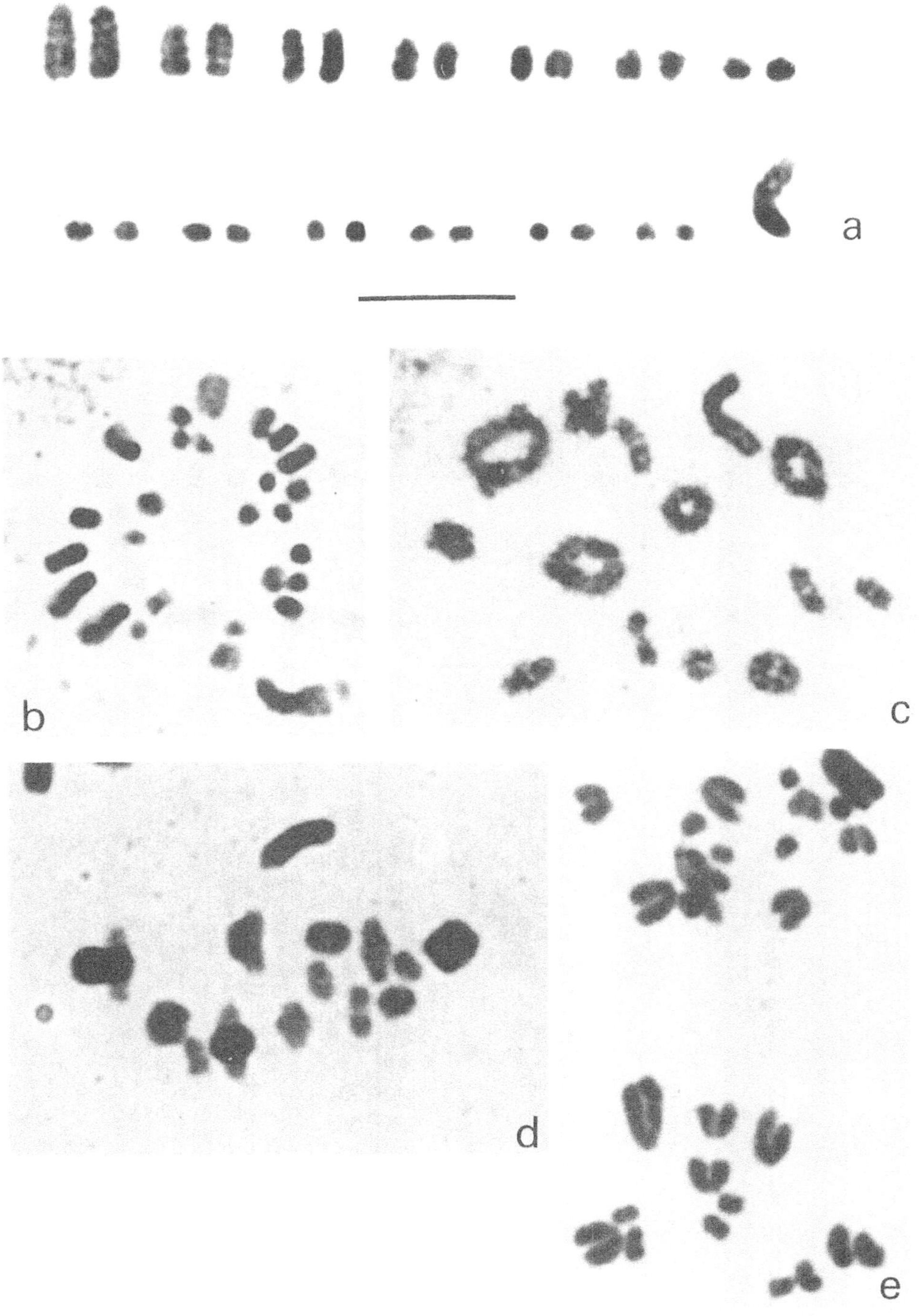

Fig. 10. *Metaballus anchigyus* Rentz. a: Male karyotype, b: Spermatogonial metaphase, c: Diplotene, d: First metaphase, and e: First anaphase.

Metaballus anchigyus Rentz (Fig. 10)

Material examined: R80-30 (holotype), R80-31 (stop 18), -32 (stop 18), R82-173, -201, -161, -179

Metaballus mesopterus Rentz (Fig. 11)

Material examined: R77-88 (holotype), R77-92, -96, -102

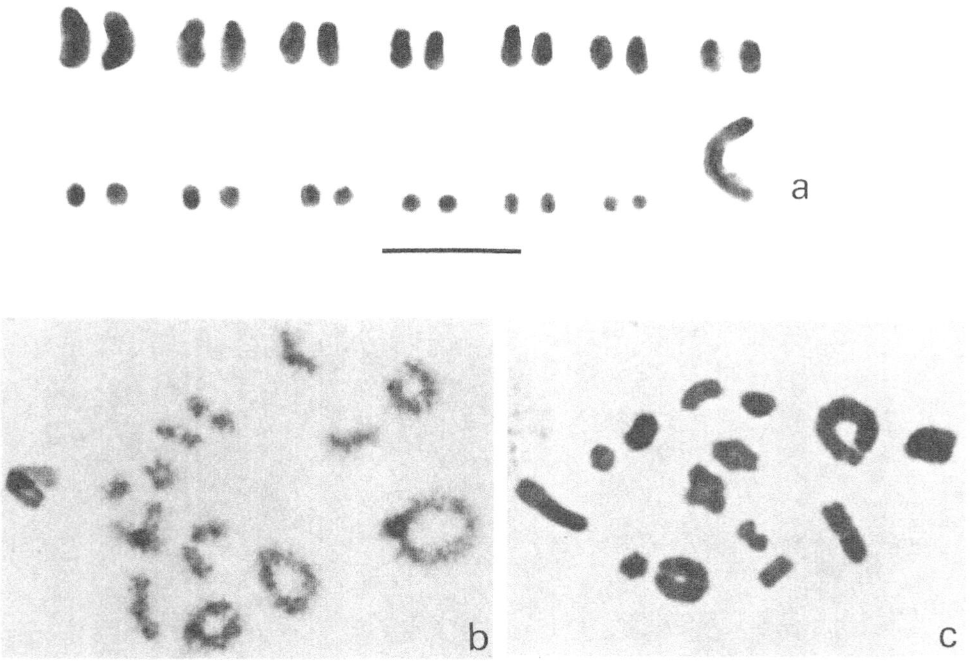

Fig. 11. *Metaballus mesopterus* Rentz. a: Male karyotype, b: Diplotene, and c: First metaphase.

Metaballus decticoides (Walker) (Fig. 12)
Material examined: R81-314, -281, -287, -308, -312, -347, R82-183, -170, -194

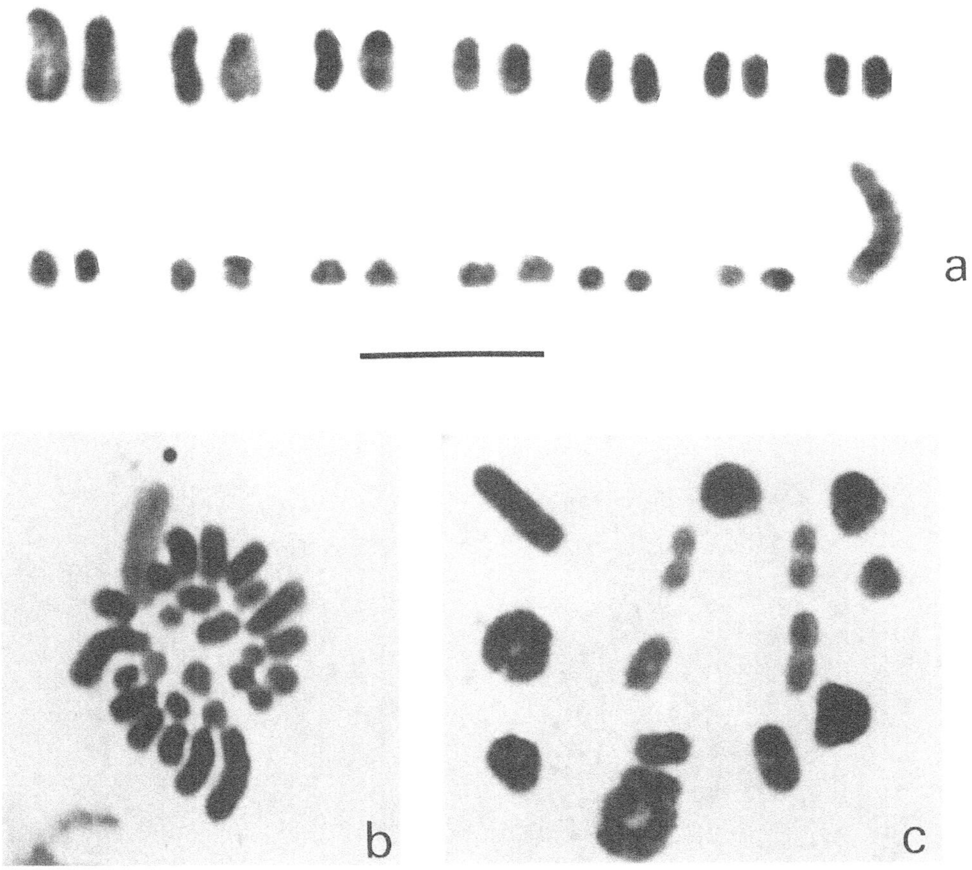

Fig. 12. *Metaballus decticoides* (Walker). a: Male karyotype, b: Spermatogonial metaphase, and c: First metaphase.

Metaballus mucronatus Rentz (Fig. 13)
Material examined: R81-309 (holotype), R81-300, -289

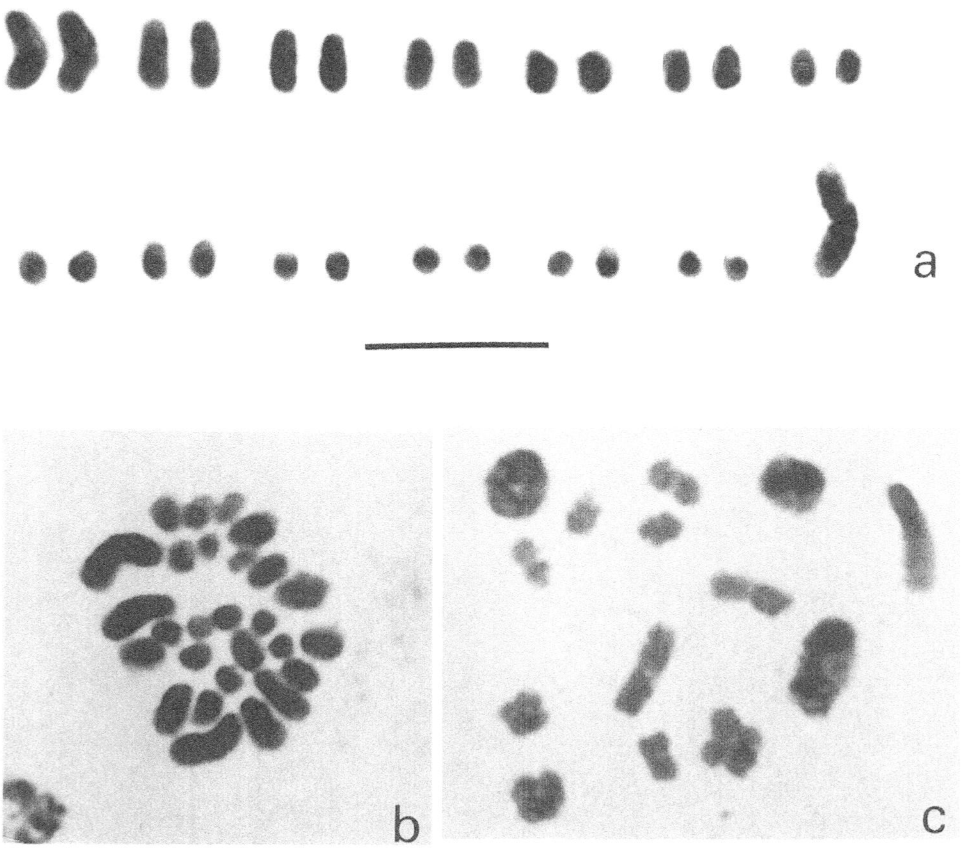

Fig. 13. *Metaballus mucronatus* Rentz. a: Male karyotype, b: Spermatogonial metaphase, and c: First metaphase.

Metaballus frontalis (Walker) (Fig. 14)

Material examined: R81-255, -265, -282, -283, -291, -293, -302, -303, -310, R82-167, -171, -193

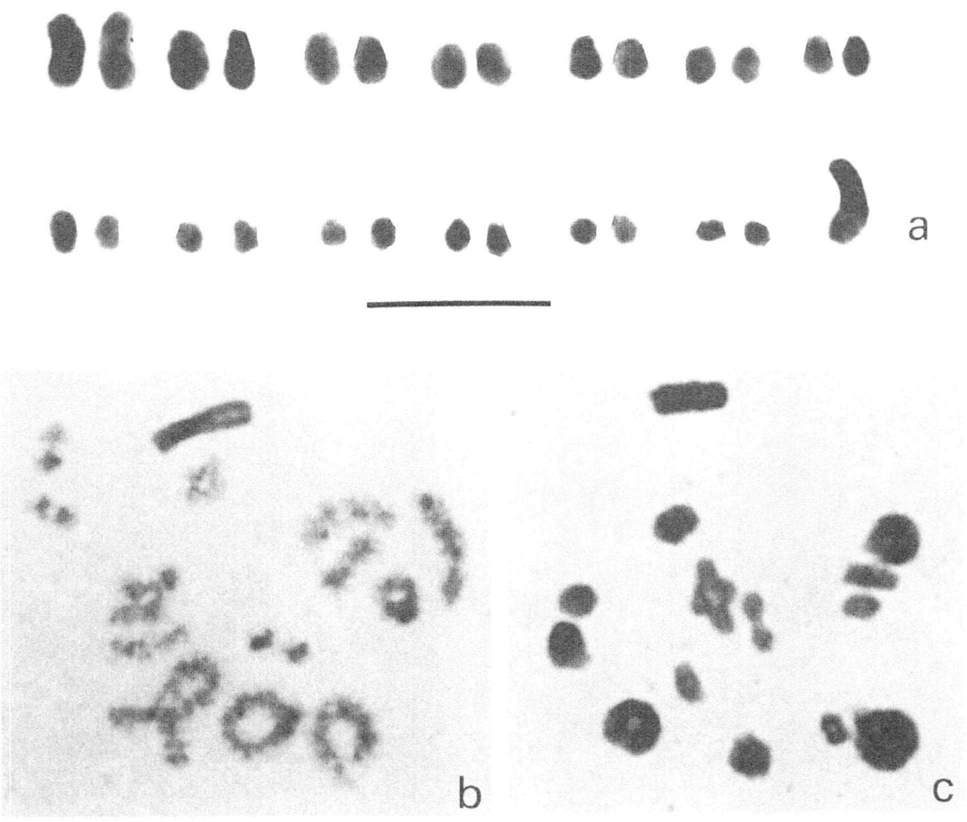

Fig. 14. *Metaballus frontalis* (Walker). a: Male karyotype, b: Diplotene, and c: First metaphase.

Metaballus litus Rentz (Fig. 15)
Material examined: R80-83 (holotype)

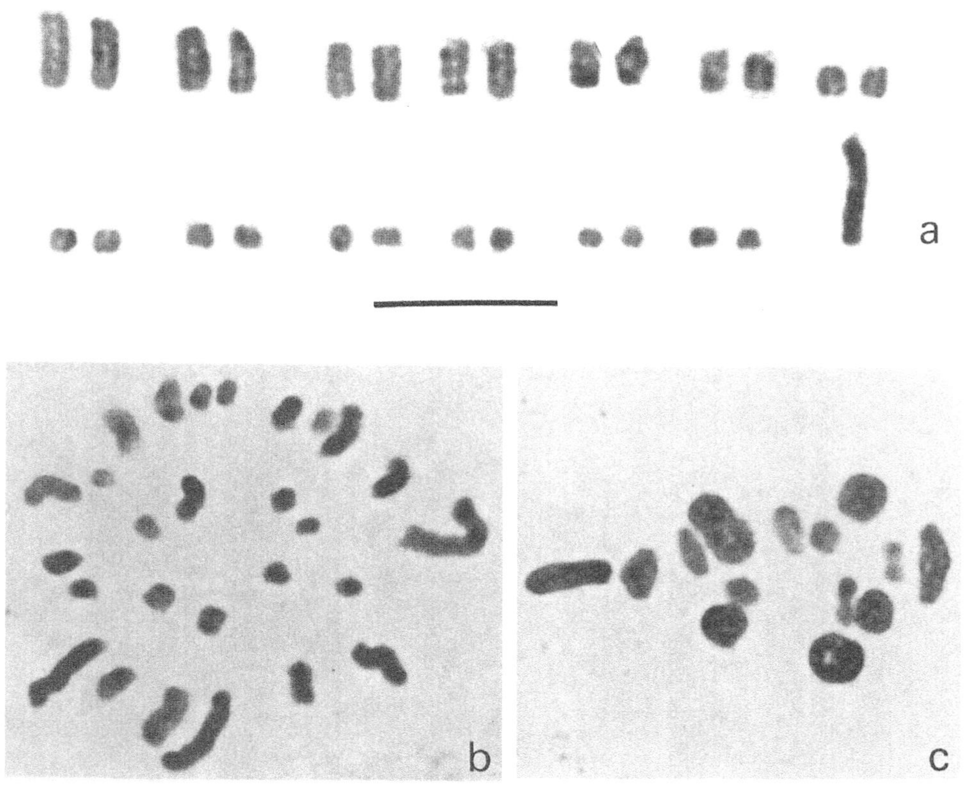

Fig. 15. *Metaballus litus* Rentz. a: Male karyotype, b: Spermatogonial metaphase, and c: First metaphase.

Metaballus bynoei Rentz (Fig. 16)
Material examined: R80-95 (holotype)

Lanciana albidicornis Walker (Fig. 17)
Material examined: R86-19, -20
Locality: 34.14S 150.33E Thirlmere Lakes, nr. Picton, N.S.W.
29.xi.1986 (D.C.F. Rentz, R. Chiang, stop 28).
2n♂ =25

The chromosome number of this species was reported by Ferreira (1969) as 25 in the male diploid. Our observations confirm his result. All 25 chromosomes are telocentric and the X is the largest component (Figs 17a, 17b). Twelve pairs of telocentric autosomes are of gradually decreasing in size.

At meiosis, autosomal bivalents each usually form a terminal chiasma, however, larger ones may form distal or proximal chiasma as well (Fig. 17c). At first metaphase, 12 autosomal bivalents and an X are present (Fig. 17d). The X moves to one pole without dividing at first anaphase (Fig. 17e). As a result of first division, two types of second spermatocytes are formed: 12A+X (Fig. 17f) and 12A (Fig. 17g).

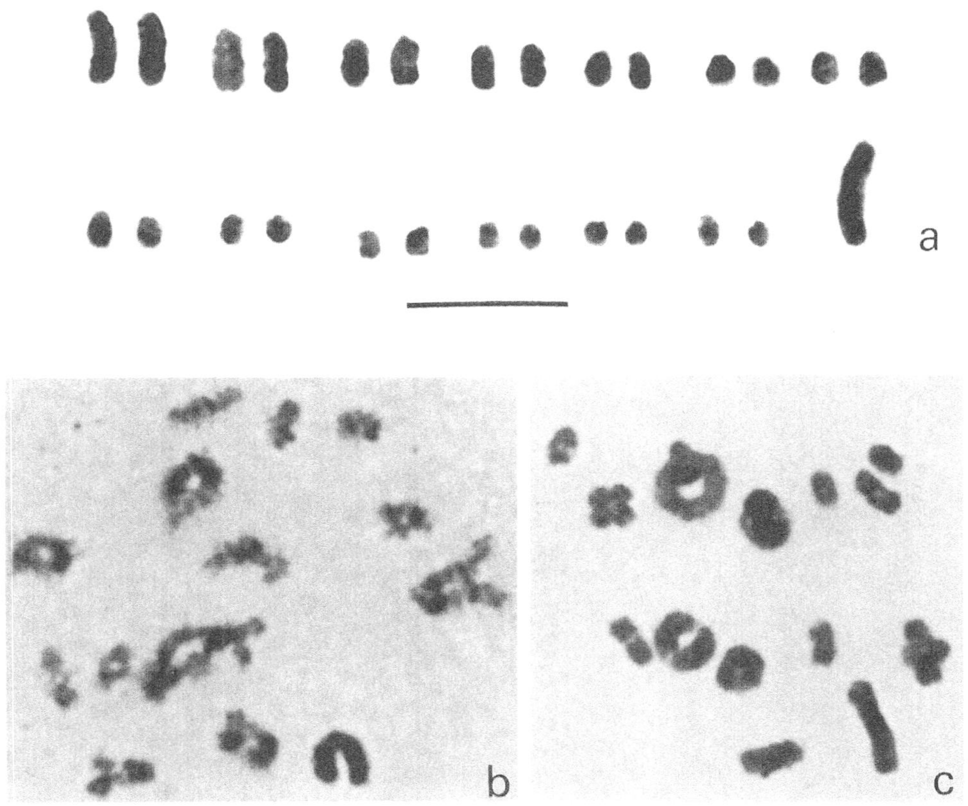

Fig. 16. *Metaballus bynoei* Rentz. a: Male karyotype, b: Diplotene, and c: First metaphase.

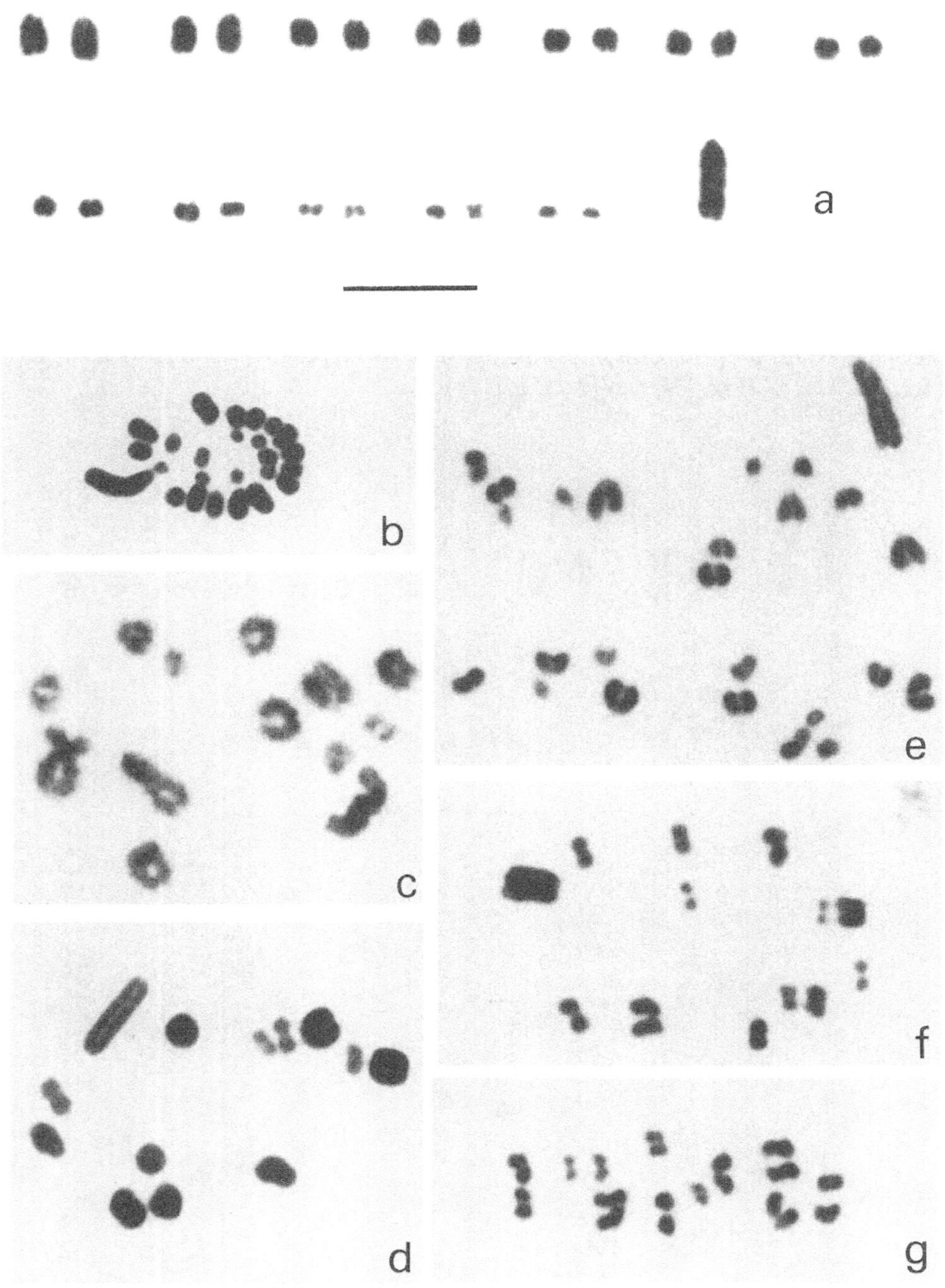

Fig. 17. *Lanciana albidicornis* Walker. a: Male karyotype, b: Spermatogonial metaphase, c: Diplotene, d: First metaphase, e: First anaphase, f: Second metaphase with X, and g: Second metaphase without X.

Lanciana semialata Rentz (Fig. 18)
Material examined: R83-149 (holotype), R83-147, -148, -162, -165
2n♂ = 25
The karyotype of this species is similar to that of *L. albidicornis*.

It consists of 24 telocentric autosomes and a large telocentric X (Figs 18a, 18b). Twelve pairs of autosomes gradually decrease from medium to small size. Twelve autosomal bivalents and an X are present at first metaphase (Fig. 18c). As the result of first division, two types of second metaphase are formed, 12A+X (Fig. 18d) and 12A (Fig. 18e).

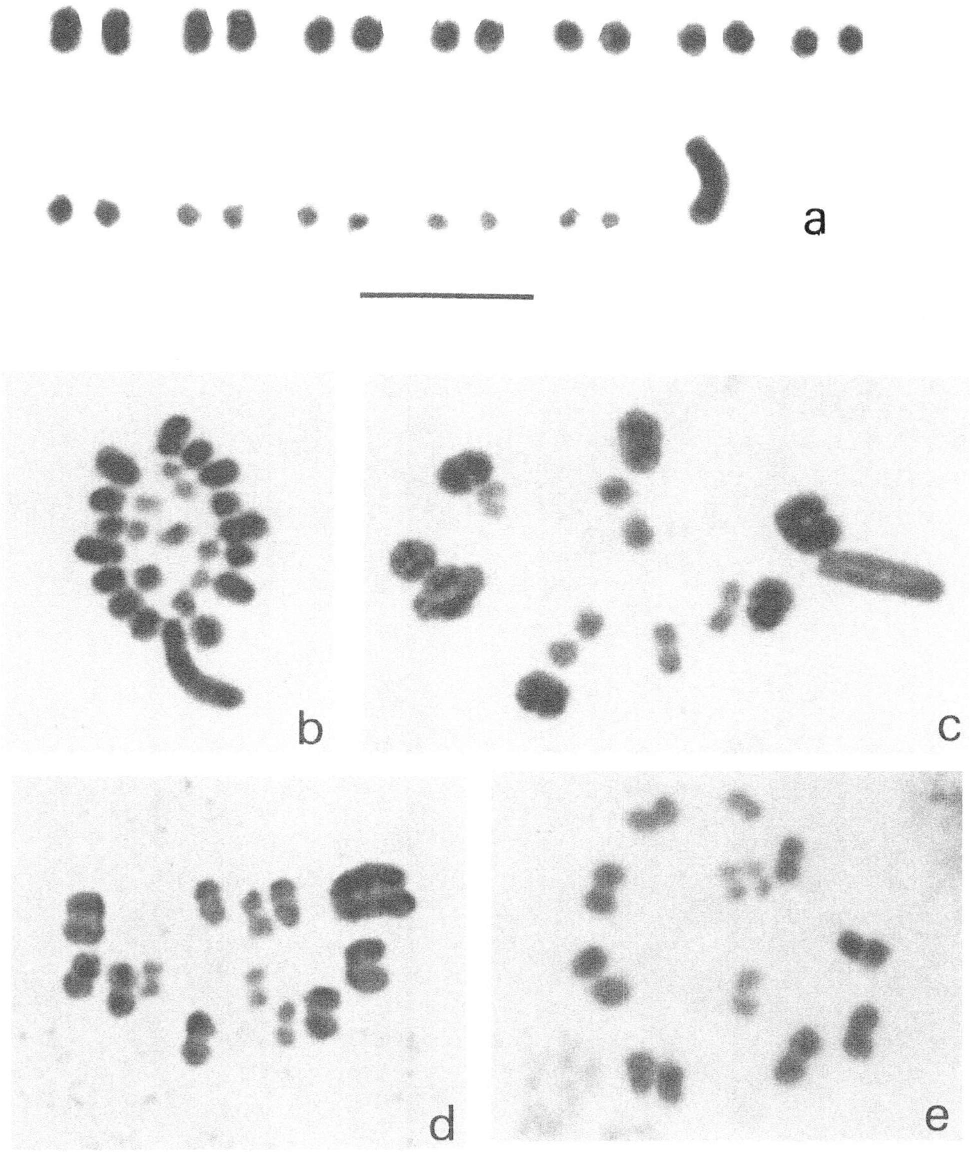

Fig 18. *Lanciana semiatata* Rentz. a: Male karyotype, b: Spermatogonial metaphase, c: First metaphase, d: Second metaphase with X and e: Second metaphase without X.

Lanciana hesperpotana Rentz (Fig. 19)

Material examined: R81-315 (holotype), R81-278

2n♂ = 21

The karyotype of this species comprises one pair of metacentrics with arms equal in length, and nine pairs of telocentric autosomes and a telocentric large X (Figs 19a, 19b). The X is the largest component. The nine pairs of telocentric autosomes can easily be arranged in two principal groups, seven pairs of medium and two pairs of small chromosomes.

At first metaphase, there are 10 autosomal bivalents and an X (Fig. 19c). As the result of first division, two types of second metaphases are formed: 10A+X (Fig. 19d) and 10A (Fig. 19e). Fig. 19f shows the second anaphases of 11A+X and 11A.

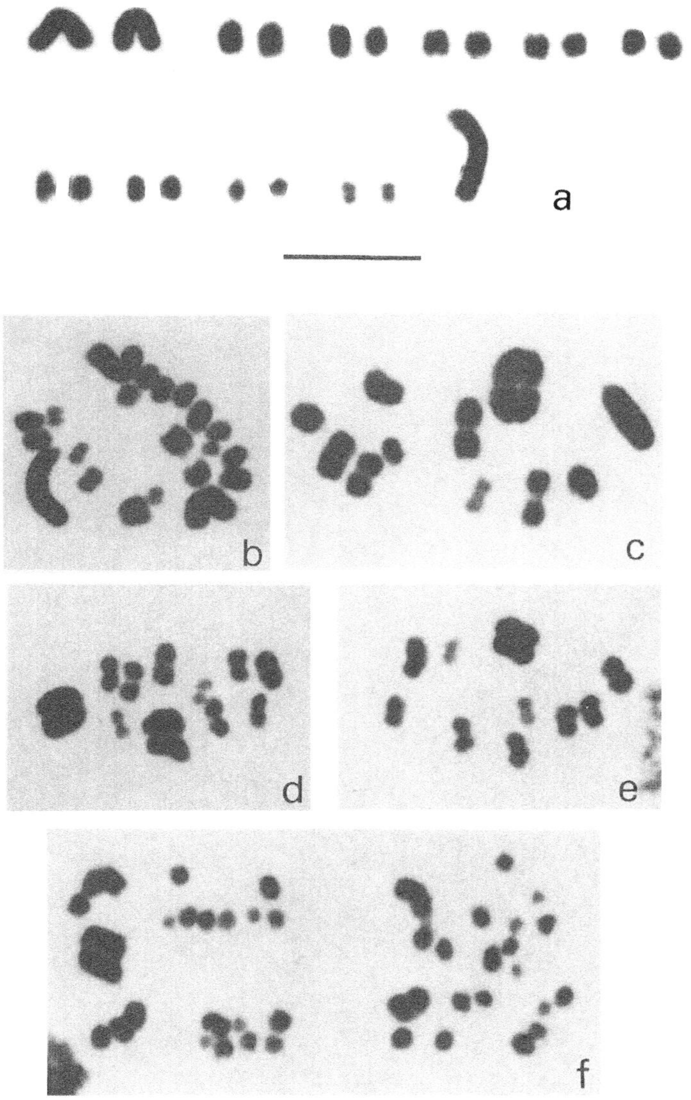

Fig. 19. *Lanciana hesperpotana* Rentz. a: Male karyotype, b: Spermatogonial metaphase, c: First metaphase, d: Second metaphase with X, e: Second metaphase without X and f: Second anaphase with two types of second spermatocytes.

Lanciana montana Rentz (Fig. 20)

Material examined: R77-129 (holotype), R79-8, R80-1 (stop 2), R82-18 (stop 17)

2n♂ =23

The male karyotype consists of one pair of metacentric autosomes with the arms equal in length, 10 pairs of telocentric autosomes, and a large telocentric X (Figs 20a, 20b).

At first metaphase, there are 11 autosomal components and an X (Fig. 20c). The X moves to one pole without dividing at first anaphase (Fig. 20d). As a result of first division, there are two types of second spermatocytes: 11A+X (Fig. 20e) and 11A (Fig. 20f).

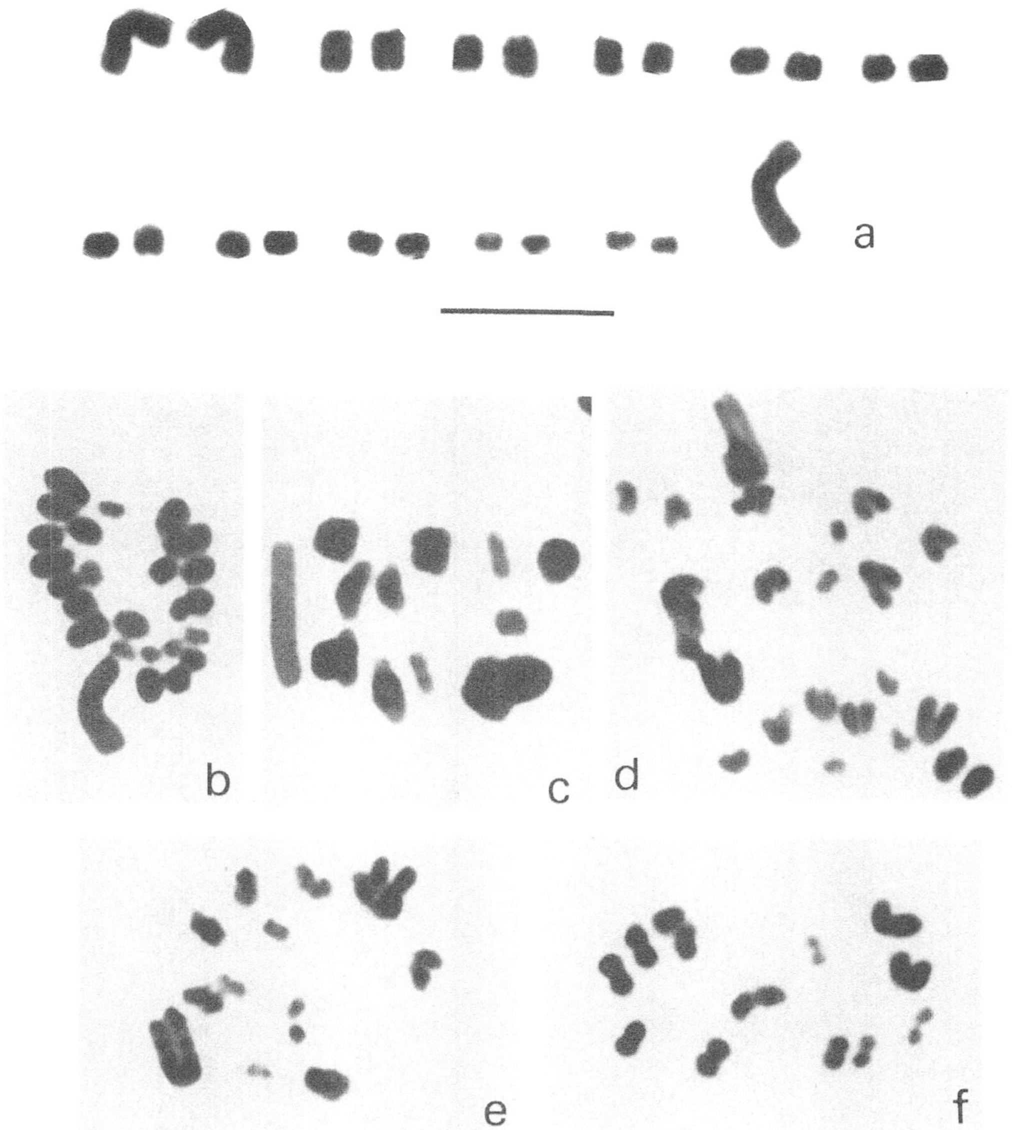

Fig. 20. *Lanciana montana* Rentz. a: Male karyotype, b: Spermatogonial metaphase, c: First metaphase, d: First anaphase, e: Second metaphase with X and f: Second metaphase without X.

Lanciana occidentalis Rentz (Fig. 21)
Material examined: R81-333 (holotype)
2n♂ = 21

The karyotype of the species is similar to that of *L. hesperpotana*. One pair of autosomes is metacentric with nearly equal arms (Fig. 21a). Nine pairs of autosomes are telocentric in a graded series, two of them very small. The X is the largest element of the telocentrics (Figs 21a, 21b).

The chiasmata are strongly localised distally (Fig. 21c). At first metaphase almost all showed terminalised chiasmata in both arms of the metacentric bivalent and one chiasma in each of the telocentric bivalents (Fig. 21d). At first anaphase, the X moves to one pole without dividing (Fig. 21e). As a result of the first division, there are two types of second metaphases: 10A+X (Fig. 21f) and 10A (Fig. 21g). Fig. 21h shows a second anaphase of 10A.

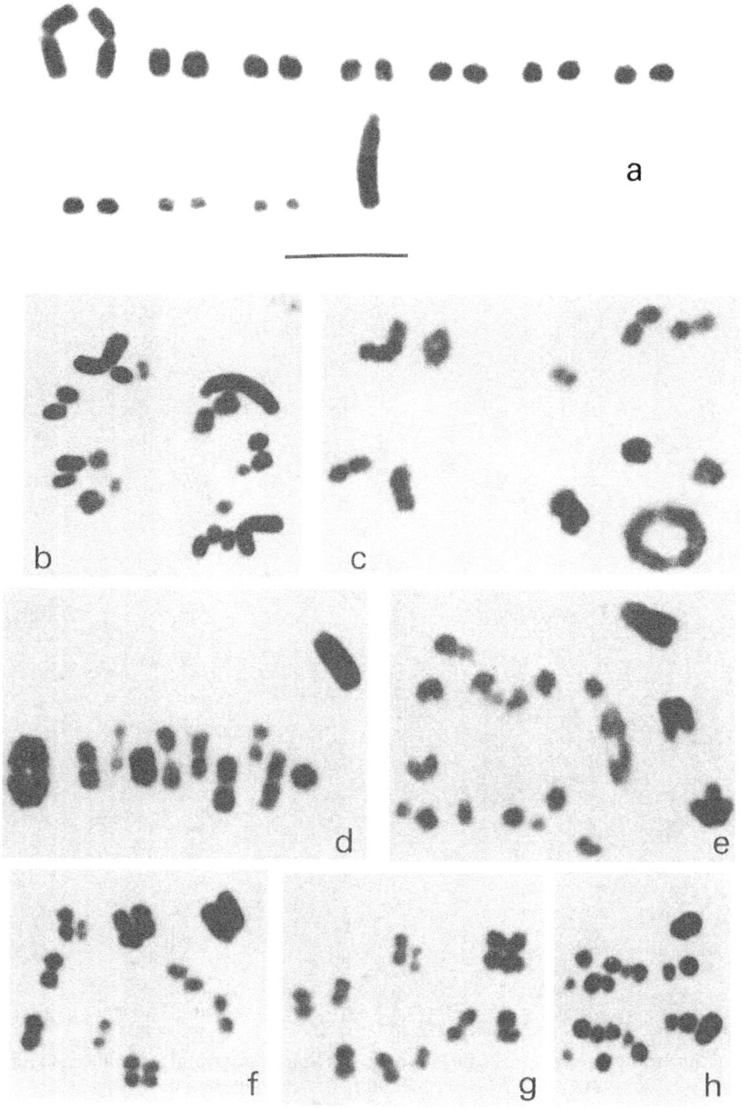

Rhachidorus semoni Krauss (Fig. 22)

Material examined: R82-30 (stop 22)

2n♂ = 31

The karyotype comprises 15 pairs of telocentric autosomes and a large X chromosome (Fig. 22a). The 15 pairs of autosomes can easily be arranged in two groups: eight pairs of medium with graded series and seven pairs of small chromosomes. In the latter group, one is a little larger than the others and the rest are nearly equal in size.

At first metaphase, 15 autosomal bivalents and an X univalent are present (Fig. 22b). The X moves to one pole without dividing at first anaphase (Fig. 22c). As the result of first division, there are two types of second metaphases: 15A+X (Fig. 22d) and 15A (Fig. 22e).

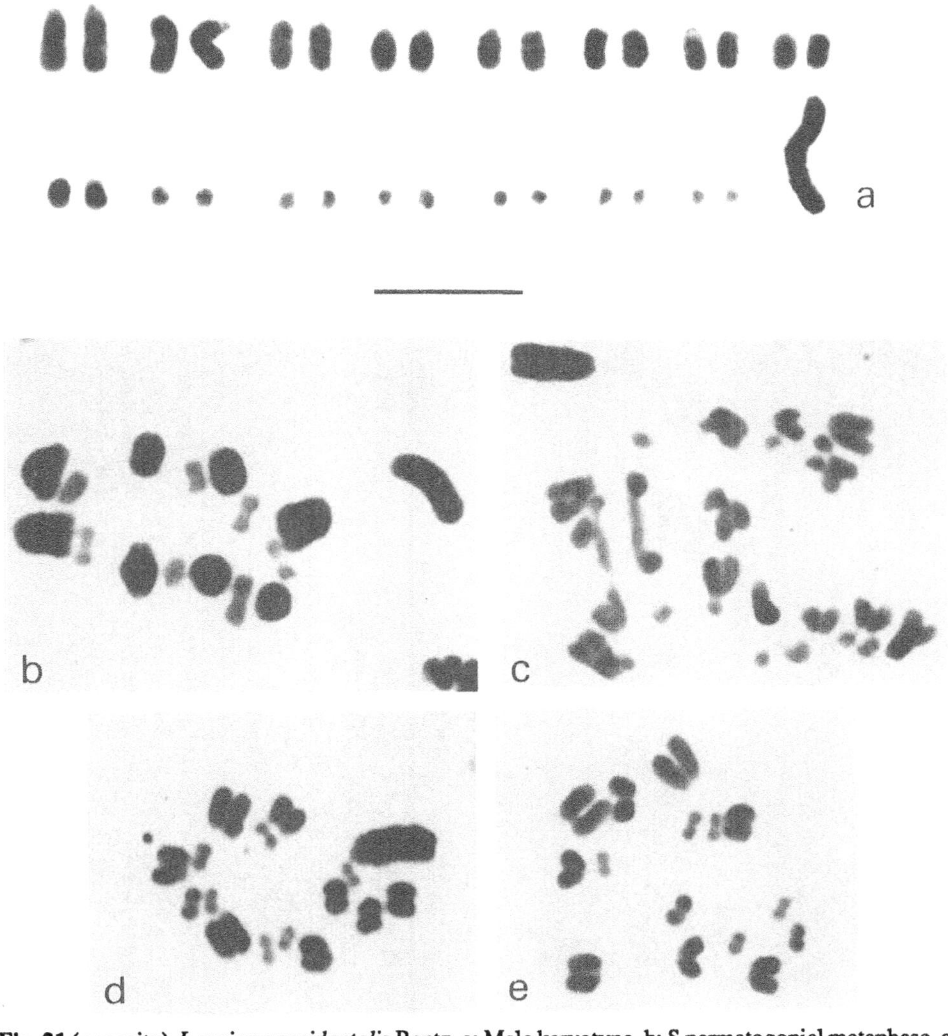

Fig. 21 (opposite). *Lanciana occidentalis* Rentz. a: Male karyotype, b: S permatogonial metaphase, c: Diakinesis, d: First metaphase, e: First anaphase, f: Second metaphase with X, g: Second metaphase with X, g: Second metaphase without X and h: Second anaphase without X.

Fig. 22. *Rhachidorus semoni* Krauss. a: Male karyotype, b: First metaphase, c: First anaphase, d: Second metaphase with X and e: Second metaphase without X.

Rhachidorus blackdownensis Rentz (Fig. 23)
Material examined: R81-22 (holotype), R81-21

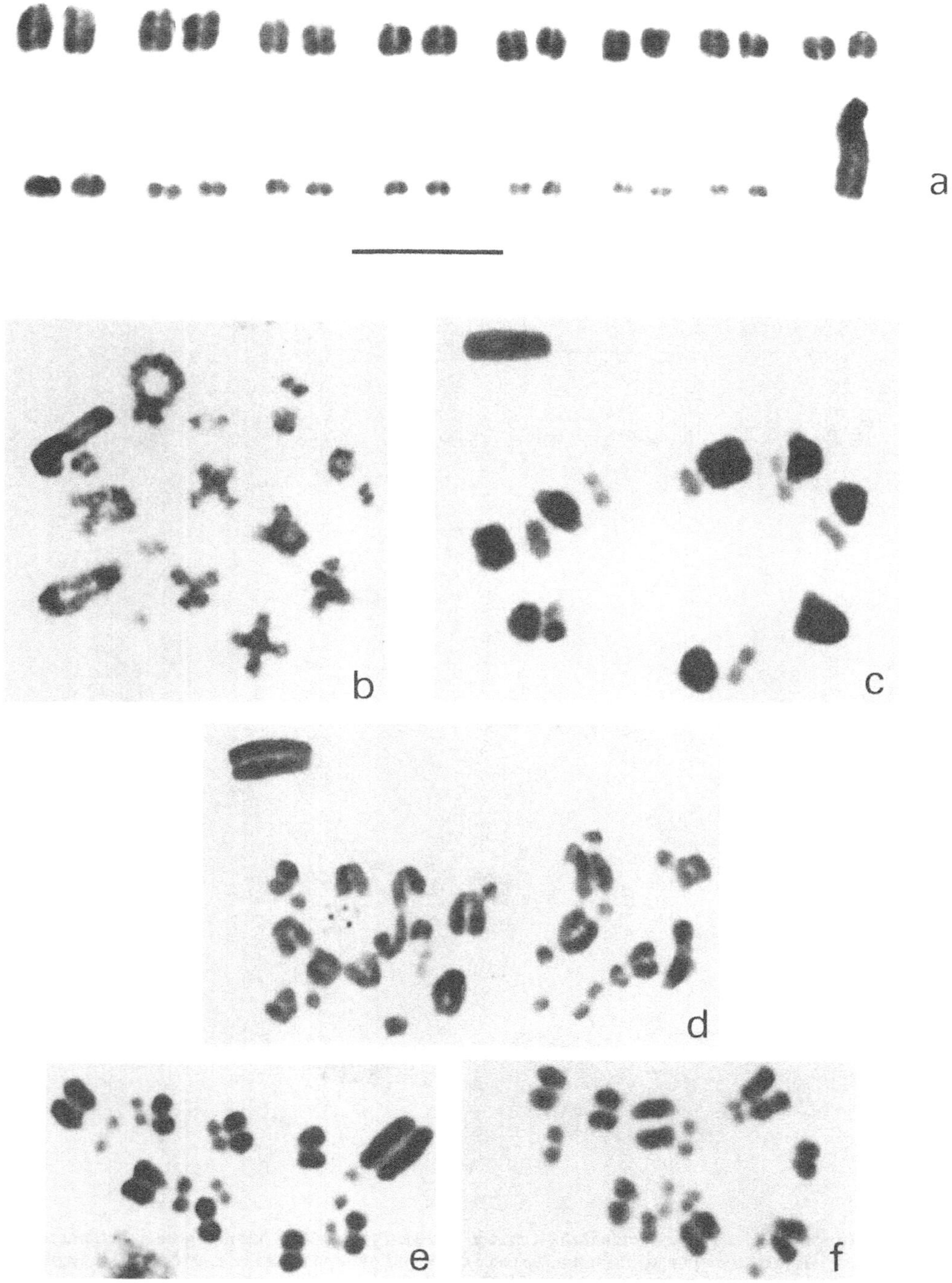

Fig. 23. *Rhachidorus blackdownensis* Rentz. a: Male karyotype, b: Diplotene, c: First metaphase, d: Fi rst anaphase, e: Second metaphase with X, and f: Second metaphase without X.

Rhachidorus longipennis Rentz (Fig. 24)

Material examined: R77-130 (holotype)

2n♂ = 31 (both species)

The chromosome cytology of these species is the same as *R. semoni*. The male karyotype consists of 15 pairs of telocentric autosomes and a large X chromosome (Figs 23a, 24a). Fifteen pairs of autosomes are again arranged in two size groups, eight medium, seven small.

During meiosis, there are 15 autosomal elements and a heteropycnotic X element (Fig. 23b). Telocentric autosomal bivalents usually form a single terminal or distal chiasma each. At first metaphase, 15 autosomal bivalents and an X univalent are present (Fig. 23c). The X moves to one pole without dividing at first anaphase (Fig. 23d). As the result of first division, there are two types of second metaphases: 15A+X and 15A (Figs 23e, 23f).

Antipodectes graminicolus Rentz (Fig. 25)

Material examined: R81-146 (holotype), R81-145

2n♂ = 29

The male karyotype consists of one pair of metacentric autosomes, 13 pairs of telocentric autosomes and a telocentric X chromosome (Fig. 25a). The arms of the metacentric autosome are almost equal in length. The X is the largest component of the telocentric chromosomes.

During diplotene, there are 14 autosomal elements and a heteropycnotic X element (Fig. 25b). At first metaphase, there are 14 autosomal bivalents, including a metacentric and the X univalent (Fig. 25c). At first anaphase, the X moves to one pole without dividing (Fig. 25d).

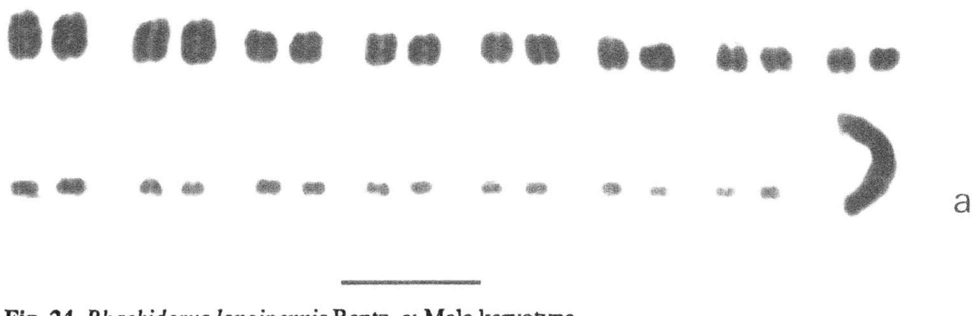

Fig. 24. *Rhachidorus longipennis* Rentz. a: Male karyotype.

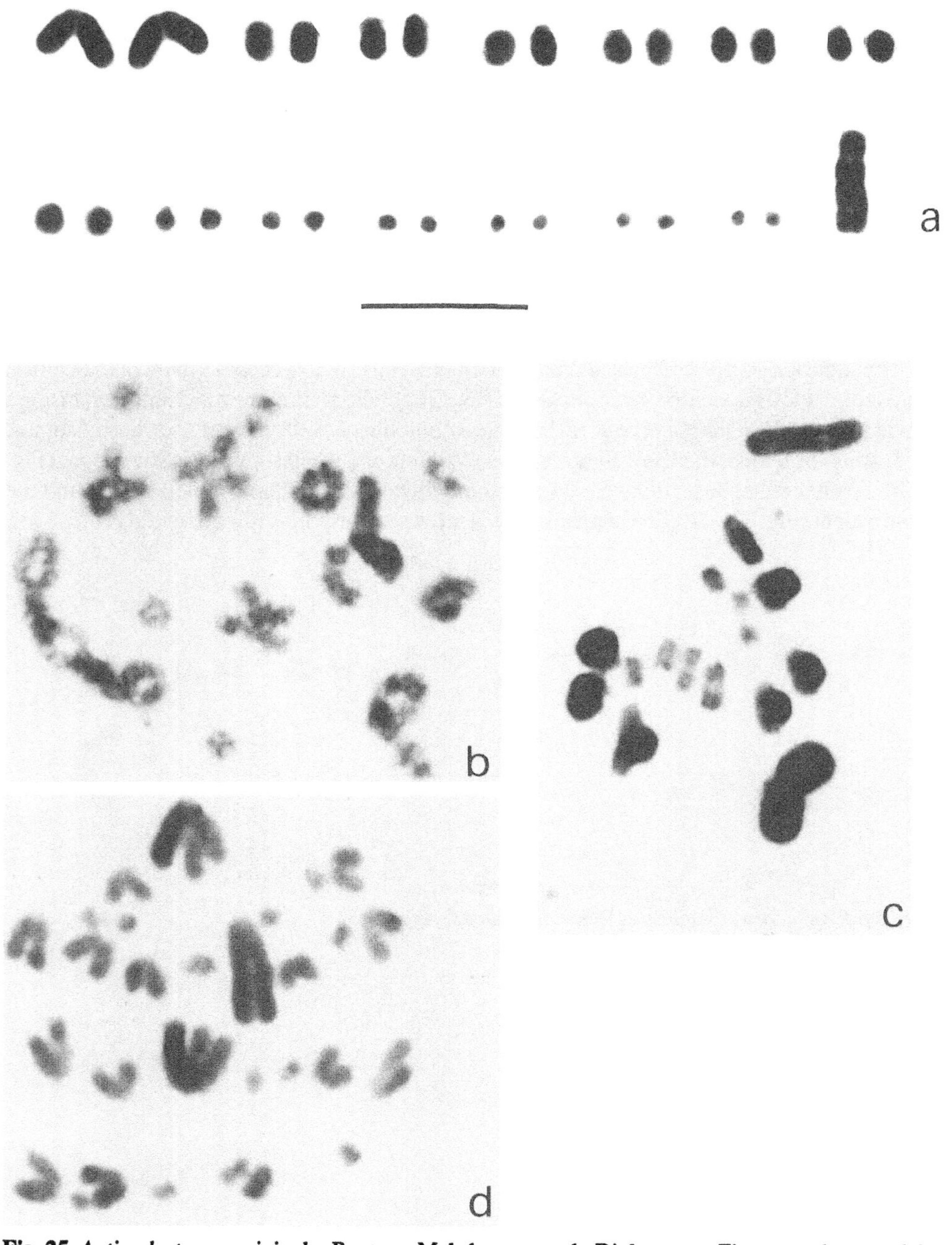

Fig. 25. *Antipodectes graminicolus* Rentz. a: Male karyotype, b: Diplotene, c: First metaphase, and d: First anaphase.

Antipodectes brevicaudus Rentz (Fig. 26)
Material examined: R82-113, -114, -115

Antipodectes giganteus Rentz (Fig. 27)
Material examined: R83-86, -87
2n♂ = 31 (both species)
The karyotype of these species comprises 15 pairs of telocentric autosomes and a large

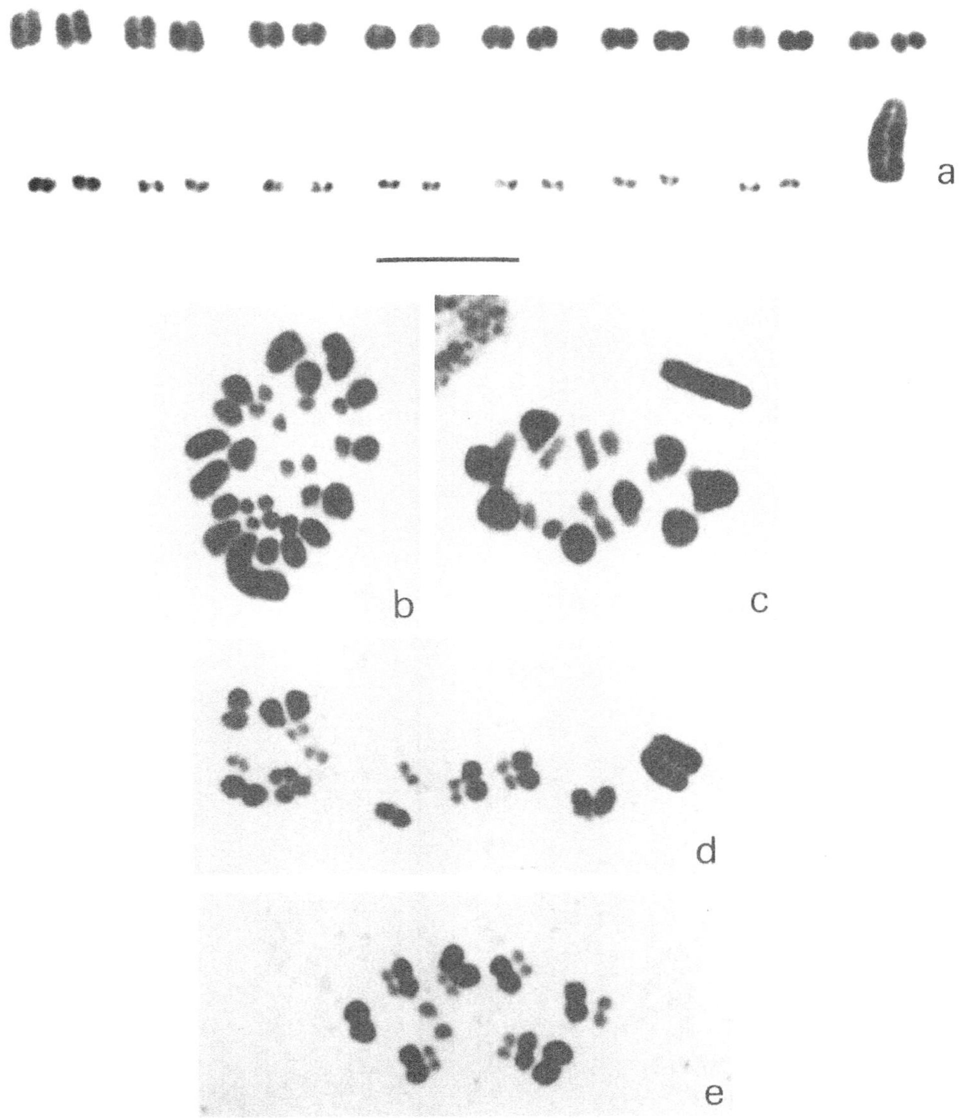

Fig. 26. *Antipodectes brevicaudus* Rentz. a: Male karyotype, b: Spermatogonial metaphase, c: First metaphase, d: Second metaphase with X, and e: Second metaphase without X.

Fig. 27 (overleaf). *Antipodectes giganteus* Rentz. a: Male karyotype, b: Spermatogonial metaphase, c: first metaphase and d: First anaphase.

telocentric X chromosome (Figs 26a, 26b, 27a, 27b). Fifteen pairs of telocentrics can be classified into two groups, eight pairs of medium chromosome in a graded series and seven pairs of small chromosomes. In the latter group, one chromosome is a little larger than the others which are nearly equal in size.

At first metaphase, there are 15 autosomal bivalents and an X univalent (Figs 26c, 27c). The X moves to one pole without dividing at first anaphase (Fig. 27d). As the result of first division, there are two types of second metaphase: 15A+X and 15A (Figs 26d, 26e).

Chinnandectes neenan Rentz (Fig. 28)

Material examined: R83-43 (holotype), R83-3, -43, -45, -51, -53, -60

2n♂ = 29

The spermatogonial metaphase of this species comprises eight pairs of medium-sized telocentric autosomes, six pairs of small telocentric autosomes and a large telocentric X

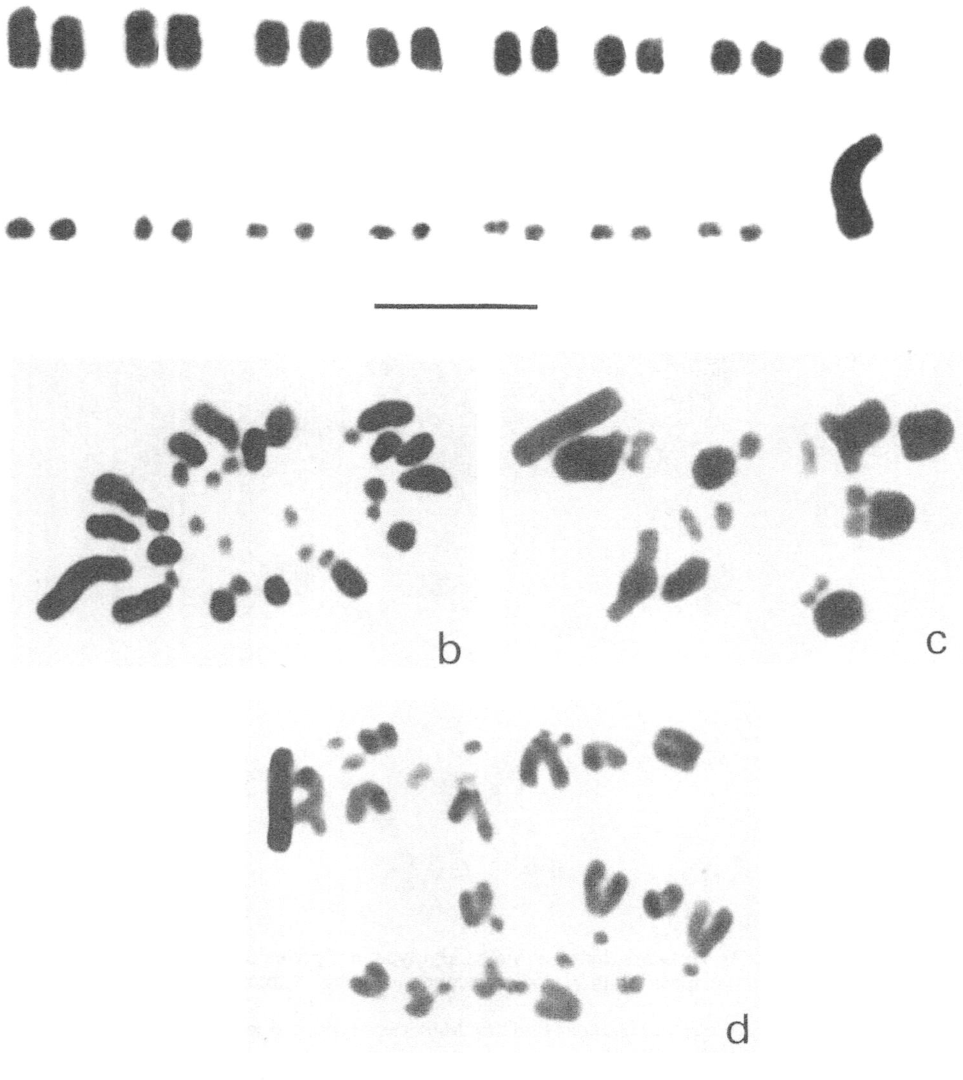

chromosome (Figs 28a, 28b).

Fourteen autosomal bivalents and an X univalent are present at first metaphase (Fig. 28c). At first anaphase, the X moves to one pole without dividing (Fig. 28d). As the result of first division, there are two types of second metaphase: 14A+X and 14A.

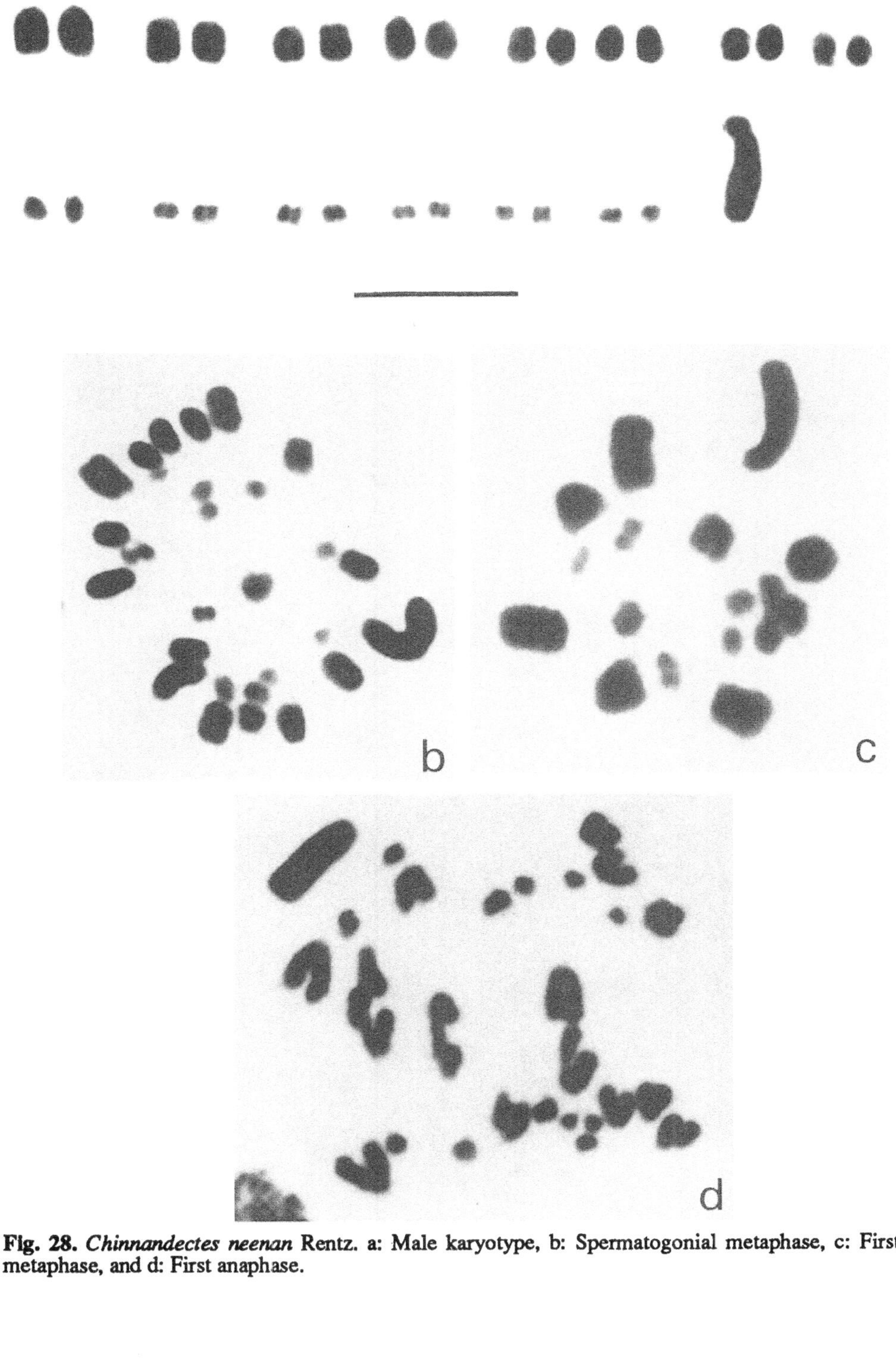

Fig. 28. *Chinnandectes neenan* Rentz. a: Male karyotype, b: Spermatogonial metaphase, c: First metaphase, and d: First anaphase.

Glenbalodectes norrisi Rentz (Fig. 29)

Material examined: R80-55 (stop 39), -59 (stop 42)

2n♂ =33

The spermatogonial metaphase comprises 16 pairs of telocentric autosomes and a large X chromosome (Figs 29a, 29b). Sixteen pairs of telocentric autosomes form two groups: eight pairs of medium, in a graded series, and eight pairs of small chromosomes. In the latter group, first two are a little larger than the others.

During meiosis, telocentric autosomes usually form a single terminal or distal chiasma each (Fig. 29c). At first metaphase, there are 16 autosomal bivalents and a telocentric X univalent (Fig. 29d).

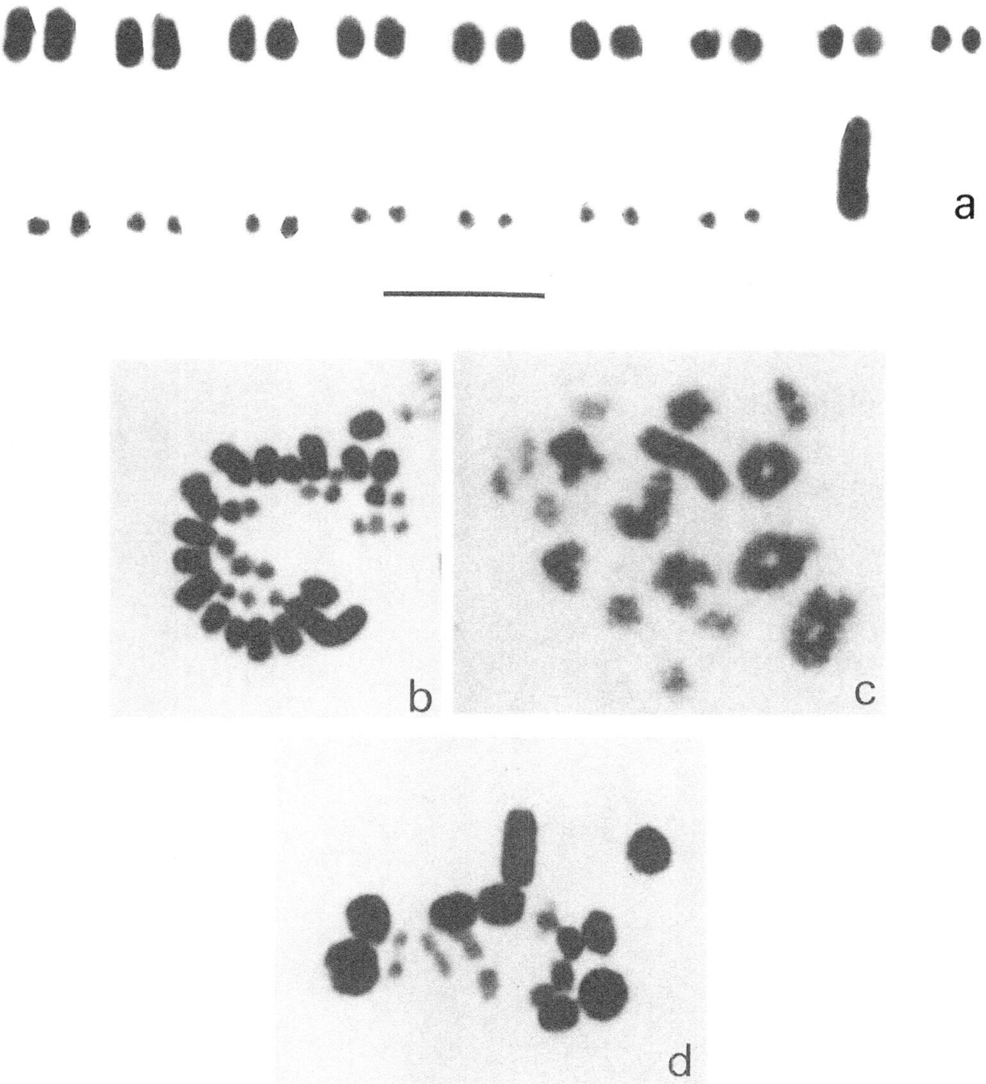

Fig. 29. *Glenbalodectes norrisi* Rentz. a: Male karyotpye, b: Spermatogonial metaphase, c: Diplotene, and d: First metaphase.

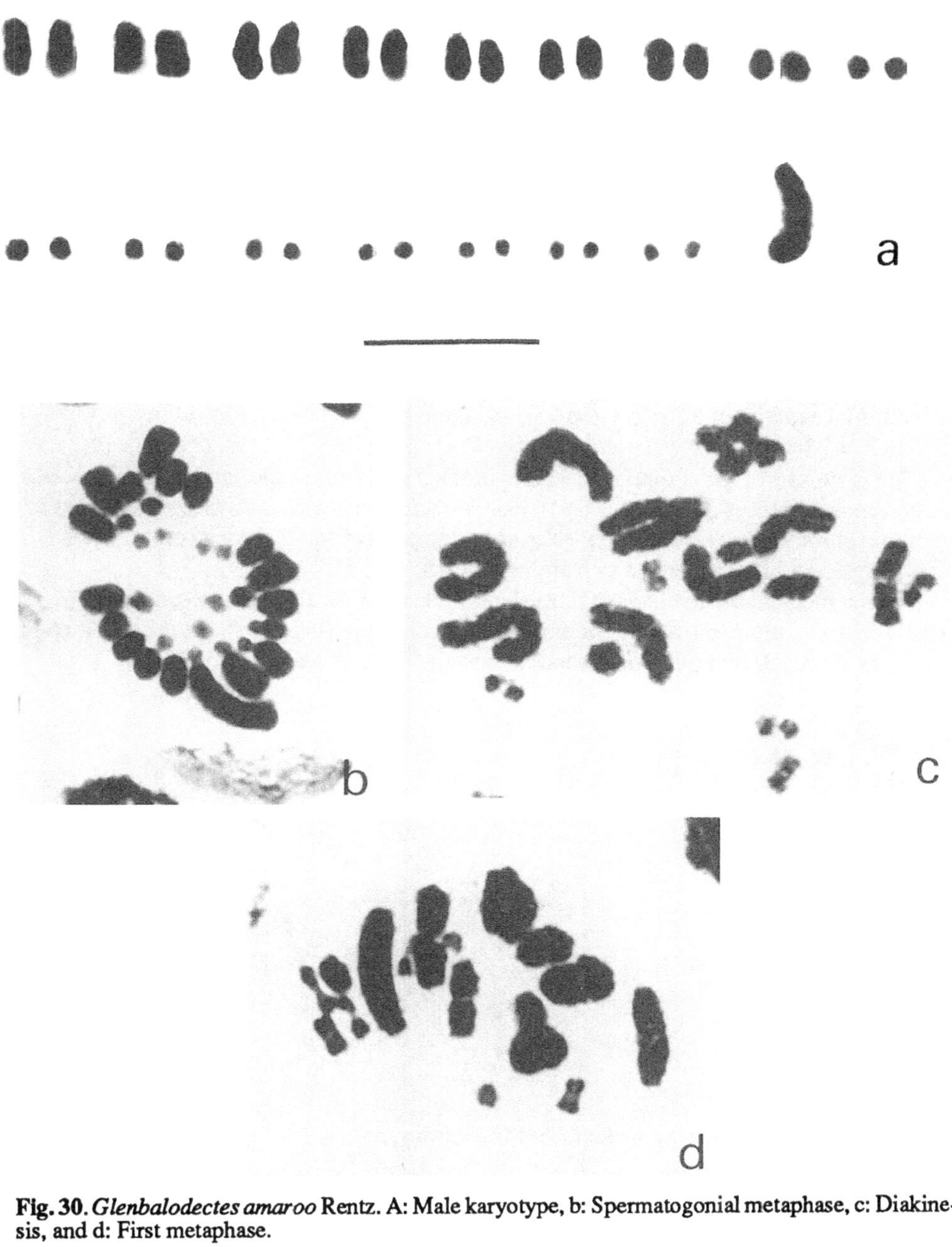

Fig. 30. *Glenbalodectes amaroo* Rentz. A: Male karyotype, b: Spermatogonial metaphase, c: Diakinesis, and d: First metaphase.

Glenbalodectes amaroo Rentz (Fig. 30)

Material examined: R81-299

2n♂ = 33

The chromosome cytology of this species is nearly identical to that of *Glenbalodectes norrisi*. The karyotype comprises 16 pairs of telocentric autosomes and a large telocentric X chromosome (Figs 30a, 30b). The 16 pairs of autosomes form two groups: eight medium and eight small chromosomes.

At meiosis, telocentric autosomal bivalents usually form a single terminal or distal chiasma each (Fig. 30c). At first metaphase, there are 16 autosomal bivalents and the telocentric X (Fig. 30d).

Glenbalodectes narranga Rentz (Fig. 31)

Material examined: R80-28 (holotype)

2n♂ = 33

The chromosome complement of this species is the same as described for *G. norrisi* and *G. amaroo*. The male has 16 pairs (eight medium and eight small) of telocentric autosomes and a large telocentric X chromosome (Fig. 31a).

Xederra barbarae Rentz (Fig. 32)

Material examined: R80-75 (stop 55), -80 (stop 54)

2n♂ = 33

The male karyotype comprises 16 pairs of telocentric autosomes and a large telocentric X chromosome (Figs 32a, 32b). The 16 pairs of autosomes form two groups: eight pairs of medium chromosomes in a graded series and eight pairs of small chromosomes. In the latter group, one is a little larger and two are very small.

At first metaphase, 16 autosomal bivalents and an X univalent are present (Fig. 32c). At first anaphase, the X moves to one pole without dividing (Fig. 32d). As the result of first division, there are two types of second metaphase: 16A+X and 16A.

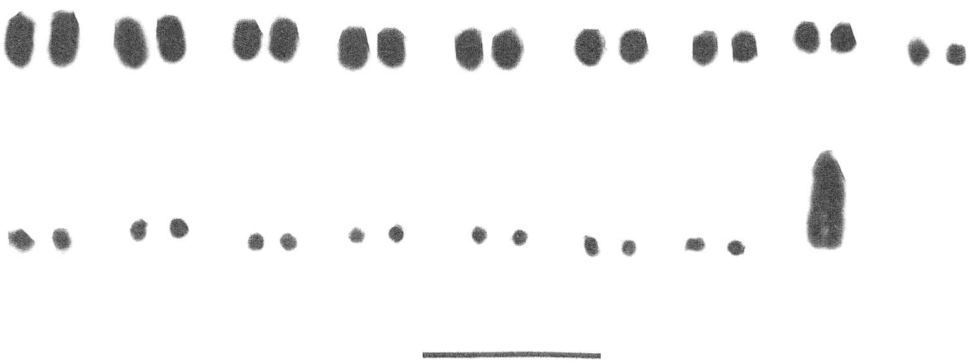

Fig. 31. *Glenbalodectes narranga* Rentz. a: Male karyotype.

Ixalodectes whitei Rentz (Fig. 33)
Material examined: R81-317 (stop 56)
2n♂ = 15
The karyotype comprises two pairs metacentric and five pairs of telocentric autosomes all of which are equal in length except for one pair which is slightly larger and a large telocentric X chromosome (Figs 33a, 33b). One arm of each of the two pairs of metacentrics is a little longer than the other.

At meiosis, metacentric bivalents usually have more than three chiasmata in each arm. Telocentric bivalents usually form a single terminal or distal chiasma (Fig. 33c). At first metaphase, two metacentric and five telocentric bivalents and an X univalent are present (Fig. 33d). During first anaphase, the X moves to one pole without dividing (Fig. 33e).

Ixalodectes uptoni Rentz (Fig. 34)
Material examined: R82-129 (holotype), R82-105, -137
2n♂ = 23
The karyotype of this species is very different from that of *I. whitei*. It comprises 11 pairs of telocentric autosomes and a large telocentric X chromosome (Figs 34a, 34b). The telocentric autosomes can easily be arranged in three groups: two large; five medium; and four small. The X chromosome is the largest component in the set.

During meiosis, large telocentric bivalents show an interstitial or distal chiasma, and the medium or small telocentric bivalents usually form a single terminal or distal chiasma (Fig. 34c). At first metaphase, there are 11 autosomal bivalents and an X univalent (Fig. 34d). As a result of first division, two types of second metaphase are formed: 11A+X (Fig. 34e) and 11A (Fig. 34f).

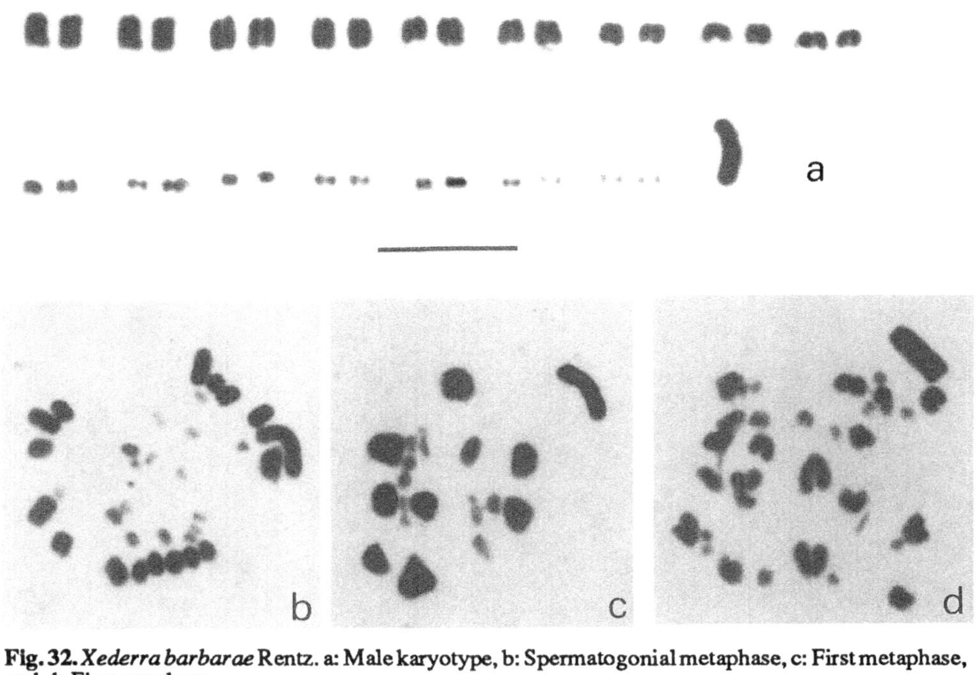

Fig. 32. *Xederra barbarae* Rentz. a: Male karyotype, b: Spermatogonial metaphase, c: First metaphase, and d: First anaphase.

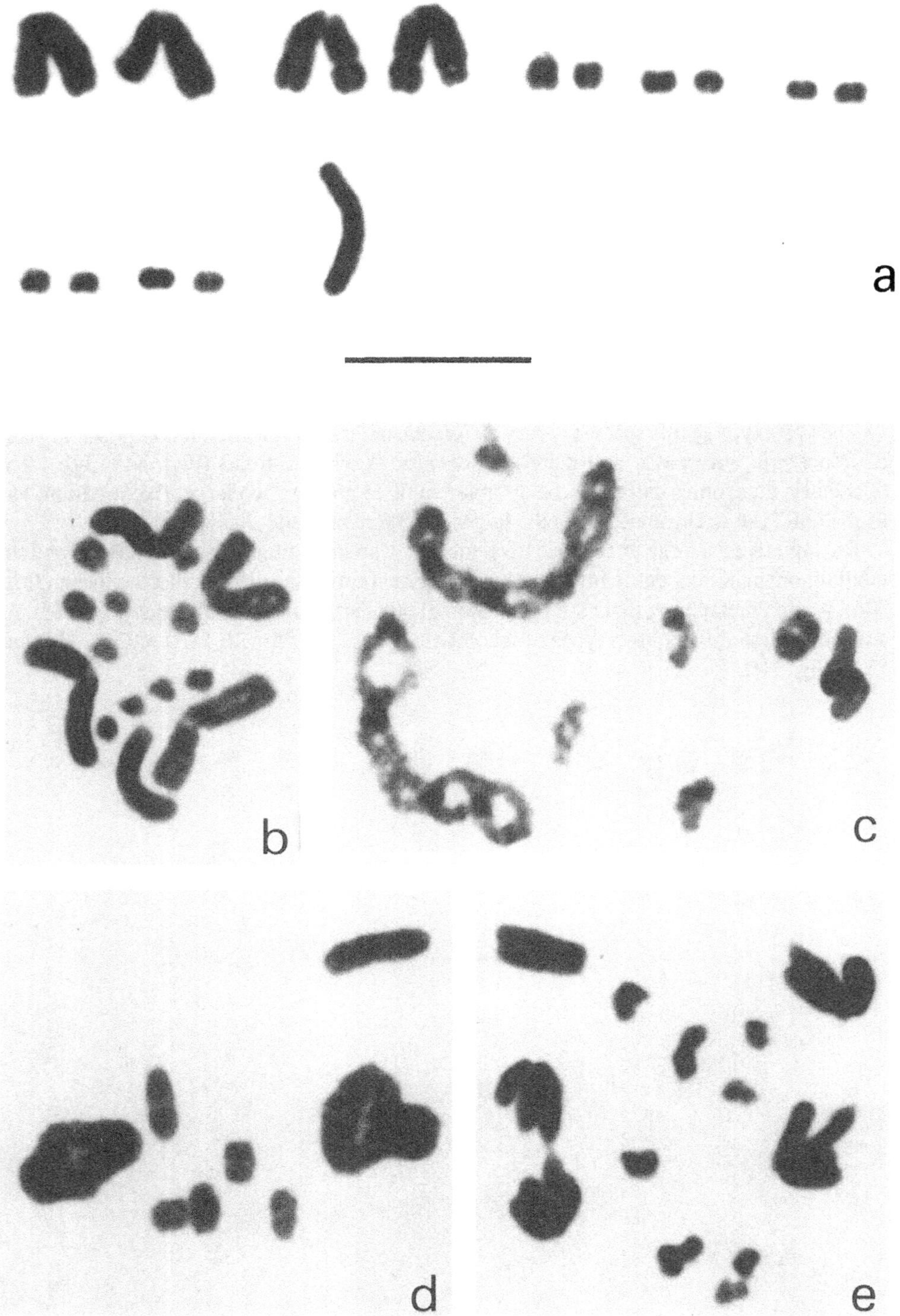

Fig. 33. *Ixalodetes whitei* Rentz. a: Male karyotype, b: S permatogonial metaphase, c: Diplotene, d: First metaphase, and e: First anaphase.

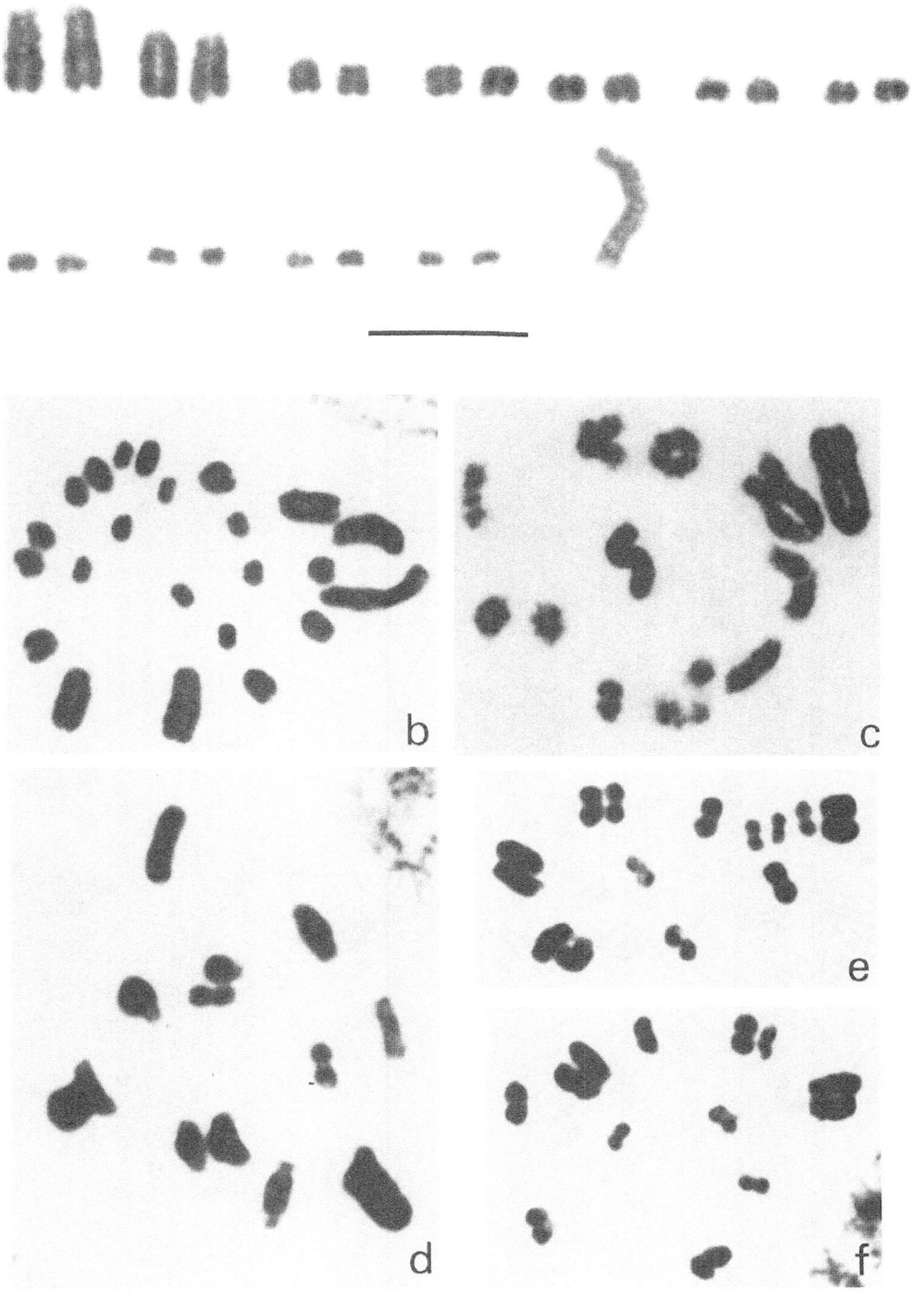

Fig. 34. *Ixalodectes uptoni* Rentz. a: Male karyotype, b: Spermatogonial metaphase, c: Diplotene, d: First metaphase, e: Second metaphase with X, and f: Second metaphase with X.

Ixalodectes nigrifrons Rentz (Fig. 35)

Material examined: R81-258, R84-96 (stop 65), -97 (stop 65)

2n♂ = 23

The karyotype of this species is the same as that of *I. uptoni*. It comprises two pairs of large, five pairs of medium and four pairs of small telocentric autosomes and a large telocentric X (Figs 35a, 35b).

At first metaphase, 11 autosomal bivalents and an X univalent are present (Fig. 35c). As the result of the first division, there are two types of second metaphase: 11A+X (Fig. 35d) and 11A (Fig. 35e).

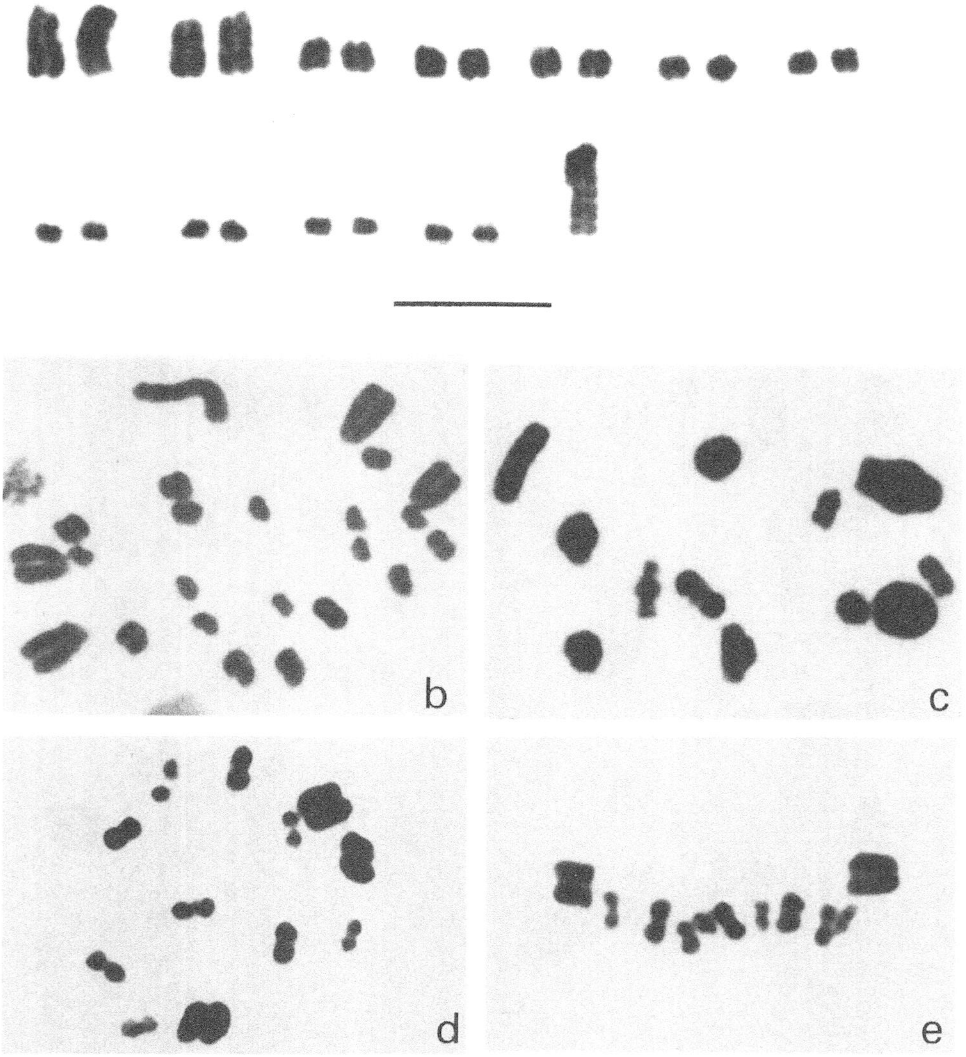

Fig. 35. *Ixalodectes nigrifrons* Rentz. a: Male karyotype, b: Spermatogonial metaphase, c: First metaphase, d: Second metaphase with X, and e: Second metaphase without X.

Ixalodectes megacercus Rentz (Fig. 36)

Material examined: R78-58 (holotype), R82-128, -134, -136, -143, -155

2n♂ = 15

The karyotype of this species is identical to that of *I. whitei* (Figs 36a, 36b).

During meiosis, there are two metacentric and five telocentric bivalents and a heteropycnotic X univalent (Fig. 36c). At first metaphase, eight chromosomes are present (Fig. 36d). As the result of first division, there are two types of second metaphase: 7A+X (Fig. 36e) and 7A (Fig. 35f).

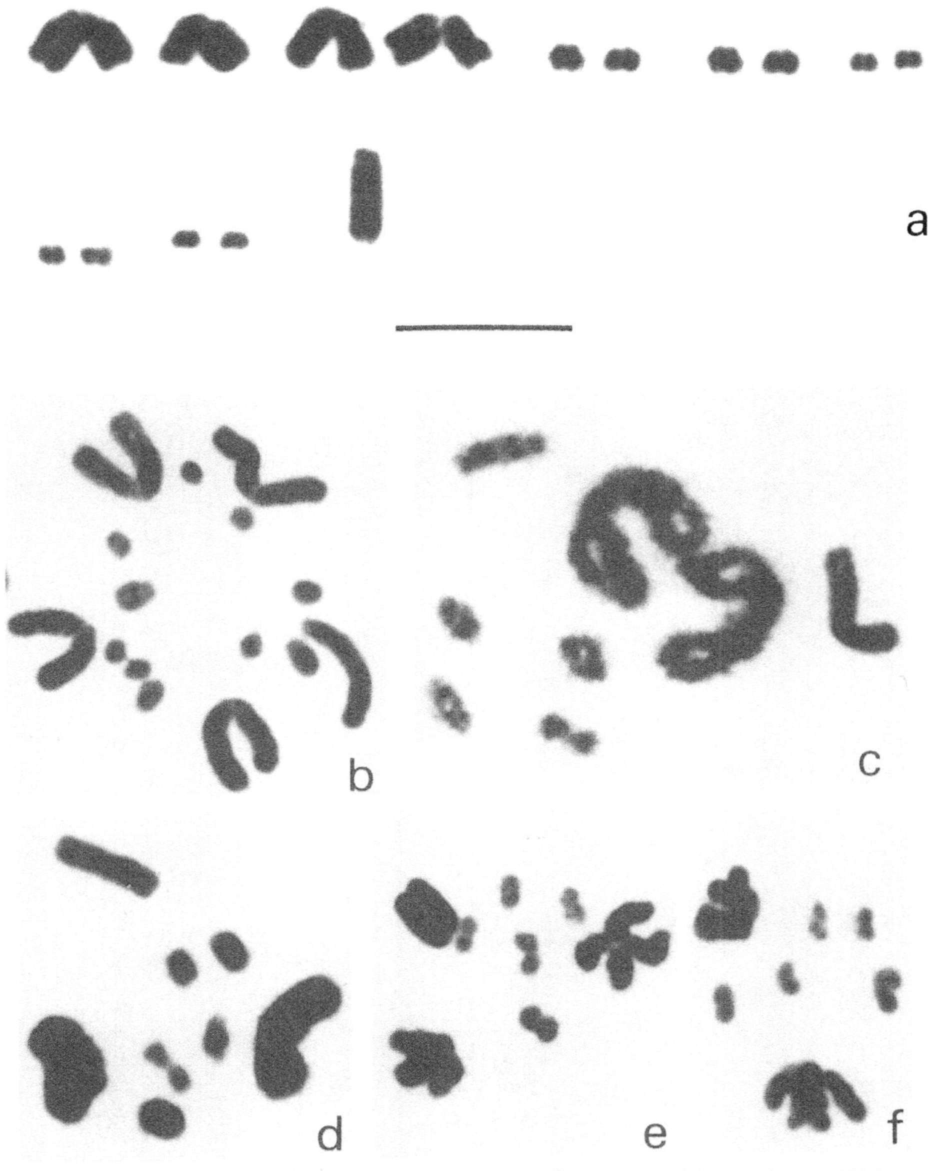

Fig. 36. *Ixalodectes megacercus* Rentz. a: Male karyotype, b: Spermatogonial metaphase, c: Diplotene, d: First metaphase, e: Second metaphase with X, and f: Second metaphase without X.

Austrodectes monticolus Rentz (Fig. 37)

Material examined: R78-6 (holotype), R78-90, -96, -97

2n♂ = 21

The karyotype comprises two pairs of large and eight pairs of medium-sized telocentric autosomes and a large telocentric X chromosome (Figs 37a, 37b).

At meiosis, telocentric bivalents usually form a single terminal or distal chiasma (Fig. 37c) and the X univalent is heteropycnotic. At first metaphase, ten autosomal bivalents and an X univalent are formed (Fig. 37d). At first anaphase, the X moves to one pole without dividing (Fig. 37e). As the result of first division, there are two types of second metaphase: 10A+X (Fig. 37f) and 10A (fGi. 37g).

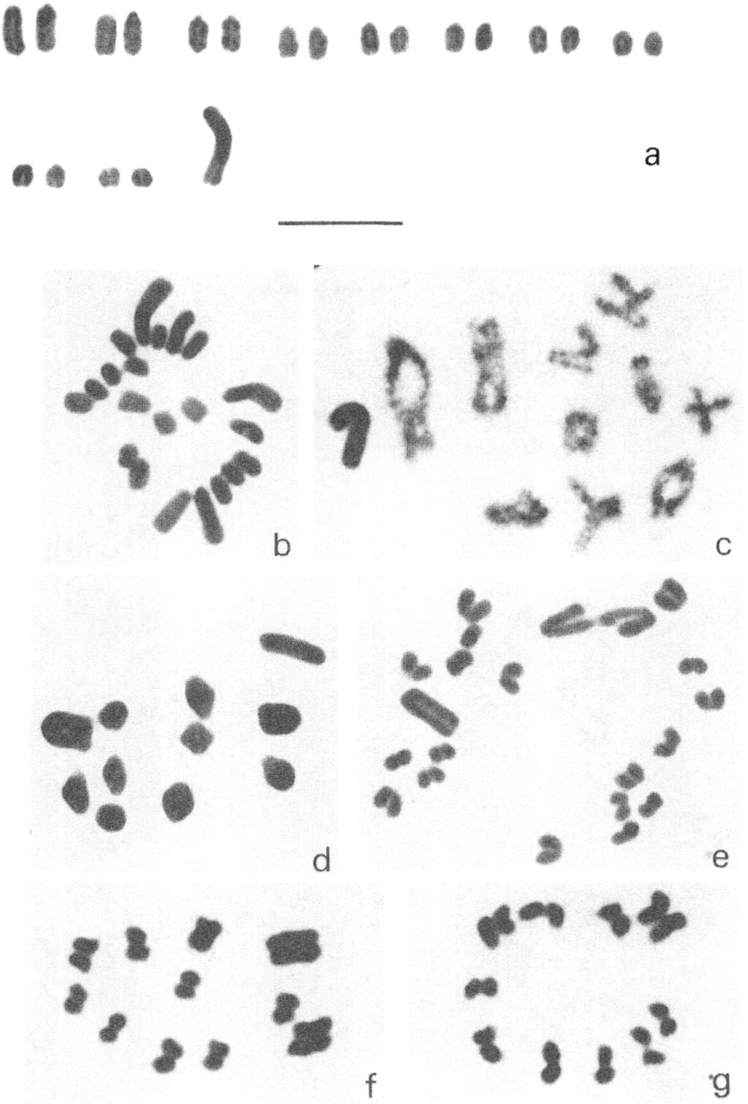

Fig. 37. *Austrodectes monticolus* Rentz. a: Male karyotype, b: Spermatogonial metaphase, c: Diplotene, d: First metaphase, e: First anaphase, f: Second metaphase with X, and g: Second metaphase without X.

Oligodectoides tindalei Rentz (Fig. 38)
Material examined: R78-54
2n♂ = 29

The karyotype comprises one pair of large, five pairs of medium (in a graded series), and eight pairs of small telocentric autosomes and a large telocentric X chromosome (Figs 38a, 38b). The X chromosome is the largest component in the set.

At meiosis, there are 14 autosomal bivalents, each usually forming a terminal or distal chiasma and a heteropycnotic X chromosome (Fig. 38c). At first metaphase, 14 autosomal bivalents and an X univalent are present (Fig. 38d).

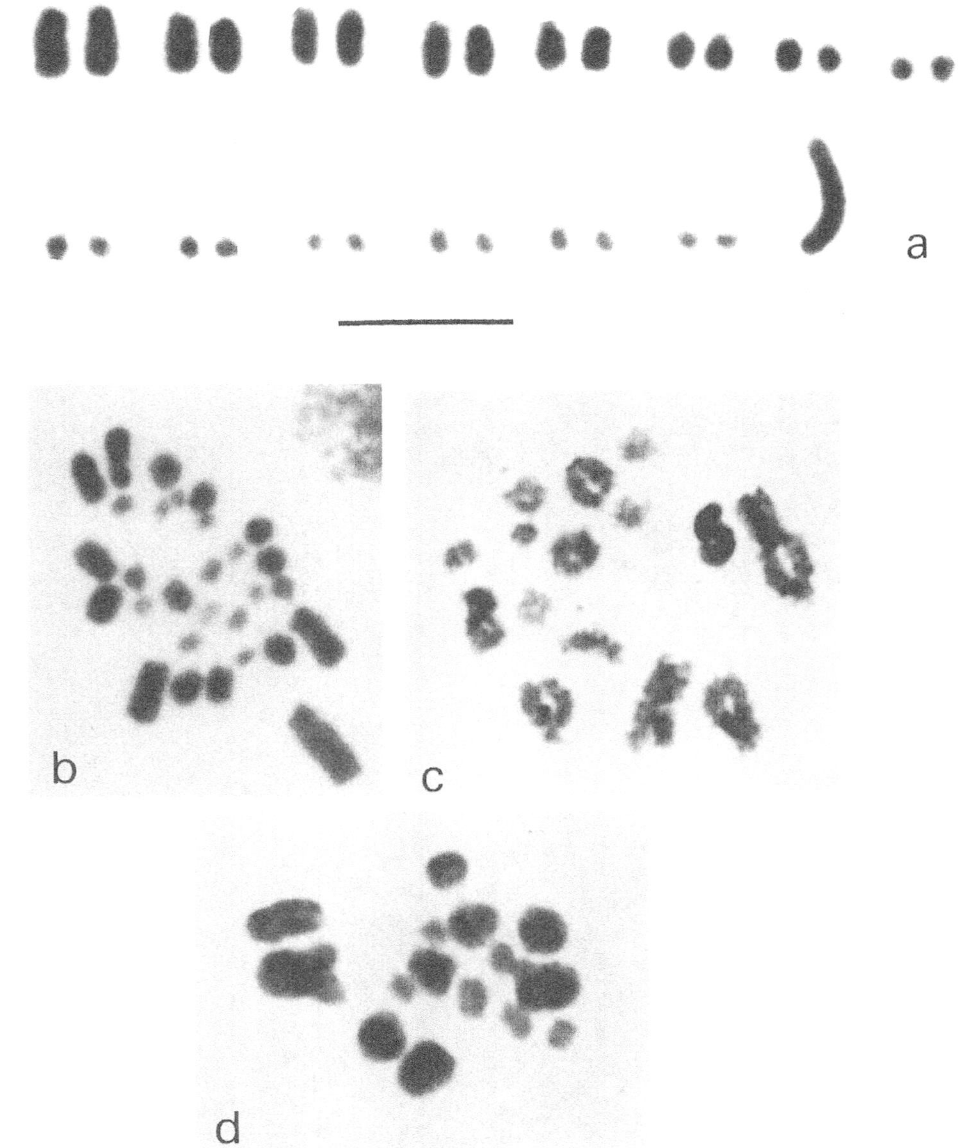

Fig. 38. *Oligodectoides tindalei* Rentz. a: Male karyotype, b: Spermatogonial metaphase, c: Diplotene, d: First metaphase.

Oligodectes longicercus Rentz (Fig. 39)

Material examined: R81-247 (holotype), R81-236, R82-113

2n♂ = 21

The karyotype comprises three pairs of large, two pairs of medium and five pairs of small telocentric autosomes with the first pair slightly larger than the others. There is a large telocentric X chromosome about the same size as the largest autosomal pair (Fig. 39a).

At meiosis, three large autosomal bivalents usually form more than two chiasmata and the rest of autosomal bivalents usually form a single terminal or distal chiasma (Fig. 39b). At first metaphase, ten autosomal bivalents and an X univalent are present (Fig. 39c). As the result of first division (Fig. 39d), there are two types of second metaphase: 10A+X and 10A (Figs 39e, 39f).

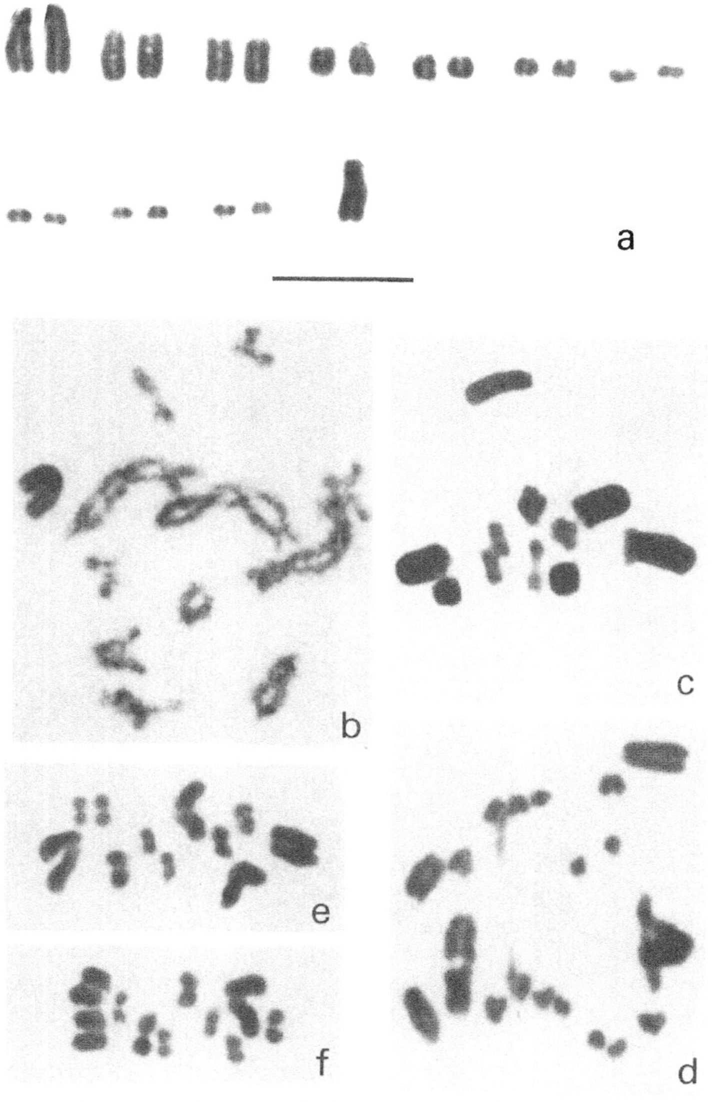

Fig. 39. *Oligodectes longicercus* Rentz. a: Male karyotype, b: Diplotene, c: First metaphase, d: First anaphase, e: Second metaphase with X, and f: Second metaphase without X.

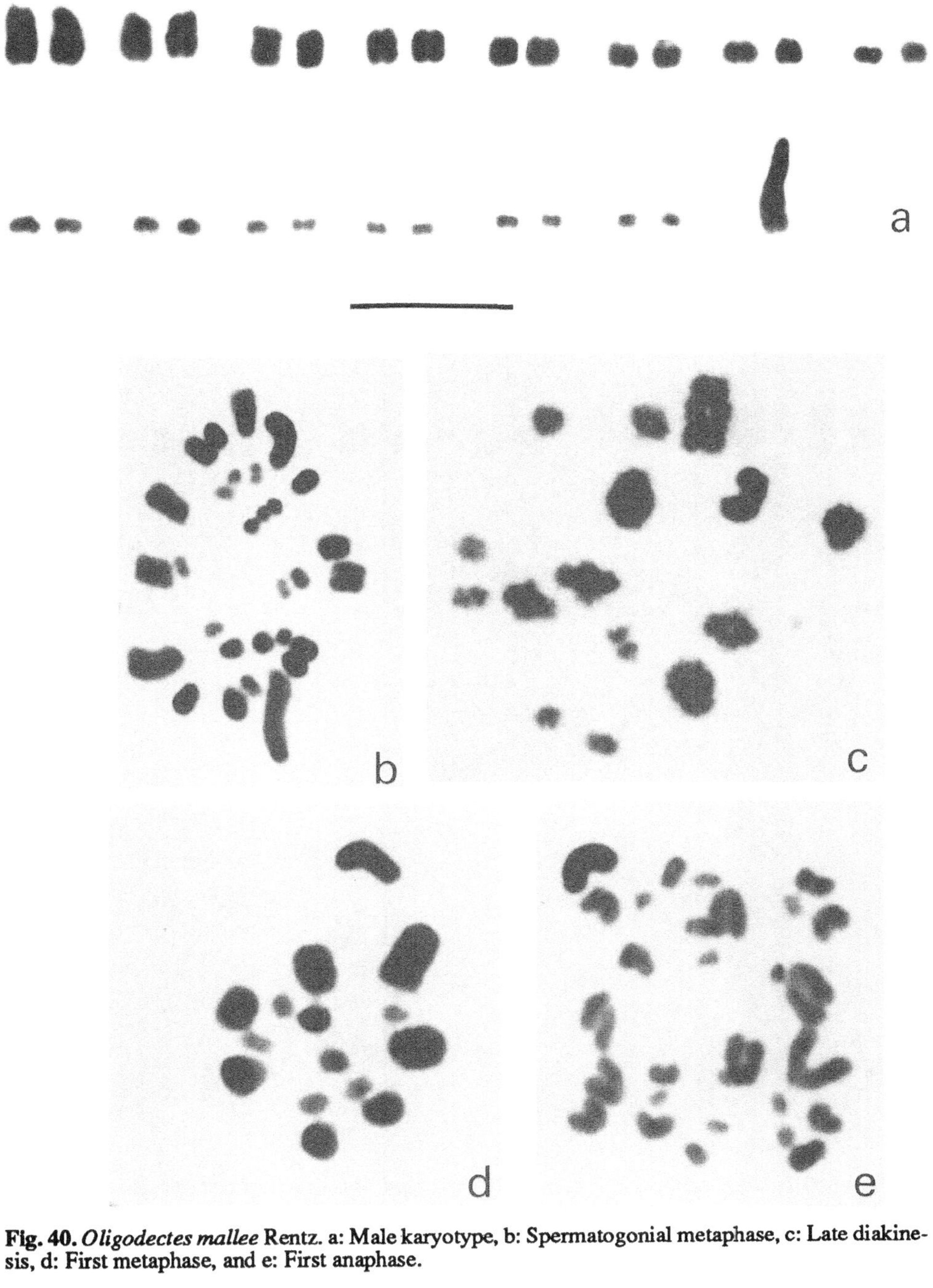

Fig. 40. *Oligodectes mallee* Rentz. a: Male karyotype, b: Spermatogonial metaphase, c: Late diakinesis, d: First metaphase, and e: First anaphase.

Oligodectes mallee Rentz (Fig. 40)

Material examined: R78-41 (stop 21), -42 (stop 21), R80-52 (stop 34)

2n♂ = 29

The karyotype comprises 14 pairs of telocentric autosomes consisting of one large, six medium (in a graded series) and seven small chromosomes. There is a telocentric X chromosome which is the largest in the set (Figs 40a, 40b).

During meiosis, 14 autosomal elements and a heteropycnotic X are present (Fig. 40c). At first metaphase, there are 14 autosomal bivalents and an X univalent (Fig. 40d). At first anaphase, the X moves to one pole without dividing (Fig. 40e). As the result of first division, there are two types of second metaphase: 14A+X and 14A.

Oligodectes pallens Rentz (Fig. 41)

Material examined: R80-56 (stop 39), -60 (stop 44)

2n♂ = 29

The karyotype is identical to that of *O. mallee* described above (Figs 41a, 41b, 41c).

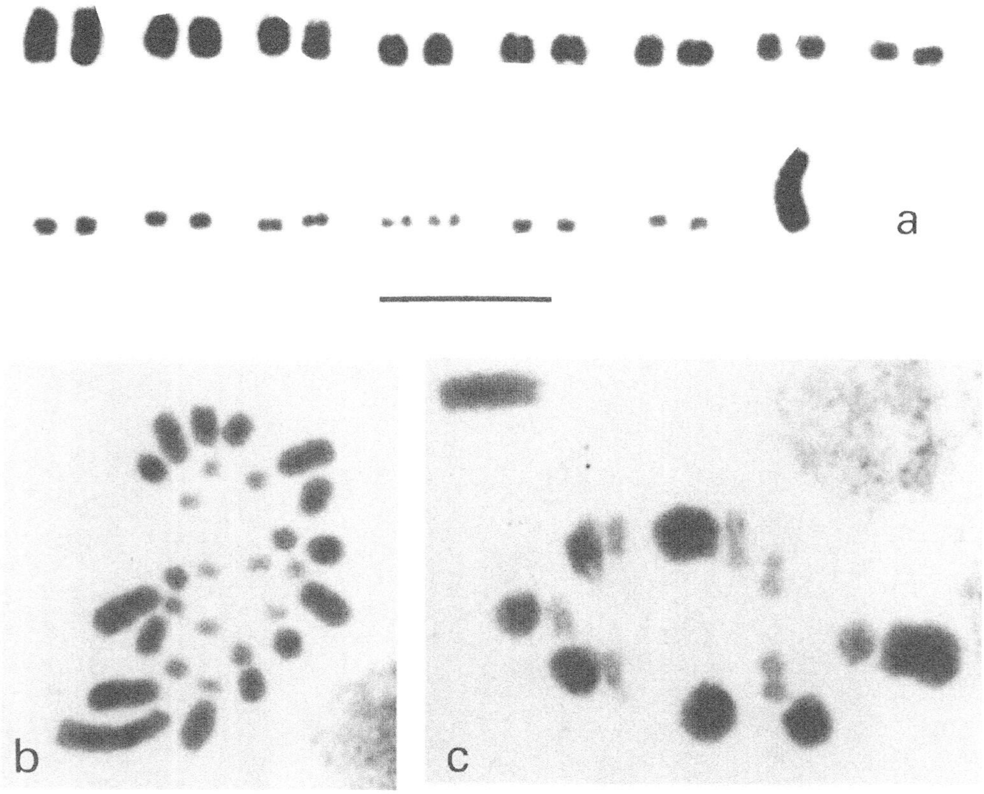

Fig. 41. *Oligodectes pallens* Rentz. a: Male karyotype, b: Spermatogonial metaphase, and c: First metaphase.

Fig. 42 (opposite). *Dexerra turpis* Walker. a: Male karyotype, b: Spermatogonial metaphase, c: First metaphase, and d: First anaphase.

Dexerra turpis Walker (Fig. 42)

Material examined: R81-316

2n♂ = 29

The karyotype comprises one pair of large, six pairs of medium (in a graded series) and seven pairs of small telocentric autosomes and a large telocentric X (Figs 42a, 42b). The X is the largest chromosome. The karyotype of the species is somewhat similar to that of both *Oligodectes mallee* and *O. pallens*.

At first metaphase, there are 14 autosomal bivalents and an X univalent (Fig. 42c). At first anaphase, the X moves to one pole without dividing (Fig. 42d).

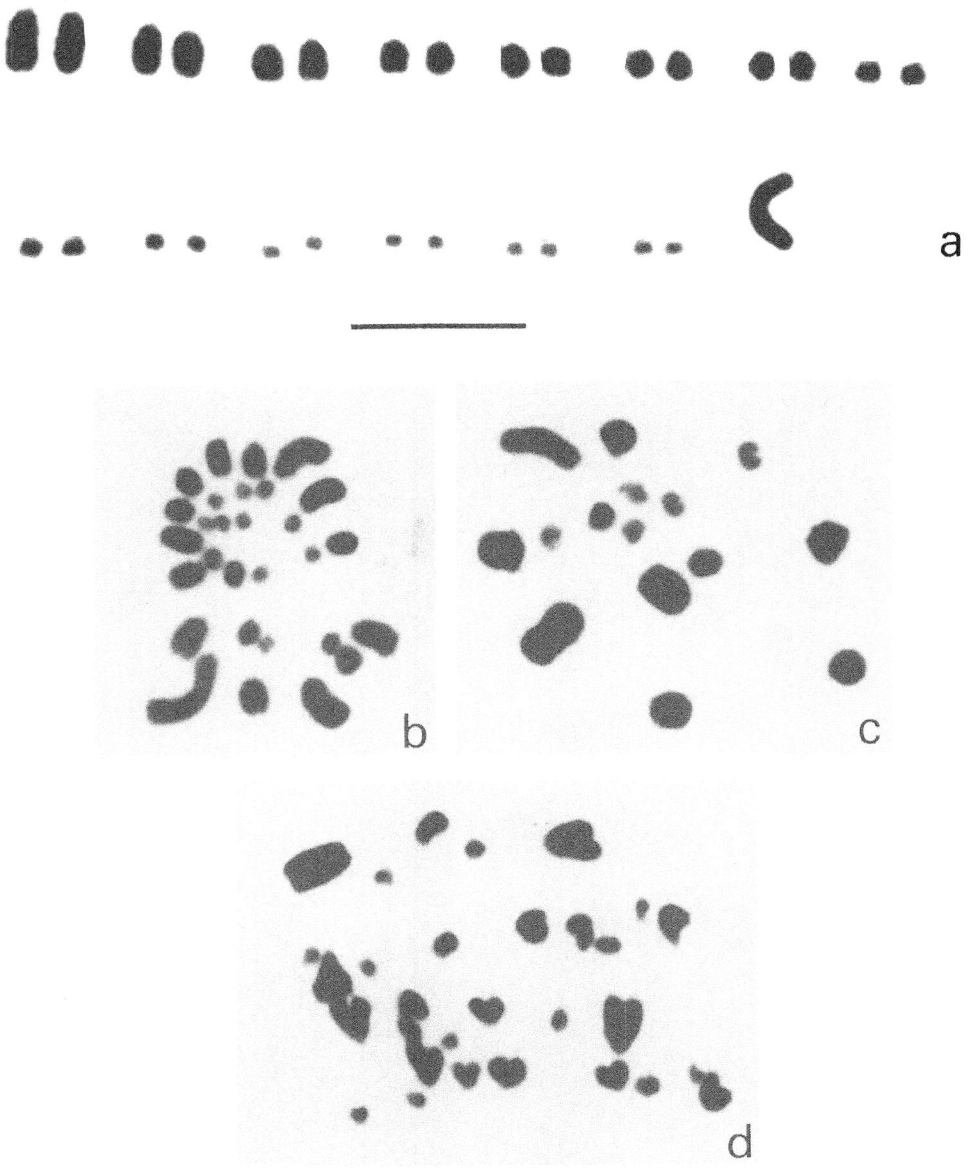

Dexerra serrata Rentz (Fig. 43)

Material examined: R80-68 (stop 46)

2n♂ = 29

The chromosome cytology of this species is the same as that described for *Dexerra turpis* (Fig. 43a).

At first metaphase, 14 autosomal bivalents and an X univalent are present (Fig. 43b). The X moves to one pole without dividing at first division. As the result of first division, there are two types of second metaphase: 14A+X and 14A (Figs 43c, 43d).

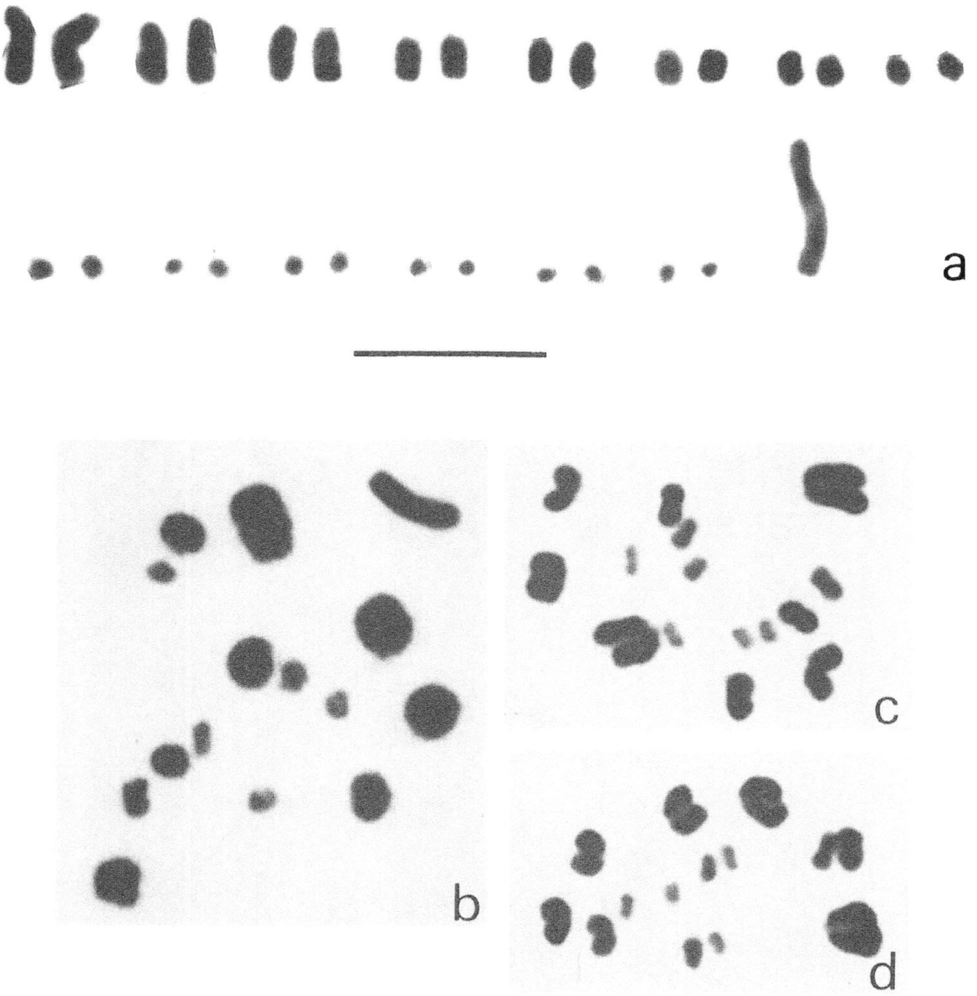

Fig. 43. *Dexerra serrata* Rentz. a: Male karyotype, b: First metaphase, c: Second metaphase with X, and d: Second metaphase without X.

Dexerra vigescens Rentz (Fig. 44)

Material examined: R80-78 (stop 52), -81 (stop 52), -86 (stop 61)

Karyotype: 2n♂ = 29

The chromosome cytology of this species is also the same as described for *D. turpis* and *D. serrata* (Figs 44a, 44b).

The 14 telocentric autosomes usually form a single terminal or distal chiasma each at diplotene. During meiosis, the X is a heteropycnotic element (Fig. 44c). At first metaphase, there are 14 autosomal bivalents and an X univalent (Fig. 44d). At first anaphase, the X moves to one pole without dividing. There are two types of second metaphase: 14A+X and 14A. Both second anaphases are shown in Figs. 44e and 44f.

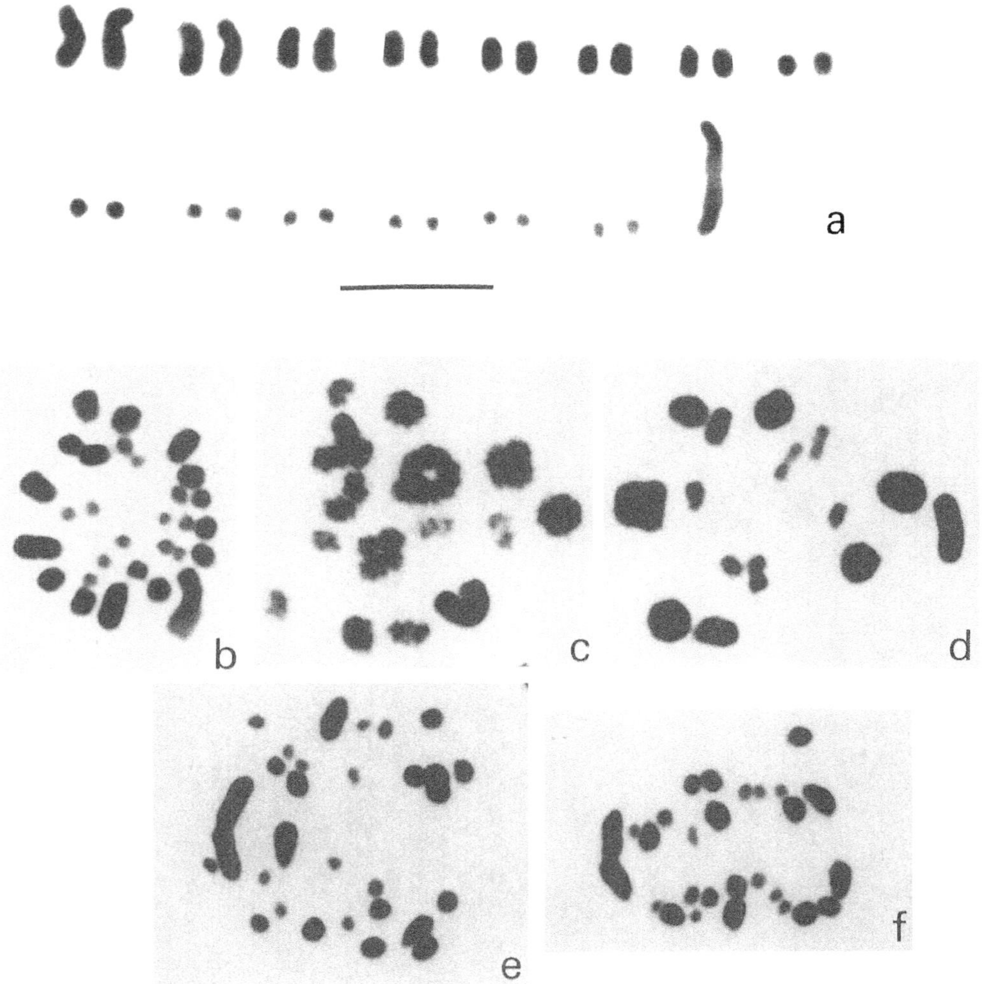

Fig. 44. *Dexerra vigescens* Rentz. a: Male karyotype, b: Spermatogonial metaphase, c: Diplotene, d: First metaphase, e: Second anaphase with X and f: Second anaphase without X.

Fig. 45 (overleaf). *Dexerra acanthiterga* Rentz. a: Male karyotype, b: Spermatogonial metaphase, c: First metaphase, and d: First anaphase.

Dexerra acanthiterga Rentz (Fig. 45)

Material examined: R80-60 (stop 44)

2n♂ = 29

The karyotype of this species is the same as the other species of *Dexerra* (Figs 45a, 45b).

At first metaphase, 14 autosomal bivalents and an X univalent are present (Fig. 45c). The X moves to one pole without dividing at first anaphase (Fig. 45d).

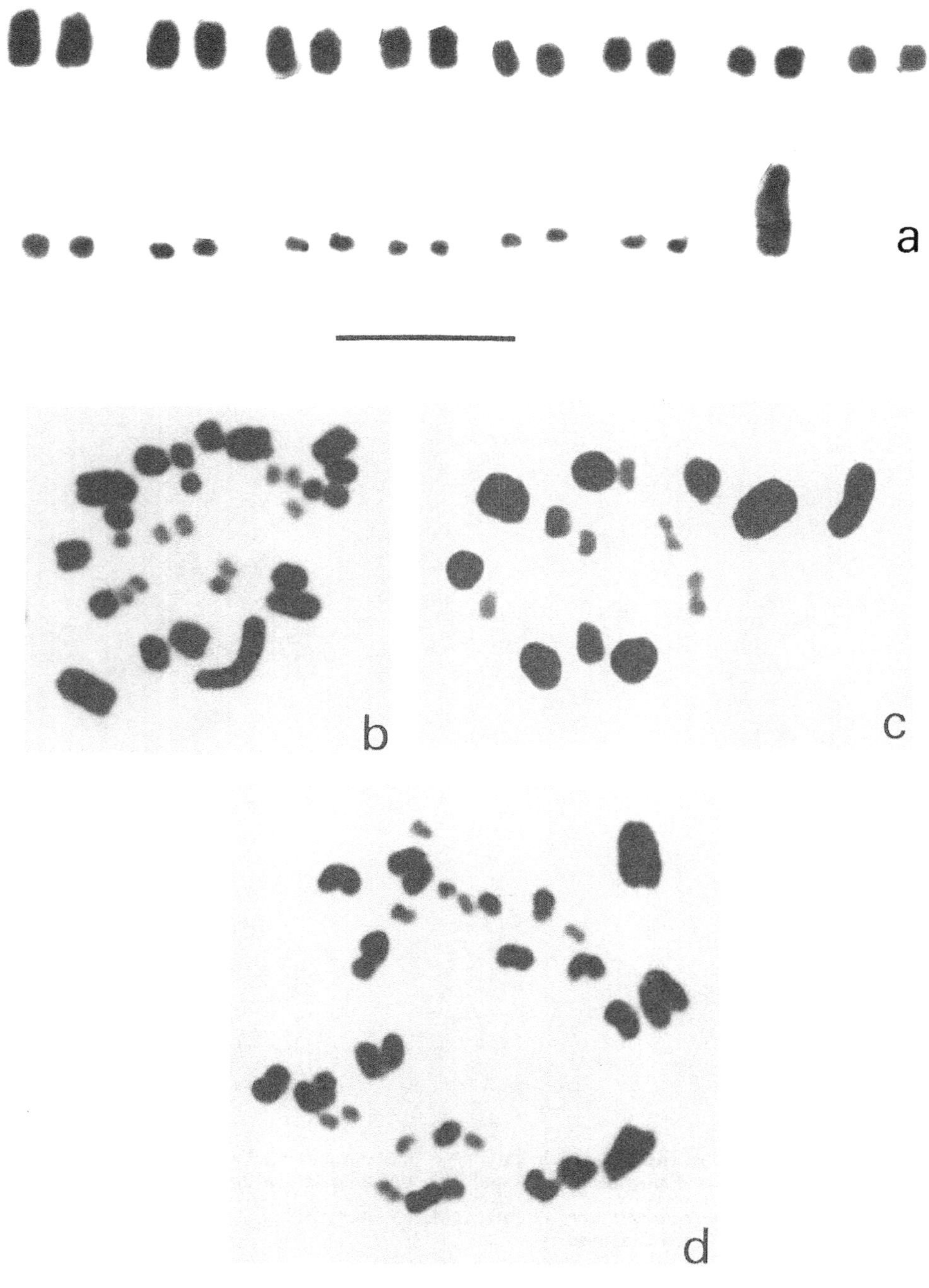

Discussion

The karyotypes of 54 of the 72 described species (in 15 of the 17 known genera) of Australian Tettigoniinae are described with the exception of the nine known species of *Nanodectes* Rentz. These latter species show different patterns of speciation from others in the subfamily and are being treated separately.

Chromosome numbers in Australian tettigoniine species range from 2n = 33 to 2n = 15 in males (Table 1). Two related genera, *Glenbalodectes* Rentz and *Xederra* Ander, share the highest number, 2n = 33. A modal number, 2n = 31, is shared by *Rhachidorus* Krauss and *Antipodectes* Rentz. Decreases in this number can be explained by fusion, as discussed by White (1973). This is illustrated among the cytologically known species of *Antipodectes*. Both *A. brevicaudus* and *A. giganteus* have 31 telocentric chromosomes; *A. graminicolus* has 29 chromosomes a pair of which are metacentric. This is most easily explained by centric fusion between two pairs telocentric autosomes. This phenomenon has been reported for other members of the subfamily by John and Hewitt (1968) and Ueshima and Rentz (1979).

A number of genera have 29 chromosome complements, all of which are telocentric. This is illustrated in *Chinnandectes*, *Oligodectes*, *Oligodectoides* and *Dexerra*. If this is considered a reduction from the modal number of 31, it can be explained by tandem fusion of two pairs of medium or small telocentric autosomes. Two species of *Oligodectes*, *O. mallee* and *O. pallens*, have 29 chromosomes, another, *O. longicercus*, has 21, none of which are metacentric. This may also be explained by multiple tandem fusions, but the pathway of evolution here is not clear. The monotypic genus *Austrodectes* also has 21 telocentric chromosomes. *Metaballus* has 27 telocentric chromosomes in all known species. If a reduction has occurred, this can be explained by tandem fusion of four pairs of medium or small autosomes independently.

Chlorodectes species exhibit three distinct karyotypes. *C. loquax* has 27 chromosomes consisting of one pair of metacentric and 12 pairs of telocentric autosomes and a metacentric X. *C. baldersoni* and *C. montanus* each have 11 pairs of telocentric autosomes instead of the 12 in *C. loquax*. The chromosome number has, therefore, been reduced to 25. Comparison of the relative sizes of telocentric autosomes between *C. loquax* (Figs 5a, 5b) and *C. baldersoni* (Figs 3a, 3b) suggests that one pair of the largest telocentrics in *C. baldersoni* may have been derived by tandem fusion of two pairs of large and small telocentrics in *C. loquax*. A further reduction in number is found in *C. ligaenus* where 2n = 23. Comparison of size relationships (Figs 3a, 3b, 4a, 4b) indicates that one pair of medium sized telocentrics in *C. ligaenus* could have been derived by the tandem fusion of two pairs of small telocentrics in *C. baldersoni*. This would account for the reduction of number to 23. The X chromosome in all *Chlorodectes* is unique for all Australian Tettigoniinae in that it is metacentric; however the relative size of both arms is different among the species (see Figs 3a, 4a, 5a). This could be the result of a pericentric inversion of a large telocentric X.

Lanciana is of interest to the student of chromosome evolution. *L. albidicornis* and *L. semialata* have 25 telocentric chromosomes. In *L. montana* the number is reduced to 23 a pair of which are metacentric. This is simply centric fusion between two pairs of telocentric autosomes in the species with 25 chromosomes (Figs 17a, 17b, 18a, 18b, 20a, 20b). Further reduction in chromosome number to 21 occurs in *L. hesperpotana* and *L. occidentalis*. The complements of these two species each have a pair of metacentrics and nine pairs of telocentric autosomes and a large telocentric X. Comparison of size relationships in the species with 21 and 23 chromosomes indicates that the reduction to 21 in *L. occidentalis* and *L. hesperpotana* is the result of tandem fusion by two pairs of small telocentric autosomes in *L. montana* (Figs 19a, 19b, 20a, 20b, 21a, 21b).

A more complex pattern of chromosome evolution seems indicated in *Ixalodectes*. Both *I. uptoni* (Fig. 34) and *I. nigrifrons* (Fig. 35) have 23 telocentric chromosomes. Surprisingly, *I. whitei* (Fig. 33) and *I. megacercus* (Fig. 36) have 15, consisting of two pairs of metacentrics and five pairs of telocentrics and a large telocentric X. A possible explanation is as follows: the metacentric autosome could be derived by centric fusion of two pairs of large and medium telocentrics in the species with 23 chromosomes. This would form a submetacentric autosome reducing the number from 23 to 21. Subsequently a tandem fusion takes place between a short arm of a metacentric and a medium sized telocentric. The submetacentric is thereby changed to a metacentric reducing the number further from 21 to 19.

The secondary constriction in the arm of the metacentric autosome (Fig. 33a), which frequently occurs, suggests that fusion is still a dynamic process. This same pattern of chromosome evolution has independently occurred in another pair of metacentrics in both *I. whitei* and *I. megacercus*. This satisfactorily explains the reduction of number from 23 to 15 in *Ixalodectes* and is illustrated diagramatically in Fig. 46.

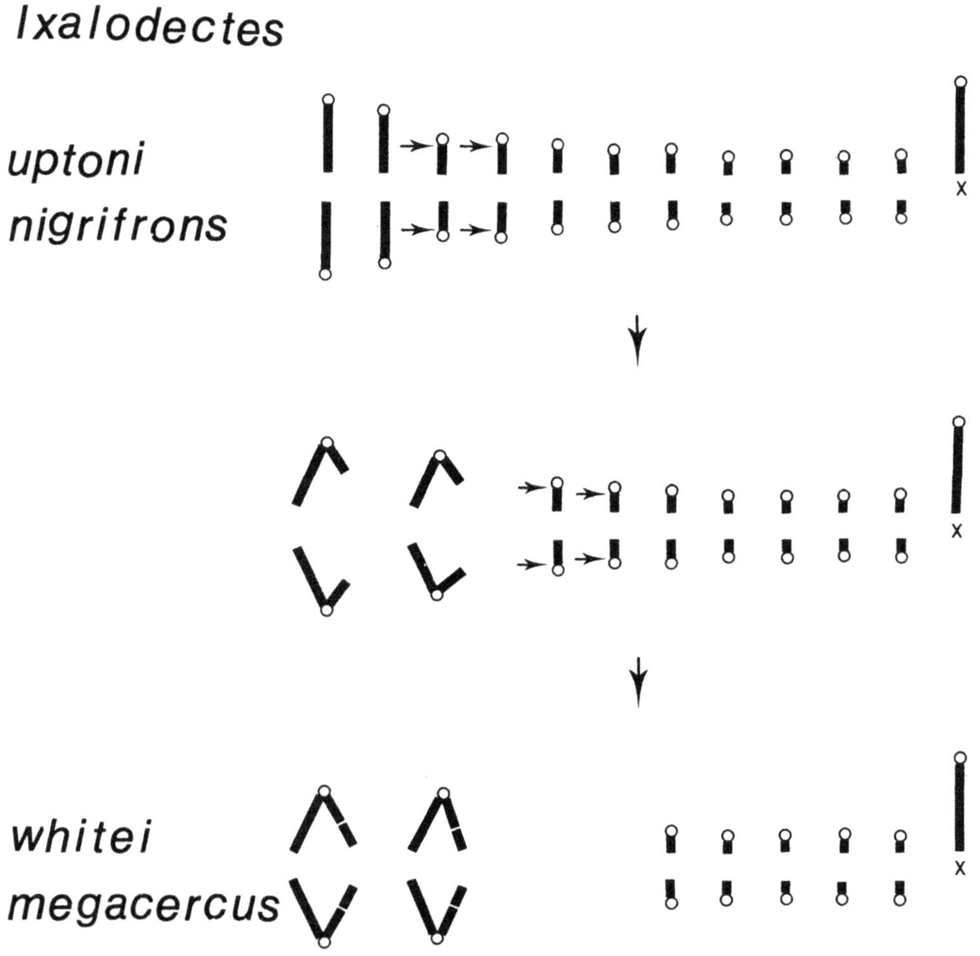

Fig. 46. Diagrammatical scheme of chromosome evolution in *Ixalodectes*.

REFERENCES

Camacho, J.P.M., Orozco, J.C., and Pascual, F. (1981). Chromosomal rearrangements and karyotypic evolution in Decticinae (Orthoptera: Tettigonioidea). *Cytologia* 46: 209-217.

Ferreira, A. (1969). Chromosome survey of some Australian Tettigoniids (Orthoptera: Tettigonioidea): Two species with neo-XY sex determining mechanism. *Animal Cytologia* 34: 511-522.

Hewitt, G.M. (1979). 'Grasshoppers and crickets'. In: *Cytogenetics*. vol. 3. Insecta 1. 170 pp. Gebruder Borntraeger, Berlin.

John, B., and Hewitt, G.M. (1968). Patterns and pathways of chromosome evolution within the Orthoptera. *Chromosoma* 25: 40-74.

Mesa, A., and Ferreira, A. (1977). The chromosomes of two species of North American tettigoniids (Orthoptera: Tettigonioidea). *Entomological News* 88: 99-103.

Rentz, D.C.F. (1985). 'A Monograph of the Tettigoniidae of Australia. vol. 1. The Tettigoniinae'. 384 pp. CSIRO, Melbourne; E.J. Brill, Leiden.

Ueshima, N. (1986). Chromosome systems of some Japanese Tettigoniidae (Orthoptera). *Journal of Matsusaka University* 4: 13-20.

Ueshima, N., and Rentz, D.C. (1979). Chromosome systems in the North American Decticinae with reference to Robertsonian changes (Orthoptera: Tettigoniidae). *Cytologia* 44: 693-714.

Warchalowska-Sliwa, E. (1984). Karyological studies on Polish species of the Tettigonioidea superfamily. II. Karyotypes of families Tettigoniidae and Decticidae. *Folia Biol.* (Krakow) 32: 311-325.

White, M.J.D. (1973). 'Animal *Cytology* and Evolution'. 3rd ed. 961 pp. Cambridge University Press, London.

17. A CLASSIFICATION OF THE SHIELD-BACKED KATYDIDS (TETTIGONIINAE) OF THE WORLD

D.C.F. Rentz and D.H. Colless

Introduction

The Tettigoniinae (shield-backed katydids) includes some of the most recognizable of tettigoniids. The greatest number of species occur in the Holarctic region but considerable numbers occur elsewhere. Although not generally considered a tropical group, many species occur within the tropics in grasslands and savannahs of Australia. A generous estimate of the number of species worldwide might place the total at around 500.

Shield-backs range in size from among the smallest of tettigoniids to those rivalling the largest. The majority of species are brachypterous or micropterous but many are fully winged. A single known species is apterous in both sexes. Most stridulate at night, but many such "nocturnal" species perform various other activities during the day. Similarly, "diurnal" species can be found actively moving about, feeding, ovipositing and even intermittently stridulating at night. Most species are to be found on or near the ground. Some are associated with grasses and their long legs enable them to successfully straddle the grass stems in high winds. Many of the long-winged species are associated with bushes and trees. Some alate species move considerable distances at night, perhaps in search of food or mates. Shield-backed katydids were once thought to occur mainly in desert or arid regions, but many exceptions are now known. A number of species are to be found in montane situations, some well above tree-line, but in situations that could be interpreted as "arid". Most tettigoniines are probably opportunistic scavengers. A number are strictly predaceous, others facultatively so; but the great majority feed on foliage, flowers, seeds, dead animal matter and insects as the situation provides. Few species are of any economic importance, but the Mormon cricket, *Anabrus simplex* (Haldeman), and the Coulee cricket, *Peranabrus scabricollis* (Thomas), of the United States cause considerable damage to crops during certain outbreak years. Similar damage to cereal crops has been caused by the Ethiopian *Decticoides brevipennis* Ragge. Other species can be numerous during years of favourable rainfall and abundant food and may cause local damage to pasture, especially when other orthopteroid species are similarly abundant.

A modern review of this subfamily for the Southern Hemisphere has only recently been completed, with the review of the Tettigoniinae of southern Africa by Rentz (1988). By contrast, the large fauna of the Northern Hemisphere, from both Palaearctic and Nearctic regions, has for long been well-documented and, for the most part, the generic classification has remained stable.

It seemed opportune here to attempt a generic classification of the Tettigoniinae on a world-wide basis. We decided to ask two questions at the outset. First, could we produce an improved classsification of the group based on numerical analyses, both phenetic and cladistic, of a large set of characters. At the same time, would the integrity of the commonly recognized taxon "Decticinae" (which continues to be used by European authors; see Rentz, 1979 and subsequent faunal revisions of the author) survive under such detailed analysis.

History

The Tettigoniinae have long been known to taxonomists. The first "overview" was that of Herman (1874) which was followed by Brunner (1882) who monographed the European forms and later (1893) proposed a worldwide classification (to be discussed later). Scudder (1894) and Caudell (1907) presented revisionary studies on the North American species and Caudell (1908) provided a synoptic treatise on the entire group, but inadvertently omitted several Australian genera. More modern interpretations have been provided by Ramme, Uvarov and Zeuner for Europe. Harz (1969) provided keys, synopses and literature references for the European species. Rentz and Birchim (1968) reviewed the North American fauna, Rentz and Gurney (1985) reviewed the South American and Rentz (1985, 1988) revised the Australian and African faunas respectively. References to all the important generic revisions can be found in the bibliographies of those publications. The cytological features of the American species were presented by Ueshima and Rentz (1979), while cytological studies of the Australian fauna may be found in the previous chapter. References to the relatively few Palaearctic species for which the cytology is known are provided in the latter works.

Zeuner (1941) proposed a large number of generally monotypic genera for certain species in *Platycleis* and *Metrioptera*. A few have been accepted as genera, but most have been treated as subgenera or synonyms of the two genera. For the sake of argument, a number of Zeuner's taxa have been included in our study but it is not proposed that they should be accorded generic rank. Along the same lines, Rentz (1972) considered *Oreopedes* as a subgenus of *Eremopedes* but for the purposes of this exercise, they are ranked as genera. Rentz (1988) raised *Aglaothorax* from subgeneric status within *Neduba* to generic rank largely because of chromosomal evidence; but this decision may change once the total cytology of the two taxa is analyzed together with morphological and acoustical data. Here they are considered as distinct. *Onconotus*, usually considered a shield-back, was accorded its own separate subfamily by Harz (1969) but is here considered a tettigoniine. *Petropedes* Tinkham proved to be identical with *Inyodectes* on the characters used in this study.

Gorochov (1988) presented a major contribution to our understanding of the tettigoniine insects, integrating fossil and extant taxa. He compared a wide range of characters over the entire family. He split the Tettigoniidae (*sens. lat.*) into a number of families and divided several of the currently accepted subfamilies. Of relevance to this study is the changed status of *Glyphonotus* (here placed in a nine-genus tribe), for which he erected a monogeneric subfamily on the basis of the fusion of the male epiproct with the tenth tergite, and the "primitive cerci" (here scored as having a simple tooth). Gorochov seemed to relate this subfamily to the Conocephalinae and the Saginae, especially the former, rather than to the Tettigoniinae. Of more interest here is his erection of a subfamily Nedubinae to contain *Neduba*, *Aglaothorax*, *Platydecticus*, *Xyrdectes*, *Falcidectes*, and "perhaps" *Calodectes*, basing his opinion on the conical or hypognathous head and the unique structure of the paraprocts. He reserved judgement on the placement of several Australian genera. The present study confirms the distinctiveness of this group of genera, including *Calodectes*, although we regard it as a tribe, not a subfamily. Incidentally, we disagree with the description of the head of *Neduba*, *Aglaothorax*, *Xyrdectes* and *Falcidectes* as "conical": it is decidedly round. However, in the genera we refer to the Nedubini the head does range from conical to round. We agree with Gorochov and earlier authors in not recognizing a subfamily Decticinae and in the tribal status accorded to *Onconotus*, with which we here associate several African genera (see Figs 1-3).

Outgroups were chosen that were either formerly considered in the subfamily or bear overall similarity to tettigoniine katydids.

Specimens of a number of genera were not available for study. They are *Drymapedes* Bey-Bienko 1967, *Evergoderes* Bolívar 1936, *Ferganusa* Uvarov 1926, *Lithoxenus* Bey-Bienko 1951, *Microdrymadusa* Bey-Bienko 1967, *Plicigastra* Uvarov 1940, and *Tadzhikia* Mistshenko, 1952. Their relations to the other genera remain unclear.

The Genera

The numbering and the bold-faced code are used in Figs 1-3 and in the listing of the character states at the end of the chapter.

ACRO1. *Acrodectes* Rehn and Hebard 1919
AFRO2. *Afrodrymadusa* Ramme 1939
AGLA3. *Aglaothorax* Scudder 1907
ALFR4. *Alfredectes* Rentz 1988
AMMO5. *Ammoxenulus* Bey-Bienko 1951
AMPH6. *Amphiestris* Fieber 1853
ANAB7. *Anabrus* Haldeman 1852
ANAD8. *Anadolua* Ramme 1939
ANAX9. *Anadrymadusa* Karabag 1961
ANAR10. *Anarytropteris* Uvarov 1924
ANAT11. *Anatlanticus* Bey-Bienko 1951
ANON12. *Anonconotus* Camerano 1878
ANTA13. *Antaxius* Brunner 1882
ANTE14. *Anterastes* Brunner 1882
ANTI15. *Antipodectes* Rentz 1985
APOT16. *Apote* Scudder 1894
ARIA17. *Ariagona* Krauss 1892
ARYT18. *Arytropteris* Herman 1874
ATEL19. *Ateloplus* Scudder 1900
ATLA20. *Atlanticus* Scudder 1894
AUST21. *Austrodectes* Rentz 1985
BERG22. *Bergiola* Stschelkanowzeff 1910
BIEN23. *Bienkoxenus* Cejchan 1968
BUCE24. *Bucephaloptera* Ebner 1926
CALL25. *Calliphona* Krauss 1892
CALO26. *Calodectes* Rentz and Gurney 1985
CALX27. *Calopterusa* Uvarov 1942
CAPN28. *Capnobotes* Scudder 1897
CERA29. *Ceraeocercus* Uvarov 1910
CHIN30. *Chinnandectes* Rentz 1985
CLIN31. *Clinopleura* Scudder 1876
CHLO32. *Chlorodectes* Rentz 1985
CTEN33. *Ctenodecticus* Bol!var 1876
CYRT34. *Cyrtophyllicus* Hebard 1908
DECT35. *Decticita* Hebard 1939

DECX36. *Decticoides* Ragge 1977
DECY37. *Decticus* Serville 1831
DEXE38. *Dexerra* Walker 1869
DRYM39. *Drymadusa* Stein 1860
ECTO40. *Ectopistidectes* Rentz 1985
EREM41. *Eremopedes* Cockerell 1898
EULI42. *Eulithoxenus* Bey-Bienko 1951
SCHI43. *Schizonotinus* Ramme 1948
EUPH44. *Eupholidoptera* Ramme 1951
FALC45. *Falcidectes* Rentz and Gurney 1985
GAMP46. *Gampsocleis* Fieber 1852
GLYP47. *Glyphonotus* Redtenbacher 1889
GLEN48. *Glenbalodectes* Rentz 1985
HYPH49. *Hyphonomus* Uvarov 1921
IDIO50. *Idionotus* Scudder 1894
IDIX51. *Idiostatus* Pictet 1881
HUBB52. *Hubbellia* Hebard 1927
INYO53. *Inyodectes* Rentz and Birchim 1968; *Petropedes* Tinkham 1972 scored identically.
IXAL54. *Ixalodectes* Rentz 1985
LANC55. *Lanciana* Walker 1869
MEDE56. *Medecticus* Uvarov 1912
META57. *Metaballus* Herman 1874
METR58. *Metrioptera* Wesmael 1838
MIRA59. *Miramiola* Uvarov 1939
NAMA60. *Namaquadectes* Rentz 1988
NANO61. *Nanodectes* Rentz 1985
NEDU62. *Neduba* Walker 1869
OLIG63. *Oligodectes* Rentz 1985
OLIX64. *Oligodectoides* Rentz 1985
ONCO65. *Onconotus* Fischer de Waldheim 1839
OREO66. *Oreopedes* Rehn and Hebard 1919
PACH67. *Pachytrachis* Uvarov 1940
PARA68. *Paratlanticus* Ramme 1939
PEDI69. *Pediodectes* Rehn and Hebard 1916
PARX70. *Paradrymadusa* Herman 1874
PERA71. *Peranabrus* Scudder 1894
PHOL72. *Pholidoptera* Wesmael 1838
PLAG73. *Plagiostira* Scudder 1876
PLAT74. *Platycleis* Fieber 1852
PLAX75. *Platydecticus* Chopard 1951
PLAY76. *Platyproctidectes* Rentz 1985
PTER77. *Pterolepis* Rambur 1839
RAGG78. *Raggeana* Pener, Broza, Ayal 1971
RHAC79. *Rhachidorus* Herman 1874
RHAX80. *Rhacocleis* Fieber 1853
SCOT81. *Pezodrymadusa* Ramme 1939
SEPI82. *Sepiana* Zeuner 1941

STEI83. *Steiroxys* Herman 1874
THOR84. *Thoracistus* Pictet 1888
THRO85. *Throscodectes* Rentz 1985
THYR86. *Thyreonotus* Serville 1839
UVAR87. *Uvarovina* Ramme 1939
UVAX88. *Uvarovistia* Ramme 1951
YERS89. *Yersinella* Ramme 1939
XEDE90. *Xederra* Ander 1939
XYRD91. *Xyrdectes* Rentz and Gurney 1985
ZACY92. *Zacycloptera* Caudell 1907
ZEUN93. *Zeuneriana* Ramme 1951
UVAY94. *Uvarovites* Tarbinsky 1932
SURE95. *Sureyaella* Uvarov 1934
RAMM96. *Rammeola* Uvarov 1934
PSOR97. *Psorodonotus* Brunner 1861
PSAL98. *Psalmatophanes* Chopard 1938
PHYT99. *Phytodrymadusa* Ramme 1948
PEZO100. *Pezodrymadusa* Karabag 1961
PARY101. *Parapholidoptera* Ramme 1951
MONG102. *Mongolodectes* Bey-Bienko 1951
IRAN103. *Iranusa* Uvarov 1942
FEST104. *Festella* Giglio-Tos 1894
EXOD105. *Exodrymadusa* Karabag 1961
ELAS106. *Elasmocercus* Chopard 1943
DRYM107. *Drymadusella* Ramme 1939
CERE108. *Ceresia* Uvarov 1928
APHO109. *Apholidoptera* Ramme 1951
AFGH110. *Afghanoptera* Ramme 1952
SEME111. *Semenoviana* Zeuner 1941 (subgenus of *Platycleis*)
SQUA112. *Squamiana* Zeuner 1941 (subgenus of *Platycleis*)
SPHA113. *Sphagniana* Zeuner 1941 (usually considered a subgenus of *Metrioptera*)
TETT114. *Tettigonia* Linnaeus 1758
ZULU115. *Zuludectes* Rentz 1988
ULUD116. *Uludaghia* Ramme 1951
PLAZ117. *Platyoplus* Tinkham 1973

Undescribed Genera

GNT-1118. (Kalbarri, Western Australia)
GNT-2119. (Kalbarri, Western Australia)
GNT-3120. (Balladonia, Western Australia)
GNT-4121. (San Jacinto Mountains, California)

Outgroups

BRAD122. *Bradyporus* Charpentier 1825 (Bradyporinae)
HEMI123. *Hemisaga* Saussure 1888 (Saginae)
PERI124. *Peringueyella* 1888 Saussure (Saginae)
NEOB125. *Neobarrettia* Rehn 1901 (Listroscelidinae)
REQU126. *Requena* Walker 1869 (Listroscelidinae)
TERP127. *Terpandrus* Stål 1874 (Listroscelidinae)

PACH128.*Pachysaga* Brunner 1893 (Saginae)
APTE129.*Apteropedetes* Rentz 1979 (originally placed in the Apteropedetinae but here transferred to Tettigoniinae)
EPHI130.*Ephippiger* Berthold 1827 (Ephippigerinae)

The Characters

The following is a list of characters and their states as used in the numerical analysis. All of the characters have been assessed by only one of us (DCFR) so those that are less objective have been, at least, comparatively rated. If one species in a genus possessed a given structure, then it was ranked as present in that genus. Full details of the data matrix are given at the end of the paper. The type of character is shown below in brackets, as: numerical, N; two-state, B; ordered multistate, OM ; unordered multistate, UM.

HEAD

1. SHAPE: conical 0, round 1. (B)
2. FASTIGIUM; extremely narrow 0, narrow 1, broad 2. (OM)
3. EYE: normal 0, disproportionately large for head 1. (B)
4. FASTIGIUM OF VERTEX; not sulcate 0, sulcate 1. (B)

THORAX

5. PROSTERNAL ARMATURE; absent 0, spiniform 1, minutely spiniform 2, tuberculate 3. (UM)
6. PRONOTUM: CAUDAL MARGIN; concave 1, truncate 2, obtuse 3, angulate 4. (OM)
7. PRONOTUM: LATERAL LOBE, VENTRAL ASPECT; truncate 0, produced 1. (B)
8. PRONOTUM: LATERAL CARINAE: absent 0, parallel 1, divergent 2. (UM)
9. PRONOTUM: MEDIAN CARINA: absent 0, present 1. (B)
10. PRONOTUM: LATERAL LOBE: not rimmed 0, rimmed 1. (B)
11. PRONOTUM: VENTRAL AND CAUDAL EDGES OF LATERAL LOBE: etched 1, not etched 0. (B)
12. PRONOTUM: SURFACE OF DISK: smooth 0, tuberculate 1, rugose 2. (OM)
13. MESOSTERNUM: unarmed 0, with two tubercles 1, with platyform projections 2, spiniform 3. (UM)
14. METASTERNUM: unarmed 0, with two tubercles 1, with platyform projections 2, spiniform 3. (UM)
15. MALE WING TYPE: macropterous 1, mesopterous 2, brachypterous 3, micropterous 4. (OM)
16. STRIDULATORY FILE: arched 0, straight 1. (B)
17. STRIDULATORY FILE: not on raised swelling 0, on raised swelling 1. (B)
18. STRIDULATORY FILE: portion with largest teeth either: proximal 1, middle 2, distal 3. (OM)
19. MIRROR AREA: without marginal venation 0, with marginal venation 1. (B)
20. FEMALE TEGMEN: absent 0, short 1, surpassing end of abdomen 2. (OM)
21. FEMALE TEGMINA: not overlapping 0, overlapping 1. (B)

LEGS

22. TIBIA I: number of spines on anterior margin of ventral surface: (N)
23. TIBIA I: number of spines on posterior margin of ventral surface: (N)
24. FEMUR I: VENTRAL SURFACE, ANTERIOR MARGIN: unarmed 0, armed 1. (B)
25. FEMUR I: VENTRAL SURFACE, POSTERIOR MARGIN: unarmed 0, armed 1. (B)
26. FEMUR II: VENTRAL SURFACE, ANTERIOR MARGIN: unarmed 0, armed 1. (B)

27. FEMUR II: VENTRAL SURFACE, POSTERIOR MARGIN: unarmed 0, armed 1. (B)
28. TARSUS II: normal development 0, slightly modified 1, somewhat modified 2, greatly modified 3. (OM)
29. FEMUR III: DORSAL SURFACE smooth 0, DORSAL SURFACE armed 1. (OM)
30. TIBIA III: SPURS ON DORSAL SURFACE: absent 0, two spurs present 1, one spur 2. (B)
31. TARSUS III: PLANTULA: less than half length of metatarsus 1, about half length 2, more than half length 3. (OM)
32. FEMUR I, GENICULAR LOBES: unarmed 0, armed on anterior margin 1, armed on posterior margin 2, armed on both sides 3. (UM)
33. FEMUR II, GENICULAR LOBES: unarmed 0, armed on anterior margin 1, armed on posterior margin 3, armed on both sides 3. (UM)
34. FEMUR III, GENICULAR LOBES: unarmed 0, armed on anterior margin 1, armed on posterior margin 2, armed on both margins 3. (UM)
35. FEMUR III: ARMATURE OF VENTRAL SURFACE, INTERNAL MARGIN: unarmed 0, armed 1. (B)
36. FEMUR III: ARMATURE OF VENTRAL SURFACE, EXTERNAL MARGIN: unarmed 0, armed 1. (B)
37. TIBIA I: DORSAL SURFACE, APICAL SPINE, ANTERIOR MARGIN: absent 0, present 1. (B)
38. TIBIA I: DORSAL SURFACE, APICAL SPINE, POSTERIOR MARGIN: absent 0, present 1 (B)
39. TIBIA I: NUMBER OF SPINES, ANTERIOR MARGIN, DORSAL SURFACE (N)
40. TIBIA I: NUMBER SPINES, POSTERIOR MARGIN, DORSAL SURFACE (N)
41. TIBIA II: NUMBER OF SPINES, ANTERIOR MARGIN, DORSAL SURFACE (N)
42. TIBIA II: NUMBER OF SPINES, POSTERIOR MARGIN, DORSAL SURFACE (N)

COLOUR

43. SEXES DICHROMATIC: no 0, yes 1. (B)
44. FEMALE BODY COLOUR: grey 1, brown 2, green 3, combination 4. (UM)
45. ANTENNAL SCAPE AND PEDICEL CONCOLOROUS: no 0, yes 1. (B)
46. ANTENNAL FLAGELLUM: concolourous 0, annulate 1. (B)
47. LATERAL PRONOTAL LOBE: CENTRAL PORTION SHINING DARK BROWN OR BLACK: no 0, yes 1. (B)
48. LATERAL PRONOTAL LOBE: MARGIN COLOUR: ventral and caudal margins whitish 1, ventral and caudal margins concolorous 2, margins variable 3, margins yellow 4. (UM)
49. VENTRAL SURFACE OF ABDOMEN: yellow 1, green 2, creamish white 3. (UM)
50. TARSI BANDED: no 0, yes 1. (B)
51. CERCUS DORSO-VENTRALLY BICOLOURED: no 0, yes 1. (B)

MALE CHARACTERS

52. TENTH TERGITE: not hairy 0, hairy 1. (B)
53. CERCUS: styliform 1, with simple tooth 2, with complex tooth 3, variable 4. (OM)
54. SUBGENITAL PLATE: without styli 0, with styli 1. (B)
55. SUBGENITAL PLATE: MEDIAN INCISION AND STYLE: style longer than the length of one side 0, style equal to length of one side 1, style shorter than length of

one side 2. (OM)

56. SUBGENITAL PLATE: MEDIAN CARINA: absent 0, present 1. (B)
57. SUBGENITAL PLATE: MEDIAN PORTION: without spiniform process 0, with spiniform process 1, with tubercle and spiniform process 2. (OM)
58. TITILLATOR: unsclerotised 1, simple unfused cones 2, simple arms 3, without arms but with simple median projection 4, median projection and arms 5, two pairs of arms 6, arms connected by bar 7, complex capsule 8. (UM)
59. TITILLATOR: AUXILLIARY SCLERITES (see Rentz and Birchim, 1968, P. 45): absent 0, present 1. (B)
60. TITILLATOR: TUBERCLES ON MEMBRANOUS PORTION: absent 0, present 1. (B)

FEMALE CHARACTERS

61. OVIPOSITOR: GENERAL SHAPE: falcate 0, ensiform 1. (B)
62. OVIPOSITOR: CURVATURE: decurved 0, straight 1, weakly curved upward 2, strongly curved upward 3. (OM)
63. OVIPOSITOR: ARMATURE AT TIP: not carinate 0, carinate 1. (B)
64. OVIPOSITOR: ARMATURE NEAR TIP: without tubercles 0, with tubercles 1. (B)
65. OVIPOSITOR: MARGIN OF TIP: not serrate 0, serrate 1. (B)
66. SUBGENITAL PLATE: MEDIAN CARINA: absent 0, present 1.(B)
67. SUBGENITAL PLATE: MEDIAN SULCUS: absent 0, present 1.(B)
68. SUBGENITAL PLATE: APEX: obtuse 1, acute 2, divided 3, emarginate 4.(UM)
69. SUBGENITAL PLATE: APICAL AND LATERAL DIGITIFORM PROCESSES: absent 0, present 1. (B)
70. SUBGENITAL PLATE: NUMBER OF LATERAL LOBES: absent 0, one pair 1, two pairs 2. (OM)

MISCELLANEOUS

71. TIBIA II: ANTERIOR MARGIN, VENTRAL SURFACE, NUMBER OF SPINES: (N)
72. TIBIA II: POSTERIOR MARGIN, VENTRAL SURFACE, NUMBER OF SPINES: (N)
73. ANTENNA: SCAPE, PEDICEL AND FLAGELLUM CONCOLOUROUS? no 0, yes 1.(B)
74. TIBIA III: NUMBER OF APICAL SPURS ON VENTRAL SURFACE:(N)
75. MALE PARAPROCTS: unmodified 0, feebly developed plates 1, a pseudocercus 2, elongate prongs 3.(UM)
76. MALE TEGMEN COMPLETELY CONCEALED BY PRONOTUM: no 0, yes 1.(B)
77. FEMALE: TUBERCLE ON STERNITE VII: absent 0, present 1.(B)
78. FEMALE: TUBERCLE ON STERNITE VI: absent 0, present 1.(B)
79. ARMATURE OF TIP OF OVIPOSITOR: not rugose 0, rugose 1.(B)
80. TIBIA II: APICAL SPINE ON ANTERIOR MARGIN, DORSAL SURFACE: absent 0, present 1.(B)
81. TIBIA II: APICAL SPINE ON POSTERIOR MARGIN, DORSAL SURFACE: absent 0, present 1. (B)
82. OVIPOSITOR, VENTRAL SUBAPICAL HOOK: absent 0, present 1.(B)

Numerical Procedures

For phenetic studies (Sneath & Sokal, 1973), a distance matrix was computed using "Gower metric" (Manhattan distance on range-coded data). An ordination was then produced using the standard technique of Principal Coordinate Analysis. The metric was disturbed by the many missing data and a relatively small amount of variation was explained by the first few coordinates (13.4%, 7.7% and 6.1% on numbers 1-3). However, the plot of coordinates 1 versus 2 made a degree of taxonomic sense (Fig. 1); see discussion below. A minimum spanning tree was also computed, but added little to our understanding and is not presented here.

The distance matrix was also processed by a battery of standard sorting strategies, including UPGMA, WPGMA, Furthest Neighbor, Flexible and Increment-in Sums-of Squares (ISS; Dallwitz, in press), the last two at several levels of "intensity". Figure 2 shows the phenogram yielded by the ISS strategy, with intensity factor = 0.5. This was amongst the best, as judged by variation in moat size (the moat of a fork is the difference in fusion level between that fork and the one that subtends it; the cooeficient of variation of moats is showing some promise as an index of utility of phenograms). There were one or two marginally better by that criterion; but, although very similar to the one selected, they made less sense taxonomically and (especially) zoogeographically.

Cladograms were produced by the PAUP package, version 2.4.0 (Swofford, 1985) using Farris optimisation, global branch-swapping, and *Pachysaga*, No. 128 (PACH) as an outgroup. Fifty equally short trees were found, of length 1660 and consistency index 0.12.

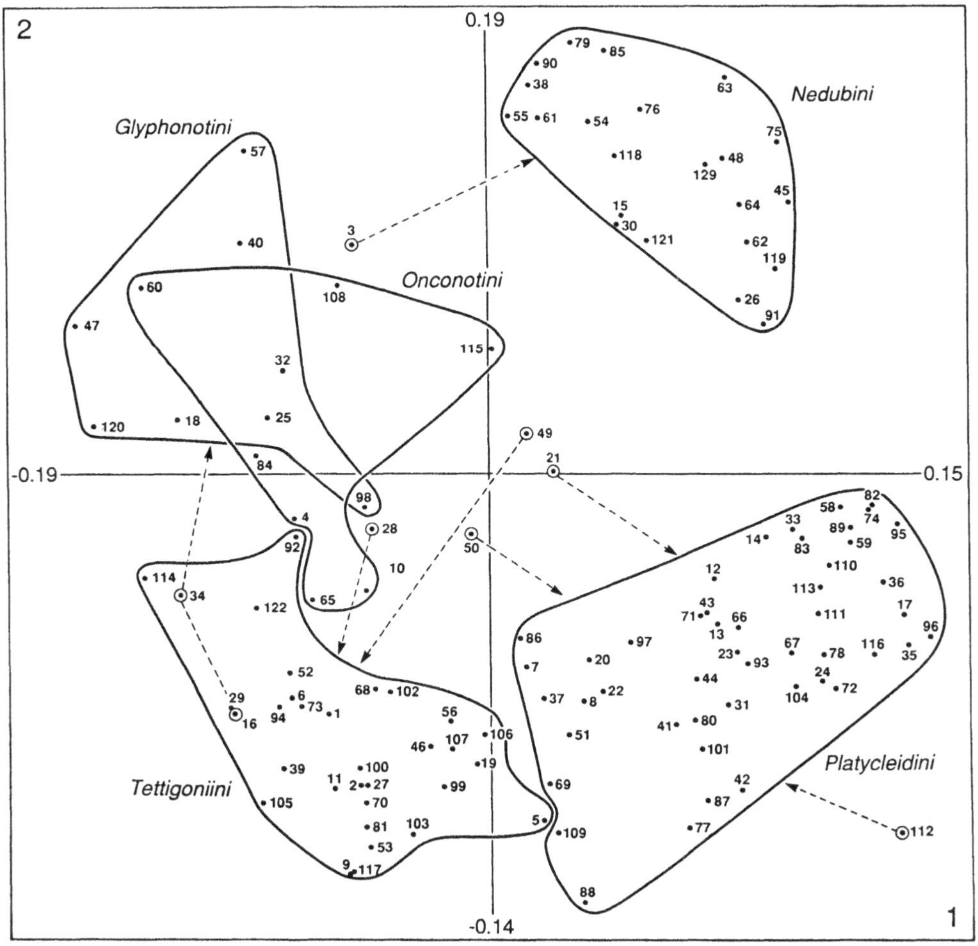

Fig. 1. Ordination of all 130 genera on axes I and II from principal coordinate analysis (see text).

Figure 3 shows the "majority" consensus tree, yielded by a local modification of Swofford's CONTREE program.

Results

No geographical parameters were entered in the character state list. However, almost every analysis, whether it be represented by the phenogram, Wagner tree or ordination plot, affirms strong zoogeographic correlation among the genera. The cladogram, as might be expected, interprets one grouping – the African – as paraphyletic, i.e., based on resemblance in primitive attributes. However, its credibility as an estimate of phylogeny (and, therefore, its utility in classification) suffers from the failure of four out of eight prima facie outgroups to appear in that position in the diagram. Indeed, three (BRAD, TERP and NEOB) are deeply embedded at the lowest level of the cladogram. Interestingly, the suggested relationships of at least some of those was once accepted, but is now generally rejected.

Five groups are apparent from the phenogram (Fig. 3) and these reflect accepted zoogeographical patterns. These groups are hereby accorded tribal status. They can be identified by using the procedures following this discussion.

The Tettigoniini contain genera from the Holarctic region. Included here are some genera previously considered as typical tettigoniines when the Tettigoniinae were distinguished from the Decticinae on the basis of relative length of the plantula of the hind tarsus. But since this character is so variable even among related genera, it has little value in distinguishing groups of genera. As an illustration of this, *Decticus*, the type genus of the latter taxon, belongs to this tribe along with *Tettigonia* further confirming that there is no justification in retaining those two genera in separate taxa.

The Nedubini consists of Australian, South American and Californian genera and further supports the hypothesis of a Gondwanan nature of these taxa by Rentz (1985). The Californian genera *Aglaothorax* and *Neduba* and GNT-4 were suggested by Rentz as being important links with those of South America and with many of the Australian genera. This is strongly reflected in our analysis (see Figs 1-3).

The Chilean *Apteropedetes*, with its single-known species which was originally placed in *Platydecticus*, was accorded its own subfamily by Rentz (1979). This genus is peculiar in many respects, not the least of which is the fact that it is apterous in both sexes. It was regarded as an outgroup at the outset of our analysis, but in every result it paired strongly with *Platydecticus*. We therefore now reject its subfamily status and regard it as an aberrant genus related to *Platydecticus* in the Tettigoniinae.

The Onconotini comprise genera from southern Africa with the single exception of the European *Onconotus*. The distant relations of *Arytropteris* with *Neduba* suggested by Rentz (1988) are not reflected in any of our results; however, the spinules borne on the sides of the abdomen in both of these genera was not used as a character in the analysis.

The Glyphonotini contain a few genera formerly considered as "typical tettigoniines". Many, if not all, are predaceous and are of similar general appearance. The grouping is difficult to explain on a zoogeographical basis but was regarded as a monogeneric tribe by Gorochov (1988, see Fig. 1 in chapter 1).

The Platycleidini, with a single exception, consist of genera that are of Holarctic origin. This group can be further subdivided into two subgroups (Fig. 3). *Austrodectes* is an interesting anomaly which was noted by Rentz (1985) as having features reminiscent of

Fig. 2 (overleaf). Phenogram (ISS Strategy; see text) for 122 genera (outgroups, other than *Apteropedetes*, not included).

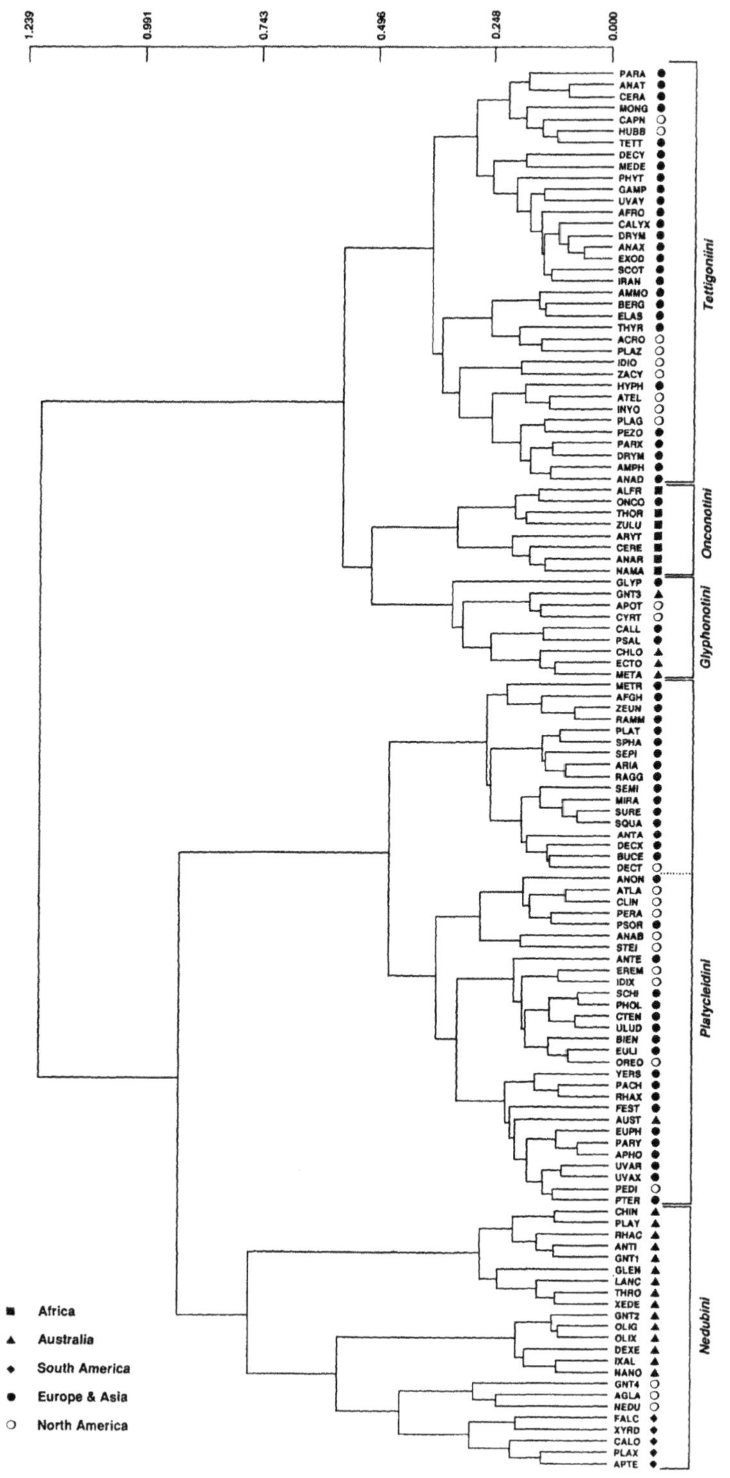

1.239
0.991
0.743
0.496
0.248
0.000
PARA
ANAT
CERA
MONG
CAPN
HUBB
TETT
DECY
MEDE
PHYT
GAMP
UVAY
AFRO
CALYX
DRYM
ANAX
EXOD
SCOT
IRAN
AMMO
BERG
ELAS
THYR
ACRO
PLAZ
IDIO
ZACY
HYPH
ATEL
INYO
PLAG
PEZO
PARX
DRYM
AMPH
ANAD
ALFR
ONCO
THOR
ZULU
ARYT
CERE
ANAR
NAMA
GLYP
GNT3
APOT
CYRT
CALL
PSAL
CHLO
ECTO
META
METR
AFGH
ZEUN
RAMM
PLAT
SPHA
SEPI
ARIA
RAGG
SEMI
MIRA
SURE
SQUA
ANTA
DECX
BUCE
DECT
ANON
ATLA
CLIN
PERA
PSOR
ANAB
STEI
ANTE
EREM
IDIX
SCHI
PHOL
CTEN
ULUD
BIEN
EULI
OREO
YERS
PACH
RHAX
FEST
AUST
EUPH
PARY
APHO
UVAR
UVAX
PEDI
PTER
CHIN
PLAY
RHAC
ANTI
GNT1
GLEN
LANC
THRO
XEDE
GNT2
OLIG
OLIX
DEXE
IXAL
NANO
GNT4
AGLA
NEDU
FALC
XYRO
CALO
PLAX
APTE
Tettigoniini
Onconotini
Glyphonotini
Platycleidini
Nedubini
Africa
Australia
South America
Europe & Asia
North America

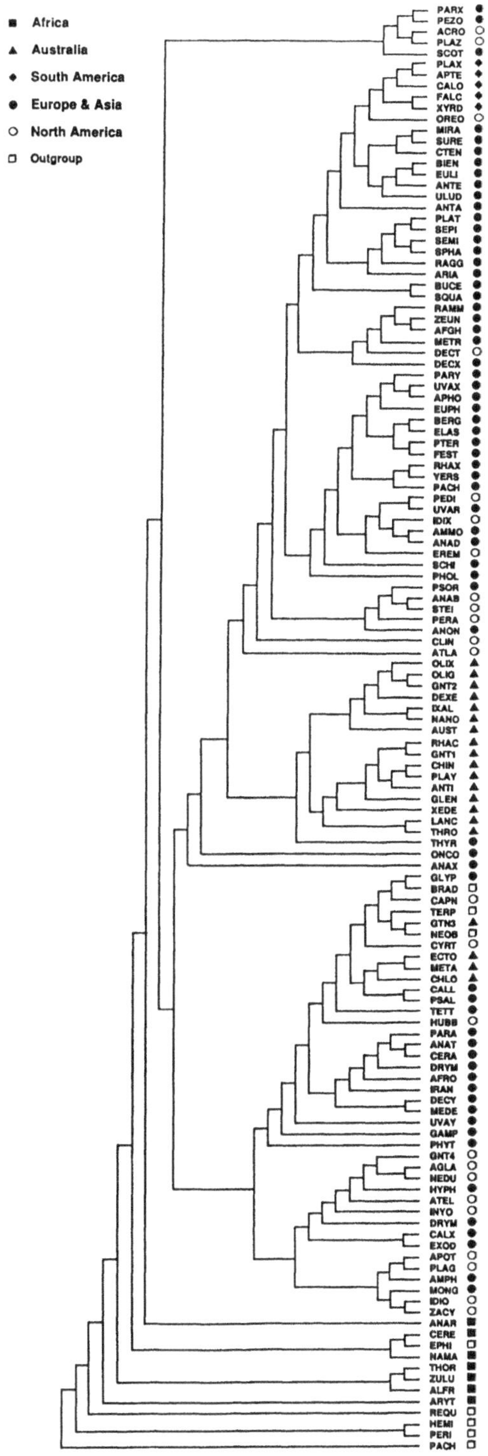

Fig. 3. Majority consensus tree on 50 equally short cladograms (see text).

European genera. This seems to be reflected in its placement here. One of the subgroups contains *Metrioptera*, *Platycleis* and all of their subgenera and relatives. All are of Palaearctic distribution with the single exception of the Nearctic *Decticita*, which has always been considered a relative of *Metrioptera*. *Afghanoptera* is placed here with some reservation. Lacking females we were unable to assess the important character of ovipositor shape (character no. 61) which is distinctive of members of the subgroup. In addition to *Austrodectes*, the second subgroup contains many of the well known Holoarctic genera formerly considered as "typically decticine". An interesting placement is that of *Atlanticus* which has been reported many times in the literature to be related to *Anatlanticus* and *Paratlanticus*. Both of these reside in the Tettigoniini in our analysis.

The Tribes

Our data confirm that there is no justifiable reason for division of the Tettigoniinae into two subfamilies as had been done prior to Rentz (1979). There is no set of characters that we know of that can be used to segregate the genera along traditional lines. In all of our analyses, the genera hitherto included in the Tettigoniinae and Decticinae are distributed quite unsystematically in our various groups. And, as noted previously, the type genera of both of these segregate into one of our groups. However, it does seem reasonable to divide the Tettigoniinae into tribes based on our results. The proposed tribes are indicated by our Groups and share a wide range of morphological characters and enjoy considerable zoogeographical congruity. The rationale and the character suites used to distinguish them follow at the end of the list. We offer this classification for criticism by the community of orthopterists.

Tribe Tettigoniini

Paratlanticus
Anatlanticus
Ceraeocercus
Mongolodectes
Capnobotes
Hubbellia
Tettigonia
Decticus
Medecticus
Phytodrymadusa
Gampsocleis
Uvarovites
Afrodrymadusa
Calopterusa
Drymadusa
Anadrymadusa
Exodrymadusa
Scotodrymadusa
Iranusa
Ammoxenulus
Bergiola
Elasmocercus
Thyreonotus
Acrodectes
Platyoplus
Idionotus
Zacycloptera
Hyphonomus
Ateloplus
Inyodectes
Petropedes
Plagiostira
Pezodrymadusa
Paradrymadusa
Drymadusella
Amphiestris
Anadolua
Nanodectes

Tribe Nedubini

Chinnandectes
Platyproctidectes
Rhachidorus
Antipodectes
GNT-1
GNT-2
GNT-4
Glenbalodectes
Lanciana
Throscodectes
Xederra
Oligodectes
Oligodectoides
Dexerra
Ixalodectes
Aglaothorax
Neduba
Falcidectes
Xyrdectes
Calodectes
Platydecticus
Apteropedetes

Tribe Onconotini

Ceresia
Arytropteris
Anarytropteris
Namaquadectes
Alfredectes
Thoracistus
Zuludectes
Onconotus

Tribe Glyphonotini

Chlorodectes
Ectopistidectes
Metaballus
Glyphonotus
GNT-3
Calliphona
Psalmatophanes
Cyrtophyllicus
Apote

Tribe Platycleidini

subgroup

Metrioptera
subgenera:
a. *Metrioptera*
b. *Sphagniana*
c. *Zeuneriana*
d. *Eobinana*
e. *Bicolorana*
f. *Vichetia*
g. *Broughtonia*
h. *Roeseliana*
Afghanoptera
Zeuneriana
Rammeola

Platycleis
subgenera:
a. *Montana*
b. *Semenoviana*
c. *Parnassiana*
d. *Squamiana*
e. *Platycleis*
f. *Alticolana*
g. *Modestana*
h. *Sepiana*
i. *Tessellana*
j. *Incertana*
k. *Decorana*
l. *Sporadiana*
m. *Eumetrioptera*

Ariagona
Raggeana
Miramiola
Sureyaella
Antaxius
Decticoides
Bucephaloptera
Decticita

subgroup

Anonconotus
Atlanticus
Clinopleura
Peranabrus
Psorodonotus
Anabrus
Steiroxys
Anterastes
Bienkoxenus
Eulithoxenus
Oreopedes
Eremopedes
Idiostatus
Schizonotinus
Pholidoptera
Ctenodecticus
Uludaghia
Yersinella
Pachytrachis
Rhacocleis
Festella
Austrodectes
Eupholidoptera
Parapholidoptera
Apholidoptera
Uvarovina
Uvarovistia
Pediodectes
Pterolepis

Disjunctive Diagnosis

The notion of a "polytypic" group – later emended to polythetic – was formally introduced by Beckner (1959), although much the same notion had been examined by Waissman (1945) in discussing the "open-texture" of empirical concepts. Hull (1965) provided what is probably the clearest exposition of the ideas involved, and the taxonomic implications were reviewed in some detail by Sneath and Sokal (1973:18-27). What is involved is the familiar situation of a "good" taxon whose members all possess a substantial fraction of the attributes in some "defining" set, but nonetheless no attribute in that set is possessed by all and only members of that taxon. Technically, no single attribute is necessary for group-membership, and it may be that no single attribute or subset of attributes is sufficient.

To the cladist such groups present the challenge of discovering an appropriate synapomorphy (or rejecting the taxon); but inevitably the best that can be done in many cases is to provide an historical "explanation": that the relevant synapomorphy is no longer recognizable in some members of the group, or that homoplasy has bestowed the synapomorphous attribute elsewhere, on non-members. While interesting, such explanations provide no useful diagnosis of group-membership. Nor, for that matter, is a rigorous "polythetic" diagnosis possible, since the requirement is simply for a large (but unspecified) number of attributes from the defining set.

There remains, however, the possibility (also discussed by Beckner, 1959), of what we shall call disjunctive diagnosis. This differs from the polythetic case simply in that the number of attributes from the defining set *is* specified. That is, we may be able to specify a set of G attributes and a number n, such that any subset of n attributes from G will provide a necessary and sufficient condition for group-membership. Such a diagnosis may seem very artificial and would quite likely be somewhat temporary and subject to frequent revision; but it does provide a formal advance upon simple "extensional" diagnosis – where, for instance, we might have to "define" the tribe X-ini simply as comprising genera X-us, Y-us, and Z-us. In any case, one of our purposes in this study has been to test the possibility of providing disjunctive diagnoses for the five groupings recognizable in the phenogram (Fig. 2) and here awarded tribal status.

With a large data matrix, such a project clearly requires assistance from the computer; but it seems equally clear that the problem is what mathematicians call "NP-complete", rigorously soluble only by testing each and every one of the vast array of possibilities. Fortunately, we were able to devise heuristic methods that, with a little ingenuity and much hard work, yielded at least one diagnosis for each tribe. No doubt many more could be found. Two separate programmes were used: one (NCOM) accepts a data matrix and the specification of one or more groups, and outputs for each character the frequency of each state in each group. Also output are coefficients that, for each pairwise comparison of groups, show how "efficient" each state (and, for ordered characters, each segment of the range of states) is in separating the groups. With practise, this output enables one to select a suite of useful candidates for the diagnostic set. Another programme (POLYDEF) then accepts the data matrix and specifications of membership of the groups to be compared and the suite of character states. It then scores the number of such states possessed by each member of each group and (interactively) provides details on the success or failure of diagnosis.

By iterating these processes (manually) we were able to home in on a successful solution for each tribe. In four cases, the tribe was diagnosible with respect to the entire study group; in the remaining one, the right-hand half of the phenogram (tribes Tettigoniini, Onconotini and Glyphonotini) had to be diagnosed first, and the Tettigoniini then distinguished from tribes Onconotini and Glyphonotini. The latter, empirical result may seem somewhat illogical; it is, we believe, due to interactions of our processes with missing data cells in our matrix. It is noteworthy that in all cases the proportion of character states required from the diagnostic set is close to 50%. The significance of this is not clear.

The diagnoses are set out below, where, for instance, "3/1" specifies state 1 of character 3, and "2+" and "2-" specify state 2 or greater and state 2 or less respectively.

Group 1+3+4: any 20 from the following 43 attributes – 1/1, 4/1, 5/1, 6/4, 7/0, 10/1, 13/

1-, 13/3, 14/1-, 14/3+, 15/3-, 16/0, 16/3, 18/2, 19/1, 21/1, 24/1, 26/1, 31/1-, 32/3, 33/1, 34/1+, 35/1, 36/1, 38/0, 44/3-, 45/1, 47/0, 48/0, 48/2+, 50/0, 55/1+, 56/0, 58/7, 62/1-, 63/1, 67/1, 71/6, 72/6, 74/4, 75/0, 81/1, 82/0.

Tettigoniini: membership in group 1+3+4, then any 7 of the following 14 attributes – 1/1, 2/1, 4/0, 5/2+, 9/0, 10/0, 12/0, 13/2, 14/2, 34/1-, 41/3, 42/4, 49/3, 63/1.

Onconotini: any 15 from the following 30 attributes – 1/0, 2/1, 4/0, 5/1, 6/4, 10/1, 13/1-, 13/3, 14/1-, 15/4, 16/1, 17/1, 21/0, 26/1, 34/3, 36/1, 38/0, 40/2-, 41/3, 42/3, 44/1, 48/2, 49/3, 56/0, 58/3, 70/0, 71/6, 72/6, 76/1, 81/0.

Glyphonotini: any 8 from the following 16 attributes – 5/2, 6/2, 8/1+, 9/1, 12/1+, 14/3, 15/2-, 19/1, 20/2, 31/1, 32/6, 34/3, 36/1, 44/3, 49/2, 56/1.

Platycleidini: any 23 from the following 43 attributes – 3/1, 4/0, 5/0, 5/2, 6/2, 8/1+, 13/2, 14/2, 15/4, 17/1, 18/2, 18/3, 19/1, 20/1, 21/0, 24/0, 25/1, 26/0, 27/0, 31/2+, 32/1, 33/2, 34/0, 35/0, 36/0, 44/2, 45/0, 47/1, 48/2, 50/1, 52/1+, 55/0, 56/1, 58/3, 62/2+, 63/0, 64/0, 65/0, 66/1, 68/3, 70/0, 71/6, 72/6.

Nedubini: any 3 from the following 6 attributes – 12/0, 12/2, 71/5, 71/7+, 72/5-, 72/7+.

Discussion

Membership in Group 1+3+4, which consists of the Tettigoniini, Onconotini and Glyphonotini, utilises character states which run the full gamut of the range of morphological characters listed. Membership in the Tettigoniini involves membership in Group 1+3+4 and then 7 of the 14 attributes stressing mainly features of the head, pronotum and thoracic regions. Similarly membership in the Platycleidini or the Onconotini requires a number of characters in all body regions but does not require prior diagnosis of membership in Group 1+3+4. Membership in the Glyphonotini places emphasis on head, pronotum and thoracic characters. It is noteworthy that no emphasis is given to the armature of any of the legs except that of the ventral surface of the fore femur; however, the relative length of the plantula and the armature of the genicular lobes is important. Membership in the Nedubini requires the least number of characters for inclusion, those being the nature of the texture of the pronotum and the number of spines on the anterior and posterior margins of the ventral surface of the middle tibia.

Acknowledgements

The nature of this project was such that it could not have been attempted without the aid of the individuals and institutions listed alphabetically below. We extend our sincere thanks to them all.

Academy of Natural Sciences of Philadelphia, Dr D. Otte, Mr D. Azuma.

British Museum (Natural History), Dept. of Entomology, Dr D. Ragge, Ms J. Marshall.

CSIRO, Division of Entomology, Canberra, Drs K. H. L. Key, R. B. Halliday, T. E. Bellas, Ms K. Pickerd, Mr C. Hunt, Mr E. Zurcher.

Dr K. Harz, Steinsfeld, West Germany.

Mr J. R. Helfer, Mendocino, California.

Museo de Ciencias Naturales, Madrid, Dr V. Llorente.

Museum d'Histoire Naturelle, Paris, Mr M. Donskoff.

Naturhisorisches Museum, Vienna, Drs U. Aspöck, A. Kaltenbach.

Zoological Institute, Academy of Sciences, Leningrad, Dr A. V. Gorochov, L. I. Podgornaja.

Zoologisches Museum, Humboldt Univertsity, East Berlin, Dr K. K. Günther.

Zoologisk Museum, Copenhagen, Dr W. Enghoff.
University of Florida, Gainesville, Dr T. J. Walker.

References

Beckner, M. 1959. The Biological Way of Thought. Columbia University Press, New York.

Brunner von Wattenwyl, C. 1882. Prodromus der Euroäischen Orthopteren. Leipzig.

Brunner von Wattenwyl, C. 1893. Revision du systeme des Orthopteres et description des especes rapportees par M. Leonardo Fea de Birmanie. Annali Museo Civico Storia Nat. Giacomo Doria: 1-230.

Caudell, A. N. 1907. The Decticinae (a group of Orthoptera) of North America. Proceedings of the United States National Museum, 32: 285-410.

Caudell, A. N. 1908. Orthoptera Fam. Locustidae Subfam. Decticinae. Genera Insectorum 72: 1-42.

Dallwitz, M. J. 1989. A flexible clustering method based on UPGMA and ISS. Australian Computing Journal, accepted for publication.

Gorochov, A. V. 1988. (In Russian) 4. The classification and phylogeny of grasshoppers (Gryllida-Orthoptera, Tettigonioidea). In: The Cretaceous Biocoenotic Crisis and the Evolution of Insects. Ed. A. Pomerenko. "Hayka", Moscow pp. 145-190.

Harz, K. 1969. "The Orthoptera of Europe. I." Series Entomoligica 5. W. Junk: The Hague.

Herman, O. 1874. Die Decticiden der Brunner von Wattenwyl 'schen Sammlung. I. Genera. Verhandlungen der Zoologische Botanischen Gesellschaft in Wien, 24: 191-210.

Hull, D. L. 1965. The effect of essentialism on taxonomy – two thousand years of stasis. British Journal of Philosophy and Science, 15: 314-326 and 16: 1-18.

Rentz, D. C. 1972. Taxonomic and faunistic comments on decticine katydids with the description of several new species (Orthoptera: Tettigoniidae: Decticinae). Proc. Acad. Nat. Sci. Phila., 124: 41-77.

Rentz, D. C. F. 1979. Comments on the classification of the orthopteran family Tettigoniidae, with a key to subfamilies and description of two new subfamililes. Australian Journal of Zoology, 27: 991-1013.

Rentz, D. C. F. 1985. A Monograph of the Tettigoniidae of Australia. Volume I: The Tettigoniinae, with an Appendix by D. H. Colless. E. J. Brill, CSIRO: Australia.

Rentz, D. C. F. 1988. The shield-backed katydids of southern Africa: Their taxonomy, ecology and relationships to the faunas of Australia and South America (Orthoptera: Tettigoniidae; Tettigoniinae). Invertebrate Taxonomy, 2: 223-335.

Rentz, D. C. & J. D. Birchim. 1968. Revisionary Studies in the Nearctic Decticinae. Memoirs of the Pacific Coast Entomological Society, 3:1-173.

Rentz, D. C. F. & A. B. Gurney. 1985. The shield-backed katydids of South America (Orthoptera: Tettgionidae, Tettigoniinae) and a new tribe of Conocephalinae with genera in Chile and Australia. Entomologica Scandinavica, 16: 70-119.

Scudder, S. H. 1894. A preliminary review of the North American Decticidae. Canadian Entomologist 26: 177-184.

Sneath, P. H. A. & R. R. Sokal. 1973. Numerical Taxonomy. W. H. Freeman, San Francisco.

Swofford, D. 1985. PAUP. Phylogenetic analysis using parsimony. Illinois Natural History Survey, unnumbered.

Ueshima, N. & D. C. Rentz. 1979. Chromosome systems in the North American Decticinae with reference to Roberstonian changes. Cytologia, 44: 693-714.

Waissman, F. 1945. Verifiability. Proceedings of the Aristotelian Society. Supplementary Volume 19: 119-150.

Zeuner, F. E. 1941. The classification of the Decticinae hitherto included in *Platycleis* Fieb. or *Metrioptera* Wesm. (Orthoptera, Saltatoria.) Trans. Ent. Soc. Lond., 91: 1-50.

THE DATA MATRIX

Character states are listed below for each genus for each of the 82 characters studied, as described above.

001ACRO	1 2 0 0 0 1 0 0 0 0 0 0 2 3 0 0 1 0 0 1 0 6 6 1 1 1 1 0 0 1
	1 3 3 2 1 1 0 1 0 3 2 4 0 2 0 0 0 2 3 0 0 0 2 0 0 0 0 1 0 0
	1 2 0 0 0 0 0 3 0 0 6 6 0 4 0 0 0 0 0 0 1 0
002AFRO	1 2 1 0 1 3 1 0 0 0 0 0 2 2 1 0 0 1 1 2 1 6 6 1 0 1 0 0 0 1
	1 1 1 1 1 1 0 1 0 3 2 4 0 4 1 0 0 2 3 0 0 0 2 1 0 0 0 3 0 0
	1 3 1 0 0 0 0 1 0 0 6 6 1 4 1 0 0 0 0 0 1 0
003AGLA	1 1 0 1 3 3 0 2 1 1 0 2 2 0 4 1 0 2 0 1 0 6 6 1 0 1 0 0 1 0
	1 1 3 1 1 1 0 1 0 1 3 4 0 4 1 1 0 2 4 1 0 0 1 1 2 1 0 7 1 1
	1 3 0 1 1 0 0 3 1 1 7 7 0 2 2 1 0 0 0 1 1 0
004ALFR	0 2 0 0 1 4 0 2 0 1 0 0 3 0 4 0 0 1 0 1 0 6 6 1 0 1 0 0 0 1
	1 1 1 2 1 1 0 0 0 1 2 3 0 2 1 0 0 2 3 0 0 0 2 1 0 0 0 3 0 0
	1 2 0 0 0 0 0 3 0 0 6 6 1 4 0 1 0 0 0 0 0 0
005AMMO	1 2 1 0 3 2 1 0 0 0 0 0 2 2 4 0 1 2 1 1 0 6 6 1 0 1 0 0 0 1
	1 2 2 ? 1 1 0 1 0 3 2 4 0 1 0 1 1 1 3 1 0 0 2 1 1 0 0 3 0 0
	1 2 1 1 1 0 1 3 0 0 6 6 0 4 0 0 0 0 1 0 1 0
006AMPH	1 1 0 1 1 2 1 0 0 0 0 0 2 2 3 0 0 2 0 1 0 6 6 1 0 1 0 0 0 1
	1 1 3 3 1 1 0 1 0 3 2 4 0 2 1 0 0 1 3 0 0 0 2 1 2 1 0 7 1 0
	1 1 1 0 0 1 0 3 0 1 6 6 0 4 0 0 0 0 0 0 1 0
007ANAB	1 2 0 0 0 2 0 0 0 0 0 0 2 0 4 0 0 2 0 1 1 6 6 0 0 0 0 0 0 1
	2 1 3 0 1 1 0 1 2 4 4 5 0 4 1 0 0 2 4 0 0 1 3 1 0 0 0 3 0 0
	1 2 0 0 0 1 0 4 1 2 6 6 0 4 0 1 0 0 0 0 1 0
008ANAD	1 2 1 1 3 1 1 0 0 0 0 0 2 2 4 0 0 2 1 1 0 6 6 0 0 0 0 0 0 1
	2 1 1 1 1 1 0 1 0 3 2 4 0 1 1 0 1 1 3 1 0 0 2 1 1 1 0 3 1 0
	1 1 1 0 0 0 0 3 0 1 6 6 0 4 0 0 0 0 0 0 1 0
009ANAX	1 2 0 0 1 3 1 0 0 0 0 0 2 2 4 0 1 2 1 1 1 6 6 1 0 1 0 0 0 1
	1 1 3 0 1 1 0 1 0 3 2 4 0 2 1 0 0 2 3 0 0 1 2 1 0 0 0 3 ? 0
	1 0 1 0 0 0 0 3 0 1 6 6 0 4 0 0 0 0 0 0 1 0
010ANAR	0 1 0 0 1 3 1 0 1 1 0 0 2 2 4 0 0 1 0 1 1 6 6 1 0 1 0 0 0 1
	0 1 3 3 1 1 0 0 0 2 2 3 0 2 0 0 0 2 3 0 0 0 2 1 2 1 0 3 0 0
	1 2 0 0 0 0 0 3 0 1 6 6 0 4 0 0 0 0 0 0 1 0
011ANAT	1 1 0 1 3 3 1 2 0 0 0 0 2 2 2 0 0 1 1 ? ? 6 6 1 0 1 0 0 0 1
	1 0 0 1 1 1 0 1 0 3 3 ? 0 2 1 0 0 2 3 1 0 1 2 1 0 1 0 3 1 0
	? ? ? ? ? ? ? ? ? ? 6 6 0 4 0 0 ? ? ? 0 1 ?
012ANON	1 2 ? 0 2 2 1 1 1 1 0 2 2 2 4 1 0 2 0 1 1 6 6 0 0 0 0 0 0 1
	1 0 0 0 0 0 0 1 0 3 2 4 0 2 1 0 1 1 3 0 0 0 2 1 2 0 0 3 0 0
	1 2 1 1 0 0 0 3 0 1 6 6 0 4 0 1 0 0 1 0 1 0
013ANTA	1 2 1 0 3 2 0 0 0 0 0 0 2 2 4 1 0 1 1 1 1 6 6 1 0 0 0 0 0 1
	3 0 0 0 0 0 0 1 0 3 2 4 0 4 0 0 0 1 3 0 0 0 2 1 0 1 0 3 0 0
	1 0 0 0 0 0 1 3 0 1 6 6 0 2 0 0 0 0 0 0 1 0
014ANTE	1 2 1 0 2 1 0 0 0 0 0 2 2 3 4 1 1 1 0 1 0 6 6 0 0 0 0 0 0 1
	3 1 3 0 0 0 0 1 0 3 2 4 0 2 0 0 1 4 2 0 0 1 2 1 2 0 0 3 0 0
	1 2 0 0 1 1 0 3 0 1 6 6 0 2 0 0 0 0 0 0 1 0

015ANTI	0 2 0 0 1 2 1 0 0 1 1 0 2 2 ? 0 1 4 0 ? 1 6 6 0 0 0 0 0 0 1 3 3 3 1 1 0 0 1 0 3 2 5 0 2 1 0 1 1 3 1 0 0 1 1 0 0 0 ? 0 0 1 0 0 1 0 0 1 3 0 1 7 7 1 2 1 0 0 0 0 0 1 0
016APOT	1 2 1 0 1 2 0 0 1 0 0 2 3 3 4 0 0 2 0 1 1 6 6 1 1 1 1 0 0 1 1 1 3 3 1 1 0 1 0 3 2 4 0 2 1 0 0 2 3 0 0 0 2 1 2 1 1 6 0 0 1 0 1 0 0 0 0 4 0 0 6 6 0 4 0 0 0 0 0 0 1 0
017ARIA	0 2 1 0 0 2 1 0 0 0 0 0 2 2 4 0 0 2 0 1 1 6 6 0 0 0 0 0 0 1 3 0 0 0 0 0 0 1 0 3 0 3 0 2 0 0 1 1 3 0 0 0 2 1 0 0 0 3 0 0 0 3 0 0 1 0 0 4 0 1 6 6 0 4 0 0 1 1 0 0 1 0
018ARYT	0 1 0 1 1 4 1 2 0 0 0 0 3 3 4 1 0 2 0 1 0 6 6 1 1 1 1 0 0 1 1 1 2 3 1 1 0 0 0 2 2 3 0 2 1 1 0 2 3 0 0 0 2 1 2 1 0 3 0 0 1 3 0 0 0 0 0 3 0 0 6 6 0 4 0 0 0 0 0 0 0 0
019ATEL	1 1 1 1 3 2 0 0 0 0 0 0 2 0 4 0 0 2 1 1 0 6 6 0 0 1 0 0 0 1 1 0 2 0 1 1 0 1 0 1 2 4 0 4 1 0 0 3 3 0 0 0 2 1 2 0 0 3 0 0 1 3 1 0 0 0 1 3 0 1 6 6 0 4 0 0 1 1 1 0 1 0
020ATLA	1 2 0 0 1 3 1 2 0 0 0 0 2 2 4 1 0 2 0 1 0 6 6 1 0 0 0 0 0 1 1 0 2 0 1 0 0 1 0 3 2 4 0 2 1 0 1 1 3 0 0 0 2 1 0 0 0 3 0 0 1 1 0 0 0 0 0 3 0 1 6 6 0 4 0 0 0 0 0 0 1 0
021AUST	0 2 1 0 1 2 1 0 0 0 0 0 2 2 3 0 0 3 0 1 0 6 6 0 0 0 0 0 0 1 2 3 3 3 1 0 0 1 0 2 2 3 0 2 1 0 0 2 3 1 0 0 2 1 0 0 0 1 0 0 1 1 1 0 0 1 0 2 0 0 6 6 0 2 0 0 0 0 0 0 1 0
022BERG	1 2 1 0 3 2 1 0 0 0 0 0 2 2 2 0 1 1 0 1 0 6 6 1 0 0 0 0 0 1 1 3 3 3 1 0 0 1 0 3 2 4 0 2 1 0 0 1 3 1 0 0 2 1 0 0 0 3 1 0 1 2 1 1 1 0 1 3 0 1 6 6 0 4 0 0 1 0 0 0 1 0
023BIEN	1 2 0 0 0 2 0 0 0 0 0 0 0 0 4 0 1 1 1 1 0 6 6 0 0 0 0 0 0 0 1 1 1 1 0 0 0 1 0 3 2 4 0 1 0 0 1 1 3 0 0 0 2 1 2 1 0 3 1 0 1 2 1 0 1 0 0 3 0 1 6 6 0 4 0 0 0 0 0 0 1 0
024BUCE	1 2 1 0 0 3 1 0 1 0 0 0 2 2 4 0 0 2 1 1 1 6 6 0 0 0 0 0 0 1 3 0 0 0 0 0 0 1 0 4 2 4 0 2 0 0 0 2 3 0 0 2 1 0 0 0 0 6 0 0 1 2 0 0 0 0 0 3 0 1 6 6 0 4 0 0 0 0 0 0 1 0
025CALL	1 1 0 1 1 2 1 0 1 1 1 2 3 3 1 0 0 2 1 2 1 6 6 1 0 0 0 0 0 1 1 3 3 3 1 1 0 1 0 3 2 4 0 3 1 0 0 0 2 0 0 1 2 1 0 0 0 3 0 0 1 1 0 0 0 0 0 4 1 1 6 6 0 4 0 0 0 0 0 0 1 0
026CALO	1 1 1 1 0 2 1 0 0 0 0 0 2 2 4 0 0 3 0 1 0 5 6 0 0 0 0 0 0 2 1 1 3 1 0 0 1 1 1 1 2 2 0 2 1 0 1 1 3 0 0 0 2 1 2 0 0 3 0 1 1 1 0 1 1 0 0 3 0 1 7 7 0 2 0 0 0 0 0 1 1 1
027CALX	1 2 1 1 0 2 1 0 0 0 0 0 2 0 4 0 0 2 0 1 1 6 6 1 0 1 0 0 0 1 1 1 3 0 1 1 0 1 0 3 2 4 0 4 1 0 0 2 3 0 0 0 2 1 1 0 0 3 0 0 1 0 1 0 0 0 0 4 0 1 6 6 0 0 0 0 0 0 0 0 1 0
028CAPN	1 1 1 1 ? 2 0 2 0 0 0 0 2 0 1 1 0 2 1 2 1 6 6 ? ? ? ? 0 0 1 1 3 3 3 ? ? 0 1 0 4 4 6 0 4 1 0 0 1 4 0 0 0 2 1 0 0 0 3 0 0 1 0 0 0 0 0 0 3 0 1 6 6 0 4 0 0 0 0 0 0 1 0
029CERA	1 1 0 1 1 3 1 2 0 0 0 0 2 2 2 1 0 1 1 2 1 6 6 1 0 1 1 0 0 1 1 1 0 0 1 1 0 1 0 3 2 4 0 2 1 0 0 2 3 1 0 0 2 1 0 0 0 3 0 0 1 0 1 0 0 1 0 4 0 0 6 6 0 4 0 0 0 0 0 0 1 0
030CHIN0	2 1 0 1 3 1 0 0 1 0 2 2 2 1 1 0 1 1 2 1 6 6 0 0 0 0 0 0 1 3 3 3 1 1 0 0 1 0 3 2 4 0 2 1 0 1 1 1 1 0 0 1 1 0 0 0 2 0 0 1 1 0 1 0 0 1 3 0 1 7 7 1 2 1 0 0 0 1 0 1 0

031CLIN	1 2 0 0 0 3 1 1 1 0 0 0 2 2 4 0 0 2 0 1 1 6 6 0 0 0 0 0 0 1
	1 0 2 0 1 0 0 1 0 4 2 4 0 2 0 0 1 1 3 0 0 0 2 1 0 0 0 3 0 0
	1 1 0 0 0 0 0 3 0 1 6 6 0 4 0 0 1 0 0 0 1 0
032CHLO	0 0 0 1 1 2 1 1 0 0 0 0 3 3 2 0 0 2 1 1 ? 6 6 1 0 0 0 0 0 1
	1 3 3 3 1 1 0 1 1 3 2 4 0 3 0 0 0 4 2 1 0 1 2 1 2 0 0 2 0 0
	1 0 0 0 0 0 0 1 0 0 6 6 0 2 0 0 0 0 0 0 1 0
033CTEN	1 1 0 0 0 3 1 0 0 0 0 0 2 2 4 0 1 3 0 1 0 6 5 0 0 0 0 0 0 1
	3 0 0 0 0 0 0 1 0 3 1 4 0 2 0 0 1 1 1 0 0 1 2 1 2 1 0 1 1 0
	1 1 0 0 0 0 0 3 0 1 6 6 0 2 0 0 0 0 0 0 1 0
034CYRT	1 0 1 1 1 2 1 1 1 0 0 2 3 3 2 0 0 1 1 1 1 6 6 1 1 1 1 0 0 1
	1 1 3 3 1 1 0 1 0 4 3 5 0 3 0 0 0 2 2 0 0 1 2 1 0 1 0 7 0 0
	1 0 0 0 0 0 0 3 0 1 7 6 0 4 0 0 0 0 0 0 1 0
035DECT	1 2 1 0 0 3 1 ? 1 0 0 0 2 2 4 0 0 3 1 1 0 6 6 0 0 0 0 0 0 1
	3 0 2 0 0 0 0 1 0 3 2 4 0 2 0 0 1 1 2 1 0 0 2 1 2 0 0 3 0 0
	0 2 0 0 1 0 0 3 0 1 6 6 0 4 0 0 0 0 0 0 1 0
036DECX	1 2 1 0 0 0 2 0 1 0 0 0 2 2 3 0 0 2 1 1 1 6 6 0 0 0 0 0 0 1
	3 0 2 0 0 0 0 1 0 3 2 4 0 2 0 0 1 1 3 1 0 0 2 1 0 1 0 0 1 1
	1 1 0 1 0 0 0 3 1 1 6 6 0 4 0 0 0 0 0 0 1 0
037DECY	1 2 0 0 0 3 1 1 1 0 0 0 2 2 2 0 0 2 1 2 1 6 6 1 0 0 1 0 0 1
	3 1 1 0 1 0 0 1 0 4 2 5 0 3 1 0 0 3 3 0 0 1 2 1 0 1 0 3 0 0
	1 2 1 1 1 0 0 3 0 1 6 6 0 4 0 0 0 0 0 0 1 0
038DEXE	1 2 0 1 1 2 1 0 0 0 0 0 3 3 4 0 0 3 0 1 0 5 5 0 0 0 0 0 0 0
	3 3 3 3 1 0 0 0 0 2 1 2 0 1 1 0 0 3 3 1 0 0 2 1 2 0 0 1 0 0
	1 1 0 1 1 0 0 3 0 1 5 5 0 2 3 1 0 0 0 0 0 1
039DRYM	1 2 0 1 1 3 0 0 0 0 0 0 2 2 1 1 0 2 1 2 1 6 6 1 0 1 0 0 0 1
	1 1 2 0 1 1 0 1 0 3 3 5 0 2 1 0 0 2 3 0 0 0 2 1 0 0 0 3 0 0
	1 0 1 0 0 0 0 3 0 1 6 6 1 4 0 0 0 0 0 0 1 0
040ECTO	0 0 0 1 1 2 1 1 0 0 0 0 3 3 1 1 0 2 1 2 1 6 6 1 0 0 0 0 0 1
	1 3 3 3 1 0 0 1 1 4 2 4 0 3 1 0 0 2 2 0 0 1 2 1 0 1 0 2 0 0
	1 0 0 0 0 1 0 2 0 0 6 6 1 2 0 0 0 0 0 0 1 0
041EREM	1 2 0 0 0 2 1 0 0 0 0 0 0 2 2 1 1 2 1 1 0 6 6 0 0 0 0 0 0 1
	2 1 2 0 ? ? 0 1 0 3 2 4 0 4 0 0 0 1 4 1 0 1 2 1 ? 0 0 3 0 0
	1 2 1 0 0 0 0 3 0 1 6 6 0 4 0 0 0 0 0 0 1 0
042EULI	1 2 1 0 0 2 0 0 0 0 0 0 0 0 4 0 1 2 0 1 0 6 6 0 0 0 0 0 0 1
	1 0 0 0 0 0 0 1 0 3 2 4 0 2 0 0 0 2 3 0 0 1 2 1 2 1 0 3 1 0
	1 2 1 0 0 0 0 3 0 1 6 6 0 4 0 0 0 0 1 0 1 0
043SCHI	1 2 0 0 0 2 1 0 0 0 0 0 2 2 4 0 1 1 1 1 0 6 6 0 0 0 0 0 0 1
	3 1 1 3 0 0 0 1 0 3 2 4 0 2 1 1 0 2 3 0 0 0 2 1 0 1 0 3 0 0
	1 2 0 0 0 0 0 3 0 1 6 6 0 4 0 0 0 0 0 0 1 0
044EUPH	1 2 1 0 0 3 1 0 0 0 0 0 2 2 4 0 0 2 0 0 0 6 6 1 0 0 0 0 0 1
	3 3 3 3 1 0 0 1 0 3 2 4 0 1 0 0 1 1 3 1 0 1 2 1 0 0 0 5 1 0
	1 2 0 0 0 0 0 3 1 1 6 6 0 4 0 0 0 0 0 0 1 0
045FALC	1 2 0 0 1 2 1 1 1 0 0 0 2 3 2 ? ? ? 1 1 1 6 6 0 0 0 0 0 0 2
	3 0 0 0 0 0 1 1 1 3 2 4 0 1 1 0 1 1 3 1 0 0 1 0 ? 0 0 6 0 1
	1 2 0 1 1 0 0 4 0 1 7 7 0 2 1 0 0 0 0 1 1 0
046GAMP	1 2 1 0 1 3 1 0 0 1 0 0 2 2 2 0 0 3 1 1 1 6 6 1 0 1 0 0 0 1
	2 1 3 0 1 0 0 1 0 3 2 4 0 3 1 0 0 2 4 0 0 1 2 1 2 0 0 6 0 0
	1 0 1 0 0 0 0 3 0 1 6 6 0 4 0 0 0 0 ? 0 1 0

047GLYP	1 1 0 1 1 4 1 1 0 1 0 1 3 3 1 1 1 2 0 2 1 9 9 1 1 1 1 0 0 1 1 3 3 3 1 1 1 1 4 6 6 8 0 3 1 0 0 2 2 0 0 0 2 1 2 1 0 2 0 0 1 2 0 0 0 0 0 3 0 1 9 9 0 4 0 0 0 0 1 1 1 0
048GLEN	0 1 0 0 3 2 0 0 0 1 1 0 2 2 3 1 1 1 0 1 1 6 6 0 0 0 0 0 0 1 3 1 1 0 0 0 0 1 0 2 2 4 1 1 0 0 0 1 3 1 0 0 1 1 1 1 0 5 0 1 1 1 0 0 0 0 0 3 0 1 7 7 0 2 1 0 0 0 0 0 1 1
049HYPH	1 1 1 1 3 2 1 0 0 0 0 1 0 0 4 0 0 2 0 1 0 6 6 0 0 0 0 0 0 0 1 0 0 0 1 1 0 1 0 1 0 0 0 2 1 0 0 ? 3 0 0 0 2 1 2 0 0 3 0 1 1 3 0 0 1 1 0 3 0 1 6 6 1 4 0 0 0 0 0 0 0 0
050IDIO	1 2 0 0 2 1 2 1 0 0 0 2 2 4 1 0 0 4 1 1 1 6 6 1 0 0 0 0 0 1 2 0 1 0 1 1 0 1 0 3 2 4 0 4 1 0 0 1 3 1 0 0 2 1 1 0 0 5 0 1 1 1 0 0 0 1 0 3 0 1 6 6 0 4 0 0 0 0 0 0 1 0
051IDIX	1 2 0 0 0 2 1 0 0 0 0 0 2 2 4 0 1 2 1 1 0 6 6 0 1 ? ? 0 0 1 1 1 ? ? 1 1 0 1 ? 3 2 4 0 4 0 0 0 3 3 1 0 ? 2 1 1 0 0 3 0 0 1 2 0 1 1 0 0 3 0 1 6 6 0 4 0 0 0 0 0 0 1 0
052HUBB	1 2 1 1 1 3 1 0 0 0 0 0 2 2 1 0 0 2 0 2 1 6 6 1 0 1 1 0 0 1 1 3 3 3 1 1 0 1 0 2 0 4 0 1 0 0 0 2 2 0 0 0 2 1 0 1 0 7 1 0 1 1 0 0 0 0 0 3 0 1 6 6 0 4 0 0 0 0 0 0 1 0
053INYO	1 2 1 1 ? 2 1 0 0 0 0 0 2 0 4 0 0 2 0 1 0 6 6 1 0 1 1 0 0 1 1 0 2 0 1 1 0 1 0 2 2 4 0 2 1 0 0 2 3 0 0 1 2 1 2 0 0 3 0 0 1 3 1 0 0 0 0 3 0 1 6 6 1 4 0 0 0 0 0 0 1 0
054IXAL	1 2 1 ? 2 2 1 0 0 0 0 0 1 1 3 0 0 3 1 1 1 5 5 0 0 0 0 0 0 1 3 3 3 3 ? 0 0 0 0 2 2 3 0 1 1 1 0 2 4 1 0 0 3 1 2 0 0 5 1 1 1 1 0 0 0 0 0 1 0 0 5 5 0 2 0 0 0 0 0 0 0 1
055LANC	0 1 0 0 1 3 1 0 0 0 0 2 1 1 ? 1 0 2 0 ? 1 6 6 0 0 0 0 0 0 1 3 3 3 3 1 0 0 1 ? 3 3 5 0 2 1 0 0 2 3 0 0 0 1 1 1 0 0 2 0 1 1 1 0 1 0 0 1 3 0 1 7 7 0 2 1 0 0 0 0 0 1 0
056MEDE	1 2 0 0 0 3 1 0 0 0 0 0 2 2 1 0 0 2 1 2 1 6 6 1 0 1 0 0 0 1 2 3 3 3 1 0 0 1 0 4 2 4 0 2 1 0 0 3 3 0 1 1 2 1 0 0 0 3 0 1 1 2 1 1 0 0 0 4 1 1 6 6 1 4 0 0 0 0 1 0 1 0
057META	0 1 0 1 1 2 1 1 0 1 0 2 3 3 ? 1 0 2 1 ? 1 6 6 1 0 0 0 0 0 1 ? 3 3 3 1 1 0 1 0 3 2 4 0 ? 1 0 0 1 4 0 0 0 2 1 1 1 0 2 2 0 1 1 0 ? ? 1 0 2 0 0 6 6 0 2 0 0 0 0 0 0 1 0
058METR	1 2 0 0 0 3 1 0 1 1 0 0 2 2 ? ? 0 2 1 ? 1 6 6 0 0 0 0 0 0 1 2 1 3 0 0 0 0 0 0 3 2 4 0 2 0 2 1 1 4 1 0 1 2 1 2 0 0 3 0 2 0 3 0 2 ? 0 0 3 1 1 6 6 0 4 0 0 ? ? 0 0 1 0
059MIRA	1 2 1 0 0 3 1 0 1 0 0 0 2 2 4 1 1 1 0 1 1 6 5 0 0 0 0 0 0 1 3 0 0 0 0 0 0 1 0 3 0 4 0 2 0 0 1 1 3 0 0 1 2 1 1 1 0 2 1 0 0 3 0 0 0 1 0 1 0 0 6 6 0 2 0 0 0 0 0 0 1 0
060NAMA	0 1 0 1 1 3 1 0 0 1 0 0 1 1 ? ? ? ? ? 1 1 6 6 1 0 1 1 0 0 1 1 3 3 1 1 1 0 0 0 1 2 3 ? 2 1 0 0 2 3 0 0 ? ? ? ? ? ? ? ? ? 1 3 0 0 0 0 0 2 0 0 6 6 0 4 ? ? 0 0 0 0 1 ?
061NANO	1 1 1 1 1 2 1 0 0 1 0 0 1 1 3 0 0 1 0 1 0 5 5 0 0 0 0 0 0 1 3 ? ? ? 1 0 0 0 0 2 2 3 0 4 1 0 0 2 4 1 0 0 2 1 ? 0 0 5 0 1 1 1 0 0 0 1 0 3 0 1 5 5 ? 2 0 0 0 0 0 0 0 1
062NEDU	1 2 0 0 ? 3 1 2 1 1 0 0 2 2 4 1 1 4 0 1 0 6 6 0 0 0 0 0 1 0 3 1 3 1 0 0 ? 1 ? 1 4 5 0 4 0 1 0 2 3 0 1 0 1 1 2 1 0 3 1 0 1 3 0 1 1 0 0 2 0 0 8 8 0 2 2 0 0 0 0 ? 1 0

063OLIG	1 1 1 1 3 1 1 0 0 0 0 0 1 1 3 0 0 3 0 1 0 5 4 0 0 0 0 0 0 0 3 0 0 1 0 0 0 0 0 ? ? ? 0 4 1 1 0 1 3 1 0 0 2 1 1 0 0 4 0 1 1 2 0 0 1 0 0 4 0 1 ? ? 0 2 0 0 0 0 0 0 0 1
064OLIX	1 2 1 0 1 1 1 0 0 0 0 0 1 1 3 0 0 2 0 1 0 5 5 0 0 0 0 0 0 0 3 0 0 3 0 0 0 0 0 0 1 0 1 1 0 2 1 3 1 0 0 2 1 2 0 0 1 0 0 1 2 0 0 0 0 0 4 0 1 5 4 0 2 0 0 0 0 0 0 0 1
065ONCO	1 2 0 0 1 3 1 1 0 1 0 1 0 0 4 1 1 2 0 1 0 6 6 1 0 1 0 0 0 1 1 0 1 1 1 1 0 1 0 3 3 5 0 2 1 0 0 2 4 0 0 0 2 1 0 0 0 3 1 0 1 0 0 0 0 0 0 3 0 0 6 6 1 4 0 0 0 0 0 0 1 0
066OREO	1 2 1 0 0 2 1 0 0 0 0 0 0 0 4 0 1 2 0 1 0 6 6 0 0 0 0 0 0 1 2 0 0 0 0 0 0 1 0 3 2 4 0 1 1 0 0 2 3 1 0 0 2 1 2 0 0 3 0 0 1 3 1 0 0 0 0 4 0 1 6 6 0 2 0 0 0 0 0 0 1 0
067PACH	1 2 1 0 0 2 0 0 0 0 0 0 2 2 4 0 1 2 1 1 0 6 6 0 0 0 0 0 0 1 3 0 2 1 1 0 0 1 0 3 2 4 0 2 0 0 1 1 1 1 1 0 1 1 0 0 0 3 0 0 1 1 0 0 0 0 0 3 0 1 6 6 0 4 0 0 0 0 0 0 1 0
068PARA	1 2 0 1 1 2 1 2 0 0 0 0 2 2 3 ? ? ? ? 1 1 6 6 1 0 1 0 0 0 1 1 1 3 3 1 1 1 1 1 3 3 4 0 2 0 0 1 3 3 1 0 0 2 1 0 0 0 3 0 0 1 0 0 0 0 0 0 3 0 1 6 6 0 4 0 0 0 0 0 1 1 0
069PEDI	1 2 0 0 1 2 1 0 0 0 0 0 2 2 4 0 0 2 1 1 0 6 6 1 0 0 0 1 0 1 1 1 3 0 1 0 0 1 1 3 2 4 0 4 0 0 0 2 4 1 0 0 2 1 ? 0 0 3 0 0 1 3 1 0 0 0 0 4 0 1 6 6 0 4 0 0 0 0 0 0 1 0
070PARX	1 2 0 0 3 1 0 0 0 0 0 0 2 2 4 1 1 1 0 1 0 6 6 1 0 1 0 0 0 1 1 1 1 1 1 1 0 1 1 3 2 4 0 1 1 0 0 2 3 1 0 0 2 1 1 0 0 3 0 0 1 0 1 0 0 1 0 3 0 1 6 6 0 4 0 0 0 0 0 0 1 0
071PERA	1 2 0 0 0 1 1 1 1 0 0 2 2 0 4 0 0 2 0 1 0 6 6 0 0 0 0 0 0 1 1 1 2 1 0 0 0 1 0 4 3 5 0 2 1 0 0 1 3 0 0 1 2 1 2 0 0 3 0 0 1 2 0 0 0 0 0 4 0 1 6 6 0 4 0 0 0 0 0 0 1 0
072PHOL	1 2 0 0 0 2 1 0 1 0 0 0 2 2 4 0 1 2 1 1 0 6 6 0 0 0 0 0 0 1 3 1 2 0 0 0 0 1 0 4 2 4 0 2 0 0 1 1 3 0 0 0 2 1 0 ? 0 3 0 0 1 2 0 0 0 0 0 3 0 1 6 6 0 4 0 0 0 0 0 0 1 0
073PLAG	1 1 1 1 ? 2 0 1 0 0 0 2 3 2 3 0 1 2 0 1 1 6 6 1 0 1 0 0 0 1 1 1 3 1 1 1 0 1 0 3 2 3 0 1 1 0 0 1 3 0 0 0 2 1 2 0 0 3 0 0 1 0 1 0 0 0 0 3 0 1 6 6 0 4 0 0 1 0 0 0 1 0
074PLAT	0 2 1 0 0 0 3 1 0 1 0 0 2 2 ? ? 1 2 ? ? 1 6 6 0 0 0 0 0 0 1 3 0 0 0 0 0 0 1 0 3 2 4 0 2 1 0 1 1 3 0 0 1 2 1 0 0 0 3 1 0 0 2 0 0 1 0 1 3 0 1 6 6 0 4 0 0 ? ? 0 0 1 0
075PLAX	1 1 1 1 0 1 1 0 0 0 0 0 2 0 4 0 0 3 1 0 0 4 4 0 0 0 0 3 0 0 3 0 ? ? 0 0 1 1 1 1 1 1 0 2 1 1 0 0 3 1 1 0 2 1 0 0 ? 8 1 1 1 3 0 1 1 0 0 1 0 0 5 5 0 2 2 0 0 0 0 1 1 1
076PLAY	0 2 1 0 1 1 3 ? 0 1 1 0 2 2 1 1 0 1 1 2 1 6 6 0 0 0 0 0 0 1 3 3 3 3 1 0 0 1 0 3 2 4 0 2 1 0 1 1 3 1 0 0 1 1 0 1 0 2 1 1 1 1 0 0 0 0 1 3 0 1 7 7 1 2 1 0 0 0 0 0 1 1
077PTER	1 2 1 0 1 2 1 0 0 0 0 0 2 2 4 0 1 2 1 1 0 6 6 1 0 0 0 0 0 1 3 1 3 0 1 0 0 1 0 4 2 4 0 2 0 0 1 1 3 1 0 1 2 1 0 1 0 3 1 0 1 2 1 0 0 0 0 4 0 1 6 6 0 4 0 0 1 0 0 0 1 0
078RAGG	0 2 1 0 0 3 1 0 0 0 0 0 2 2 1 0 0 3 1 2 1 6 6 0 0 0 0 0 0 1 3 1 3 0 0 0 0 1 0 3 2 4 0 2 0 0 1 1 3 0 0 1 2 1 0 1 0 3 0 0 1 2 0 0 0 0 0 4 0 1 6 6 0 4 0 0 1 1 0 0 1 0

079RHAC	0 1 0 0 1 4 1 1 0 1 1 0 3 2 0 0 1 0 2 1 1 6 6 0 0 0 0 0 0 1
	3 3 3 0 1 0 0 1 2 3 3 4 0 2 1 0 1 1 3 1 1 0 1 1 0 1 0 2 0 1
	1 0 0 2 0 0 0 2 0 0 7 7 0 2 1 0 0 0 0 0 1 ?
080RHAX	1 2 1 0 1 3 1 0 1 0 0 0 2 2 4 0 1 1 1 1 0 6 6 1 0 0 0 0 0 1
	3 1 2 1 1 0 0 1 2 4 2 4 0 2 0 0 0 3 3 1 1 0 2 1 0 0 0 3 0 0
	1 2 0 0 0 ? 0 3 0 1 6 6 0 2 0 0 ? ? 0 0 1 0
081SCOT	1 2 0 0 3 3 1 0 0 0 0 0 2 2 4 0 1 1 0 1 1 6 6 1 0 1 0 0 0 1
	1 1 3 1 1 1 0 1 0 3 2 4 0 2 1 0 0 0 3 0 0 0 2 1 0 0 0 1 0 0
	1 2 1 0 0 0 0 3 0 1 6 6 0 4 0 2 0 0 1 0 1 0
082SEPI	0 2 1 0 0 3 1 0 1 1 0 0 2 2 2 0 0 1 1 1 1 6 6 0 0 0 0 0 0 1
	3 0 2 0 0 0 0 1 0 3 2 4 0 2 1 0 1 1 3 0 0 1 2 1 0 1 0 3 1 1
	0 3 0 0 1 0 1 4 0 1 6 6 1 4 0 0 1 1 0 0 1 0
083STEI	1 2 1 0 0 0 2 1 1 1 0 0 2 2 4 0 0 2 0 1 0 6 6 0 0 0 0 0 0 1
	3 1 3 0 0 0 0 1 0 3 2 4 0 4 1 0 0 1 4 0 0 1 3 1 0 1 0 3 1 0
	1 2 0 1 1 1 0 4 1 1 6 6 0 4 0 0 0 0 0 0 1 0
084THOR	1 2 0 0 1 4 1 ? ? 1 0 2 3 3 4 1 1 1 0 1 0 6 6 1 0 1 0 0 0 1
	2 3 3 1 1 1 0 0 0 2 2 3 0 3 1 0 0 0 2 2 0 0 2 1 0 0 0 3 0 1
	1 2 1 0 0 0 0 3 1 1 6 6 1 4 0 1 0 0 0 0 0 0
085THRO	0 1 0 0 3 2 1 0 0 0 0 0 1 1 3 1 1 3 1 1 ? 7 7 0 0 0 0 0 0 1
	3 3 3 3 1 0 0 1 0 3 1 4 0 ? 1 1 ? ? 3 0 0 0 1 1 0 0 0 5 0 1
	? ? ? ? ? ? ? ? ? ? ? 7 7 0 2 1 ? ? ? 0 1 ? ?
086THYR	1 2 0 0 1 3 1 0 0 0 0 0 3 2 4 0 1 2 0 1 0 6 6 1 0 1 0 0 0 1
	3 0 3 0 1 0 0 1 0 3 2 4 0 2 0 0 0 2 3 0 1 0 2 1 1 0 0 5 0 0
	1 1 0 0 0 0 1 4 0 1 6 6 0 2 0 0 0 0 0 0 1 0
087UVAR	1 2 1 0 0 2 1 0 0 0 0 0 2 2 4 0 0 2 0 1 0 6 6 0 0 0 0 0 0 1
	1 1 1 3 1 0 0 1 0 3 2 4 0 2 0 0 0 1 3 1 0 0 2 1 0 1 0 3 0 0
	1 3 1 0 0 0 0 4 0 1 6 6 0 4 0 0 0 0 1 0 1 0
088UVAX	1 2 1 0 0 3 1 0 0 0 0 0 2 2 4 0 0 1 0 1 0 6 6 1 0 1 0 0 0 1
	2 1 3 1 1 0 0 1 0 3 2 4 0 2 0 0 1 4 3 0 0 0 2 1 0 1 0 6 1 0
	1 2 1 0 0 0 0 3 0 1 6 6 0 4 0 ? 0 0 1 0 1 0
089YERS	0 2 1 0 0 2 1 0 0 0 0 0 1 0 4 1 0 1 1 1 0 6 6 0 0 0 0 0 0 1
	3 1 2 1 0 0 0 1 0 3 2 4 0 2 0 0 1 1 3 1 1 0 2 1 0 1 0 3 1 0
	1 3 0 0 0 0 0 3 0 1 6 6 0 2 0 0 1 0 0 0 1 0
090XEDE	0 1 0 1 3 2 1 1 0 1 0 0 2 2 3 1 1 2 1 1 1 6 6 0 0 0 0 0 0 1
	3 3 3 3 1 0 0 1 0 3 2 4 0 2 1 0 ? 1 3 0 0 0 1 1 0 0 0 4 0 1
	1 0 0 0 0 0 0 3 0 ? 7 7 0 2 1 0 0 0 0 0 1 1
091XYRD	1 2 1 1 1 4 1 0 0 0 0 0 2 2 1 1 0 2 0 1 1 6 6 0 0 0 0 0 0 0
	0 0 2 0 0 0 1 1 1 3 2 4 0 2 0 1 1 1 3 1 0 0 2 0 ? 0 1 4 0 1
	1 1 0 1 1 0 0 4 0 0 7 7 0 2 0 0 0 0 0 0 1 0
092ZACY	1 1 0 0 2 2 1 0 0 0 0 2 3 0 2 1 0 2 0 1 1 6 6 1 1 1 1 0 0 1
	1 0 0 0 1 1 0 1 0 3 2 4 0 4 1 0 1 1 3 0 0 0 2 1 2 1 0 7 0 1
	1 1 0 0 0 1 0 4 0 0 6 6 0 4 0 0 0 1 1 0 1 0
093ZEUN	1 2 1 0 0 3 1 0 0 0 0 0 2 2 3 0 1 1 0 1 1 6 6 0 0 0 0 0 0 1
	2 1 3 0 0 0 0 1 0 3 2 4 0 2 1 0 1 2 ? 0 0 0 2 1 1 0 0 3 0 0
	0 3 0 0 0 1 0 3 1 1 6 6 0 4 0 0 0 0 0 0 1 0
094UVAY	1 2 0 0 1 3 1 0 0 0 0 0 3 3 2 0 0 2 1 1 0 6 6 1 0 1 0 0 0 1
	2 1 3 3 1 1 0 1 0 3 2 4 0 3 1 0 0 4 2 0 0 1 2 1 0 0 0 1 1 0
	1 1 1 0 0 0 0 4 0 0 6 6 0 4 0 0 0 0 0 0 1 0

095SURE	1 2 1 0 0 2 1 0 1 0 0 0 1 2 4 0 1 2 1 1 1 6 6 0 0 0 0 0 0 1
	3 0 0 0 0 0 0 1 0 3 1 4 0 2 0 0 1 1 3 0 0 1 2 1 1 1 0 3 0 1
	0 3 0 1 1 0 0 4 0 0 6 6 0 2 0 0 0 0 1 0 1 0
096RAMM	1 2 1 0 0 3 1 0 0 0 0 0 2 2 ? ? ? ? ? ? 1 1 6 6 0 0 0 0 0 0 0 1
	3 1 1 0 0 0 0 1 0 3 2 4 0 2 1 0 1 1 3 1 0 ? ? ? ? ? ? ? ? ? ?
	0 3 0 0 0 0 0 4 1 1 6 6 0 4 ? ? 0 1 0 0 1 0
097PSOR	1 2 0 0 0 1 1 2 1 0 0 2 2 2 4 0 0 2 0 1 0 6 6 1 0 0 0 0 0 1
	3 0 3 0 0 0 0 1 0 5 4 6 0 4 1 0 0 2 4 0 0 1 2 1 0 0 0 3 1 0
	1 1 0 0 0 0 0 3 0 1 6 6 0 4 0 0 0 0 0 0 1 0
098PSAL	1 1 0 1 1 1 1 1 1 0 0 2 2 2 1 0 1 3 1 2 1 6 6 0 0 0 0 0 0 1
	1 1 3 3 1 1 0 1 0 3 2 4 0 3 1 0 0 2 2 0 0 1 2 1 0 0 0 3 1 0
	1 0 0 0 0 0 1 3 0 1 6 6 0 4 0 0 0 0 0 0 1 0
099PHYT	1 2 0 0 3 3 1 0 0 0 0 0 2 2 4 0 0 2 1 1 1 6 6 1 0 1 0 0 0 1
	1 1 1 0 0 0 0 1 0 3 2 4 0 3 1 0 0 2 2 0 0 0 2 1 1 1 0 3 1 0
	1 0 1 0 0 1 0 3 0 0 6 6 1 4 0 0 1 0 0 0 1 0
100PEZO	1 1 1 1 3 1 0 0 0 0 0 0 2 2 4 0 1 1 0 1 0 6 6 1 0 1 0 0 0 1
	1 1 3 1 1 1 0 1 1 3 2 4 0 2 1 0 0 2 3 0 0 0 1 1 1 0 0 3 1 0
	1 0 1 0 0 0 0 3 0 1 6 6 0 0 0 0 0 1 1 0 1 0
101PARY	1 2 1 0 0 3 1 0 0 0 0 0 2 2 4 0 0 2 0 1 0 6 6 1 0 0 0 0 0 1
	3 1 3 1 ? ? 0 1 0 3 2 4 0 2 0 0 1 1 3 1 0 0 2 1 0 0 0 ? 0 0
	1 1 0 0 0 0 0 3 0 1 6 6 0 4 0 1 0 0 0 0 1 0
102MONG	1 1 1 1 2 3 1 0 0 0 0 0 1 2 3 0 0 2 0 1 1 6 6 1 0 1 1 0 0 1
	1 0 0 0 1 1 0 1 0 3 3 4 0 1 0 0 0 1 3 0 0 1 2 1 2 0 0 3 1 0
	1 0 0 0 0 0 0 3 0 0 6 6 0 4 1 0 0 0 0 0 1 0
103IRAN	1 2 1 0 0 3 1 0 0 0 0 0 2 2 1 0 1 1 0 2 1 6 6 1 0 1 0 0 0 1
	1 1 3 3 1 1 0 1 0 3 2 4 0 2 1 ? 0 2 3 0 0 1 2 1 0 0 0 3 0 0
	1 0 0 0 0 0 0 3 0 1 6 6 0 4 0 0 1 0 0 0 1 0
104FEST	0 2 1 0 0 2 1 0 0 0 0 0 2 2 4 1 0 4 0 1 0 6 6 0 0 0 0 0 0 1
	3 1 1 0 1 0 0 1 0 4 2 4 0 2 0 0 1 1 3 0 0 1 2 1 0 1 0 0 0 1
	2 2 1 0 0 0 0 3 1 1 6 6 0 4 0 0 1 1 0 0 1 0
105EXOD	1 2 0 1 1 ? 1 0 0 0 0 0 2 0 ? ? ? ? ? ? 1 1 6 6 1 0 1 0 0 0 1
	1 1 3 0 1 1 0 1 0 3 2 4 0 2 1 0 0 2 3 0 ? ? ? ? ? ? ? ? ? ? ? ?
	1 0 1 0 0 0 1 4 0 1 6 6 0 4 ? ? 0 0 0 0 1 0
106ELAS	1 2 1 0 1 2 1 0 0 1 0 0 2 2 4 0 1 1 0 1 1 6 6 1 0 1 0 0 0 1
	3 3 3 0 1 1 0 1 0 3 2 4 0 2 1 0 1 1 3 1 0 1 2 1 0 0 0 0 0 0
	1 2 1 0 0 0 1 3 0 1 6 6 0 4 0 0 0 0 0 0 1 0
107DRYM	1 2 0 1 0 2 1 0 0 0 0 0 0 0 4 1 0 2 0 1 0 6 6 1 0 0 0 0 0 1
	1 0 0 0 1 1 0 1 0 3 2 4 0 2 0 0 0 2 3 1 0 0 2 1 2 0 0 3 0 0
	1 0 1 0 0 1 0 3 0 1 6 6 0 4 0 0 0 0 0 0 1 0
108CERE	0 1 0 1 1 4 0 0 0 1 0 2 2 2 4 0 1 2 0 ? ? 6 6 1 0 0 0 0 0 ?
	? 3 3 3 1 0 0 0 0 1 2 4 0 ? 1 0 0 2 3 0 0 0 2 1 2 ? 0 7 0 0
	? ? ? ? ? ? ? ? ? ? 6 6 0 ? 0 1 ? ? ? 0 1 ?
109APHO	1 2 1 0 0 3 1 0 0 0 0 0 2 2 4 0 0 3 0 1 0 6 6 1 0 1 0 0 0 1
	3 1 3 1 1 1 0 1 0 3 2 4 0 2 0 0 0 4 3 1 0 0 1 1 1 0 0 3 0 0
	1 2 0 0 0 0 0 3 0 ? 6 6 0 4 0 1 1 0 0 0 1 0
110AFGH	1 2 1 0 0 3 1 0 0 1 0 0 2 2 4 0 0 3 1 1 ? 6 6 0 0 0 0 0 0 1
	1 1 2 0 0 0 0 1 0 3 2 4 0 2 1 0 1 2 3 0 0 0 2 1 2 1 3 ? ? ?
	? ? ? ? ? ? ? ? ? ? 6 6 0 4 1 0 ? ? ? 1 0 ?

111SEMI	1 2 1 0 0 3 1 1 1 0 0 0 2 2 1 0 1 2 0 1 1 6 6 0 0 0 0 0 0 1
	1 0 0 0 0 0 0 1 0 3 2 4 0 2 1 0 1 1 3 1 0 1 2 1 2 1 0 3 0 1
	0 1 0 0 0 0 1 4 0 1 6 6 0 4 0 0 1 0 0 0 1 0
112SQUA	1 2 1 0 0 3 1 0 1 0 0 0 2 2 4 0 0 2 0 1 1 6 6 0 0 0 0 0 0 1
	1 0 0 0 0 0 0 1 0 3 2 4 0 2 0 0 1 1 3 0 0 1 2 1 0 1 0 3 0 0
	? ? ? ? ? ? ? ? ? ? 6 6 0 4 0 0 ? ? ? 0 1 0
113SPHA	0 2 1 0 0 2 1 0 0 0 0 0 2 2 2 0 0 2 0 1 1 6 6 0 0 0 0 0 0 1
	1 0 0 0 0 0 0 1 0 3 2 4 0 2 1 0 1 1 3 0 0 1 2 1 2 0 0 3 1 0
	0 2 0 0 1 0 0 4 0 2 6 6 0 4 0 0 0 0 0 0 1 0
114TETT	1 1 0 1 1 3 1 0 0 0 0 0 3 2 1 0 0 2 1 2 1 6 6 1 1 1 1 0 0 1
	0 3 3 3 1 1 0 1 0 3 2 4 0 3 1 0 0 2 2 0 0 1 2 1 0 0 0 7 0 0
	1 ? 0 0 0 0 1 4 0 1 6 6 0 4 0 0 0 0 0 0 1 0
115ZULU	1 2 0 0 1 3 1 0 0 1 0 0 1 1 4 1 1 3 0 1 0 6 6 0 0 0 0 0 0 1
	2 3 3 1 1 1 0 1 0 2 3 3 0 2 1 1 0 2 4 1 0 0 2 1 2 0 0 3 0 1
	1 3 0 0 0 0 0 3 0 1 6 6 0 4 0 1 0 0 0 0 0 0
116ULUD	1 2 1 0 0 2 1 0 0 0 0 0 2 2 4 0 1 2 0 1 0 6 6 0 0 0 0 0 0 1
	3 1 2 0 0 0 0 1 0 3 2 4 0 ? 0 0 1 1 3 1 0 1 2 1 2 1 0 3 1 0
	1 2 0 0 0 0 0 3 0 1 6 6 0 2 0 0 0 0 0 0 1 0
117PLAZ	1 2 0 0 0 1 0 0 0 0 0 0 0 0 4 0 1 1 1 1 0 6 6 1 1 1 1 0 0 1
	1 3 3 0 1 1 0 1 0 3 2 4 0 2 0 0 0 2 3 1 0 1 2 1 1 0 0 3 0 0
	1 2 1 0 0 1 1 4 0 1 6 6 0 4 0 0 0 0 1 0 1 0
118GNT	10 1 0 0 1 3 1 2 1 1 1 0 2 2 1 0 1 1 0 1 1 6 6 0 0 0 0 0 0 1
	3 3 3 1 1 0 0 1 0 3 2 3 0 2 1 0 1 1 3 1 0 0 1 1 0 1 0 1 3 1
	1 1 0 0 0 0 0 3 0 1 7 7 0 2 1 0 0 0 0 0 1 0
119GNT	21 2 1 0 1 1 1 0 0 0 0 0 1 1 4 0 0 2 1 1 0 5 5 0 0 0 0 0 0 0
	3 1 2 1 0 0 0 0 0 2 1 2 0 4 0 0 0 1 3 1 0 0 2 0 ? 0 0 4 0 1
	1 3 0 0 0 0 0 3 0 1 5 4 0 2 0 0 0 0 0 0 0 1
120GNT	30 0 1 1 1 3 1 0 1 0 0 1 3 3 1 0 0 3 1 2 1 6 6 1 1 1 1 0 0 2
	1 3 3 3 1 1 0 1 0 5 3 5 0 3 1 0 0 0 2 0 0 0 2 1 0 1 0 1 0 0
	1 1 1 0 0 1 0 2 0 0 6 6 0 2 0 0 0 0 0 0 1 0
121GNT	41 0 0 1 3 1 0 0 1 0 0 2 0 0 4 0 1 2 0 1 0 6 6 0 0 0 0 0 0 0
	1 1 2 0 0 0 0 1 0 1 2 4 0 4 1 1 0 1 3 0 0 0 1 0 ? 0 0 5 0 0
	1 2 0 1 1 0 1 3 0 1 7 7 0 2 2 1 0 0 0 0 1 0
122BRAD	1 2 0 0 1 1 1 1 1 0 0 0 2 2 4 1 0 3 ? 1 1 7 7 1 0 1 1 0 0 1
	1 3 3 3 1 1 1 1 5 4 9 7 0 2 1 0 0 2 ? 0 0 0 2 0 ? 0 0 7 0 0
	1 2 1 0 0 0 0 3 0 1 9 9 1 4 0 1 0 0 0 1 1 0
123HEMI	0 0 0 0 1 2 0 0 0 0 0 1 3 3 4 0 1 2 0 1 1 6 6 1 1 1 1 0 1 0
	2 1 3 3 1 1 0 0 0 0 2 3 0 3 1 0 0 1 1 0 0 0 2 1 2 1 0 3 0 0
	1 1 0 0 0 1 0 2 0 0 6 6 1 2 0 0 0 0 0 0 0 0
124PERI	0 0 0 1 0 0 0 0 0 0 0 1 3 3 4 0 1 2 0 1 1 7 7 1 1 1 1 0 0 0
	1 2 3 3 1 1 0 0 0 0 0 0 0 3 1 0 0 1 1 0 0 0 2 1 0 1 0 4 0 0
	1 2 1 1 1 0 0 3 1 0 7 7 1 2 0 0 0 0 0 0 0 1
125NEOB	1 0 0 0 1 1 1 0 0 1 0 0 3 3 4 0 1 2 1 1 1 6 6 1 1 1 1 0 0 2
	1 3 3 3 1 1 0 1 0 5 4 6 0 3 1 0 0 1 1 0 0 0 2 1 1 1 0 1 1 0
	1 0 1 0 0 1 0 2 0 0 6 6 0 4 0 0 1 0 0 0 1 0
126REQU	1 0 0 0 1 3 1 0 0 0 0 0 3 3 4 0 1 2 0 1 0 6 6 1 1 1 1 0 0 2
	3 0 2 3 1 1 0 0 0 0 0 2 0 2 0 1 0 2 3 0 0 0 2 1 1 1 0 1 0 0
	1 1 0 0 1 0 0 3 0 0 6 6 0 4 0 0 0 0 0 0 0 0

127TERP	1 0 0 1 1 3 1 1 0 0 0 0 3 3 1 0 1 2 1 2 1 6 6 1 1 1 1 0 0 2
	1 3 3 3 1 1 0 0 0 0 4 6 0 3 1 0 0 1 2 0 0 1 2 1 0 1 0 4 0 1
	1 0 1 0 0 0 0 2 0 0 6 6 0 2 0 0 0 0 0 0 0 0
128PACH*	1 0 0 1 1 2 1 1 0 0 0 2 1 1 4 0 1 2 1 1 0 6 6 1 1 1 1 0 0 0
	2 0 0 0 1 1 0 0 0 0 1 2 0 4 1 1 1 1 3 0 0 0 2 1 2 1 0 1 0 0
	1 1 1 0 0 1 0 1 0 0 6 6 0 2 0 0 0 0 0 0 0 0
129APTE	1 1 0 1 0 1 1 0 0 0 0 0 0 0 ? ? ? ? ? 0 ? 5 5 0 0 0 0 0 0 0
	1 0 0 0 0 0 1 1 1 1 1 2 0 4 0 0 0 2 3 1 0 0 2 1 0 0 ? 1 1 1
	1 2 0 1 1 0 0 2 0 0 5 5 0 2 0 ? 0 0 0 1 1 1
130EPHI	1 0 0 1 0 1 1 0 0 1 1 2 0 0 4 0 1 1 0 1 1 6 6 0 0 0 0 0 0 2
	1 3 3 3 1 1 0 1 0 ? 0 ? 0 3 1 0 0 2 2 0 0 0 2 1 1 0 0 3 1 1
	1 2 0 0 0 0 0 2 0 0 6 6 0 2 0 0 1 1 1 0 1 0

APPENDIX
The Publications of Dr David Ragge

1954. A new grasshopper from the central Himalayas (Orthoptera: Acrididae). Proc. R. Entomol. Soc. Lond. (B) 23: 188-190.

1954. The distribution of *Chorthippus vagans* (Eversmann) in Dorset and Hampshire (Orth., Acrididae). Entomologist 87: 230-231.

1955. A note on female stridulation in the British Acridinae (Orthoptera, Acrididae). Anim. Behav. 3: 70.

1955. Le problème de la stridulation des femelles Acridinae (Orthoptera, Acrididae). Annales des Epiphyties fascicule spècial de 1954 (L'acoustique des Orthoptères): 171-174.

1955. Recent records of the mole-cricket from Hampshire and Surrey. Entomol. Rec. 67: 161.

1955. The wing-venation of the Orthoptera Saltatoria. London: Br. Mus. (Nat. Hist.). vi + 159 pp.

1955. The wing-venation of the order Phasmida. Trans. R. Entomol. Soc. 106: 375-392.

1955. Rediscovery of *Decticus verrucivorus* (L.) in Dorset (Orth., Tettigoniidae). Entomologist 88: 260-261.

1956. The male of *Cotylosoma dipneusticum* Wood-Mason, 1878 (Phasmida: Phylliidae). Proc. R. Entomol. Soc. Lond. (B) 25: 1-2.

1956. A revision of the genera *Phaneroptera* Serville and *Nephoptera* Uvarov (Orthoptera: Tettigoniidae), with conclusions of zoogeographical and evolutionary interest. Proc. Zool. Soc. Lond. 127: 205-283.

1956. A revision of the genus *Oxyecous* Chopard, 1935 (Orthoptera: Tettigoniidae). Proc. R. Entomol. Soc. Lond. (B) 25: 183-190.

1956. Some notes on the field-cricket, *Gryllus campestris* L. (Orth., Gryllidae). Entomologist 89: 300-301.

1957. A revision of the genus *Tropidonotacris* Chopard, 1954 (Orthoptera: Tettigoniidae). Proc. R. Entomol. Soc. Lond. (B) 26: 119-122.

1957. A new species of Phaneroptera Serville from Formosa (Orthptera: Tettigoniidae). Proc. R. Entomol. Soc. Lond. (B) 26: 123-126.

1957. The tracheation of the nymphal wing-pads of *Schizodactylus monstrosus* (Drury), with a revised interpretation of the adult venation (Orthoptera: Schizodactylidae). Proc. Zool. Soc. Lond. 129: 301-304.

1959. Lectotype selection for *Chorthippus parallelus* (Zetterstedt) (Orthoptera: Acrididae). Opusc. Entomol. 24: 35.

1959. A new species of *Nephoptera* Uvarov from Persia, with a revised key to the species of this genus (Orthoptera: Tettigoniidae). Eos 35: 425-428.

1959. *Mantis religiosa* L. (Dictyoptera, Mantidae) in Sussex. Entomol. Mon. Mag. 95: 273.

1960. The Acrometopae of the Ethiopian Region: a revision, with notes on the sexual dimorphism shown by the group (Orthoptera: Tettigoniidae). Bull. Br. Mus. (Nat. Hist.) (Entomol.) 8: 267-333.

1960. Further notes on the genus *Phaneroptera* Serville, with a revised key to the species (Orthoptera: Tettigoniidae). Proc. Zool. Soc. Lond. 134: 237-250.

1961. A revision of the genus *Ducetia* Stål (Orthoptera: Tettigoniidae). Bull. Br. Mus.(Nat. Hist.) (Entomol.) 10: 169-208.

1961. Further notes on the genus *Lamecosoma* Ragge, with a description of a new species (Orth.: Tettigoniidae). Eos 37: 215-219.

1961. (jointly with R. Roy).Le Parc National du Niokolo-Koba. Fasc. II. VII. Orthoptera: Tettigonioidea. Mèmoires de l'Institut Francais d'Afrique Noire 62: 105-107.

1962. A revision of the genera *Phlauocentrum* Karsch, *Buettneria* Karnsch and *Leiodontocerus* Chopard (Orthoptera: Tettigoniidae). Bull. Br.Mus. (Nat. Hist.) (Entomol.) 13: 1-17.

1962. A revision of the genera *Drepanophyllum* Karsch and *Stenamblyphyllum* Karsch (Orth.: Tettigoniidae). Eos 38: 299-309.

1963. First record of the grasshopper *Stenobothrus stigmaticus* (Rambur) (Acrididae) in the British Isles, with other new distribution records and notes on the origin of the British Orthoptera. Entomologist 96: 211-217.

1963. The nymphal wing-pad tracheation and adult axillary sclerites of the Pneumoridae (Orthoptera: Acridoidea). Ann. Mag. Nat. Hist. (13) 6: 185-191.

1963. The male of *Apoecides madagascariensis* Bolivar (Orth.: Tettigoniidae). Eos 39: 445-447.

1964. A revision of the genus *Tylopsis* Fieber (Orthoptera: Tettigoniidae). Bull. Br. Mus. (Nat. Hist.) (Entomol.) 15: 295-322.

1965. (jointly with R. Roy). Contribution à l'etude de la faune de la basse Casamance (Sènégal). XII. Orthopteres Tettigonioidea. Bull. 'Inst. Fr. 'Afr. Noire 27: 185-188.

1965. Grasshoppers, crickets and cockroaches of the British Isles. London: Warne. xii + 299 pp.

1965. Ortopteros y *Dermápteros colectados* en la Peninsula Iberica durante los anos de 1962 y 1963 por misiones del British Museum (Natural History). Graellsia 21: 95-119.

1967. (jointly with R. Roy). A review of the praying mantises of Ghana (Dictyoptera Mantodea). Bull. 'Inst. Fondam. 'Afr. Noire 29: 586-644.

1967. Contribution à la faune du Congo (Brazzaville). Mission A. Villiers et A. Descarpentries. LVI. Orthoptères Tettigoniidae (première note). Bull. 'Inst. Fondam. 'Afr.Noire 29: 1270-1277.

1968. An index-catalogue of African Phaneropterinae (Orthoptera: Tettigoniidae). Bull. Br. Mus. (Nat. Hist.) Entomol. 22: 75-108.

1968. Contributions a la connaissance de la faune entomologique de la Cote d'Ivoire (J. Decelle, 1961-1964). III. Orthoptera: Tettigoniidae. Annales, Musée Royal de l'Afrique Centrale 8vo (Sciences Zoologiques) 165: 23-26.

1969. A revision of the African species of *Pseudorhynchus* Serville (Orthoptera: Tettigoniidae). Bull. Br. Mus. (Nat. Hist.) (Entomol.) 23: 169-190.

1969. Recording insect sounds. Recorded Sound 34: 456-457.

1970. La faune terrestre de l'lle de Sainte-Helene.7.Orthoptera. a. Tettigonioidea. Annales, Musée Royal de l'Afrique Centrale 8vo (Sciences Zoologiques) 181: 185-190.

1971. (jointly with R. Roy). Le massif des Monts Loma (Sierra Leone). Fasc. I. IX. Orthoptera Tettigoniidae. Mémoires de l'Institut Fondamental d'Afrique Noire 86: 249-259.

1972. Further notes on the genus *Tylopsis* Fieber (Orthoptera: Tettigoniidae), with a description of a new species. J. Nat. Hist. 6: 121-124.

1972. An unusual case of mass migration by flight in *Gryllus bimaculatus* Degeer [Orthoptera: Gryllidae]. Bull. 'Inst Fondam. 'Afr. Noire 34: 869-878.

1973. The British Orthoptera: a supplement. Entomol. Gaz. 24: 227-245.

1973. Dictyoptera (cockroaches and praying mantises). In: Insects and other Arthropods of Medical Importance. Ed. K.G.V. Smith. London: Br. Mus. (Nat. Hist.). pp. 399-403.

1974. Tettigonioidea. In: Status of the Taxonomy of the Hexapoda of Southern Africa. Ed. W.G.H. Coaton. Entomology Memoirs, Department of Agriculture, Union of South Africa 38: 37-38.

1974. A new species of *Tylopsis* Fieber from southern Israel (Orthoptera: Tettigoniidae). Isr. J. Zool. 22: 63-66.

1976. A putative hybrid in nature between *Chorthippus brunneus* and *C. biguttulus* (Orthoptera: Tettigoniidae). Syst. Entomol. 1: 71-74.

1977. A new genus of bush-cricket causing damage to cereal crops in Ethiopia (Orthoptera: Tettigoniidae). J. Nat. Hist. 11: 509-513.

1977. Classification of Tettigonioidea. In: Kevan, D.K. McE. (Ed.), The Higher Classification of the Orthopteroid Insects. Ed. D.K. McE. Kevan. Memoirs, Lyman Entomological Museum and Research Laboratory 4: 44-46.

1979. (jointly with W.J. Reynolds). Insect sound recording in the Natural History Museum. Recorded Sound 74-75: 80-81.

1980. A review of the African Phaneropterinae with open tympana (Orthoptera: Tettigoniidae). Bull. Br. Mus. (Nat. Hist.) (Entomol.) 40: 67-192.

1981. A new leaf-resembling genus of bush-cricket from West Africa (Orthoptera: Tettigoniidae). J. Nat. Hist. 15: 327-330.

1981. An unusual song-pattern in the *Chorthippus mollis* group (Orthoptera: Acrididae): local variant or hybrid population? J. Nat. Hist. 15: 995-1002.

1984. (jointly with W.J. Reynolds). The taxonomy of the western European grasshoppers of the genus *Euchorthippus*, with special reference to their songs (Orthoptera: Acrididae). Bull. Br. Mus. (Nat. Hist.) (Entomol.) 53: 213-249.

1984. The Le Broc grasshopper population: further evidence of its hybrid status (Orthoptera: Acrididae). J. Nat. Hist. 18: 921-925.

1986. The songs of the western European grasshoppers of the genus *Omocestus* in relation to their taxonomy (Orthoptera: Acrididae). Bull. Br. Mus. (Nat. Hist.) (Entomol.) 53: 213-249.

1987. The songs of the western European grasshoppers of the genus *Stenobothrus* in relation to their taxonomy (Orthoptera: Acrididae). Bull. Br. Mus. (Nat. Hist.) (Entomol.) 55: 393-424.

1987. Speciation and biogeography of some southern European Orthoptera, as revealed by their songs. In: Evolutionary Biology of Orthopteroid Insects. Ed. B. Baccetti. Chichester: Ellis Horwood. pp. 418-426.

1988 (jointly with W.J. Reynolds). The songs and taxonomy of the grasshoppers of the *Chorthippus biguttulus* group in the Iberian Peninsula. J. Nat. Hist. (in press).

SYSTEMATIC INDEX

SUBJECT INDEX

GPSR Compliance

The European Union's (EU) General Product Safety Regulation (GPSR) is a set of rules that requires consumer products to be safe and our obligations to ensure this.

If you have any concerns about our products, you can contact us on ProductSafety@springernature.com

In case Publisher is established outside the EU, the EU authorized representative is:

Springer Nature Customer Service Center GmbH
Europaplatz 3
69115 Heidelberg, Germany

Batch number: 10838292

Printed by Printforce, the Netherlands